# Communications
# in Computer and Information Science 574

*Commenced Publication in 2007*
Founding and Former Series Editors:
Alfredo Cuzzocrea, Dominik Ślęzak, and Xiaokang Yang

More information about this series at http://www.springer.com/series/7899

Ana Fred · Hugo Gamboa
Dirk Elias (Eds.)

# Biomedical Engineering Systems and Technologies

8th International Joint Conference, BIOSTEC 2015
Lisbon, Portugal, January 12–15, 2015
Revised Selected Papers

 Springer

*Editors*

Ana Fred
Instituto de Telecomunicações, IST
University of Lisbon
Lisbon
Portugal

Dirk Elias
FhP-AICOS
Porto
Portugal

Hugo Gamboa
New University of Lisbon
Lisbon
Portugal

ISSN 1865-0929        ISSN 1865-0937   (electronic)
Communications in Computer and Information Science
ISBN 978-3-319-27706-6        ISBN 978-3-319-27707-3   (eBook)
DOI 10.1007/978-3-319-27707-3

Library of Congress Control Number: 2015956367

Printed on acid-free paper

This Springer imprint is published by SpringerNature
The registered company is Springer International Publishing AG Switzerland

# Preface

This book contains the extended and revised versions of a set of selected papers from the 8th International Joint Conference on Biomedical Engineering Systems and Technologies (BIOSTEC 2015), held in Lisbon, Portugal, during March 12–15, 2015.

BIOSTEC was sponsored by the Institute for Systems and Technologies of Information, Control and Communication (INSTICC), held in cooperation with the ACM Special Interest Group on Bioinformatics, Computational Biology, and Biomedical Informatics (SIGBio), the ACM Special Interest Group on Artificial Intelligence (SIGAI), the ACM Special Interest Group (SIG) on Management Information Systems (MIS), EUROMICRO, the International Society for Telemedicine and eHealth (ISf-TeH), the Association for the Advancement of Artificial Intelligence (AAAI), and the European Association for Signal Processing (EURASIP); it was technically co-sponsored by the Biomedical Engineering Society (BMES), the European Society for Engineering and Medicine (ESEM), and the IEEE Engineering in Medicine and Biology Society (IEEE EMBS). INSTICC is member of the Workflow Management Coalition (WfMC), Object Management Group (OMG) and Foundation for Intelligent Physical Agents (FIPA). The BMC Medical Informatics and Decision Making was the media partner and the Science and Technology Events (SCITEVENTS) the logistics partner.

The hallmark of this conference is to provide a point of contact for researchers and practitioners interested in both theoretical advances and applications of information systems, artificial intelligence, signal processing, electronics, and other engineering tools in knowledge areas related to biology and medicine.

BIOSTEC is composed of five complementary and co-located conferences, each specialized in at least one of the aforementioned main knowledge areas; namely:

- BIODEVICES - International Conference on Biomedical Electronics and Devices
- BIOIMAGING - International Conference on Bioimaging
- BIOINFORMATICS - International Conference on Bioinformatics Models, Methods and Algorithms
- BIOSIGNALS - International Conference on Bio-inspired Systems and Signal Processing
- HEALTHINF - International Conference on Health Informatics

The purpose of the International Conference on Biomedical Electronics and Devices (BIODEVICES) is to bring together professionals from electronics and mechanical engineering, interested in studying and using models, equipment, and materials inspired from biological systems and/or addressing biological requirements. Monitoring devices, instrumentation sensors and systems, biorobotics, micro-nanotechnologies, and biomaterials are some of the technologies addressed at this conference.

The aim of the International Conference on Bioimaging (BIOIMAGING) is to get together experts and to foster discussion on acquisition, processing, and visualization of image information in the context of medicine and biology. This multifaceted field covers many available image modalities, such as X-ray, computed tomography (CT), magnetic resonance imaging (MRI), and so on. The image acquisition and processing techniques are used to create images of the human body, anatomical areas, tissues, and so on, down to the molecular level, either for clinical purposes, seeking to reveal, diagnose, or examine diseases, or for medical science, including the study of normal anatomy and physiology.

The authors are encouraged to submit papers describing methods, techniques, advanced prototypes, applications, systems, tools, or survey papers. The International Conference on Bioinformatics Models, Methods and Algorithms (BIOINFOR-MATICS) intends to provide a forum for discussion to researchers and practitioners interested in the application of computational systems and information technologies to the field of molecular biology, including, for example, the use of statistics and algo-rithms in understanding biological processes and systems, with a focus on new developments in genome bioinformatics and computational biology. Areas of interest for this community include sequence analysis, biostatistics, image analysis, scientific data management and data mining, machine learning, pattern recognition, computa-tional evolutionary biology, computational genomics, and other related fields.

The goal of the International Conference on Bio-inspired Systems and Signal Processing (BIOSIGNALS) is to bring together researchers and practitioners from multiple areas of knowledge, including biology, medicine, engineering, and other physical sciences, interested in studying and using models and techniques inspired from or applied to biological systems. A diversity of signal types can be found in this area, including image, audio, and other biological sources of information. The analysis and use of these signals is a multidisciplinary area including signal processing, pattern recognition, and computational intelligence techniques, among others.

The International Conference on Health Informatics (HEALTHINF) aims to be a major meeting point for those interested in understanding the human and social implications of technology, not only in health-care systems but in other aspects of human–machine interaction such as accessibility issues and the specialized support to persons with special needs.

The eighth edition of the joint conference BIOSTEC received 375 paper submis-sions from 62 countries. From these, 61 papers were published as full papers, 92 were accepted for short presentation, and another 73 for poster presentation. These numbers, leading to a "full-paper" acceptance ratio of about 16 % and an oral paper acceptance ratio close to 41 %, show the intention of preserving a high quality forum for the next editions of this conference.

We sincerely thank the authors, whose research and development efforts are recorded here.

September 2015

Alberto Cliquet Jr.
Mário Forjaz Secca
Jan Schier
Oscar Pastor
Christine Sinoquet
Harald Loose
Marta Bienkiewicz
Christine Verdier
Ana Fred
Hugo Gamboa
Dirk Elias

# Organization

## Conference Co-chairs

Ana Fred       Instituto de Telecomunicações/IST, Portugal
Hugo Gamboa       CEFITEC/FCT - New University of Lisbon, Portugal
Dirk Elias       University of Porto/Fraunhofer, Portugal

## Program Co-chairs

### BIODEVICES

Alberto Cliquet Jr.       University of São Paulo and University of Campinas, Brazil

### BIOIMAGING

Mário Forjaz Secca       CEFITEC, FCT/UNL, Portugal
Jan Schier       The Institute of Information Theory and Automation of the Czech Academy of Sciences, Czech Republic

### BIOINFORMATICS

Oscar Pastor       Universidad Politécnica de Valencia, Spain
Christine Sinoquet       University of Nantes, France

### BIOSIGNALS

Harald Loose       Brandenburg University of Applied Sciences, Germany

### HEALTHINF

Marta Bienkiewicz       TUM, Technische Universität München, Germany
Christine Verdier       LIG - Joseph Fourier University of Grenoble, France

## BIODEVICES Program Committee

Elli Angelopoulou       University of Erlangen-Nuremberg, Germany
Nizametin Aydin       Yildiz Technical University, Turkey
Mohammed Bakr       CCIT-AASTMT, Egypt
Steve Beeby       University of Southampton, UK
Egon L. van den Broek       University of Twente/Radboud UMC Nijmegen, The Netherlands
Jan Cabri       Norwegian School of Sport Sciences, Norway
Wenxi Chen       The University of Aizu, Japan

Seonghan Ryu                Hannam University, Korea, Republic of
Chutham Sawigun            Mahanakorn University of Technology, Thailand
Michael J. Schöning        FH Aachen, Germany
Rahamim Seliktar           Drexel University, USA
Mauro Serpelloni           University of Brescia, Italy
Anita Lloyd Spetz          Linköpings Universitet, Sweden
João Paulo Teixeira        Polytechnic Institute of Bragança, Portugal
Renato Varoto              University of São Paulo, Brazil
Pedro Vieira               Universidade Nova de Lisboa, Portugal
Bruno Wacogne              FEMTO-ST UMR CNRS 6174, France
Xintong Wang               Vanderbilt University, USA
Huikai Xie                 University of Florida, USA
Hakan Yavuz                Çukurova Üniversity, Turkey
Stefan Zappe               Carnegie Mellon University, USA
Aladin Zayegh              Victoria University, Australia

## BIODEVICES Additional Reviewers

Simone Benatti             University of Bologna, Italy
Jie Ding                   Macquarie University, Australia
Jie Zhou                   Macquarie University, Australia

## BIOIMAGING Program Committee

Jesús B. Alonso            Universidad de Las Palmas de Gran Canaria, Spain
Sameer K. Antani           National Library of Medicine, National Institutes of
                           Health, United States, USA
Radu Badea                 University of Medicine and Pharmacy Iuliu Hatieganu
                           Cluj Napoca, Romania
Peter Balazs               University of Szeged, Hungary
Ewert Bengtsson            Centre for Image Analysis, Uppsala University,
                           Sweden
Abdel-Ouahab Boudraa       Ecole Navale, France
Tom Brown                  University of St. Andrews, UK
Begoña Calvo               University of Zaragoza, Spain
Rita Casadio               University of Bologna, Italy
Tony Cass                  Imperial College London, UK
Chung-Ming Chen            National Taiwan University, Taiwan
Christos E. Constantinou   Stanford University, USA
Giacomo Cuttone            INFN – Laboratori Nazionali del Sud Catania, Italy
Alain Dagher               Montreal Neurology Hospital Institute, Canada
Xiuzhen Dong               Fourth Military Medical University, China
Dimitrios Fotiadis         University of Ioannina, Greece
Esteve Gallego-Jutglà      University of Vic, Spain
Xiaoping Hu                Emory University, USA
Ioannis S. Kandarakis      Technological Education Institute of Athens, Greece

# BIOINFORMATICS Program Committee

| | |
|---|---|
| Mohamed Abouelhoda | Nile University, Egypt |
| R. Acharya | The Pennsylvania State University, USA |
| Tatsuya Akutsu | Kyoto University, Japan |
| Hesham Ali | University of Nebraska at Omaha, USA |
| Jens Allmer | Izmir Institute of Technology, Turkey |
| Rui Alves | University of Lleida, Spain |
| Paul Anderson | College of Charleston, USA |
| Joel Arrais | Universidade de Coimbra, Portugal |
| Kiyoshi Asai | University of Tokyo, Japan |
| Rolf Backofen | Albert-Ludwigs-Universität, Germany |
| Tim Beissbarth | University of Göttingen, Germany |
| Shifra Ben-Dor | Weizmann Institute of Science, Israel |
| Ulrich Bodenhofer | Johannes Kepler University Linz, Austria |
| Carlos Brizuela | Centro de Investigación Científica y de Educación Superior de Ensenada, Baja California, Mexico |
| Egon L. van den Broek | University of Twente/Radboud UMC Nijmegen, The Netherlands |
| Conrad Burden | Australian National University, Australia |
| Heorhiy Byelas | University Medical Center Groningen, The Netherlands |
| João Carriço | Universidade de Lisboa, Portugal |
| Claudia Consuelo Rubiano Castellanos | Universidad Nacional de Colombia - Bogota, Colombia |
| Ching-Fen Chang | NYMU, Taiwan |
| Wai-Ki Ching | The University of Hong Kong, Hong Kong, SAR China |
| Mark Clement | Brigham Young University, USA |
| Ana Conesa | Centro de Investigaciones Príncipe Felipe, Spain |
| Francisco Couto | Universidade de Lisboa, Portugal |
| Antoine Danchin | AMAbiotics SAS, France |
| Thomas Dandekar | University of Würzburg, Germany |
| Sérgio Deusdado | Instituto Politecnico de Bragança, Portugal |
| Julie Dickerson | Iowa State University, USA |
| Eytan Domany | Weizmann Institute of Science, Israel |
| Richard Edwards | University of Southampton, UK |
| George Eleftherakis | CITY College, International Faculty of the University of Sheffield, Greece |
| André Falcão | Universidade de Lisboa, Portugal |
| Fabrizio Ferre | University of Rome Tor Vergata, Italy |
| António Ferreira | Universidade de Lisboa, Portugal |
| Gianluigi Folino | Institute for High Performance Computing and Networking, National Research Council, Italy |
| Bruno Gaeta | University of New South Wales, Australia |
| Max H. Garzon | The University of Memphis, USA |

| | |
|---|---|
| Igor Goryanin | University of Edinburgh, UK |
| Julian Gough | University of Bristol, UK |
| Reinhard Guthke | Hans Knoell Institute, Germany |
| Arndt von Haeseler | Center for Integrative Bioinformatics Vienna, Austria |
| Joerg Hakenberg | Medical School at Mount Sinai, USA |
| Christopher E. Hann | University of Canterbury, New Zealand |
| Ronaldo Fumio Hashimoto | University of São Paulo, Brazil |
| Artemis Hatzigeorgiou | University of Thessaly, Greece |
| Shan He | University of Birmingham, UK |
| Volkhard Helms | Universität des Saarlandes, Germany |
| Hailiang Huang | Massachusetts General Hospital, USA |
| Yongsheng Huang | University of Michigan, USA |
| Daisuke Ikeda | Kyushu University, Japan |
| Costas S. Iliopoulos | King's College London, UK |
| Seiya Imoto | University of Tokyo, Japan |
| Sohei Ito | National Fisheries University, Japan |
| Bo Jin | Sigma-aldrich, USA |
| Giuseppe Jurman | Fondazione Bruno Kessler, Italy |
| Ed Keedwell | University of Exeter, UK |
| Sami Khuri | San José State University, USA |
| Inyoung Kim | Virginia Tech, USA |
| Toralf Kirsten | University of Leipzig, Germany |
| Jirí Kléma | Czech Technical University in Prague, Czech Republic |
| Sophia Kossida | Academy of Athens, Greece |
| Malgorzata Kotulska | Wroclaw University of Technology, Poland |
| Ivan Kulakovskiy | VIGG RAS, EIMB RAS, Russian Federation |
| Yinglei Lai | George Washington University, USA |
| Matej Lexa | Masaryk University, Czech Republic |
| Leo Liberti | Ecole Polytechnique, France |
| Antonios Lontos | Frederick University, Cyprus |
| Shuangge Ma | Yale University, USA |
| Paolo Magni | Università degli Studi di Pavia, Italy |
| Elena Marchiori | Radboud University, The Netherlands |
| Majid Masso | George Mason University, USA |
| Petr Matula | Masaryk University, Czech Republic |
| IvanMerelli | ITB CNR, Italy |
| Luciano Milanesi | ITB-CNR, Italy |
| Saad Mneimneh | Hunter College CUNY, USA |
| Pedro Tiago Monteiro | INESC-ID, Portugal |
| Shinichi Morishita | University of Tokyo, Japan |
| Vincent Moulton | University of East Anglia, UK |
| Chad Myers | University of Minnesota, USA |
| Radhakrishnan Nagarajan | University of Kentucky, USA |
| Arcadi Navarro | Universitat Pompeu Fabra, Spain |
| Jean-Christophe Nebel | Kingston University, UK |
| José Luis Oliveira | Universidade de Aveiro, Portugal |

| | |
|---|---|
| Hakan S. Orer | Koc University, Turkey |
| Gülay Özcengiz | METU - Middle East Technical University Ankara, Turkey |
| Patricia Palagi | Swiss Institute of Bioinformatics, Switzerland |
| Gustavo Parisi | Universidad Nacional de Quilmes, Argentina |
| Oscar Pastor | Universidad Politécnica de Valencia, Spain |
| Florencio Pazos | National Centre for Biotechnology, Spain |
| Marco Pellegrini | Consiglio Nazionale delle Ricerche, Italy |
| Matteo Pellegrini | University of California, Los Angeles, USA |
| Horacio Pérez-Sánchez | Catholic University of Murcia, Spain |
| Guy Perrière | Université Claude Bernard - Lyon 1, France |
| Francisco Pinto | Universidade de Lisboa, Portugal |
| Olivier Poch | ICube UMR7357 CNRS-Université de Strasbourg, France |
| Alberto Policriti | Università degli Studi di Udine, Italy |
| Giuseppe Profiti | University of Bologna, Italy |
| Junfeng Qu | Clayton State University, USA |
| Mark Ragan | The University of Queensland, Australia |
| Laura Roa | University of Seville, Spain |
| Miguel Rocha | University of Minho, Portugal |
| David Rocke | University of California, Davis, USA |
| Simona E. Rombo | Università degli Studi di Palermo, Italy |
| Eric Rouchka | University of Louisville, USA |
| Carolina Ruiz | WPI, USA |
| J. Cristian Salgado | University of Chile, Chile |
| Armindo Salvador | Universidade de Coimbra, Portugal |
| Alessandro Savino | Politecnico di Torino, Italy |
| Jaime Seguel | University of Puerto Rico at Mayaguez, USA |
| Joao C. Setubal | Universidade de São Paulo, Brazil |
| Ugur Sezerman | Sabanci University, Turkey |
| Christine Sinoquet | University of Nantes, France |
| Neil R. Smalheiser | University of Illinois Chicago, USA |
| Pavel Smrz | Brno University of Technology, Czech Republic |
| Gordon Smyth | Walter and Eliza Hall Institute of Medical Research, Australia |
| Peter F. Stadler | Universität Leipzig – IZBI, Germany |
| Kristel Van Steen | University of Liège, Belgium |
| David Svoboda | Masaryk University, Czech Republic |
| Peter Sykacek | BOKU: University of Natural Resources and Life Sciences, Austria |
| Silvio C.E. Tosatto | Università di Padova, Italy |
| Jyh-Jong Tsay | National Chung Cheng University, Taiwan |
| Allegra Via | Università di Roma La Sapienza, Italy |
| Juris Viksna | University of Latvia, Latvia Thomas Werner, University of Michigan, Germany |
| Yanbin Yin | Northern Illinois University, USA |

| | |
|---|---|
| Jingkai Yu | Institute of Process Engineering, Chinese Academy of Sciences, China |
| Nazar Zaki | United Arab Emirates University (UAEU), UAE |
| Alexander Z. Zelikovsky | Georgia State University, USA |
| Erliang Zeng | University of South Dakota, USA |
| Bao-Hong Zhang | East Carolina University, USA |
| Louxin Zhang | National University of Singapore, Singapore |
| Leming Zhou | University of Pittsburgh, USA |

## BIOINFORMATICS Additional Reviewers

| | |
|---|---|
| Huy Dinh | University of Southern California, USA |
| Artem Kasianov | VIGG, Russian Federation |

## BIOSIGNALS Program Committee

| | |
|---|---|
| Aybar Acar | Middle East Technical University, Turkey |
| Jean-Marie Aerts | M3-BIORES, Katholieke Universitëit Leuven, Belgium |
| Jesús B. Alonso | Universidad de Las Palmas de Gran Canaria, Spain |
| Sergio Alvarez | Boston College, USA |
| Julián David Arias | Universidad de Antioquia, Colombia |
| Luis Azevedo | Anditec, Portugal |
| Ofer Barnea | Tel Aviv University, Israel |
| Peter Bentley | UCL, UK |
| Jovan Brankov | Illinois Institute of Technology, USA |
| Egon L. van den Broek | University of Twente/Radboud UMC Nijmegen, The Netherlands |
| Tolga Can | Middle East Technical University, Turkey |
| Guy Carrault | University of Rennes 1, France |
| Maria Claudia F. Castro | Centro Universitário da FEI, Brazil |
| M. Emre Celebi | Louisiana State University in Shreveport, USA |
| YangQuan Chen | University of California at Merced, USA |
| Joselito Chua | Monash University, Australia |
| Jan Cornelis | VUB, Belgium |
| Justin Dauwels | NTU, Singapore |
| Gordana Jovanovic Dolecek | Institute INAOE, Mexico |
| Pier Luigi Emiliani | Italian National Research Council (CNR), Italy |
| Pedro Encarnação | Universidade Católica Portuguesa, Portugal |
| Luca Faes | Università degli Studi di Trento, Italy |
| Dimitrios Fotiadis | University of Ioannina, Greece |
| Esteve Gallego-Jutglà | University of Vic, Spain |
| M. Ghogho | University of Leeds, UK |
| Inan Güler | Gazi University, Turkey |
| Md. Kamrul Hasan | Bangladesh University of Engineering and Technology (BUET), Bangladesh |
| Thomas Hinze | Friedrich Schiller University Jena, Germany |

James R. Hopgood    University of Edinburgh, UK
Bart Jansen    Vrije Universiteit Brussel, Belgium
Visakan Kadirkamanathan    The University of Sheffield, UK
Shohei Kato    Nagoya Institute of Technology, Japan
Natalya Kizilova    ICM Warsaw University, Poland
Dagmar Krefting    Hochschule für Technik und Wirtschaft Berlin – University of Applied Sciences, Germany
Lenka Lhotska    Czech Technical University in Prague, Czech Republic
Chin-Teng Lin    National Chiao Tung University, Taiwan
Ana Rita Londral    Universidade de Lisboa, Portugal
Harald Loose    Brandenburg University of Applied Sciences, Germany
Wenlian Lu    Fudan University, China
Hari Krishna Maganti    Horowitz Biometrics Limmited, UK
Jan Mares    Institute of Chemical Technology, Czech Republic
Pina Marziliano    Nanyang Technological University, Singapore
G.K. Matsopoulos    National Technical University of Athens, Greece
Des McLernon    University of Leeds, UK
Paul Meehan    The University of Queensland, Australia
Pramod Kumar Meher    Nanyang Technological University, Singapore
Vojkan Mihajlovic    Holst Centre/IMEC the Netherlands, The Netherlands
Mihaela Morega    University Politehnica of Bucharest, Romania
Nicoletta Nicolaou    Imperial College London, UK
Giandomenico Nollo    Università degli Studi di Trento, Italy
Michael Ochs    The College of New Jersey, USA
Krzysztof Pancerz    University of Information, Technology and Management in Rzeszow, Poland
George Panoutsos    The University of Sheffield, UK
Joao Papa    UNESP – Universidade Estadual Paulista, Brazil
Gennaro Percannella    University of Salerno, Italy
Vitor Pires    Escola Superior de Tecnologia de Setúbal – IPS, Portugal
Octavian Postolache    Institute of Telecommunications, Portugal
Ales Prochazka    Institute of Chemical Technology, Czech Republic
José Joaquín Rieta    Universidad Politécnica de Valencia, Spain
Marcos Rodrigues    Sheffield Hallam University, UK
Heather Ruskin    Dublin City University, Ireland
Carlo Sansone    University of Naples, Italy
Gerald Schaefer    Loughborough University, UK
Christian Schmidt    University of Rostock, Germany
Reinhard Schneider    Fachhochschule Vorarlberg, Austria
Lotfi Senhadji    University of Rennes 1, France
Tapio Seppänen    University of Oulu, Finland
David Simpson    University of Southampton, UK
Jordi Solé-Casals    University of Vic – Central University of Catalonia, Spain
John J. Soraghan    University of Strathclyde, UK

| | |
|---|---|
| Olga Sourina | Nanyang Technological University, Singapore |
| Deborah Stacey | University of Guelph, Canada |
| Alan A. Stocker | University of Pennyslvania, USA |
| Yannis Stylianou | University of Crete, Greece |
| Hiroki Takada | University of Fukui, Japan |
| Asser Tantawi | IBM, USA |
| Wallapak Tavanapong | Iowa State University, USA |
| João Paulo Teixeira | Polytechnic Institute of Bragança, Portugal |
| Kasim Terzic | University of the Algarve, Portugal |
| Ana Maria Tomé | University of Aveiro, Portugal |
| Vicente Traver | ITACA, Universidad Politécnica de Valencia, Spain |
| Carlos M. Travieso | University of Las Palmas de Gran Canaria, Spain |
| Ahsan Ahmad Ursani | Mehran University of Engineering and Technology, Pakistan |
| Bart Vanrumste | Katholieke Hogeschool Kempen/Katholieke Universiteit Leuven, Belgium |
| Michal Vavrecka | Czech Technical University, Czech Republic |
| Giovanni Vecchiato | IRCCS Fondazione Santa Lucia, Italy |
| Pedro Gómez Vilda | Universidad Politécnica de Madrid, Spain |
| Andreas Voss | University of Applied Sciences Jena, Germany |
| YuanyuanWang | Fudan University, China |
| Quan Wen | University of Electronic Science and Technology of China, China |
| Kerstin Witte | Otto von Guericke University Magdeburg, Germany |
| Didier Wolf | Research Centre for Automatic Control - CRAN CNRS UMR 7039, France |
| DongruiWu | GE Global Research, USA |
| Pew-Thian Yap | University of North Carolina at Chapel Hill, USA |
| Nicolas Younan | Mississippi State University, USA |
| Rafal Zdunek | Wroclaw University of Technology, Poland |
| Li Zhuo | Beijing University of Technology, China |

## BIOSIGNALS Additional Reviewers

| | |
|---|---|
| Virgine Lerolle | INSERM 1099, France |
| Hervé Saint-Jalmes | LTSI, INSERM U1099, France |

## HEALTHINF Program Committee

| | |
|---|---|
| Francois Andry | Philips HealthCare, USA |
| Wassim Ayadi | LERIA, University of Angers, France and LaTICE, University of Tunis, Tunisia |
| Turgay Ayer | Georgia Tech, USA |
| Philip Azariadis | University of the Aegean, Greece |
| Elarbi Badidi | United Arab Emirates University, UAE |
| Omar Badreddin | Northern Arizona University, Canada |

| | |
|---|---|
| Michele Luglio | University of Rome Tor Vergata, Italy |
| Pierre Maret | Université de Saint Etienne, France |
| Alda Marques | University of Aveiro, Portugal |
| José Luis Martínez | Universidade Carlos III de Madrid, Spain |
| Paloma Martínez | Universidad Carlos III de Madrid, Spain |
| Maria Di Mascolo | University of Grenoble, France |
| Sally Mcclean | University of Ulster, UK |
| Marilyn McGee-Lennon | University of Glasgow, UK |
| Gianluigi Me | LUISS University, Italy |
| Gerrit Meixner | Heilbronn University, Germany |
| Mohyuddin Mohyuddin | King Abdullah International Medical Research Center (KAIMRC), Saudi Arabia |
| Christo El Morr | York University, Canada |
| Roman Moucek | University of West Bohemia, Czech Republic |
| Radhakrishnan Nagarajan | University of Kentucky, USA |
| Hammadi Nait-Charif | Bournemouth University, UK |
| Tadashi Nakano | Osaka University, Japan |
| Goran Nenadic | University of Manchester, UK |
| José Luis Oliveira | Universidade de Aveiro, Portugal |
| Rui Pedro Paiva | University of Coimbra, Portugal |
| Danilo Pani | University of Cagliari, Italy |
| Guy Pare | HEC Montréal, Canada |
| Rosario Pugliese | Università di Firenze, Italy |
| Juha Puustjärvi | University of Helsinki, Finland |
| Arkalgud Ramaprasad | University of Illinois at Chicago, USA |
| Marcos Rodrigues | Sheffield Hallam University, UK |
| Roberto J. Rodrigues | eHealthStrategies, USA |
| Valter Roesler | Federal University of Rio Grande do Sul, Brazil |
| Elisabetta Ronchieri | INFN, Italy |
| George Sakellaropoulos | University of Patras, Greece |
| Ovidio Salvetti | National Research Council of Italy – CNR, Italy |
| Akio Sashima | AIST, Japan |
| Jacob Scharcanski | UFRGS – Universidade Federal do Rio Grande do Sul, Brazil |
| Bettina Schnor | Potsdam University, Germany |
| Carla Simone | Università degli studi di Milano-Bicocca, Italy |
| Irena Spasic | University of Cardiff, UK |
| Jan Stage | Aalborg University, Denmark |
| Abdel-Rahman Tawil | University of East London, UK |
| Francesco Tiezzi | IMT – Institute for Advanced Studies Lucca, Italy |
| Ioannis G. Tollis | University of Crete, Greece |
| Vicente Traver | ITACA, Universidad Politécnica de Valencia, Spain |
| Gary Ushaw | Newcastle University, UK |
| Aristides Vagelatos | CTI, Greece |
| Christine Verdier | LIG – University Joseph Fourier Grenoble, France |
| Francisco Veredas | Universidad de Málaga, Spain |

JustinWan                     University of Waterloo, Canada
Rafal Wcislo                  AGH – University of Science and Technology in
                              Cracow, Poland
Janusz Wojtusiak              George Mason University, USA
Lixia Yao                     University of North Carolina at Charlotte, USA
Serhan Ziya                   University of North Carolina, USA
André Zúquete                 IEETA/IT/Universidade de Aveiro, Portugal

## HEALTHINF Additional Reviewers

Michael Bauer                 Heilbronn University, Germany
Giovanni Bottazzi             University of Rome Tor Vergata, Italy
Juan Pablo Suarez Coloma      University of Grenoble, France
Dirk-Tassilo Hettich          University of Tübingen, Germany
Marius Koller                 Hochschule Heilbronn, Germany
Thomas Kübler                 University of Tübingen, Germany
Silvia Macis                  University of Cagliari, Italy
Sebastian Rauh                Heilbronn University, Germany

## Invited Speakers

Erik Meijering                Erasmus University Medical Center, The Netherlands
Lionel Pazart                 CHU, France
David Rose                    MIT Media Lab, USA

# Contents

## Bioinformatics Models, Methods and Algorithms

## Bio-inspired Systems and Signal Processing

**Health Informatics**

# Invited Paper

# How to Cross the Border from R to D? The Example of Conception of New Medical Devices

Lionel Pazart[✉]

INSERM CIC 1431, Besançon University Hospital,
Place Saint Jacques, 25030 Besançon cedex, France
lpazart@chu-besancon.fr

**Abstract.** The border between Research and Development for a new medical device is often unclear due to non-linear process of its development and frequent feedbacks from trials in clinical settings to a new conception of the product. Sometimes researchers under-estimate these translational studies since they do not lead to an increase of fundamental knowledge. However, and especially in the field of medical devices, users have to face specific difficulties due to the variability of the biological systems under study. Results obtained in translational research often depend on this variability and new questions or scientific obstacles arise from the confrontation to the real world. In order to address these new challenges, reverse translational research is required. Fundamental research is then fuelled by the results of translational research. A useful model of medical device development is presented here through several examples of translational research.

**Keywords:** Translational research · Medical device · Innovation · R&D - clinical trials

## 1 Introduction

Fundamental research rarely knows what discovery will be of use and disruptive innovations in health mostly come from basic research whose authors have not expected the consequences (for instance, the discovery of electron spin in 1922 to the MRI in 70's). Conversely, applied research is primarily directed towards a specific practical objective (for instance the long story to capture and preserve images began with the Egyptians some ten thousand years ago when they noticed the ability of light to transmit images). Between these enemy brothers, experimental development is a systematic work, using knowledge gained from basic research and/or practical experience, which is directed to produce new products or to improve substantially those already existing. In order to transform fundamental research results into practical innovation, translational research implies difficult efforts to define future applications, and requires thinking differently and changing the mindset of researchers. Research and Development for a new medical device is often unclear since the process of development of a new medical device remains non-linear and requires feedbacks from trials in clinical setting to a new conception of the product. Sometimes researchers neglect these further studies because it is thought that, although essential to create innovative technologies, they do not lead to an increase

© Springer International Publishing Switzerland 2015
A. Fred et al. (Eds): BIOSTEC 2015, CCIS 574, pp. 3–13, 2015.
DOI: 10.1007/978-3-319-27707-3_1

of scientific knowledge. However, and especially in the field of medical devices, users have to face specific difficulties due to the variability of the biological systems under study. Variability from one patient to the other is easily understood. But there is also the variability of a single patient whose metabolism evolves naturally over time and additionally with the therapeutic actions. Translational research has to understand this variability and new fundamental research questions arise from the confrontation to the real world. In order to address these new challenges, reverse translational research is required. Fundamental research is then fuelled by the results of translational research. In this overview, we would like to present a useful model of translational research for medical device development through several examples in order to illustrate the adequacy of research to bring fundamental research results as closely as possible to the patients.

## 2    Classification of Research Activities

The Frascati Manual [1] defines R&D as *"creative work undertaken on a systematic basis in order to increase the stock of knowledge"*. The term R&D covers in fact three activities: basic research, applied research and experimental development.

- *Basic research* is experimental or theoretical work undertaken primarily to acquire new knowledge of the underlying foundation of phenomena and observable facts, without any particular application or use in view.
- *Applied research* is also original investigation directed primarily towards a specific practical aim. In technical fields, applied research could be associated with the '*industrial research*' to develop new products or processes.
- *Experimental development* is systematic work, drawing on existing knowledge gained from research and/or practical experience, which is directed to producing new materials, products or devices, to installing new processes, systems and services, or to improving substantially those already produced.

For example, the determination of the amino acid sequence of an antibody molecule would be basic research. Investigations undertaken to identify the antibody specifically binding to a viral membrane protein would be applied research. Experimental development would then consist in designing a biochip functionalized with the appropriate antibody specific to the disease. This is done on the basis of the knowledge of the antibody structure and after clinically testing it with a biological liquid of interest (blood, urine, etc.) in order to make the right diagnosis.

*Translational Research* involves, for the National Institutes of Health (NIH), *"the extensive body of work required moving a discovery from bench to bedside"* and the Wikipedia definition insists on the capacity of translational research to shorten the time-frame to be placed on the market. According to these definitions, translational research spans from applied research and experimental development to the market launch of the product. In healthcare fields, others add to these definitions the "bedside-to-bench" feedback loop.

## 3  Challenges of R&D on Medical Devices

With the great diversity of medical devices ranging from crutches to programmable pacemakers, it is not feasible to subject all medical devices to the same development scheme. Much specificity of medical devices vs. drugs should be taken into account [2] and methodological adaptations should be done to offer a feasible methodology:

- clinical investigation is particularly needed to get CE mark for class IIb and III medical devices [3]. First tests in humans are needed to ensure that the device fulfils the essential requirements and to demonstrate its safety by in vitro, bench and in vivo (animals) tests. For other types of devices (ex: gloves, eyes occlusion plasters, conductive gels, non-invasive electrodes, image intensifying screens) predictability of performance could be a useful manner to answer the question.
- the product may be of interest in various fields such as treatment, diagnosis, or compensation of disability and the methodology of clinical assessment should therefore be adapted to the main objective of the study. For instance, clinical trials of diagnostic tests are sometimes divided into exploratory phases, challenge phases and advanced phases to see how effective and how accurate the tests are [4]. In all cases, a distinction should be made between the clinical proof of efficacy and safety in order to get the market approval and the place in the diagnostic or therapeutic armamentarium in order to define the price or the reimbursement of a medical device. Finally, randomized controlled trials are generally conducted to compare a new intervention or strategy to the classical one.
- the level of innovation; should the new medical device be considered as an incremental evolution or as a rupture evolution? Minor or incremental changes on an existing medical device are the most frequent type of innovation activity in companies. Activities leading to minor, incremental changes or adaptations should in principle not be counted as R&D activities, unless they are part of, or result from, a formal R&D project in the firm.
- the equivalence with the legally marketed device known as the predicate; substantial equivalence means that the new device is at least as safe and effective as the predicate [5]. This concept can be applied to many products including high-risk products, such as coronary stents or hip prostheses. In order to prove the equivalence, technical bench tests and preclinical studies may be done. Production of specific clinical data may be limited to a cohort study to obtain similar results to the predicate. However, this applies only if the equivalence criteria are not affected claim, clinical and technical data and environment.
- the operator/medical device interaction; the clinical benefit may depend not only on the medical device itself but also on the performance of the medical team (operator dependent nature, learning curve) and the technical platform. This organizational dimension is an element which must be taken into account in the early investigations of a new medical device; trials should incorporate this learning curve by providing a first acquisition phase, in the number of subjects required for example, and/or any interim analyzes. Another possibility is to use a sequential adaptive two-stage design (i.e. Fleming methods)

- the diversity of use; one or more studies are needed to develop the implementation of a new medical device and to describe various operators (medical staff or the patient himself), operating times, technical facilities and personnel skill required for the success of the procedure.
- the reduced life cycle. The clinical assessment should be realized in short-term monitoring, on technical and clinical intermediate parameters. Nevertheless, a long term monitoring should be performed till failure occurrence for all patients who were implanted with an old version of medical device (particularly for implantable devices like cardiac prostheses, breast implants, cochlear implants etc.).
- the small size of target population. Particular methodological solutions can be offered: conducting multicentre clinical trials in Europe (within ECRIN network for instance), or exhaustive surveys of patients through national or international registers.
- the short track of development; a lot of medical devices could be developed with few technical tests to get the Proof of Concept without clinical tests, like for instance dental impression materials, tubes used to pump the stomach, urinary catheters intended for transient use etc. For other medical devices categories, the absence of an animal model to test preclinical medical device and the impossibility to test it on healthy volunteers contribute to a rapid transfer to the patient, for instance for hip prostheses or implantable analgesic pumps.

## 4    Practical Situations of Translational Research

### 4.1    Optical Biopsy

Nowadays invasive biopsy is still the reference diagnostic technique of a lot of skin or mucosa diseases (inflammation, tumours). Nevertheless, several diagnosis methods should be kept as minimally invasive as possible. Consequently, non-invasive imaging techniques (ultrasound, computed tomography, magnetic resonance imaging) have been developed for clinical use. Based on the principle of white-light interferometry and developed initially in 1991 for in-vivo imaging of the human eye [6], OCT was investigated by a large number of groups worldwide. With regards to penetration depth and resolution, OCT could be a perfect trade-off between ultrasound and confocal microscopy. The use of optically based on specific swept sources for OCT was first demonstrated in 2011 but since that time, the threshold towards the use of low-cost electrically-pumped devices has still not been reached.

How to translate the basic knowledge to a practical application in healthcare?

A first way could be to define the possible application field (for instance skin biopsy) and ask the specialists (here university dermatologists) about the potential clinical use in accordance with the technical characteristics of the future device (e.g.: spatial resolution, field of view and imaging magnification). The design parameters will be selected according to the system specifications and technological constraints, for instance a reduced size ($<15cm^3$), low cost OCT imager providing cross-sectional 3-D tomograms with a depth around 0.5 mm, axial and transverse resolutions of 5 μm and imaging field of $5 \times 5$ mm$^2$. Of course, specialists could envision possible clinical applications [7] such as superficial baso-cellular cancer, follow-up of healing after an injury or surgery, assessment of new wound dressings or graft, determination of the degree of skin burns, the local efficacy and tolerance of topical treatments etc. But the usefulness of such an OCT imager remains questionable, since pictures interpretation may be difficult for clinicians who have no experience feedbacks about such imaging. The learning curve is currently very slow with those new techniques (new images, new colours, new field of view...), which is currently a real limiting factor for the diffusion of those technologies.

A second approach could the specifications based on clinical use. First of all, dermatologists are invited to express their needs (Fig. 1). To this end, they describe a new device able to provide detection of early skin cancer by differentiating abnormal and healthy skin, and able to help the practitioner to accurately determine the margins for resection. This can usually be done by the examination of the overall architecture of epidermis and counting the number of atypical cells per unit of area.

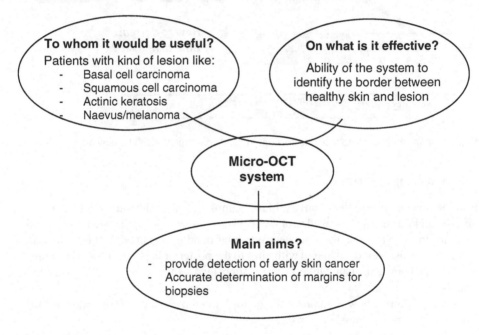

**Fig. 1.** Model to express the clinical needs for a new Micro-OCT system for dermatology.

Thus, the parameters of new OCT microsystem have to be determined by examining the biomedical application requirements as well as the instrumental characteristics of selected interferometric architecture including array-type as well as high-speed camera requirements. The Medical ISO13485 methodology requires also a Risk Analysis of the final product. It must be initiated with all participants and especially the future users. A Functional Analysis can then describe what is expected from the medical device and split it in building blocks.

Current works are trying to improve the accuracy, resolution, penetration depth of these devices. Manufacturers and researchers should focus their insights on the easiness of recording, measuring and analyzing, the daily practice in doctor office, the reliability and the price.

To summarize, the best approach could be: analysing the constraints of available techniques, defining the needs from the end-user (medical) point of view and adapting research program to conciliate both requirements. The diagram in Fig. 2 tries to represent this approach.

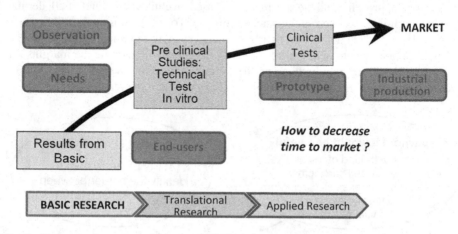

**Fig. 2.** Main components to take into account in the development plan of a new medical device.

## 4.2 Screening at Birth

Routine screening tests by capillary blood sample at birth aims at detecting several diseases in France. The lateral edge of the feet was chosen as sampling area by scientific societies in order to avoid the main neurovascular bundles and the risk of osteomyelitis of the calcaneus, previously found with bites to the posterior heel. The method is painful for the newborn and quantitative failures often lead to the need for a second blood sampling.

The question is: how to improve the quality and the capacity of screening at birth, particularly to reduce newborns' pain?

***The First Way consists in Searching for Available Techniques on the Shelves, then to try to adapt them to the Need.*** Micro-needles array appears to be a good solution to replace the lancet (Fig. 3). This matrix would be applied on the heel as a patch, the multitude of micro-needle (deemed not painful) replacing the wide blade of the lancet.

But a lot of questions emerge to adapt this technology to the heel of newborns:

- How deep to prick?
- Which spatial density of needles should be considered?
- Which size for the internal channel, if any?

**Fig. 3.** Micro needles array, (source FEMTO-ST).

A better understanding of the distribution of capillary networks could improve the specifications of a new device based on micro-needles array. To collect these data, we conducted a clinical study [8] using ultrasound (device Dermacup Atys) and videocapillaroscopy (device Moritex, MS-500C Micro-Scopeman) on both sides (lateral and medial edges) of the heel of 62 newborns according to gestational age at birth. The parameters of the microcirculation were obtained by ultrasonography (depth of dermis) and capillaroscopy (capillary density and distribution, inter-capillary distance and average diameter of the capillaries). The results show that on average, the density of the capillary network is 60 capillaries/mm$^2$, the inter-capillary distance of 155 μm and a diameter of 22 μm. Another result shows that the capillary network is oriented mainly parallel to the lateral edge of the foot and is less present on the medial edge.

From our study, the results of capillaroscopy and skin ultrasound will help determining the right micro-needles array configuration as follows:

- The area provided for the needle plane is 25 mm$^2$, and the number of micro-needles depends on the density of capillaries and of the inter-capillary distance
- The depth of the dermis specified the maximum depth of the micro-needles.

After some prototypes adaptations, we performed several tests on animals. The results show that a network of 8 micro-needles could be acceptable and avoid any "fakir effect" of the skin. But these micro-needles must penetrate about 1 mm in the heel of the newborn and three applications of the matrix are needed to achieve a 96 % probability of blood collection. Under these conditions, it is difficult to talk about painless sampling.

*A Second Way to Solve the Problem of pain consists firstly in understanding the mechanisms of this painful process* in order to provide inputs for improvement of the medical device and also to explore new avenues for screening at birth.

A systematic clinical observation of blood collection steps at the 72th hour of life was conducted on a sample of 50 newborns (58 stings). The purpose of this observation was on the one hand to get confirmation of the frequency of pain, when it appears and on the other hand to understand the factors behind its occurrence on which further action could be taken. The anatomical data from videocapillaroscopy and ultrasound were also collected to be correlated with results of observation of the act of screening in terms of

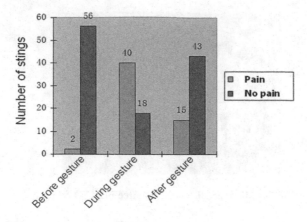

**Fig. 4.** Pain before, during and after the screening gesture in newborns (Source Inserm CIC 1431).

pain (DAN scale) and quantity of blood obtained. 89 % of newborns have expressed pain. It appeared that pain is mainly observed when pressure is applied on the newborn's heel to collect blood on the blotter paper and secondarily at the heel prick with the lancet (Fig. 4). As a consequence, the most painful is not the bite on the foot but the pressure itself. The pain at the sting of the lateral edge appears less important than the sting of the medial edge, but this result should be confirmed by a more powerful study with more cases. No correlation was found between pain and deep dermis, or density, or the diameter of the capillary.

Considering these results, we were committed to finding new methods of blood collection, particularly to avoid pressure on the heel and the proliferation of bites to get the blood in sufficient quantity on the blotter paper (Fig. 5).

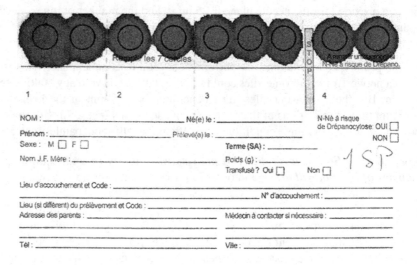

**Fig. 5.** The blotter paper to collect blood from newborn's heel (Source Inserm CIC 1431).

We have therefore developed a system provided with a micro-machined nozzle [9]. This tip is applied to the heel, after bite of the lancet. The blood viscosity properties and the geometry of the tip make possible to maintain the blood captive inside a reservoir. This tip is then stamped on the blotter paper. Tests have shown that a volume of less than 800 μl of blood is sufficient to properly soak the spaces provided on the blotter paper.

To go on with the development of the new device, it is planned to conduct clinical trials on parallel groups for ethical reasons (not to stress newborn twice), with prototypes and classic screening methods.

In summary, knowledge acquired from these practical experiences change our mind of thinking about the way to cross the border from basic research to applied research. The Fig. 6 illustrates the chronological steps from idea to market and replace the fundamental research as a provider of solutions, either existing or developing.

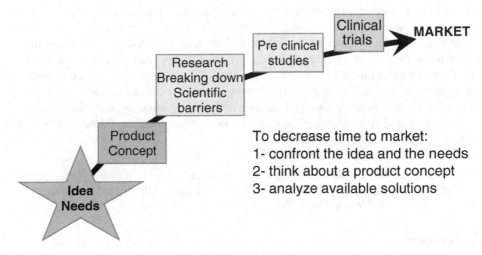

**Fig. 6.** Chronological steps toward the development plan of a new medical device.

In such a scheme, the first challenge is to confront the new idea to the needs in terms of practice (unsatisfied clinical needs), market (awaiting bankable innovation) and society (uncovered social needs). This confrontation mostly concerns end-users, stakeholder and industrials more than researchers. The second challenge should be to transform an idea in a product concept. These engineering and design tasks involve mainly engineer either from labs or companies, but also end-users in order to specify the right technical characteristics for a useful product. Then, an interaction with fundamental researchers could analyze available solutions or define the way to search. This part is crucial in order to get sense to the development of a new medical product. The forwarded steps are more common with pre-clinical and clinical trials before launching the product on the market. More accurately, a global model [10] should to take into account, in parallel, research aspects, regulatory constraints, economics and industrials approach.

## 5  Conclusion

To conclude, in this conference, we considered examples of the conception of medical devices either based on technology availability or on clinical use requirements. In both approaches, identification of clinical useful technical characteristics is a critical issue to accelerate the development of new medical devices. As mentioned in the introduction, translational research activities are sometimes under-estimated and postponed because they do not lead to an increase of scientific knowledge.

However, translational research covers applied research and experimental development and it is essential to create new tools especially in the field of medical devices. It is not only a question of semantics as illustrated few years ago in a similar congress, but a way of thinking.

If we design a clinical study to describe the capillary network of newborns without any objective other than knowledge, this study should fall into the basic research category. If the same study is intended to complete technical specifications for a new screening device, it becomes applied research. The translation is there, behind the aims of the study protocol: Have the study results any practical application? Basic researchers should be aware of this new paradigm, even if they do not have to focus their attention on application. Louis Pasteur perfectly summarized this necessary connection: *"There is no such thing as basic research on one side and applied research on the other. There are research and applications thereof, united to each other as the fruit of the tree is joined to the branch that bears it."*

**Acknowledgements.** The studies presented in this paper are funded by public grants, under the European Commission's 7th Framework Program, or French Health Ministry and French National Agency for Research. We would like to thank all researchers and patients implied in these studies.

## References

1. OECD: Frascati Manual 2002: Proposed Standard Practice for Surveys on Research and Experimental Development, The Measurement of Scientific and Technological Activities, OECD Paris (2002)
2. Parquin, F., Audry, A., Giens, X.X.V.I.I.: Constraints and specificities of the clinical evaluation of medical devices. Thérapie **67**(4), 301–309 (2012)
3. Annex IX from 93/42/EEC concerning medical devices as amended by Directive 2007/47/EC
4. ISO 14155:2011 Clinical investigation of medical devices for human subjects - Good clinical practice (2011)
5. Food and Drugs Administration, Premarket Notification (510 k)
6. Fujimoto, J., et al.: Biomedical Optical Imaging. Oxford University Press, Oxford (2009)
7. Sattle, E., Kästle, R., Welzel, J.: Optical coherence tomography in dermatology. J. Biomed. Opt. **18**(6), 061224 (2013)
8. Robieux, C., Sainthilier, J.M., Vidal, C., et al.: Study of the correlation between the structure of the capillary network of the heel of newborns and pain during screening at birth. J. Invest. Dermatol. **131**, 2144 (2011)

9. Wacogne, B., Pieralli, C., Cabodevila, G., Baron, N., Marioli, S., Pazart, L.: Pain and efficiency in neonatal blood sample screenings, New devices for reducing pain and improving blood sample quality. In: Proceedings of the International Conference on Biomedical Electronics and Devices. Biodevices, Porto, Portugal, pp. 290–295 (14–17 January 2009). doi:10.5220/0001776902900295
10. Moreau-Gaudry, A., Pazart, L.: Development of an innovation in health: the cycle CREPS Concept - Research - Evaluation - Product – Care. IRBM **31**(1), 12–21 (2010)

# Biomedical Electronics and Devices

# Low Power Programmable Gain Analog to Digital Converter for Integrated Neural Implant Front End

Amir Zjajo[✉], Carlo Galuzzi, and Rene van Leuken

Delft University of Technology, Delft, The Netherlands
a.zjajo@tudelft.nl

**Abstract.** Integrated neural implants interface with the brain using biocompatible electrodes to provide high yield cell recordings, large channel counts and access to spike data and/or field potentials with high signal-to-noise ratio. By increasing the number of recording electrodes, spatially broad analysis can be performed that can provide insights on how and why neuronal ensembles synchronize their activity. However, the maximum number of channels is constrained by noise, area, bandwidth, power, thermal dissipation and the scalability and expandability of the recording system. In this chapter, we characterize the noise fluctuations on a circuit-architecture level for efficient hardware implementation of programmable gain analog to digital converter for neural signal-processing. This approach provides key insight required to address signal-to-noise ratio, response time, and linearity of the physical electronic interface. The proposed methodology is evaluated on a prototype converter designed in standard single poly, six metal 90-nm CMOS process.

## 1 Introduction

Bio-electronic interfaces allow the interaction with neural cells by both recording, to facilitate early diagnosis and predict intended behavior before undertaking any preventive or corrective actions [1], and stimulation devices, to prevent the onset of detrimental neural activity, such as the one resulting in tremor. Monitoring large scale neuronal activity and diagnosing neural disorders has been accelerated by the fabrication of miniaturized micro-electrode arrays, capable of simultaneously recording neural signals from hundreds of channels [2]. By increasing the number of recording electrodes, spatially broad analysis of local field potentials can be performed to provide insights into how and why neuronal ensembles synchronize their activity. Studies on body motor systems have uncovered how kinematic parameters of movement control are encoded in neuronal spike time-stamps [3] and inter-spike intervals [4]. Neurons produce spikes of nearly identical amplitude near to the soma, but the measured signal depends on the position of the electrode relative to the cell. Additionally, the signal quality in neural interface front-end, beside the specifics of the electrode material and the electrode/tissue interface, is limited by the nature of the bio-potential signal and its biological background noise, dictating system resources.

For any portable or implantable device, micro-electrode arrays require miniature electronics locally to amplify the weak neural signals, filter out noise and out-of band

E. Tambouris et al. (Eds.): EGOV 2015, LNCS 9248, pp. 17–32, 2015.
DOI: 10.1007/978-3-319-27707-3_2

interference and digitize for transmission. A single-channel [5] or a multi-channel integrated neural amplifiers and converters provide the front-line interface between recording electrode and signal conditioning circuits and, thus, face critical performance requirements. Multi-channel, fully differential designs allow for spatial neural recording and stimulation at multiple sites [6–8]. The maximum number of channels is constrained by noise, area, bandwidth, power [9], which has to be supplied to the implant from outside, thermal dissipation e.g. to avoid necrosis of the tissues even with a moderate heat flux [10], and the scalability and expandability of the recording system.

The block diagram of a typical neural recording system architecture is illustrated in Fig. 1(a). When a neuron fires an action potential, the cell membrane becomes depolarized by the opening of voltage-controlled neuron channels leading to a flow of current both inside and outside the neuron. Since extracellular media is resistive [11], the extracellular potential is approximately proportional to the current across the neuron membrane [12]. The membrane roughly behaves like an $RC$ circuit and most current flows through the membrane capacitance [13]. The data acquired by the recording electrodes is conditioned using analog circuits. As a result of the small amplitude of neural signals and the high impedance of the electrode tissue interface, amplification and bandpass filtering of the neural signals is performed before the signals can be digitized by a successive approximation register (SAR)-based analog to digital converter (ADC) [14]. To avoid the large capacitive DACs found in the SAR ADC, and to lower demands on driving capabilities of the amplifier and relax power, noise and cross-talk requirements, in the alternative architecture illustrated in Fig. 1(b), the programmable gain amplifier (PGA) and ADC are combined and embedded in every recording channel.

The programmable gain analog to digital converter (PG ADC) implements simultaneously both signal acquisition and amplification, and data conversion. As illustrated in Fig. 2 [15], the schematic incorporates a fully-differential operational transconductance

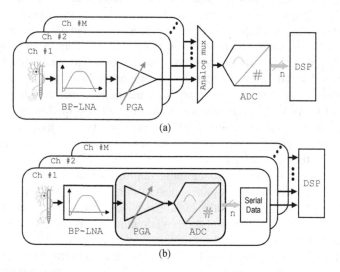

(a)

(b)

**Fig. 1.** Multichannel neural interfaces: (a) An ADC is multiplexed between M channels, (b) An ADC per channel with serial interfacing.

amplifier (OTA), a comparator and circuitry for control of the acquisition and amplifi-
cation operation set by the clock phases $\varphi_{s1}$, $\varphi_{s2}$ and $\varphi_{s3}$ and output generation data
conversion operation, controlled by the clock phases $\varphi_1$ and $\varphi_2$. The recorded signals are
capacitively coupled to the input of the amplifier to reject the dc polarization. The differential input signal is sampled, amplified by the capacitance ratio (gain $G^A$ is adjustable
by implementing $C_3$ as a programmable capacitor array, $G^A = C_3/C_4$), and transferred to
the integration capacitors $C_4$ at the feedback loop of the OTA. At data conversion operation, the differential signal stored in $C_4$ is converted to digital domain by successively
adding or subtracting binary-scaled versions of the reference voltage to the integration
capacitors. Voltage addition or subtraction is implemented by means of the four cross-
coupled switches controlled by the signals $\varphi_{2p}$ and $\varphi_{2n}$. Digital representation of the output
signals are then sequentially stored in the SAR register for further processing.

A low-power monolithic digital signal processing (DSP) unit provides additional
filtering and executes spike discrimination and sorting algorithms [16]. The relevant
information is then transmitted to an outside receiver through the transmitter or used for
stimulation in a closed-loop framework. Understanding the role of noise in such systems
is one of the central challenges in the heterogeneous neural simulation and neural rehabilitation [17]. In this chapter, we try to characterize the noise fluctuations on a circuit-
architecture level for efficient hardware implementation of neural signal-processing
circuitry. This approach provides the key insight required to address signal-to-noise ratio
(SNR), response time, and linearity of the physical electronic interface (i.e., saturation
level).

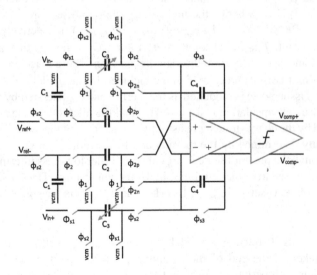

**Fig. 2.** Schematic of a programmable gain ADC.

# 2  Noise Characterization

## 2.1  Noise Models

**Neural Cell Noise Model.** In the Hodgkin and Huxley framework, the configuration of the neural channel is determined by the states of its constituent subunits, where each subunit can be either in an open or closed state [18]. Adding a noise term $\chi_x(V,t)$ ($x = m,h$, or $n$) to the deterministic ordinary differential equation (ODE) of Hodgkin and Huxley is consistent with the behavior of the Markov process for channel gating [19]. Such process can be contracted to a Langevin description (via a Fokker–Planck equation) and expressed as delta-correlated noise processes $\Gamma_{neuron}(t + \tau,t) = 1/\kappa[\alpha_x(1-x) + \beta_x x]\delta(\tau)$, where $\kappa$ is the total number of neural channels, and the transition rates $\alpha_x(t)$ and $\beta_x(t)$ are instantaneous functions of the membrane potential $V(t)$. Dirac's delta function $\delta$ designates that the noise at different times is uncorrelated and the variables $m$, $h$, and $n$ represent the aggregated fraction of open subunits of different types, aggregated across the entire cell membrane.

**Electrode-Tissue Interface Noise Model.** The overall noise of an electrode-tissue interface has contributions from the tissue/bulk thermal noise, the electrode-electrolyte interface noise, and the electronic noise. The most important types of electrical noise sources (thermal, shot, and flicker noise) in passive elements and integrated circuit devices have been investigated extensively, and appropriate models derived [20] as stationary and in [21] as nonstationary noise sources. We adapt model descriptions as defined in [21], where thermal and shot noise are expressed as $\Gamma_{thermal}(t + \tau,t) = 2kTG(t)\delta(\tau)$ and $\Gamma_{shot}(t + \tau,t) = qI_D(t)\delta(\tau)$, respectively, where $k$ is Boltzmann's constant, $T$ is the absolute temperature, $G$ is the conductance, $q$ is the electron charge, and $I_D$ is the current through the junction. These noise processes correspond to the current noise sources, which are included in the models of the integrated circuit devices. Tissue noise is modelled as the thermal noise generated by the solution/spreading or tissue/encapsulation resistance [13] and the electrode noise is the thermal noise generated by the charge transfer resistor [22]. The noise of the recording electronic circuits is mainly determined by the thermal and flicker noise generated by the input amplifier. Although the preamplifier can provide first-order low-pass filtering, dedicated low-pass filters are used to further minimize high-frequency noise. The cut-off frequency of the low-pass filters is set to $f_{Neuron} = 10$ kHz, where $f_{Neuron}$ is the signal bandwidth of the action potential.

**A/D Converter Noise Model.** Sampled data systems operate on the series of discrete-time samples taken at the end of the sampling period. Although the details of the processing during each period result in nonstationary noise voltages and currents, the same operation is performed each clock cycle, leading to the same signal statistics at each cycle. Consequently, such stochastic process can be described as wide-sense cyclostationary.

The special case of a white noise input source is of particular importance since the majority of the noise sources can be traced back to white noise generated in circuit components. For a white noise step input, the autocorrelation is a delta function, where $S_{xo}$ is the one-sided white noise power spectral density (PSD) of the underlying noise process. By using Parseval's theorem, the variance of the output as a function of the autocorrelation simplifies to $\Gamma(t + \tau,t) = 1/2S_{xo}(t)\delta(\tau)$ [23]. The one-sided noise PSD of the sampled output can then be found from the sum of the filtered and shifted two-sided input noise PSD $S_x(f)$ [24]. Measurements of the output codes for a dc input signal to the A/D converter can be used to obtain an input-referred noise PSD estimate, $S_{ADC}(f)$. The noise of the input sampler and the converter quantization noise add to the input-referred noise PSD to give the total input noise PSD $S_{total}(f) = S_{sample}(f) + S_{ADC}(f) + S_q(f)$, where $S_{sample}(f) = (kT/C_s)/(f_s/2)$ is the noise PSD from the input sampler over the Nyquist range ($0 \leq f_{Neuron} \leq f_s/2$) and $S_q(f) = (V_{LSB}^2/12)/(f_s/2)$ is the A/D converter quantization noise.

## 2.2  Noise Analysis of Programmable Gain Analog to Digital Converter

A fundamental technique to reduce the noise level, or to increase the signal-to-noise ratio of a programmable gain ADC, is to increase the size of the sampling capacitors, or by either over-sampling or with calibration. However, for a fixed input bandwidth specification, the penalty associated with these techniques is the increased power consumption. Consequently, a fundamental trade-off exists between noise, speed, and power dissipation. For discrete-time analog signal-processing circuits, analog signals are acquired and processed consecutively, and a sample of the signal is taken periodically, according to a clock signal. As the sampling circuit cannot differentiate the noise from the signal, part of this signal acquisition corresponds to the instantaneous value of the noise at the moment the sampling takes place. In this context, when the sample is stored as charge on a capacitor, the root-mean-square total integrated thermal noise voltage is $kT/C_4$, where $kT$ is the thermal energy. This noise usually comprises two major contributions - the channel noise of the switches, which is a function of the channel resistance and the OTA noise.

The OTA output noise is, in most cases, dominated by the channel noise of the input transistors, where the thermal noise and the $1/f$ noise both contribute. If the input transistors of the OTA are biased in saturation region to derive large transconductance $g_m$, impact ionization and hot carrier effect will enhance their thermal noise level [25]. Similarly, the $1/f$ noise increases as well, due to the reduced gate capacitance resulted from finer lithography and, therefore, shorter minimum gate length. As a consequence, an accurate consideration of the intrinsic noise sources in such a circuit should have the thermal noise of switches and all amplifier noises readily included. Nevertheless, the input-referred noise $v_n$ (the total integrated output noise as well) still takes the form of $kT/C$, with some correction factor $\chi_1$,

$$\overline{v_n^2} = \chi_1 kT/C_4 \tag{1}$$

**kT/C Noise.** During the acquisition process, the *kT/C* noise is sampled on the capacitors $C_4$ along with the input signal. To determine the total noise charge sampled onto the capacitor network, the noise charge $Q_{ns}$ is integrated over all frequencies

$$\overline{Q_{ns}^2} = \int_0^\infty \left| \frac{V_{ns}(C_4 + C_p + C_{OTA})}{1 + j\omega R_{on}(C_4 + C_p + C_{OTA})} \right|^2 d\omega = kT(C_4 + C_p + C_{OTA}) \qquad (2)$$

where $R_{on}$ is the resistance of the switch, $V_{ns}$ is the noise source, $C_p$ is the parasitic capacitance and $C_{OTA}$ is the input capacitance of the OTA. Then, in the conversion mode, the sampling capacitor $C_4$, which now contains the signal value and the offset of the OTA, is connected across the OTA. The total noise charge will cause an output voltage of

$$\overline{v_{ns(out)}^2} = \frac{\overline{Q_{ns}^2}}{C_4^2} = kT\frac{(C_4 + C_p + C_{OTA})}{C_4^2} = \frac{1}{\beta}\frac{kT}{C_4} \qquad (3)$$

where $\beta$ is the feedback factor. For differential implementation of the circuit, the noise power of the previous equation increases by a factor of 2, assuming no correlation between positive side and negative side, since the uncorrelated noise adds in power. Thus, input referred noise power, which is found by dividing the output noise power by the square of the gain ($G^A = C_3/C_4$), is given by

$$\overline{v_{ns(in)}^2} = \frac{\overline{v_{ns(out)}^2}}{(G^A)^2} = \frac{1}{\beta(G^A)^2}\frac{kT}{C_4} \qquad (4)$$

**OTA Noise in Conversion Mode.** The resistive channel of the MOS devices in OTA has also thermal noise and contributes to the input referred noise of the PG ADC circuit. The noise power at the output is found from

$$\overline{v_{ns(out)}^2} = \int_0^\infty \left( \left| H\left(s|_{j\omega}\right) \right|^2 \times \overline{i_{ns}^2} \right) d\omega = \frac{kT\gamma}{C_{LT}}\frac{G_m R_o}{(1 + G_m R_o \beta)} = \frac{\gamma}{\beta}\frac{\cdot kT}{C_{LT}} \qquad (5)$$

where $R_o$ is the output resistance and $C_{LT}$ is the capacitance loading at the output

$$C_{LT} = C_L + \beta \times (C_p + C_{OTA}) \qquad (6)$$

The thermal noise coefficient $\gamma$ depends on the effective mobility and channel length modulation [26]; it is 2/3 for older technology nodes and between 0.6 and 1.3 for submicron technologies [27]. Assuming $G_m R_o \beta \gg 1$, and gain of the conversion operation $G^C = C_2/C_4$, the input referred noise variance is

$$\overline{v_{ns(in)}^2} = \frac{\gamma}{\beta(G^C)^2}\frac{kT}{C_{LT}} \qquad (7)$$

**Total Input Referred Noise.** The noise from acquisition and conversion mode can be added together to find the total input referred noise assuming that the two noise sources are uncorrelated. Using the results from (4) and (7), the total input referred noise power for differential input is given by

$$\overline{v^2_{ns(in)}} = \frac{2\gamma}{\beta(G^C)^2}\frac{kT}{C_{LT}} + \frac{2}{\beta(G^A)^2}\frac{kT}{C_4} = 2\gamma\frac{1}{\beta}\left(\frac{1}{(G^C)^2C_{LT}} + \frac{1}{\gamma(G^A)^2C_4}\right)\cdot kT \qquad (8)$$

The optimum gate capacitance of the OTA is proportional to the sampling capacitor $C_{OTA,opt} = \chi_2 C_4$, where $\chi_2$ is a circuit-dependent proportionality factor. The drain current $I_D$ yields

$$I_D = \frac{\chi_1^2 L^2 \omega_1^2 C_4}{\mu\chi_2} \qquad (9)$$

where $\mu$ is the carrier mobility, $\omega_1$ is the gain-bandwidth product, and $L$ is the channel length. For a noise dominated by $kT/C$, the total power consumption $P_T$ is found as

$$P_T \propto I_D V_{DD} = \frac{\chi_1^2 L^2 \omega_1^2 SNR \cdot 8kT}{\mu\chi_2}\frac{V_{DD}}{V^2_{max}} \qquad (10)$$

For a given speed requirement and signal swing, a two times reduction in noise voltage requires a four times increase in the sampling capacitance value and the OTA size. This means that the PG ADC circuit power *quadruples* for every additional bit resolved for a given speed requirement and supply voltage as illustrated in Figs. 3 and 4. Notice that for a small sampling capacitor values, the thermal noise limits the SNR, while for a large sampling capacitor, the SNR is limited by the quantization noise and the curve flattens out. Improving the power efficiency beyond topological changes of the OTA and supply voltage reduction require smart allocation of the biasing currents.

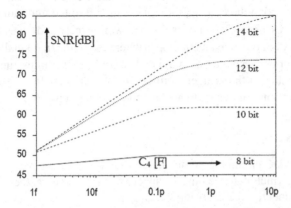

**Fig. 3.** Maximum achievable SNR for different sampling capacitor values and resolutions.

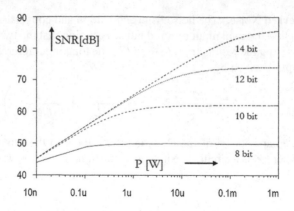

**Fig. 4.** SNR vs. Power dissipation.

Hence, techniques such as current reuse [28, 29], time multiplexing [4, 29] and adaptive duty-cycling of the entire analog front end [30, 31] can be used to improve power efficiency by exploiting the fact that neurons spikes are irregular and have low frequency. Choosing the OTA bandwidth too high increases the noise and, additionally, demands unnecessarily low on-resistance of the switches and, thus, large transistor dimensions. The optimum time constant remains constant regardless of the circuit size (or $I_D$) because $C_L$ scales together with $C_4$ and the parasitic capacitance $C_p$. The choice of the hold capacitor value is a trade-off between noise requirements on one hand and speed and power consumption on the other. The sampling action adds $kT/C$ noise to the system, which can only be reduced by increasing the hold capacitance $C_4$.

A large capacitance, on the other hand, increases the load of the operational amplifier and, thus, decreases the speed for a given power. The OTA size and its bias current for a given speed requirement and minimum power dissipation are determined using $\tau$-vs.-$C_4$ curves, as in Fig. 5. Note that for low frequency operation (where $\tau/\tau_t$ is large), the $C_{OTA}$ that achieves the minimum power dissipation for given settling time and noise requirements, usually, does not correspond to the minimum time constant point. This is a consequence of setting the $C_4/C_{OTA}$ ratio of the circuit to the minimum time constant point, which requires larger $C_{OTA}$ and results in power increase and excessive bandwidth. Near the speed limit of the given technology (where the ratio $\tau/\tau_t$ is small), however, the difference in power between the minimum power point and the minimum time constant point becomes smaller, as the stringent settling time requirement forces the $C_4/C_{OTA}$ ratio (Fig. 6) to be at its optimum value to achieve the maximum bandwidth.

## 2.3    The OTA Noise

The OTA in PG ADC circuit has some unique requirements; the most important is the input impedance, which must be purely capacitive so as to guarantee the conservation of charge. Consequently, the OTA input has to be either in the common source or the source follower configuration. Another characteristic feature is the load at the OTA output, which is typically purely capacitive and, as a result, the OTA output impedance

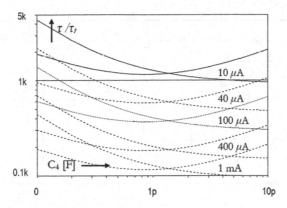

**Fig. 5.** Closed loop normalized time constant vs. Hold capacitance $C_H$ for different biasing conditions; case for $C_4 = 3C_L$, $C_L = C_p$. The time constant is normalized to the $\tau_t$ ($= 1/f_{t,intrinsic}$) of the device, which is approximately $(C_G/g_m)$.

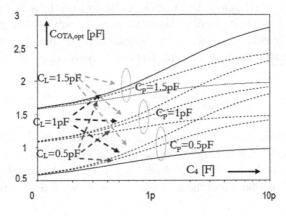

**Fig. 6.** Optimum gate capacitance $C_{OTA,opt}$ vs. Hold capacitance $C_4$ for different loading and parasitic conditions.

must be high. The benefit of driving solely capacitive loads is that no output voltage buffers are required. The implemented folded-cascode OTA and dynamic latch are illustrated in Fig. 7. The input stage of the OTA is provided with two extra transistors $T_{10}$ and $T_{11}$ in a common-source connection, having their gates connected to a desired reference common-mode voltage at the input, and their drains connected to the ground [32]. The advantage of this solution is that the common-mode range at the output is not restricted by a regulation circuit, and can approach a rail-to-rail behavior very closely.

The transistors of the output stage have two constrains: (*i*) the $g_m$ of the cascading transistors $T_{5,6}$ must be high enough, in order to boost the output resistance of the cascode, allowing a high enough *dc* gain and (*ii*) the saturation voltage of the active loads $T_{3,4}$ and $T_{7,8}$ must be maximized, in order to reduce the extra noise contribution of the output stage. These considerations underline a tradeoff between fitting the saturation

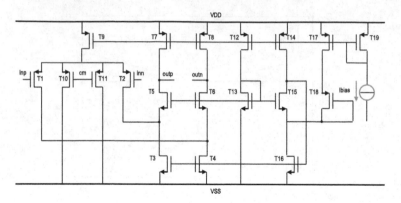

**Fig. 7.** OTA schematic.

voltage into the voltage headroom and minimizing the noise contribution. A good compromise is to make the cascading transistors larger than the active loads: in such a way, the $g_m$ of the cascading transistors is maximized, boosting the *dc* gain, while their saturation voltage is reduced, allowing for a larger saturation voltage for the active loads, without exceeding the voltage headroom. The output SNR is equal to

$$SNR_{out} = \frac{C_L}{2\gamma kT} \frac{A^2 \times g_{m1,2}R_{out}}{g_{m1,2} + g_{m3,4} + g_{m7,8}} \tag{11}$$

where $A$ is the amplitude of the input signal and $R_{out}$ denotes the open-loop output resistance of the OTA. From (11), it can be concluded that in order to maximize the output SNR, $C_L$ must be maximized, which means that the bandwidth must be minimized. The noise contribution of the individual transistors are shown in Fig. 8.

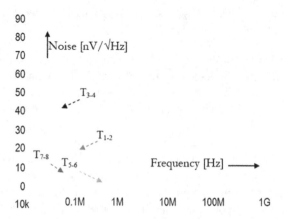

**Fig. 8.** Noise contribution of the individual transistors in the OTA.

The input-referred noise of the OTA input pair is reduced by increasing the $g_m$, increasing the current, or increasing the aspect ratio of the devices. The effect of the last method, however, is partially canceled by the increase in the noise excess factor. When referred to the OTA input, the noise voltages of the transistors used as current sources (or mirrors) in the first stages are multiplied by the $g_m$ of the device itself and divided by the $g_m$ of the input transistor, which again suggests that maximizing input pair $g_m$ minimizes noise. It can be further reduced by decreasing the $g_m$ of the current sources. Since the current is usually set by other requirements, the only possibility is to decrease the aspect ratio of the device. This leads to an increase in the gate overdrive voltage, which, as a positive side effect, also decreases $\gamma$. Increasing $L$ to avoid short channel effects is also possible, although with a constant aspect ratio it increases the parasitic capacitances.

### 2.4   The Comparator Noise

The dynamic latch illustrated in Fig. 9 consists of the pre-charge transistors $T_{14}$ and $T_{17}$, the cross-coupled inverter $T_{12\text{-}13}$ and $T_{15\text{-}16}$, the differential pair $T_{10}$ and $T_{11}$ and the switch $T_9$, which prevent the static current flow at the resetting period [33]. When the latch signal is low (resetting period), the drain voltages of $T_{10\text{-}11}$ are $V_{DD}\text{-}V_T$ and their source voltage is $V_T$ below the latch input common mode voltage. Therefore, once the latch signal goes high, the $n$-channel transistors $T_{11\text{-}13}$ immediately go into the active region. As each transistor in one of the cross-coupled inverters turns off, there is no static power dissipation from the latch, once the latch outputs are fully developed. A large portion of the total comparator current is allocated to the input branches to boost the input $gm$. Similarly, the noise from the non-gain element, i.e. the load transistor, is minimized, by applying a small biasing current. Additionally, small width and a large length for their gate dimensions are chosen.

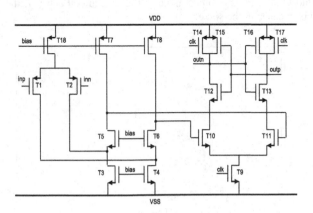

**Fig. 9.** Comparator schematic.

## 3   Experimental Results

The time series representation of neuron signal (Fig. 10) are composed of a spike burst, plus additive Gaussian white noise (grey area with 1000 randomly selected neural channel compartments, Fig. 11). In typical electrode-tissue interface, we are relying on the current measurement to sense these neural signals. Hence, by maintaining a constant current density, the relative uncertainty of the current becomes inversely proportional to the square of the interface area. The electrode noise spectral density has an approximate dependence of $-10$ dB/dec for small frequencies. However, for frequencies higher than 1–10 kHz, capacitances at the interface form the high-frequency pole and shape both the signal and the noise spectrum; the noise is low-pass filtered to the recording amplifier inputs. After band-pass filtering and amplification, the noisy neural signal is further processed with programmable gain analog to digital converter. The fluctuation of the voltage on the sampling capacitor is inversely proportional to the capacitance ( the variance of the capacitor voltage is $kT/C$ at any given time). This implies that with scaling, the uncertainty of the sampled voltage increases.

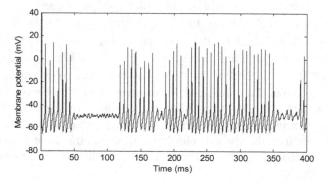

**Fig. 10.** Nominal (without noise) voltage trace of the neuron cell activity; the complex spike burst is followed by a pause in the spike activity.

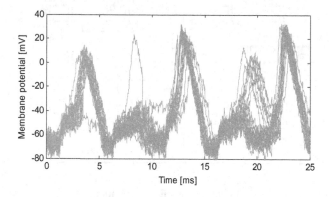

**Fig. 11.** Statistical voltage trace of the neuron cell activity; grey area - voltage traces from 1000 randomly selected neural channel compartments.

It can be seen that in both scenarios, in electrode-tissue interface and in PG ADC, the noise in the neural interface front end greatly increases, as the interface size reduces. The interface's input equivalent noise voltage decreases, as the gain across the amplifying stages increase, e.g. the ratio of the square of the signal power over its noise variance can be expressed as

$$SNR = A^2 / \left[ \overline{v_{ns(neural)}^2} + \overline{v_{ns(electrode)}^2} + \sum_i \left( \prod_j G_j^{-1} \right) \overline{v_{ns(amp,i)}^2} \right] \tag{12}$$

where $v_{ns(amp,i)}^2$ represents the variance of the noise added by the $i$th amplification stage with gains $G_j$. The variance of the electrode is denoted as $v_{ns(electrode)}^2$ and $v_{ns(neural)}^2$ is the variance of the biological neural noise. The observed SNR of the system also increases as the system is isomorphically scaled up, which suggests a fundamental trade-off between the SNR and the speed of the system. This lower bound on the speed in a converter loop is primarily a function of the technology's gate delay and the $kT/C$ noise multiplied by the number of SAR cycles necessary for one conversion. All PG ADC simulations were performed with a 1.2 V supply voltage at room temperature (25 °C). Spectral signature of PG ADC is illustrated in Fig. 12. The circuit offers a programmable amplification of 0–18 dB by digitally scaling the input capacitance $C_3$.

As shown in Fig. 13, the Signal-to-Noise and Distortion Ratio (SNDR), Spurious-Free Dynamic Range (SFDR) and Total Harmonic Distortion (THD) remain constant at different gain settings. The THD in the range of 10–100 kS/s is above 54 dB for a $f_{in}$ of 5 kHz (Fig. 14). Within the bandwidth of neural activity of up to 5 kHz, SNDR is above 44 dB and SFDR more than 57 dB. The degradation with a higher input signal is mainly due to the parasitic capacitance, the clock non-idealities and the substrate switching noise. The parasitic capacitance decreases the feedback factor resulting in an increased settling time constant. The non-idealities of the clock such as the clock jitter, the non-overlapping period time, the finite rising and the fall time, and unsymmetrical duty cycle are other causes for this degradation. The three latter errors reduce the time allocated for the setting time.

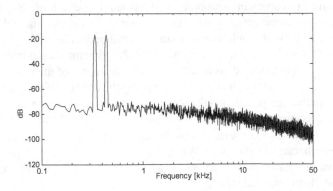

**Fig. 12.** Spectral signature of programmable gain A/D converter-two tone test.

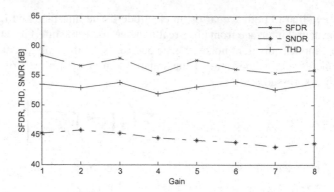

**Fig. 13.** SFDR, SNDR and THD vs. Gain settings.

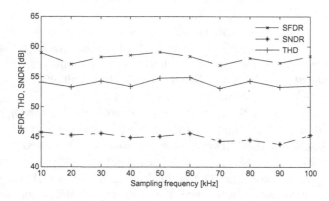

**Fig. 14.** SFDR, SNDR and THD vs. Sampling frequency, with $f_{in}$ = 5 kHz and gain set to one.

## 4    Conclusions

The high density of neurons in neurobiological tissue requires a large number of electrodes for accurate representation of neural activity. To develop neural prostheses capable of interfacing with single neurons and neuronal networks, multi-channel neural probes and the electrodes need to be customized to the anatomy and morphology of the recording site. The increasing density and the miniaturization of the functional blocks in these multi-electrode arrays, however, presents significant circuit design challenge in terms of area, bandwidth, power, and the scalability, programmability and expandability of the recording system. In this chapter, for one such functional block, programmable analog to digital converter, we evaluate the trade-off between noise, speed, and power dissipation and characterize the noise fluctuations on a circuit-architecture level. This approach provides the key insight required to address SNR, response time, and linearity of the physical electronic interface.

**Acknowledgements.** This research was supported in part by the European Union and the Dutch government as part of the CATRENE program under Heterogeneous INCEPTION project.

# References

1. Nicolelis, M.A.L.: Actions from thoughts. Nature **409**, 403–407 (2001)
2. Frey, U., et al.: An 11 k-electrode 126-channel high-density micro-electrode array to interact with electrogenic cells. IEEE International Solid-State Circuits Conference Digest of Technical Papers, pp. 158–159 (2007)
3. Georgopoulos, A.P., Schwartz, A.B., Kettner, R.E.: Neuronal population coding of movement direction. Science **233**(4771), 1416–1419 (1986)
4. Chae, C., et al.: A 128-channel 6 mw wireless neural recording IC with spike feature extraction and UWB transmitter. IEEE Trans. Neural Syst. Rehabil. Eng. **17**(4), 312–321 (2009)
5. Yin, M., Ghovanloo, M.: A low-noise preamplifier with adjustable gain and bandwidth for bio potential recording applications. In: IEEE International Symposium on Circuits and Systems, pp. 321–324 (2007)
6. Shahrokhi, F., et al.: The 128-channel fully differential digital integrated neural recording and stimulation interface. IEEE Trans. Biomed. Circuits Syst. **4**(3), 149–161 (2010)
7. Gao, H., et al.: HermesE: a 96-channel full data rate direct neural interface in 0.13um CMOS. IEEE J. Solid-State Circuits **47**(4), 1043–1055 (2012)
8. Han, D., et al.: A 0.45 V 100-channel neural-recording IC with sub-$\mu$W/channel comsumption in 0.18 $\mu$m CMOS. IEEE Trans. Biomed. Circuits Syst. **7**(6), 735–746 (2013)
9. Chae, M.S., Liu, W., Sivaprakasham, M.: Design optimization for integrated neural recording systems. IEEE J. Solid-State Circuits **43**(9), 1931–1939 (2008)
10. Seese, T.M., Harasaki, H., Saidel, G.M., Davies, C.R.: Characterization of tissue morphology, angiogenesis, and temperature in the adaptive response of muscle tissue to chronic heating. Lab. Invest. **78**(12), 1553–1562 (1998)
11. de Zeeuw, C.I., et al.: Spatiotemporal firing patterns in the cerebellum. Nat. Rev. Neurosci. **12**(6), 327–344 (2011)
12. Kölbl, F., et al.: In vivo electrical characterization of deep brain electrode and impact on bio-amplifier design. In: Proceedings of IEEE Biomedical circuits and Systems Conference, pp. 210–213 (2010)
13. West, A.C., Newman, J.: Current distributions on recessed electrodes. J. Electrochem. Soc. **138**(6), 1620–1625 (1991)
14. Harpe, P., Cantatore, E., van Roermund, A.: A 10b/12b 40 kS/s SAR ADC with data-driven noise reduction achieving up to 10.1b ENOB at 2.2 fJ/conversion-Step. IEEE J. Solid-State Circuits **48**(12), 3011–3018 (2013)
15. Rodríguez-Pérez, L., et al.: A 64-channel inductively-powered neural recording sensor array. In: Proceedings of IEEE Biomedical Circuits and Systems Conference, pp. 228–231 (2012)
16. Harrison, R., et al.: A low-power integrated circuit for a wireless 100-electrode neural recording system. IEEE J. Solid-State Circuits **42**(1), 123–133 (2007)
17. Harrison, R.: The design of integrated circuits to observe brain activity. Proc. IEEE **96**(7), 1203–1216 (2008)
18. Hodgkin, A., Huxley, A.: A quantitative description of membrane current and its application to conduction and excitation in nerve. J. Physiol. **117**, 500–544 (1952)
19. Fox, R.F., Lu, Y.-N.: Emergent collective behavior in large numbers of globally coupled independently stochastic ion channels. Phys. Rev. E. **49**, 3421–3431 (1994)

20. Gray, P.R., Meyer, R.G.: Analysis and Design of Analog Integrated Circuits. Wiley, New York (1984)

21. Demir, E., Liu, A., Sangiovanni-Vincentelli, A.: Time-domain non-Monte Carlo noise simulation for nonlinear dynamic circuits with arbitrary excitations. In: Proceedings of IEEE International Conference on Computer Aided Design, pp. 598–603 (1994)

22. Yang, Z., Zhao, Q., Keefer, E., Liu, W.: Noise characterization, modeling, and reduction for in vivo neural recording. Advances in Neural Information Processing Systems, pp. 2160–2168 (2010)

23. Fischer, J.H.: Noise sources and calculation techniques for switched capacitor filters. IEEE J. Solid-State Circuits 17(4), 742–752 (1982)

24. Sepke, T., Holloway, P., Sodini, C.G., Lee, H.-S.: Noise analysis for comparator-based circuits. IEEE Trans. Circuits Syst.-I 56(3), 541–553 (2009)

25. Enz, C., Cheng, Y.: MOS transistor modeling for RF IC design. IEEE J. Solid-State Circuits 35(2), 186–201 (2000)

26. Jindal, R.P.: Compact noise models for MOSFETs. IEEE Trans. Electron Devices 53(9), 2051–2061 (2006)

27. Ou, J.: $g_m/I_D$ based noise analysis for CMOS analog circuits. In: Proceedings of IEEE International Midwest Symposium on Circuits and Systems, pp. 1–4 (2011)

28. Song, S., et al.: A 430nW 64nV/VHz current-reuse telescopic amplifier for neural recording application. In: Proceedings of IEEE Biomedical Circuits and Systems Conference, pp. 322–325 (2013)

29. Zou, X., et al.: A 100-channel 1-mW implantable neural recording IC. IEEE Trans. Circuits Syst. I Regul. Pap. 60(10), 2584–2596 (2013)

30. Lee, J., Rhew, H.-G., Kipke, D.R., Flynn, M.P.: A 64 channel programmable closed-loop neurostimulator with 8 channel neural amplifier and logarithmic ADC. IEEE J. Solid-State Circuits 45(9), 1935–1945 (2010)

31. Abdelhalim, K., Genov, R.: CMOS DAC-sharing stimulator for neural recording and stimulation arrays. In: Proceedings of IEEE International Symposium on Circuits and Systems, pp. 1712–1715 (2011)

32. Bult, K., Geelen, G.: A fast-settling CMOS op amp for SC circuits with 90-dB DC gain. IEEE J. Solid-State Circuits 25(6), 1379–1384 (1990)

33. Kobayashi, T., Nogami, K., Shirotori, T., Fujimoto, Y.: A current-controlled latch sense amplifier and a static power-saving input buffer for low-power architecture. IEEE J. Solid-State Circuits 28(4), 523–527 (1993)

# Integrated Chip Power Receiver for Wireless Bio-implantable Devices

Vijith Vijayakumaran Nair and Jun Rim Choi$^{(\boxtimes)}$

School of Electronics Engineering, Digital Convergence Laboratory,
Kyungpook National University, Daegu 702-701, South Korea
vijith133@knu.ac.kr, jrchoi@ee.knu.ac.kr

**Abstract.** Wireless bio-medical devices employ inductive link as medium for transfer of energy between the external source and the implant. But, the inductive power picked by receiver results in high voltage, that may largely exceed the voltage compliance of low voltage integrated chips. The high voltage at the receiver is due to high load impedance offered by electrodes within the implant. To limit the magnitude of induced voltage, majority of the low voltage circuits use power inefficient methods like voltage clippers and shunt regulators. Therefore, to overcome voltage limitation and to enhance power efficiency, a power receiver topology based on step-down approach is designed and implemented for input voltage as high as 30 V. The implemented design consists of rectifier and series voltage regulator. In addition a battery charger circuit that ensures safe and reliable charging of the implant battery is designed and tested. The proposed design is fabricated in 0.35 $\mu$m high voltage BCD foundry. Rectifier and regulator power efficacy are analyzed based on simulation and measurement results.

**Keywords:** Bio-medical implant · Power receiver · High voltage IC · Rectifier · Series voltage regulator · Charger circuit · Power efficiency

## 1 Introduction

In implantable biomedical devices such as in neural stimulators, energy is transmitted through near–field inductive links. The wireless power links provides sufficient energy for the device excitation and to recharge the implant battery, thus extending the operational life time of the battery. [9,12,14,15] Spiral coils used as inductive links transmit energy and transmitted energy is recovered at the receiver (Rx) end, with the aid of Rx coils that resonates at 13.56 MHz (ISM – Industrial Scientific Medical band) carrier frequency. The Rx voltage is then rectified, filtered and supplied to regulators to meet the system voltage requirements. For neural stimulators, the power link transfer the necessary energy to meet the high current stimulation requirements and addresses the large cuff-nerve electrode interface impedance.

In general low-voltage (LV) integrated chips (ICs) are employed for power recovery at the Rx end in bio-medical devices. However, during periods of low

© Springer International Publishing Switzerland 2015
A. Fred et al. (Eds): BIOSTEC 2015, CCIS 574, pp. 33–48, 2015.
DOI: 10.1007/978-3-319-27707-3_3

current stimulation, the recovered voltage at the Rx may largely exceed the compliance of the LV ICs. To protect LV systems, conventional zener diode voltage regulators, on-chip voltage limiters [1,9,19] or shunt regulators [25] are employed. These systems restricts the supply voltage to below 5.5 V and also degrades the power efficiency of the total system, as most of the excess power is grounded or heat dissipated. Moreover, for implantable stimuli generator, the impedance characteristics of the cuff-nerve electrode interface implicates a gradual increase in magnitude over the course of time after the implant [11,12,14,20]. As a result, low supply voltages turn out to be insufficient to generate the necessary stimulus current. The typical stimulation current for neural stimulation is between 1–2 mA and the cuff-nerve electrode interface impedance varies from 0.3–10 $K\Omega$. In consideration to the worst case scenario, for maximum stimulus current (2 mA) at maximum impedance (10 $K\Omega$), the Rx voltage is 20 $V_{peak}$ in ideal conditions. Therefore, prior LV power recovery schemes [1,3,7,9,10,19,25] become unsuitable for the implementation . The neural stimulator consists of two characteristic blocks: (1) High voltage (HV) block to directly provide the high current stimulus, (2) LV block that provides power to the rest of the system.

Considering the rectified voltage is limited by a shunt regulator, the possible approaches to provide dual supplies are voltage step-up and step-down methods. The step-up voltage is preferred over step-down in most of the systems as minimized inductive voltage improves power efficiency. However, considering the total system and operating conditions an appropriate topology has to be determined. Taking in consideration the actual stimulator conditions (more than 800 $\mu$A system current from 4.2 V implant battery, up to 2 mA high stimulation pulses and secondary coil equivalent source resistance $(R_S)$ = 12 $K\Omega$ [23]) and applying theoretical study published in [17], an approach based on voltage step-down proves to be more efficient if the converter efficiency is higher than 45 % and its output impedance is lower than 10 $K\Omega$.

The possible solution can be an approach based on IC that would not limit the received voltage at the Rx end. Power recovery circuit when realized in HV technology could withstand high operating voltages. Further, it will essentially serve the advantage of storing the excess power in filter capacitor (proceeding rectification), leading to higher rectified voltage. Where as in voltage limiting structures, excessive power is heat dissipated or grounded. Hence, limiter [1] and shunt regulator [25] could be successively replaced by a step-down buck converter. But buck converter requires large inductor, practically not feasible to embed on a chip. Moreover to maintain the regulation quality and efficiency for high operating frequency is a difficult task [21]. A possible alternative to the above topologies is a high supply input series voltage regulator. For a HV series regulator, the input voltage is only limited by the available power at Rx and the power consumption that includes the series regulator dissipation.

In [14,15], IC based on HV technology for power recovery in implantable device showed improved power efficiency based on simulation results. A HV IC power recovery system for contactless memory card [21] reported 50 % power efficiency at higher input power level. But, the system feature size and its low power efficiency made it unsuitable for biomedical applications. This paper proposes the design and fabrication of on-chip power recovery analog front end in HV technology. The design is based on voltage step down approach. The system is detailed in the following sections. Section 2 describes the system design, Sect. 3 details the circuit design methods, Sect. 4 presents the simulation and measurement results and Sect. 5 the conclusion. The IC is fabricated in 5060BD35BA bipolar-CMOS-DMOS 0.35 μm 60 V Mixed Technology in Dongbu Hi-tek process. The prototype fabricated can be used for the purpose of generating stimulus current and recharging battery in biomedical implants.

## 2  System Design

### 2.1  Architecture Work Flow

The wireless neurostimulator topology comprises of two modules, an external transmitter (Tx) module and an internal Rx module. Figure 1 shows the system architecture of the neurostimulator. The internal and external modules are embedded with data modules to monitor and control the sequential power transmission and reception. At Tx side the user interface, data transceiver and data modulator/demodulator aid the user to control the power transmission and operation within the implant. Power amplifier module holds responsible for the amount of power generated for transmission. Impedance matching circuit and power coils builds up the resonance link so as to transfer energy with maximum efficiency [24]. At the Rx end (Fig. 1) the coil picks up the power and transferred to the HV receiver IC for further transformations. HV IC, then supplies the power to either charge the implant battery through a charge control circuit or to the high current stimulation stages for neural stimulation based on the user control. The data coil and Rx module collects the command signals, conditions the signal and is transferred to the micro–controller and FPGA for interpretation of the commands. The control and FPGA constantly monitors the operations

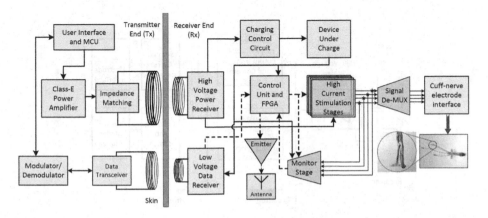

**Fig. 1.** Block diagram of wireless power and data transfer chain for neurostimulator.

and directs the system towards different stimulation stages or battery recharging process. The user communicates with the implant based on the feedback signal transmitted by the antenna in the Rx side.

## 2.2   Topology and System Specifications

The power recovering topology comprises of HV bridge rectifier, current and voltage reference generator, linear regulator (low drop–out (LDO) regulator) and a charging control circuit [23]. The rectifier translates the AC power to DC power with the help of bridge structure and filter capacitor. The output DC from the rectifier is supplied to the voltage reference generator and the regulator. The reference circuit generates the necessary reference voltage and bias current for LDO and charging control circuit. The regulator provides stable output voltage that is, responsible for providing the necessary charging and stimulation current. The block schematic of the power recovering topology is depicted in Fig. 2

As power efficiency in biomedical implants is of high significance, the design constraints are more critical. The total power efficiency ($\eta_{total}$) of the wireless power system is given by,

$$\eta_{total} = \eta_{supply} + \eta_{coils} + \eta_{recti} + \eta_{LDO} + \eta_{ctrl/sti} \tag{1}$$

where $\eta_{supply}$ is efficiency of the radio frequency (RF) power amplifier/generator, $\eta_{coils}$ is the power transmission efficiency of the wireless link, $\eta_{recti}$ is the power conversion efficiency (PCE) of the rectifier, $\eta_{LDO}$ is the LDO regulator efficiency, $\eta_{ctrl}$ is the power efficiency of the charging control circuit and $\eta_{sti}$ is the power efficiency of stimulus generator.

The biomicrosystem generates stimulus currents in the form of pulses to regenerate the nerve cells. Typically it is in the order of 1–2 mA [14,15]. In general, low voltage supply favors the neural stimulation when the cuff-nerve impedance is smaller in magnitude. But, few months after the implantation,

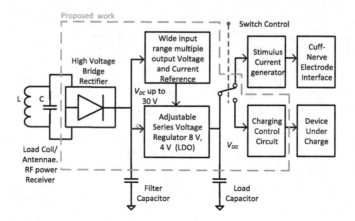

**Fig. 2.** Schematic block of power recovery topology in neurostimulator.

magnitude of the cuff–nerve interface's impedance rises to $0.5$–$8\,K\Omega$ [11,12,14, 20]. Under these circumstances to generate the necessary simulation currents, the supply voltage must be increased. Typically, it rises from 3.3–5 V to 10– 20 V. Rest of the system functions with typical low voltage supply from the battery (3.7 V in proposed system). The continuous stimuli generator is powered by battery and essentially uses lesser than $30\,\mu A$ current pulse at 100 Hz [14]. In practice, batteries rating of 3.7 V and 20 mAh are used in biomedical implants due to their smaller size and low power requirements. According to our design specifications, a charging current of 10 mA is required to enable quick charging of the battery.

# 3  Circuit Design Methods

## 3.1  Semi–Active HV Bridge Rectifier

Full wave bridge rectifiers are generally implemented with diode connected transistors in CMOS technology. To overcome the leakage losses and improve PCE threshold cancellation techniques [16] and cross-coupled transistors based bridge rectifiers are introduced [13]. Though these schemes provided efficient rectification, the PCE reported is $< 70\,\%$. Further to compensate for the low PCE, several works based on comparator techniques are proposed and implemented [3,7,9]. But, the PCE response curve at high frequencies (above 10 MHz), indicated rapid decline. Prior rectifier design schemes are incompatible for the intended application due to voltage constraints, low operational frequencies, comparably low PCE and complicated circuit calibration. The HV design library provides limited parametric properties and less flexible design functionality for transistors in comparison to LV design library. In addition, the main constraint that limits the implementation of earlier work in our HV design approach is the gate-to-source voltage ($V_{GS}$) limitation (maximum $V_{GS}$ Dongbu BCD process = 13.2 V). Design of active rectifier in this technology requires level shifters that

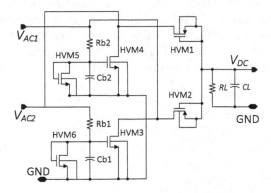

**Fig. 3.** Circuit schematic of the proposed high voltage rectifier.

affects the total efficiency. Apart from this, level shifters are generally designed for specific input voltages. In [14], a HV rectifier design based on simulation result reports 93 % PCE, but the design is unsuitable for technology with $V_{GS}$ limitations.

This section presents, the proposed semiactive HV bridge rectifier based on partial and adaptive threshold cancellation techniques. The schematic diagram of the proposed rectifier is depicted in Fig. 3. The gate terminal of the HV LDN-MOS transistors, HVM3 and HVM4 are biased with cross-coupled resistors (Rb1, Rb2), diode connected transistors (HVM5, HVM6), and bias capacitors (Cb1, Cb2). The dynamic bias voltage tracks the threshold voltage ($V_{th}$) of HVM5 and HVM6 through various temperature and process conditions. The average value of the bias voltage is lower than the $V_{th}$ of transistors HVM3 and HVM4. The desired bias voltage is obtained by appropriately sizing the bias transistors, resistors, and capacitors. Increasing the bias voltage in the conduction phase decreases the on-resistance of the power transistors, but increases the leakage current through the off transistor and thus decreasing the PCE. Therefore, the complete $V_{th}$ cancellation is not a viable option for the design. In our design, the average bias voltage about 65 % of $V_{th}$ of power transistors is maintained. The bias voltage adaptively increases to 90 % of the $V_{th}$ in the conduction phase and decreases to 40 % in the non-conduction phase providing higher PCE. For p-type switches, HV P-type DMOS transistor with source terminal shorted with gate terminal is employed to obtain diode connected transistor configuration. In comparison prior designs [3,7,9,13,14,16], the proposed design does not limit the supply voltage and moreover the average voltage drop is less than two diode voltage drops. In our technology, by default the source to bulk terminals are connected internally and thus restricts the implementation of bulk biasing scheme that would have reduced the leakage current.

## 3.2  HV Current and Voltage Reference

The voltage references and bias current generators are implemented using sub-threshold MOSFET techniques to reduce the power consumption in the circuit [8,22]. Moreover the designs help to improve the temperature immunity of generated voltage over the range of $-10°C - 150°C$ [22]. To accommodate higher supply voltages a topology based on LV–HV transistor cascading scheme is employed [2]. The transistors LVM11 and LVM12 operating in subthreshold regime generate a proportional to absolute temperature (PTAT) current $I_A$ independent of supply voltage variations. Further, LVM4 is also operated in subthreshold regime, which generates a complementary to absolute temperature (CTAT) current $I_B$ independent of variations in input voltage. Figure 4 shows the schematic circuit of the designed voltage reference and current generator with the essential startup circuitry. Currents $I_A$ and $I_B$ are summed to generate the supply voltage independent and zero temperature coefficient current $I_{REF}$. Therefore, the reference voltage is expressed as:

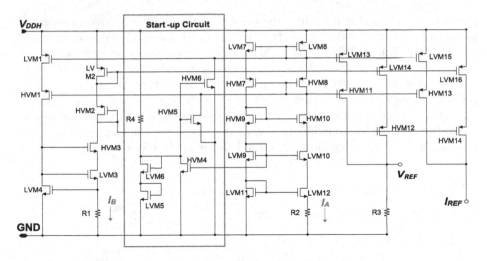

**Fig. 4.** Circuit schematic of voltage and current reference generator.

$$V_{REF} = \left( \frac{K_{LVM15}}{K_{LVM10}} I_A + \frac{K_{LVM16}}{K_{LVM6}} I_B \right) R_a \tag{2}$$

where,

$$I_A = \frac{V_{GSLVM13} - V_{GSLVM14}}{R2} = \frac{\zeta V_T ln\left( \frac{K_{LVM13}}{K_{LVM14}} \right)}{R2} \tag{3}$$

$$I_B = \frac{V_{GSLVM4}}{R1} \tag{4}$$

where, $K_{LVM}$ is aspect ratio of the transistors, R is the resistor values, $V_{GS}$ is the gate to source voltage, $\zeta > 1$ is a non-ideal factor and $U_T = KT/q$ is the thermal voltage.

## 3.3   High Voltage Series Regulator

The series voltage regulator provides 4 V and 10 V output, when delivering current in the range of 0.03 – 100 mA. The LDO comprises of pass device, an error amplifier with a output buffer stage, resistor feedback network and output capacitor. The fabricated HV series regulator utilizes cascode compensation technique to attain optimum optimum phase margin and loop gain. The scheme employs Miller compensation with common gate transistor as current buffer that effectively replaces RHP zero with LHP zero providing improved stability [6]. For the error amplifier design as in [6], the LV differential pair transistors and common gate transistor in buffer stage are replaced by HV n-type LDMOS transistors with floating source to accommodate for high input voltage conditions (up to 30 V). The circuit schematic of the fabricated LDO is shown in Fig. 5. Cascaded PMOS transistors (LVM 1–2, HVM 7–8) in differential pair stage are employed to increase the error amplifier gain. An n-type LDMOS transistor is used as pass transistor to reduce the area of the pass device three times compared to the p-type device in our technology. The total current consumption of the design is limited to less than 170 A. The over current protection circuit is composed of resistor and a PMOS transistor (Q1) [18]. Q1 is off during normal operations. But as the output current hikes above the maximum limit, the voltage drop across $R_{OC}$ is approximately 0.7 V, which is equal to base-emitter voltage ($V_{be}$) and Q1 starts to conduct. As Q1 starts to conduct the gate terminal of the pass device is pulled up restricting excess current and protecting the pass device. The stability analysis of the LDO is analyzed as in [6].

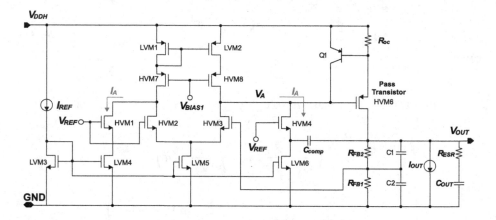

**Fig. 5.** Schematic diagram of the implemented HV LDO.

The on-chip capacitor used for the compensation could withstand voltage up to 35 V (break-down voltage), ensured safe operation within the design specifications. Transistor HM4 is a cascode transistor, that helps in reducing the drain to source voltage ($V_{DS}$) mismatch between LM3 and LM5, thus effectively decreasing the DC bias current mismatch. Current $I_A$ is a scaled version of the bias current of the error amplifier, so as to reduce the total quiescent current. To suppress the high frequency noise adequately, capacitors C1 and C2 are connected in parallel to the feedback resistors RFB2 and RFB1 respectively. The values of the feedback resistors are determined based on the reference voltage, output voltage and the resistor bias current [18]. Further, an additional circuit to limit the output current is placed before the pass transistor that provides over current protection. The HV Dongbu Hi-tek technology offers limited functionality in transistor sizing. Consequently, optimization of the error amplifier in the LDO is difficult as the transistor parameters are fixed. Particularly when the input variations are large, the stability and performance must be ensured.

### 3.4 Charging Control Circuit

In biomedical implants, battery management circuits are inevitable to prolong the battery life time. Various circuit architectures are proposed and implemented in prior works based on digital and analog control for the purpose of charging the batteries [4, 10]. These designs may utilize expensive precision sense resistors or analog-to-digital converters that occupy large area for end of charge (EOC) detection.

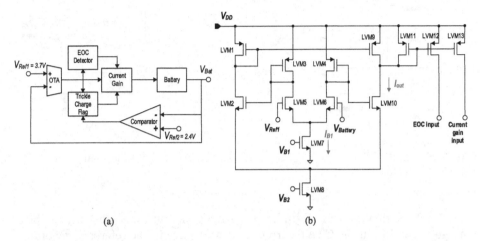

(a)                                                                          (b)

**Fig. 6.** (a) Simplified block diagram of charger. (b) Circuit schematic of the OTA.

In [5], a compact lithium-ion (Li-ion) battery charger based on analog control is reported. This design overcame the demerits of prior designs and addressed to all the four (Trickle charging, constant current (CC), constant voltage (CV) and EOC) charging regimes. But, the designed subthreshold operational transconductance amplifier (OTA), that determines the linear voltage range (approximately 120 mV) is directly dependent on the temperature. Temperature fluctuations vary the linear voltage range of tanh function, that results in either overcharging or undercharging of the implant battery. This irregularity may decrease battery life time. To overcome the temperature artifacts on linear voltage range, we propose a charger based on adaptive bias OTA, where transistors are operating in saturation region. The simplified block diagram of the charging is shown in Fig. 6(a). The OTA in Fig. 6(b), is designed to operate in saturation region where the output current is a function of tanh. The output current is generated based on comparison of the battery voltage to reference. By operating the OTA in saturation regime, the linear voltage range is controlled adaptively by the bias voltage $V_{B1}$. The linear voltage range of the OTA is given by

$$V_L = \frac{V_{B1} - V_{th}}{k_s \sqrt{2}} \qquad (5)$$

where $V_{B1}$ is the bias voltage and $k_s$ is the subthreshold exponential slope parameter . The bias voltage was fixed to be 810 mV ($V_{th}$ = 700 mV) ensuring the saturation region operation of the OTA and the derived linear range was

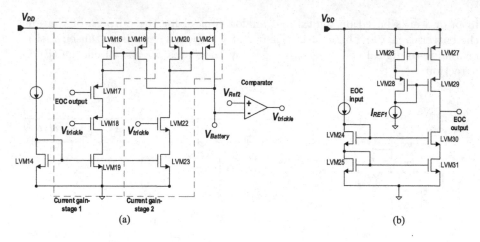

**Fig. 7.** Schematic diagram of (a) Current gain stages and trickle charge detector, (b) EOC detector.

160 mV. Though $V_{th}$ in the Eq. 5 has negative temperature coefficient, the bias voltage derived from the CTAT current (Eq. 4 ) source in the reference generater compensates for the decrease in threshold voltage, thus maintaining the constant bias current $I_{B1}$ and linear range. The linear range signifies that for battery voltages less than 3.54 V, the OTA output is saturated allowing maximum output current and for battery voltages greater than 3.54 V the OTA operates in the linear region implying decrease in output current. The output current of the OTA $I_{OTAout}$ is given by

$$I_{out} = I_{B1}tanh(\frac{k_s(V_{in+} - V_{in-})}{2U_T})$$ (6)

where $I_{B1}$ is the bias current of OTA, $k_s$ is the subthreshold exponential slope parameter, $V_{in+}$, $V_{in-}$ are the input voltages and $U_T$ is the thermal voltage. The voltage reference circuit described in earlier section generates zero temperature coefficient supply independent voltage (3.7 V) in relative to the battery voltage used in the biomedical implant. The current gain circuit uses two separate multiplier transistors. First multiplier is turned on during the CC regime and has higher current gain and the second multiplier operates at the time of trickle charging providing lower gain. In CC regime the charging current is designed to be 10 mA and 1 mA during trickle charging. Since the OTA is operated in the saturation regime (bias current 20 μA) the current gain stage is compact and shows good stability in high power designs. In CC regime, LVM17 and LVM18 (Fig. 7(a)) are conducting, which enables first current gain stage and 10 mA current charges the battery. The Trickle charge detector circuit comprises of a simple comparator that compares the battery voltage to a reference voltage. The comparator output is high when the reference voltage ($V_{Ref2} = 2.5$ V) is higher than the battery voltage. This output turns on transistor LVM22 and turns off

LVM18 (Fig. 7(a)), thus enabling current through second gain stage implying trickle charging. The EOC detector shown in Fig. 7(b) is developed as in [4]. It compares the EOC input to a reference current. The schematic of the current comparator is shown in the Fig. 7(b).

## 4  Simulation and Experimental Results

The designed system is implemented in 0.35 μm BCD process and simulations are performed in Cadence Spectre environment. The AC input voltage is set below 20 V peak maximum for load currents less than 2 mA. But, when the system is directed to charge the battery, the input voltage is varied between 4.5 − 7 V peak with load current range from 9–11 mA and filter capacitor being 0.5 μF.

The bias voltage varies in the range of 0.48–1.06 V in non-conduction and conduction phase ensures that the HVNMOS transistors are operated in linear region (Dongbu 60V LDNMOS $V_{th}$ = 1.2 V). Figure 8(a) shows the experimental

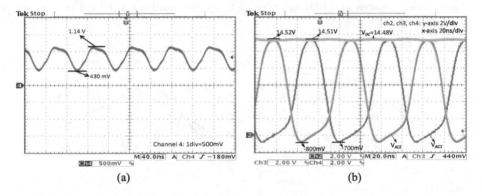

(a)                              (b)

**Fig. 8.** Oscilloscope capture of (a) Rectifier bias voltage, (b) Transient response of rectifier input nodes and output voltages

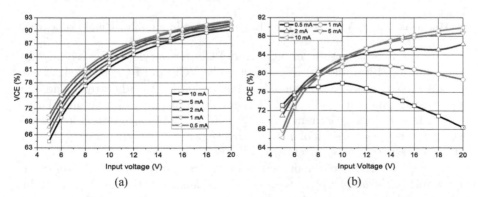

(a)                              (b)

**Fig. 9.** Simulated (a) VCE and (b) PCE with varying input voltage and load current

result of the gate bias voltage of HVNMOS transistor for an input voltage of 15 $V_{peak}$. Rectifiers are often characterized with their PCE and voltage conversion efficiency (VCE). VCE is calculated based on the equation VOUT/VIN. The PCE in all measurement and simulation results in this paper is calculated based on the Eq. (7).

$$\eta_{PCE} = \frac{V_{OUT} * I_{OUT}}{P_{in}} = \frac{V_{OUT}^2 * T}{R_L \int_0^T V_{in}(t) * I_{in}(t)\, \mathrm{d}t}. \tag{7}$$

Figure 9(a) and (b) shows the post-layout simulation results of VCE and PCE of the rectifier respectively for various load currents with varying input voltage. The VCE plot indicates increasing trend for increased supply voltage. In the PCE plot, a similar increasing trend is followed with increase in supply voltage but, after a particular range the PCE starts to decrease. This is due to lack of bulk biasing scheme, that results in increased leakage current that affects the PCE. The IC is tested with different loads employing discrete resistors ranging from 0.5–8 K$\Omega$. Discrete resistors were used to replicate the increasing cuff-nerve electrode interface impedance. Oscilloscope capture of the rectifier input and output node voltages at 13.56 MHz input frequency and output load current of 1 mA is shown in Fig. 8(b).

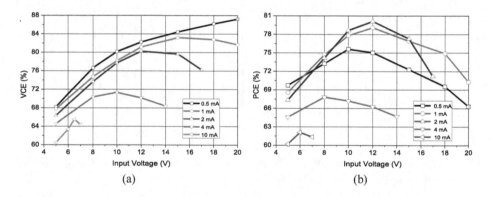

**Fig. 10.** Measured (a) VCE and (b) PCE with varying input voltage and load current

The experimental results of VCE and PCE for varying input voltage are depicted in Fig. 10(a) and (b) respectively. The VCE and PCE at low load currents resembles closely to the simulation results. At high load currents and high supply voltage, measured VCE and PCE deviate from simulation results. This is due to substrate leakage current and results in latch-up which destroys the chip. The existence of parasitic bipolar transistors that is not modelled during the simulation process, conducts current when the reverse voltage exceeds certain limit [26]. Though, modelling of the parasitic bipolar transistors have been explored in prior works (for example: [26]), an accurate modelling is still not developed due to the process variations in different technologies. This is a major limitation for the implementation of bulk HV process and is beyond the scope of this paper. The results are based on operating points where latch-up does not occur.

**Table 1.** Performance comparison with prior rectifiers.

| Publication | Guo | Lee | Mounaim | Cha | This work |
|---|---|---|---|---|---|
| Technology | 0.35 $\mu$m | 0.5 $\mu$m | DALSA | 0.18 $\mu$m | **0.35 $\mu$m** |
| | CMOS | CMOS | C08G-C08E | CMOS | **BCD** |
| $V_{in,peak}$(V) | 2.4 | 3.8 | 16.8 | 1.5 | **20.0** |
| $V_{REC}$ (V) | 2.28 | 3.12 | 15.5 | 1.33 | **17.32** |
| Frequency (MHz) | 0.2-1.5 | 13.56 | 13.56 | 13.56 | **13.56** |
| $R_{Load}$ (K$\Omega$) | 0.1 | 0.5 | 5.0 | 1.0 | **0.5 to 8** |
| Load current $I_L$(mA) | 20 | NA | 10 | NA | **0.5 to 10** |
| PCE (%) maximum | 82-87 | 80.20 | 93.10 | 81.90 | **80.2** |
| | | | (Simulated) | | |

From the Fig. 8(b)., it can be observed that the substrate leakage current is negligible since the rectifier input node voltage never crosses the output DC voltage. At high input voltages and load currents ($I_L > 2$ mA and $V_{inpeak} > 16$ V), the rectifier becomes abnormally loaded and the input AC voltages become higher than the output DC. However, in our application for stimulus generator, the current required is 2 mA with input voltages ranging from 10–20 $V_{peak}$, the design proves to be more effective indicating higher PCE compared to prior artworks. Meanwhile, for recharging the implantable battery at low input voltages (4.5–7 V) and high load currents (9–11 mA), the maximum efficiency can be achieved as 63.2 %. A comparison of proposed work to prior works is shown in Table 1. The voltage and current references generate the essential bias current and reference voltage for the LDO. The reference voltages of 3 V, 3.7 V, 2.4 V are generated for LDO and charger circuit. The regulator is designed to generate adjustable output voltage of 4 V and 8 V for an output load range 0.03–100 mA. Load capacitor of 1 $\mu$F is used for the measurement analysis of LDO. Though, on-chip resistor option is available for the technology, we used discrete resistor divider network in our prototype to generate the different output voltage. The LDO is tested for 4 V and 10 V output for our specific application. Oscilloscope capture of transient response of the LDO at start-up for 4 V output is shown in the Fig. 11(a). The output voltage is indicated by channel 3 and input voltage by channel 4. The regulator is stable for large input voltage variations. The summary of the experimental results are shown in Table 2.

The battery is charged under typical Li-ion battery charging profile. The battery under test is a 3.7 V, 20 mAh Li-ion battery for implantable devices. It is charged with 1.2 mA and 9.8 mA current during trickle and CC charging regimes respectively. The supply voltage was set to 4 V,from the LDO. The result of the battery management circuit during the process of charging is depicted in Fig. 11(b). The CC, CV and EOC regime are depicted in the Fig. 11(b). As the battery reaches 3.54 V the CC regimes slowly translates to CV regime, driving the OTA into linear region. In the CV regime, the current slowly decreases from

Table 2. Performance summary of the HV-Regulator.

| Parameter | Value | Conditions |
|---|---|---|
| $V_{OUT}$ | $4 \pm .07$ V | $V_{IN} = 6{-}30$V; $I_{OUT} = 0.03{-}100$ mA |
| $V_{Dropout}$ | 164 mV | $I_{OUT} = 1$ mA, $V_{OUT} = 4$ V; |
| Mean | 242 mV | $I_{OUT} = 50$ mA, $V_{OUT} = 4.03$ V |
| Line Regulation | $37 \pm 20$ mV | $V_{IN} = 5{-}30$ V |
| Load Regulation | 58 mV | $V_{IN} = 6$V; $I_{OUT} = 1{-}50$ mA |
| PSRR | $-54.2$ to $-46.35$ dB | $100$ Hz$-1$ KHz |

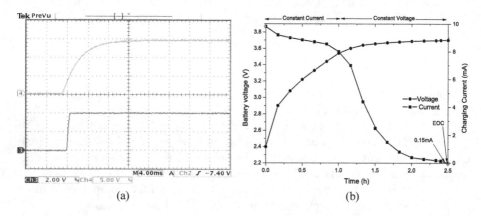

(a)                                    (b)

Fig. 11. (a) LDO at start-up (b) Experimental charging profile of the implant battery.

9.8 mA to 0.15 mA based on the tanh curve. The voltage increases to 3.703 V at the end of the charging profile. The circuit attained about 84 % power efficiency at room temperature.

## 5   Conclusion

This paper reports a energy efficient integrated chip power recovery scheme for bio-medical implants for neurostimulator. The topology is based on step-down approach and integrated on 2.5 mm x 5 mm chip for the purpose of stimulus generation and battery recharging application. Experimental results indicates a maximum of 20 $V_{peak}$ AC which is four times higher to earlier works. A maximum PCE of 80.2 % is obtained for 2 mA load current at 12 $V_{peak}$. Successful measurement results indicated that HV regulators can provide stable output for input variations as high as 30 V, and output currents varying from 0.03–100 mA. Even though, the design was focused to protect rectifier from substrate leakage current and latch-up at high input voltages and load currents, the efficiency deviated from the post-layout simulation results indicate the increased substrate leakage current. However, for typical operating voltage conditions (15 $V_{peak}$) and load currents ($\leq 2$ mA), the PCE proved to be effective. The work also provides a detailed analysis of charging control circuit for implant battery, based on experimental results. Future works include study of temperature artifacts on the integrated chip and their compensation.

**Acknowledgements.** This research was supported by Basic Science Research Program through the National Research Foundation of Korea (NRF) funded by the Ministry of Education (NRF-2013R1A1A4A01012624)

# References

1. Balachandran, G., Barnett, R.: A 110 na voltage regulator system with dynamic bandwidth boosting for rfid systems. IEEE J. Solid-State Circ. **41**, 2019–2028 (2006)
2. Ballan, H., Declercq, M., Krummenacher, F.: Design and optimization of high voltage analog and digital circuits built in a standard 5v cmos technology. In: IEEE Custom Integrated circuits Conference, pp. 574–577 (1994)
3. Cha, H.K., Park, W.T., Je, M.: A CMOS rectifier with a cross-coupled latched comparator for wireless power transfer in biomedical applications. IEEE Trans. Circ. Syst.-ll:Briefs **59**, 409–413 (2012)
4. Chen, M., Rincon-Mora, G.: Accurate, compact, and power-efficient li-ion battery charger circuit. IEEE Trans. Circ. Syst.-ll:Briefs **53**, 1180–1184 (2006)
5. Do Valle, B., Wentz, C., Sarpeshkar, R.: An area and power-efficient analog li-ion battery charger circuit. IEEE Trans. Biomed. Circ. Syst. **5**, 131–137 (2011)
6. Garimella, A., Furth, P.M., Surkanti, P.R., Thota, N.R.: Current buffer compensation topologies for ldos with improved transient performance. Analog Integr. Circ. Sig. Process. **73**, 131–142 (2011)
7. Guo, S., Lee, H.: An effiency-enhanced cmos rectifier with unbalanced-biased comparators for transcutaneous-powered high-current implants. IEEE J. Solid-State Circ. **44**, 1796–1804 (2009)
8. Huang, P., Lin, H., Lin, Y.T.: A simple subthreshold cmos voltage reference circuit with channellength modulation compensation. IEEE Trans. Circ. Syst.-ll:Briefs **59**, 882–885 (2006)
9. Lee, H.M., Ghovanloo, M.: An integrated power-efficient active rectifier with offset-controlled high speed comparators for inductively powered applications. IEEE Trans. Circ. Syst. I: Regul. Pap. **58**, 1749–1760 (2011)
10. Li, P., Bashirullah, R.: A wireless power interface for rechargeable battery operated medical implants. IEEE Trans. Circ. Syst. II: Express Briefs **54**, 912–916 (2007)
11. Li, Y.T., Chang, C.H., Chen, J.J.J., Wang, C.C., Liang, C.K.: Development of implantable wireless biomicrosystem for measuring electrodetissue impedance. J. Med. Biol. Eng. **25**, 99–105 (2005)
12. Li, Y.T., Peng, C.W., Chen, L.T., Lin, W.S., Chu, C.H., Chen, J.J.J.: Application of implantable wireless biomicrosystem for monitoring nerve impedance of rat after sciatic nerve injury. IEEE Trans. Neural Syst. Rehabil. Eng. **1**, 121–128 (2013)
13. Mandal, S., Sarpeshkar, R.: Low-power CMOS rectifier design for RFID applications. IEEE Trans. Circ. Syst. I: Regul. Pap. **54**, 1177–1188 (2007)
14. Mounaim, F., Sawan, M.: Integrated highvoltage inductive power and data-recovery front end dedicated to implantable devices. IEEE Trans. Bio-Med. Circ. Syst. **5**, 283–291 (2011)

15. Mounaim, F., Sawan, M.: Toward a fully integrated neurostimulator with inductive power recovery front-end. IEEE Trans. Bio-Med. Circ. Syst. **6**, 309–318 (2012)
16. Nakamoto, H., Yamazaki, D., Yamamoto, T., Kurata, H., Yamada, S., Mukaida, K., Ninomiya, T., Ohkawa, T., Masui, S., Gotoh, K.: A passive UHF RF idetification CMOS tag IC using ferroelectric RAM in 0.35 $\mu$m technology. IEEE J. Solid-State Circ. **42**, 101–110 (2007)
17. Nicolson, S., Phang, K.: Step-up versus stepdown DC/DC converters for RF-powered systems. In: Proceedings of the 2004 International Symposium on Circuits and Systems. ISCAS 2004, pp. 900–903 (2004)
18. Rincon-Mora, G.A.: Analog IC Design With Low-dropout Regulators. McGraw-Hill, New York (2009)
19. Su, C., Islam, S.K., Zhu, K., Zuo, L.: A hightemperature, high-voltage, fast response linear voltage regulator. Analog Integr. Circ. Sig. Process. **72**, 405–417 (2012)
20. Thil, M.A., Gerard, B., Jarvis, J.C., Vince, V., Veraart, C., Colin, I.M., Delbeke, J.: Tissue-electrode interface changes in the first week after spiral cuff implantation: preliminary results. In: Annual Conference of the International FES Society (2004)
21. Tomita, K., Shinoda, R., Kuroda, T., Ishikuro, H.: 1-W, 3.3-16.3-V boosting wireless power transfer circuits with vector summing power controller. IEEE J. Solid-State Circ. **47**, 2576–2585 (2012)
22. Ueno.K.: CMOS voltage and current reference circuits consisting of subthreshold MOSFETs- micropower circuit components for power-aware LSI applications- solid state circuits technologies. In:Tech (2010)
23. Nair, V.V., Youn J.H., Choi, J.R.: High voltage integrated chip power recovering topology for implantable wireless biomedical devices.In: International conference on Biomedical Electronics and Devices, BOIDEVICES (2015)
24. Nair, V.V., Nagakarthik, T., Choi, J.R.: Efficiency enhanced magnetic resonsnce wireless power transfer sysytem and high voltage integrated chip power recovery scheme. In: IEEE International Conference on Electronics, Computing and Communication Technologies, IEEE CONECCT (2014)
25. Wang, G., Liu, W., Sivaprakasam, M., Kendir, G.A.: Design and analysis of an adaptive transcutaneous power telemetry for biomedical implants. IEEE Trans. Circ. Syst. I: Regul. Pap. **52**, 2109–2117 (2005)
26. Zou, V., Larsen, T.: Modeling of substrate leakage currents in a high-voltage CMOS rectifier. Analog Integr. Circ. Sig. Process. **71**, 231–236 (2012)

# Research on a Novel Three-Channel Self-pressurized Wrist Pulse Acquisition System

Zhou Kan-heng[1], Qian Peng[2], Xia Chun-ming[1(✉)], and Wang Yi-qin[2]

[1] School of Mechanical and Power Engineering,
East China University of Science and Technology, Shanghai, China
kanhengzhou@gmail.com, cmxia@ecust.edu.cn
[2] School of Basic Medicine, Shanghai University of Traditional Chinese Medicine,
Shanghai, China
qqpp2000439@126.com, wangyiqin2380@vip.sina.com

**Abstract.** This paper proposed a novel three-channel self-pressurized wrist pulse acquisition system based on the principle of wrist pulse diagnosis in Traditional Chinese Medicine (TCM). This proposed acquisition system could not only sample, display, analyze and store the wrist pulses in Cun, Guan and Chi region of the wrist simultaneously under the best pulse taking force, but also simulate the process of wrist pulse diagnosis in TCM clinical science called "three regions and nine pulse takings". It is believed that this proposed system solved the problem that most of the current pulse acquisition systems could only detect the pulse in Guan region of the wrist. The result shows that the proposed acquisition system provided a user-friendly flexible sample collection platform, laid the foundation for further analysis of multi-channel wrist pulses and pushed forward the development of the standardization of wrist pulse waveforms and the objectification of wrist pulse diagnosis in TCM.

**Keywords:** Three-channel · Piezoresistive pressure sensors · The best pulse taking force · Three regions and nine Pulse-takings · Automatic self-pressurized control · Wrist pulse classification

## 1 Introduction

In TCM theory, the diagnosis and the treatment of diseases are mainly based on wrist pulse diagnosis, which is pretty hard to learn. Therefore, modern people long for a kind of objective pulse detection and description method to systematically explain the principle of wrist pulse diagnosis in TCM [1].

Recently, with the development of sensor technology, domestic and foreign researchers have already invented a series of pulse detectors to acquire and analyze wrist pulses, such as MX-811 pulse detector made by Nanchang Radio Instrument Factory, BSY-14 pulse detector made by Beijing Medical Factory, MX-3 pulse detector made by Shanghai Medical Research Institute, MTY-A pulse detector made by Tianjin Medical Research Institute, ZM-III pulse detector made by Shanghai University of Traditional Chinese Medicine and CMB-3000/2000 pulse detector made by Japan Colin

© Springer International Publishing Switzerland 2015
A. Fred et al. (Eds): BIOSTEC 2015, CCIS 574, pp. 49–59, 2015.
DOI: 10.1007/978-3-319-27707-3_4

company. Although these pulse detectors have the able to describe and analyze wrist pulses from various aspects, most of them just have a single-probe sensor, which can only detect the wrist pulse in one region of Cun, Guan and Chi of the wrist, as shown in the Fig. 1 below, and mainly detect the pulse of Guan region. These pulse detectors simplified the precisely multi-dimensional sensor of doctor's finger with a single-probe sensor. Therefore, the wrist pulses sampled by these detectors does not carry complete information of wrist pulse [2]. In addition, the pulse sampling way of these pulse detectors differed from the pulse feeling of "three regions and nine pulse-takings" in TCM clinical science. The pulse feeling of "three regions and nine pulse-takings" was a traditional way of pulse feeling, in which the wrist pulses in the region of Cun, Guan, and Chi were felt respectively and the wrist pulse in each region was felt under the floating, medium and sinking pulse taking forces. Therefore, there were totally nine pulse takings.

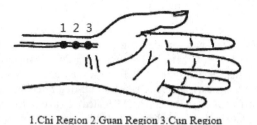

1.Chi Region 2.Guan Region 3.Cun Region

**Fig. 1.** The region of Cun, Guan and Chi.

In order to conform to the pulse feeling of "three regions and nine pulse-takings" in TCM clinical science, our research group developed a novel three-channel self-pressurized pulse acquisition system. This acquisition system can not only acquire three-channel wrist pulses under the best pulse taking force through the automatic pressure control module composed of stepper motors and ball screw pairs, but also simulate the process of the three regions and nine pulse-takings wrist pulse diagnosis in TCM. In addition, this acquisition system can real-timely display, analyze and store three-channel wrist pulses (Fig. 2).

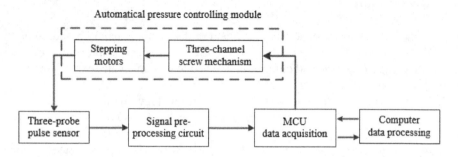

**Fig. 2.** The block diagram of the hardware of three-channel self-pressurized pulse acquisition system.

## 2  Design of Hardware

The three-channel self-pressurized pulse acquisition system consisted of three-probe pulse sensor, signal pre-processing circuit, automatic pressure controlling module and data acquisition and processing system. Figure 2 shows the framework of the hardware of the whole system.

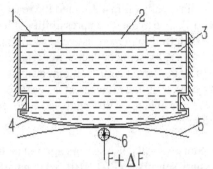

1.aluminium alloy case 2.piezoresistive pressure sensor chip 3.liquid
4.aluminium alloy film 5.skin 6.blood vessel

**Fig. 3.**  Physical model of piezoresistive pressure sensor.

### 2.1  Design of Three-Probe Pulse Sensor

The three-probe pulse sensor which was composed of three independent piezoresistive pressure sensors in the region of Cun, Guan and Chi, as shown in the Fig. 4 below, can not only realize increasing pressure automatically but also simulate the process of three regions and nine pulse-takings wrist pulse diagnosis in TCM.

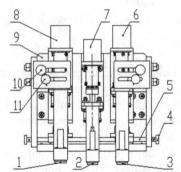

1.sensor of Chi region 2.sensor of Guan region 3.sensor of Cun region 4.x-asis position
adjustment knob 5.x-axis threaded shaft 6. pressure controlling module of Cun 7.pressure
controlling module of Guan 8. pressure controlling module of Chi 9.fixed barcket
10.y-axis fixed konb 11.y-axis position adjustment knob

**Fig. 4.**  The structure of three-probe pulse sensor.

**Principle of Pulse Detection.**   Among many kinds of pulse detection methods, ergography, which is more in line with the pulse feeling in TCM, is one of the most widely used method in the field of pulse detection. Therefore, three-probe pulse sensor applied the basic principle of ergography. Figure 3 shows the physical model of each piezoresistive pressure sensor with the same design parameters.

When aluminum alloy film 4 applied pressure onto "skin 5 and blood vessel 6" system round the radial artery (equal to the floating, medium and sinking pulse taking force with fingers), the aluminum alloy film 4 detected the reaction force of pulse taking force $F$ and pulse force $\Delta F$ [3]. Due to the incompressibility of liquid, the resultant force of these two forces was transferred to piezoresistive sensor chip 2 and made the resistivity of the sensor chip 2 made by semiconductor material changed. The signal preprocessing circuit, attached by the sensor chip 2, outputed signals in different levels with the changes of resistivity caused by the changes of $F + \Delta F$, so different pulse taking forces and different pulses were able to be distinguished according to different signals, so as to realize the detection of wrist pulse.

**Parameters of the Pulse Sensor.**   All pulse sensors applied in three-probe pulse sensor were piezoresistive pressure sensors (NPI-12-101GH) manufactured by GE. Each sensor applied silicon elastic thin film as pressure sensitive element and measures pressure through the pressure electricity conversion circuit, the Wheatstone bridge circuit, which was composed of 4 same resistance resistors fabricated by MEMS technology. Compared with the pressure sensor applied traditional metal strain foil, the sensitivity of this sensor was 50–80 timers higher. In addition, there was no mechanical linkage, therefore, the measurement accuracy of this sensor was relatively higher and the repeatability error and pressure hysteresis effect of this sensor were much lower. According to the datasheet, the original static capability indexes of the sensor are as follows:

- Sensitivity: 0.2 mV/V/psi;
- Linear range: 0–1000 gram-force;
- Accuracy: ±0.1 %;
- Working temperature: 10 °C–40 °C;
- Composite error of linearity, pressure hysteresis effect, repeatability and temperature error: <4 % FSO.

**Structure of Three-probe Pulse Sensor.**   In order to simulate the process of the three regions and nine pulse-takings wrist pulse diagnosis in TCM, the structure of three-probe pulse sensor not only realized acquiring wrist pulses simultaneously in the region of Cun, Guan and Chi under various pulse taking forces, but also realized the adjustment of the axial relative distance and the radial relative distance of these three separate pulse sensors. Figure 4 shows the structure of three-probe pulse sensor and Fig. 5 shows the structure of three-channel self-pressurized pulse acquisition system.

As can be seen form Fig. 4, three separate pulse sensors were attached to their own pressure controlling modules and these modules were attached to the fixed bracket by various directions adjustment mechanisms which were composed of screw transmission mechanisms. The sensor of Guan used as location basis can only adjust the position

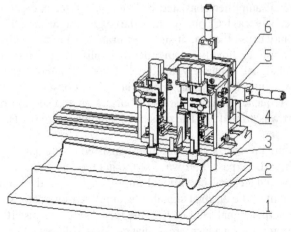

1.base 2.wrist bracket 3.sliding guide table 4.connecting plate
5.x-y axes precise manual adjustment platform 6.three-probe pulse sensor

**Fig. 5.** The structure of three-channel self-pressurized pulse acquisition system.

along z-axis, while the sensor of Cun and Chi can not only adjust the position along z-axis via automatic pressure controlling module, but also adjust the distance relative to the sensor of Guan along x-axis via the knobs on both sides so as to adapt to the different patients with different relative distance among the region of Cun, Guan and Chi. In addition, the sensor of Cun and Chi can also adjust the distance relative to the sensor of Guan along y-axis via the knobs on the front panel so as to adapt to the inconsistency of the axial physiological curve of radial artery. The feature of the three-probe pulse sensor, which can adjust the position of three separate sensors flexibly along the axes of x, y and z, made it had the ability to simulate the process of pulse feeling in TCM.

As can be seen from Fig. 5, compared with most pulse sensors that fixed with wristlet, three-probe pulse sensor of our system was fixed on a precise manual adjustment platform. Patients can lay their wrists on the wrist bracket under the three-probe pulse sensor when doing the wrist pulse acquisition test, which can not only avoid the measurement error caused by the excessive deformation of soft tissue because of fixed with wristlet, but also be more in line with the form of pulse feeling in TCM clinical science.

## 2.2 Signal Pre-processing Circuit

The signal pre-processing circuit consisted of mixed amplifier, pulse signal amplifier, pulse taking pressure signal amplifier and output circuit.

Each mixed amplifier was used to amplify the pulse taking pressure and the pulse signal that detected by the pulse sensor. In order to keep the amplified signal stable and low-noise, instrumentation operational amplifier, which had advantages of low noise, low drift, low power consumption, high consistency and strong anti-interference ability, was applied. And the gain of the amplifier was 51.

Each pulse signal amplifier consisted of band-pass filter and general operational amplifier. Because there was DC component in the original signal amplified by the mixed amplifier, which may cause the output signal saturated and distorted if this signal was put into the main amplifier directly, so, in order to amplify the pulse signal (AC component of the original signal) separately, band-pass filter was applied to remove this DC component in the signal. Meanwhile, appropriate time constant should be calculate according to the frequency characteristics of pulse signal. The gain of the amplifier applied in this part was around 50. The pass band here was from 0.1 Hz to 34.8 Hz and the time constant was 1.551 s.

Each pulse taking pressure signal amplifier consisted of low-pass filter and general operational amplifier. Low-pass filter was used to get the low-frequency component of the original signal and general operational amplifier was used to amplify the low-frequency component of the original signal. The amplification factor should not be too large to cause the output signal saturated, so the factor here was just 10.

Each output circuit was an adder circuit that consisted of general operational amplifier. And the function of the output circuit was to synthesize the completed wrist pulse waveform involving the pulse taking force and the pulse signal. Then, this completed wrist pulse waveform was transmitted to the data acquisition and processing system to real-timely display, analyze and store.

### 2.3   Automatic Pressure Controlling Module

The automatic pressure controlling module, which can not only realize automatic compression in the region of Cun, Guan and Chi, but also simulate the process of the three regions and nine pulse-takings wrist pulse diagnosis in TCM, was composed of three independent screw mechanisms and step-ping motors. Data acquisition and processing system calculated the pulse taking force of each channel and analyzed the peak-to-peak value of wrist pulses to find the best pulse taking force. Meanwhile, it sent commands to automatic pressure controlling module to adjust the position of each sensor along z-axis to ensure these sensors acquired wrist pulses under the best pulse taking force. Some researchers had studied on the range of the pulse taking force that human body can endure and found the upper limit was 261.19 gram-force [4]. Therefore, the maximum setting of the pressure range of the automatic pressure controlling module limited by the software and hardware was 0 ~ 250 gram-force.

### 2.4   Data Acquisition and Processing System

After A/D conversion, three-channel pulse data, which acquired under the best pulse taking force, was uploaded to the host computer through serial communication. Data acquisition and processing system mainly realized real-timely displaying the wrist pulses, real-timely calculating the pulse taking force and analyzing and storing the pulse data dynamically, which laid the foundation for further analysis of multi-channel wrist pulses.

# 3    Design of Software

The software of three-channel self-pressurized pulse acquisition system was composed of Micro Control Unit (MCU) data acquisition program and PC data processing software.

## 3.1    MCU Data Acquisition Program

The MCU, applied in three-channel self-pressurized pulse acquisition sys-tem to acquire three-channel pulse data, was a STM32 series chip, whose kernel architecture is ARM Cortex-M3. MCU data acquisition program up-loaded the three-channel wrist pulse samples acquired under various pulse taking forces to the host computer. At the same time, the host computer calculated the pulse taking forces and the peak-to-peak value of wrist pulse samples to search for the best pulse taking force. After finding the best pulse taking force, the host computer sent it back to MCU to regulate the pulse taking force to the best one via automatic pressure controlling module. And then, MCU filtered the signals with sliding filter, sampled the wrist pulses under the best pulse taking force and uploaded the best wrist pulses to the host computer through serial port.

## 3.2    PC Data Processing Software

Three-channel self-pressurized pulse acquisition system applied personal computer as the host computer to receive the three-channel pulse data sent from MCU and do corre-sponding processing. This computer data processing software, which was written by C#, was based on the Microsoft.NET Frame-work 4. The interface of the software is shown in Fig. 9.

This data processing software can not only real-timely display three-channel wrist pulses and their corresponding average single cycle pulse wave, but also real-timely display current pulse taking force as well as the classification result of wrist pulse on the left side. In addition, this data processing software can also search for the best pulse taking force through comparing the peak-to-peak value of two adjacent cycles of the wrist pulse. Figure 6 shows the variation tendency of peak-to-peak value with pulse taking force.

Taking floating as an example, the data processing software firstly sent the command to automatic pressure controlling module to start the searching process. Then, the step-ping motors started to slowly compressing. Mean-while, the processing software recorded the peak-to-peak value $PP[n]$ and the corresponding pulse taking force $F[n]$ separately and compared the difference of adjacent peak-to-peak value. If the difference of $PP[n + 2]$ and $PP[n + 1]$ and the difference of $PP[n + 1]$ and $P[n]$ were both smaller than zero, then the largest peak-to-peak value, $PP[n]$, was found and the corresponding pulse taking force $F[n]$ was just the best pulse taking force. In addition, this data processing software can also regulate the pulse taking pressure manually through pres-sure regulating button. All three-channel pulse data acquired can be stored into the database automatically according to the patient's name, sampling date, time and other information and can be loaded form the database for viewing and analysis at any time.

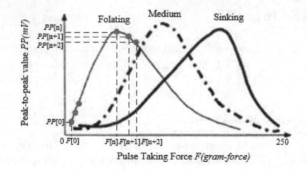

**Fig. 6.** Peak-to-peak value changes with pulse taking force.

## 3.3 Wrist Pulse Classification

In order to study the theory of wrist pulse diagnosis in TCM, the function of wrist pulse classification was added into the data processing software. The wrist pulse classification was based on the six pulse principles' classification method in TCM. So the acquired wrist pulse data could be classified into six pulse principles which are superficial, deep, slow, rapid, deficient and excess. Figure 7 shows the framework of six pulse principles' classification method.

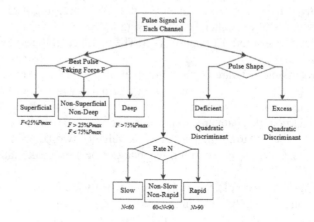

**Fig. 7.** Six pulse principles' classification method.

As can been seen in Fig. 7, to the signal of each channel, the data processing software firstly extracted three kinds of features from them, which were the best pulse taking force, the rate and the pulse shape. Then, through comparing the value of this extracted features, the sampled signal could be classified into corresponding six pulse principles.

# 4    Results

## 4.1    Result of Pulse Sensor Test

The static test of the pulse sensor was carried out between 0–250 gram-force. Figure 8 shows the result of the linear regression, whose y-axis represents output voltage (mV) and x-axis represents input pressure (gram-force). In this test, the input pressure was generated by a series of different quality weights. The R-square of the linear regression was 0.9983, which indicated the linearity of the wrist pulse sensor was excellent in the range of normal pulse taking force and could detect the pulse taking force accurately.

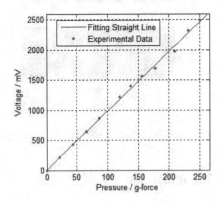

**Fig. 8.**  The result of the linear regression of the pulse sensor.

## 4.2    Result of Wrist Pulse Acquisition and Classification

Figure 9 shows the three-channel wrist pulses and their corresponding pulse taking force acquired by the three-channel self-pressurized pulse acquisition system, which indicated the acquisition system could get clear three-channel pulse data and accurate pulse taking pressure. In addition, the single-cycle average pulse signal on the right side and the result of wrist pulse classification at the lower left corner could help the doctor of traditional Chinese medicine make a correct diagnosis.

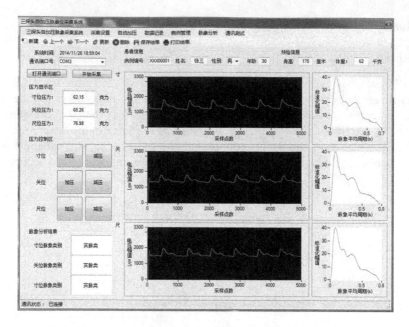

**Fig. 9.** The interface of the software with three-channel wrist pulses.

## 5    Discussions and Conclusions

All pulse sensors applied in three-probe pulse sensor were piezoresistive pressure sensors with silicon elastic thin film, which were fabricated by MEMS technology. Compared with the pressure sensor applied traditional metal strain foil, the sensitivity of this sensor was much higher. At the same time, because there was no mechanical linkage conversion, the measurement accuracy of this sensor was relatively higher and the repeatability error and pressure hysteresis effect of this sensor were much lower.

The three-probe pulse sensor avoided being fixed with wristlet, while applied wrist bracket to fix the wrist, which could not only avoid the measurement error caused by the excessive deformation of soft tissue because of fixed with wristlet, but also be more in line with the form of pulse feeling in TCM clinical science.

The PC data processing software of the three-channel self-pressurized pulse acquisition system acquired wrist pulse samples under various pulse taking forces through slowly continually compressing and then searched for the pulse taking pressure with the largest peak-to-peak value of the wrist pulses as the best pulse taking force, which was more accurate than that searching for the best pulse taking force form several setting forces and could acquire better and clearer wrist pulse signals.

The three-channel self-pressurized wrist pulse acquisition system could not only simulate the process of the three regions and nine pulse-takings wrist pulse diagnosis in TCM and acquire the wrist pulses in the region of Cun, Guan and Chi under the best pulse taking force automatically, but also real-timely display the pulse taking forces, the pulse waves, corresponding single cycle pulse waves and the result of classification, which laid the foundation for further analysis of multi-channel wrist pulses and improving the development of the standardization of wrist pulses and the objectification of wrist pulse diagnosis in TCM.

# References

1. Jingtang, L.: The objective detection and description of the types of pulse based on the Chinese traditional medical science (中医脉象的客观描述和检测). Chin. J. Med. Instrum. **25**(6), 318–323 (2001)
2. Zhaofu, F.: Contemporary Sphygmology in Traditional Chinese Medicine (现代中医脉诊学). People's Medical Publishing House Co., Ltd, Beijing (2003)
3. Weichang, T., Rui, L.: Research on the Cun-Guan-Chi pulse detecting system (三部脉象检测系统的研究). Chin. J. Med. Instrum. **29**(3), 164–166 (2005)
4. Jingjing, W., Congying, L., Xinhong, J., Yuping, Z., Weichang, T.: Research for the largest and safe pressure in the pulse testing (脉象检测中最大取脉压力和人体耐受安全值的研究). Liaoning J. Tradit. Chin. Med. **37**(4), 582–584 (2010)

# A Wearable Device for High-Frequency EEG Signal Recording

Lorenzo Bisoni$^{(\boxtimes)}$, Enzo Mastinu, and Massimo Barbaro

Department of Electrical and Electronic Engineering,
University of Cagliari, Piazza dArmi, 09123 Cagliari, Italy
lorenzo.bisoni@diee.unica.it, barbaro@unica.it

**Abstract.** The recording of high-frequency oscillations (HFO) through the skull has been investigated in the last years highlighting interesting new correlations between the EEG signals and common mental diseases. Therefore, since most of the commercially available EEG acquisition systems are focused on the low frequency signals, a wide-band EEG recorder is here presented. The proposed system is designed for those applications in which a wearable and user-friendly device is required. Using a standard Bluetooth (BT) module to transfers the acquired signals to a remote back-end, it can be easily interfaced with the nowadays widely spread smartphones or tablets by means of a mobile-based application. A Component Off-The-Shelf (COTS) device was designed on a $19\,cm^2$ custom PCB with a low-power 8-channel acquisition module and a $24-bit$ Analog to Digital Converter (ADC). The presented system, validated through in-vivo experiments, allows EEG signals recording at different sample rates, with a maximum bandwidth of $524\,Hz$, and exhibits a maximum power consumption of $270\,mW$.

**Keywords:** EEG Recorder · Wearable EEG · Wide-Band EEG · Wireless · Bluetooth EEG

## 1 Introduction

The electroencephalogram (EEG) is a common technique for detecting symptoms of neurological diseases such as epilepsy, sleep disorders, anxiety and learning disabilities. These pathologies have a great impact on people common life and are quite common. For example anxiety disorders affects approximately 13.6 % of the European population [1], and, in 2010, its overall cost in Europe was €74.4 billion [2]. Most of the mentioned mental disorders require long-term EEG monitoring to follow the course of the disease and sometimes to prevent further degradations of the patient condition such as epileptic discharges. In these cases the longer is the EEG measurements period the higher is the probability of a successful event detection. Moreover EEG acquisition during daily life activities is highly recommended to better reveal some pathologies.

Traditional ambulatory EEG systems do not satisfy these requirements. In fact patients can be continuously observed for only a few hours because of the

© Springer International Publishing Switzerland 2015
A. Fred et al. (Eds): BIOSTEC 2015, CCIS 574, pp. 60–74, 2015.
DOI: 10.1007/978-3-319-27707-3_5

costs and resource overheads. Moreover there are some inconveniences such as forcing people to take time off work and moving them from their natural environment. As a consequence, patients often feel uncomfortable and, depending on their pathology, this can affect the EEG acquisition, introducing undesired artifacts. As a result of recent technology innovations, new outpatient EEG systems were introduced. Such mobile solutions overcome some of these limitations reducing the overall patient monitoring costs and increasing the effectiveness of the measurements [3]. Despite their benefits such systems are still cumbersome and/or too complex to be used outside hospitals and require expert assistance.

Wearable EEG is aimed to overcome these issues, allowing the recording of a longer temporal window that includes all stages of sleep and wakefulness and increasing the likelihood of recording typical symptoms. Many efforts have been already put on the realization of wearable EEG systems. Some examples are presented in [4,5] in which respectively a semi-custom and a completely custom CMOS EEG recorder was realized, whereas a 4-channel BCI-cap based on off-the-shelf components is described in [6]. Other examples are the Epoc [7], the Imec's headset [10] and the Quasar's [8] DSI 10/20. They all use proprietary radio link for data transmission resulting in a reduced power consumption though they require specific hardware to interconnect a remote back-end. Only a few systems have been developed using a standard communication link such as the ThinkGear [9] and the Starfast [18]. Furthermore currently available EEG systems mainly operate with a bandwidth under 100 Hz that may be enough to cover the most common diagnostic purposes, but a wider bandwidth, up to 600 Hz, is required to investigate some pathologies [11]. As fully discussed in [11], many improvements are still required in order to get a promising solution, easy to use by non-expert users in a completely uncontrolled environment. Some critical aspects that refer to the electrode-skin adherence, the battery life time and the quality of the acquired EEG signals have to be solved.

According to the idea of spreading the use of wearable EEG recorders, the system requires a low impact on people daily life. In other words the device should not imply the use of additional special equipments and it should be very practical. These requirements and the increasing spread of smartphones and tablets among a wide variety of users, from teen to elders [16], suggest these new generation mobiles as the best solution to control the wearable EEG recorders. Furthermore this choice is supported by recent researches on moving the telemedicine toward mobile platforms [12–15].

The EEG system proposed in this paper is based on a custom PCB with off-the-shelf components (COTS) and uses a standard BT link to transmit the acquired signals. As a consequence it exhibits a higher power consumption compared to those solutions, previously presented, that use custom radio link but it has the advantage of being easily interfaced with any BT-based terminal and integrated with such new healthcare systems. In addition, todays technology allows mobile devices with high computational power, huge storage memory and fully programmable. In this way they can store and elaborate the EEG signals allowing a wide-range of applications. Anyway our EEG recorder can also be connected to a traditional desktop PCs for which a simple Microsoft Windows-based application for testing purposes was developed. Being a wearable device, special efforts were made in reducing its power consumption and in device miniaturization. As a result, only the essential components were included in the project: an amplifying/filtering block, an analog to digital converter, a microcontroller, a BT transceiver and a power management module. In the following Sect. 2 the system architecture details will be described, Sect. 3 contains a brief description of a possible remote interface whereas the validation test results will be presented in Sect. 4. Finally systems performance and future development will be shortly summarized in Sect. 5.

## 2    System Architecture

The designed system, named BlueThought by joining the implemented transmission link (Bluetooth) with the nature of the acquired signals (the human thought), is based on a differential 8 channel recording unit. The EEG signals detected with a standard EEG cap are first amplified and then converted into digital signals by an *ADS1299* component from Texas Instrument. Once acquired, digital signals are transmitted to a remote back-end by means of a *Microchip Bluetooth RN-42* module. Moreover a USB connection was introduced to charge the EEG recorder battery and as additional channel for data transfer. A *Microchip PIC18F46J50* coordinates data exchange between ADC and BT or USB external controller. The system architecture is depicted in Fig. 1.

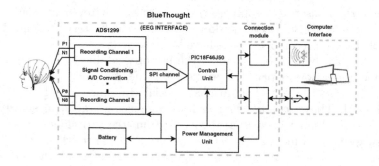

**Fig. 1.** BlueThought: System Architecture.

In addition, a power management unit generates all digital and analog voltage supplies for the ADC, the microprocessor and the BT transceiver from a 3.7 V − 950 mAh LiPo battery. Even the battery charging circuit was implemented on the board. The EEG recorder was realized on the 5.5 cm × 3.5 cm double face board depicted in Fig. 2. In the following further details on the main modules of the EEG recorder will be described.

## 2.1   Signal Conditioning and Digital Conversion

Before being converted into digital format, the input signals are filtered and amplified. To reduce power consumption and PCB area we selected the *ADS1299* A/D converter which includes the signal conditioning block avoiding the need of any additional component. Moreover, exhibiting a low power consumption of 5 mW in stand-by mode that is increased up to 40 mW during signal recording, the *ADS1299* is suitable for our application. This device contains eight independent differential channels allowing simultaneous acquisition. As depicted in Fig. 3, an internal multiplexer allows to select the P and N input signals among various sources and depending on the selected signals different recording modes are possible: normal recording, test and impedance monitoring mode. The normal recording mode is the default working set-up in which EEG signals are acquired in both single-ended and differential configuration. In single-ended measurements the N signals are internally shorted to the external reference (typically mid-supply voltage) or to the bias signal generated by the internal bias module on the base of a desired input signal combination. Whereas in differential measurements both P and N signals come from the EEG cap. To reduce the number of interconnections between the EEG recorder and the bonnet all N input lines are shorted together getting only one common reference electrode conveniently placed on the patient body. In test mode, different internally-generated test signals can be selected as input allowing the signal acquisition chain to be tested out. Another important feature provided by the *ADS1299* is the lead-off detection. It consists in a continuous patient electrode impedance monitoring to verify if a suitable connection is present or not.

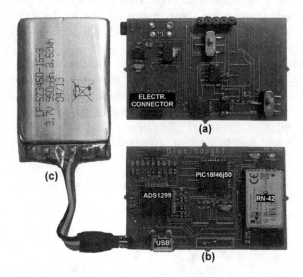

**Fig. 2.** EEG Interface prototype: power management circuit on bottom side(a); *ADS1299* (ADC), *PIC18F46J50* (microprocessor), USB and *RN-42* (BT transceiver) on top face(b) and a 3.7 V − 950 mAh LiPo battery(c).

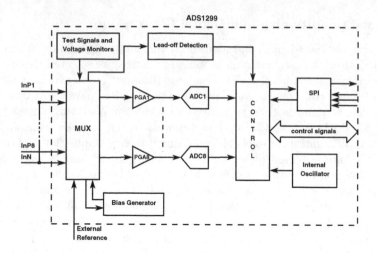

**Fig. 3.** *ADS1299* main architecture: Signal conditioning and analog to digital conversion.

The first stage of each acquisition channel is a differential low-noise programmable gain amplifier (PGA). It offers seven gain settings (1, 2, 4, 6, 8, 12, and 24) that can be set-up by writing the channel-setting registers (one per channel) of the *ADS1299*. As mentioned in Sect. 1, our EEG recorder can acquire signals with a bandwidth wider then standard EEG monitor. In fact, as reported in Table 1, the system supports different sample rates from $250 SPS$ up to $2000 SPS$ resulting in a maximum bandwidth of 524 Hz. This makes our device suitable for a wide range of applications even those requiring the analysis of signals out of standard EEG frequency. After being amplified, the signal is digitalized by a $24 - bit$ $\Sigma\Delta$ converter.

The ADC operates in two different modes: continuous mode (default) and single-shot mode. In the first modality, when a start command is sent, it continuously converts the input signal. The conversion ends when a stop command is received. Whereas, if the device is in single-shot mode it generates only one sample per received start command. This means that to begin a new conversion, a new start command has to be sent. Regardless of the operating mode, as a

**Table 1.** EEG Recorder −3 dB bandwidth with the required sample frequency and the correspondent output data rate.

| −3 dB Bandwidth [Hz] | Sample Frequency [SPS] | Output Data Rate [Kbps] |
| --- | --- | --- |
| 65 | 250 | 6.75 |
| 131 | 500 | 13.50 |
| 262 | 1000 | 27.00 |
| 524 | 2000 | 54.00 |

single sample conversion ends, a data-ready signal (DRDY) is pulled down to notify the microprocessor that a new sample is ready. After being converted, the eight samples (each per channel) are packed and sent to the micro-controller over a 3 MHz SPI connection. In the following the control unit is described. It forwards the samples received from the ADC to the BT transceiver or to the USB controller depending if a wireless or a wired connection is being used.

## 2.2  Control Unit

The *Microchip PIC18F46J50* is based on a new *nanoWatt XLP (eXtreme Low Power)* technology that hugely reduce its power consumption with respect to other micro-controllers with the same features. The microprocessor is used as control unit to serve two main tasks: system set-up and data exchange. The PIC is powered at 3.3 V with a CPU clock frequency of 48 MHz generated by an on-chip oscillator. At system power-up, the PIC is used to setup the EEG interface defining both recording and connection parameters such as acquisition gain and bandwidth, ADC SPI clock frequency, BT data rate and communication protocol parameters. All values are tuned to find a good compromise between the acquired EEG signal quality and the power consumption. Once that the system started to acquire the EEG signals, the control unit coordinates data exchange among the

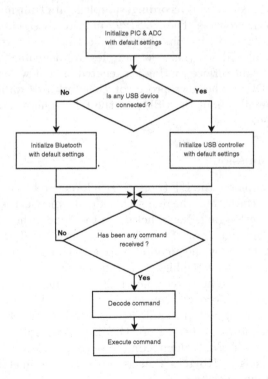

**Fig. 4.** Main steps of the control unit firmware.

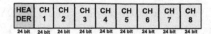

**Fig. 5.** Single data packet transmitted by the ADC.

converter and BT or USB remote back-end. The microprocessor provides several internal peripherals that, if not used, can be disabled to save power. In particular we are interested in using the USB and the UART in/out ports to respectively connect the PIC to a remote USB controller or to the BT transceiver.

Although the system communication mode can be on-line modified by the user, if, on power-up, any device is connected to the USB port, the PIC automatically enables the USB controller otherwise the BT transceiver is turned on. Once defined the connection mode, the microprocessor starts a polling cycle waiting for data coming from the remote controller. The received commands are decoded and executed. Such commands, generally are aimed at controlling the ADC or the BT transceiver but they can also be addressed to the same microprocessor for example to setup the USB controller. The main steps of the firmware are described in the flow chart of Fig. 4. Regardless of the back-end connection mode (via USB or BT) and only in single-shot recording, the same polling cycle is used by the PIC to send the sampled data to the remote controller. Otherwise, in continuous recording, sampled data transmission is handled by an interrupt service routine. The *ADS1299* data-ready signal (DRDY) is connected to an interrupt sensitive pin of the PIC acting as an external interrupt. When DRDY is pulled down (i.e. new samples are available) an exception is raised and the interrupt service routing is executed. The new samples are transferred from the ADC to the microprocessor that forwards data to the remote controller. Some details about the USB and the BT connection are given in the following paragraph.

### 2.3   Data Transmission

In normal operation mode, the BT transceiver allows wireless data transmission between the EEG recorder and the remote back-end, whereas the USB controller is used for battery recharging. Nevertheless, in test mode, the wired connection is quite useful for both data transfer and system powering. The communication link specifications are set by the amount of data to transfer during the continuous data acquisition mode at the highest frequency allowed by the system. From Table 1, in order to enable high frequency EEG recording, a sample frequencies of $2000 SPS$ is required. Moreover, considering that the *ADS1299* acquires eight channels per time and that each sample is converted into a 24 bit word, 192 bit of payload with an additional header of 24 bit has to be transferred each 0.5 ms (Fig. 5). As computed in Eq. 1, the maximum output data rate (ODR) required in worst conditions is 54 Kbyte/s that is a critical parameter to define the data-exchange channel specifications.

Single data packet: 216 bit

$$\text{ODR(@2000SPS)}: \frac{2000 * 216}{8} = 54\,\text{Kbyte/s} \tag{1}$$

The USB port is directly handled by an on-chip USB controller and can operate in two different modalities: CDC and HID mode. To make the USB suitable for our application, a standard HID protocol was implemented. Working at full speed (48 MHz) with $64 - byte$ data packet size, the data transfer speed is limited to 64 KBytes/s. In addition, to make the transmission more efficient, two sampled data packets $(2 \times 216\,\text{bit} = 432\,\text{bit})$ are grouped in the same USB frame. As a result, in worst conditions (i.e. with the maximum sample rate, $2000SPS$), the required data transfer rate amounts to 27 KByte/s that is below the USB transfer rate limit. In contrast to the USB HID protocol, the BT transceiver does not require fixed size packets, but their length is adapted to the amount of transferred data. The $RN\text{-}42$ is a small form factor, low power, class 2 BT radio with on-chip antenna. It delivers up to a 3 Mbps data rate for distances up to 20 m. It uses an UART port to communicate with the control unit and operates in two modes: data mode (default) and command mode. In data mode, the module works as a data pipe. When the module receives data, it strips the BT headers and forwards the data to the UART port. When data is written to the UART port, the module constructs the BT packet and sends it out over the BT wireless connection. Thus, the entire process of sending/receiving data to the host is transparent to the PIC. The command mode is used to defining the BT operating mode, the UART baud rate and others control flow parameters. Moreover the $RN\text{-}42$ operates in slave mode so that other BT devices (PC, tablet or smartphone) can discover and connect to the module.

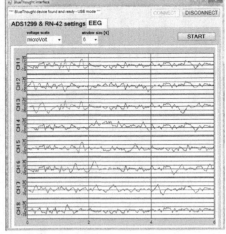

(a) System setup window

(b) On-line plotting window in a $\mu V$ amplitude scale and with a 6s temporal window

**Fig. 6.** Visual C++ application for connecting and controlling the developed EEG interface.

## 3   Remote Interface

The designed system is a general purpose EEG recorder and depending on the treated pathology a specific software can be developed. At this first stage of the project a Visual C++ application was written, implementing only the essential features for the hardware debugging. The ADC module can be completely configured in terms of PGA gain and sample rate and both continuous and single-shot modes are selectable. The eight recorded signals can be plotted together in the same graph or on separate sub-window for a real-time view and stored in a text file for off-line data computing. Figure 6(a) and (b) show the two main window of the developed interface. The first refers to the system settings and the second to the plotting of the eight recorded signals. For all channels is possible to setup the amplitude scale ($\mu V$, mV or V) and the temporal window size.

## 4   Experimental Results

All system features were first characterized and than compared with a standard laboratory equipment. To start with its static electrical characterization, Table 2 collects the EEG interface power consumption in different working conditions. In idle state (only the microprocessor is on) it has a minimum power consumption of about 119 mW whereas in worst conditions (i.e. all devices are on, sample rate of 2000$SPS$ and active BT data transmission) it absorbs a maximum of 270 mW. Under this conditions and with the chosen battery (3.7 V − 950 mAh LiPo) the system can continuously work for about 13 h. Further experiments

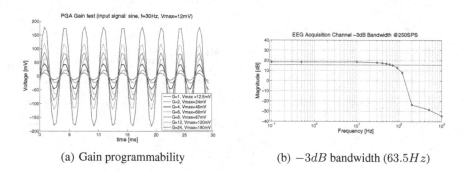

(a) Gain programmability          (b) −3$dB$ bandwidth (63.5$Hz$)

**Fig. 7.** The EEG interface gain programmability and bandwidth validation by recording a 12 mV–30 Hz sine at 250$SPS$.

**Table 2.** EEG Recorder Power Consumption in different working conditions.

| Power Consumption [mW] | ADC (ON) | ADC (OFF) |
|---|---|---|
| Bluetooth (OFF) | 119 | 158 |
| Bluetooth (ON) | 230 | 270 |

were performed to study the dynamic behaviour of the EEG acquisition channel. Its gain programmability from 1 V/V up to 24 V/V was confirmed acquiring a 12 mV–30 Hz sine as depicted in the above plot of Fig. 7(a). The device
showed a 63.5 Hz−3 dB bandwidth at sample rate of $250SPS$ and the magnitude
bode diagram of the recording channel transfer function is depicted in Fig. 7(b).

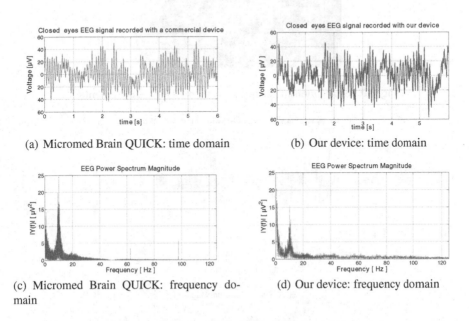

(a) Micromed Brain QUICK: time domain

(b) Our device: time domain

(c) Micromed Brain QUICK: frequency domain

(d) Our device: frequency domain

**Fig. 8.** Closed-eyes EEG signal, in time and frequency domain, recorded at $250SPS$.

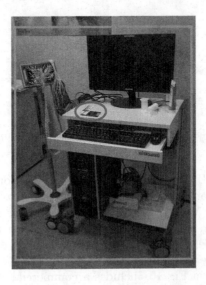

**Fig. 9.** Comparison between our wearable EEG interface (red circle) and a cumbersome
commercial device (green box) (Color figure online).

**Fig. 10.** Experimental setup for in-vivo EEG measurements.

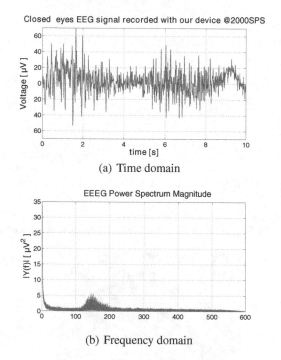

(a) Time domain

(b) Frequency domain

**Fig. 11.** Closed-eyes EEG signal, in time and frequency domain, recorded at $2000SPS$.

Moreover both wired (USB) and wireless (BT) connections were tested. Once the system main functions have been proved, some in-vivo EEG measurements, on one human subject, were performed. To evaluate the signal quality of the designed EEG recorder, the system was compared with a commercial device (Brain QUICK, [17]) depicted in Fig. 9 where the huge difference in terms of dimensions between the two devices is also highlighted. Moreover, the experimental setup, depicted in Fig. 10, includes a commercial EEG cap (*KIT-CAP-SPEXT61* from Micromed) with 61 electrodes used to acquire the neural signals.

To better compare the two devices, they were connected to adjacent electrodes and simultaneous recordings were performed in different patient conditions. During the first test, the human subject was in resting state with closed eyes to avoid any kind of artefact. Figure 8 shows the EEG signals acquired by the Brain Quick (Fig. 8(a) and (c)) and by our EEG recorder (Fig. 8(b) and (d)). The signals recorded at $250SPS$ are quite similar in both time and frequency domains.

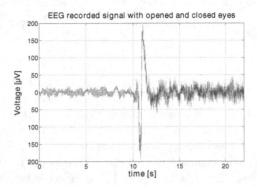

**Fig. 12.** EEG recorded signal with opened-eyes (from 0 s to 10 s) and closed-eyes (from 12 s to 22 s).

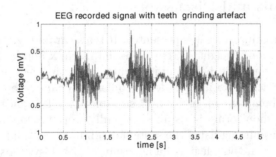

**Fig. 13.** EEG recorded signal with teeth-grinding artifacts.

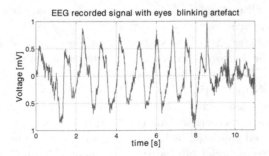

**Fig. 14.** EEG recorded signal with eyes-blinking artifacts.

**Table 3.** Comparison between some state-of-the-art EEG Recorders.

| | Our device | Quasar | Imec | Emotiv Epoc | NeuroSky | Brown L. | Enobio |
|---|---|---|---|---|---|---|---|
| CMRR | $>110dB$ | $>120dB$ | | | | | $115dB$ |
| Input Impedance | $1G\Omega$ | $47G\Omega$ | | | | | |
| Bandwidth | $0.01-524Hz$ | $0.02-120Hz$ | $0.3-100Hz$ | $0.2-45Hz$ | $3-100Hz$ | $0.5-375Hz$ | $0-250(500)Hz$ |
| Channel number | 8 | 12 | 12 | 14 | 1 | | $8-20(32)$ |
| Noise | $<2\mu Vpp$ | $3\mu Vpp$ | $4\mu Vpp$ | | | $1\mu Vpp$ | $<1\mu Vrms$ |
| Bit number | 24 | 16 | 12 | 16 | | 11 | 24 |
| Wireless protocol | BT | Proprietary | Nordic RF | Proprietary | BT | Proprietary | BT |
| Power consumption | $270mW$ | | $42mW$ | | $130mW$ | $12mW$ | |
| Run time | $13h$ | $24h$ | | $12h$ | $10h$ | $30h$ | $16h$ |
| Technology | COTS | | ASIC / COTS | | | ASIC / COTS | |

The system capability to record signals above standard EEG bandwidth was also proved by acquiring some signals at $2000 SPS$ from a patient with closed eyes as depicted in Fig. 11. Moreover, in Fig. 12 it is possible to appreciate the differences between an open-eyes (on the left) and a closed-eyes (on the right) EEG signal perfectly recorded by our device at $250 SPS$. Finally, to further validate our system, some typical EEG artefacts such as the teeth-grinding signal (Fig. 13) and the eyes-blinking effect (Fig. 14) were recorded. They respect the typical shapes and amplitudes of such signals.

# 5    Conclusions and Discussion

A COTS based EEG interface was presented. It is a wearable system that, thanks to its small dimensions (height: 5.5 cm x width: 3.5 cm x depth: 1.0 cm), can be easily placed on the patient head and integrated with the electrodes framework. The developed device has 8 independent acquisition channels and was designed with the aim of being user-friendly and suitable for all applications in which a long-time EEG monitoring is required. In fully working condition (i.e. when acquiring and transmitting data) the system exhibits an overall power consumption of 270 mW. Even-though it is higher than of other systems (Table 3), the device allows 13 h of continuous signal recording that is in line with other wearable devices. The higher power consumption is mainly due to the choice of using a COTS solution and a standard BT link to connect a remote controller. However it gives the device the great advantage to easily connect any BT-based end-terminal in contrast to other systems that, using a proprietary wireless link, require specific hardware. Moreover, compared to others state-of-the-art equipments, our EEG recorder has a wider bandwidth, up to 524 Hz, allowing high-frequency EEG monitoring. This can be very useful to deeper understand and investigate a certain number of pathologies. In addition, a Windows-based Visual C++ software was written for the EEG recorder testing purpose. The system was completed validated by in-vivo measurements on human patient and compared with a commercial laboratory equipment. Moving towards a device that can easily become part of everyday life for all people improving their living conditions from both health and entertainment points of view and with the least economical and daily-activity impact is our main goal. Therefore, being a wearable device, next developments are the reduction of the power consumption and the developing of smartphone-based application to respectively increase the battery life and to make the system completely portable. In particular, some future improvements include the use of a new generation BT called Bluetooth 4.0 Low Energy (BTLE) that drastically reduce the power transmission and a review of the control unit strategy turning off, time by time, all on-board unused devices. Moreover the possibility to optionally expand the number of input channels by plugging in an additional acquisition module and the introducing of on-board data storage capabilities might be considered. Finally, a custom chip solution for signal conditioning and converting will might be investigated to further reduce both power consumption and system dimensions.

**Acknowledgements.** The authors would like to thank Dr. Matteo Fraschini and Matteo Demuru from the University of Cagliari for their support on EEG recording in-vivo experiments. L. Bisoni gratefully acknowledges Sardinia Regional Government for the financial support of his PhD scholarship (P.O.R. Sardegna F.S.E. Operational Programme of the Autonomous Region of Sardinia, European Social Fund 2007-2013 - Axis IV Human Resources, Objective 1.3, Line of Activity l.3.1.).

# References

1. Alonso, J., Angermeyer, M.C., Bernert, S., et al.: Prevalence of mental disorders in Europe: results from the European Study of the Epidemiology of Mental Disorders (ESEMeD) project. Acta Psychiatr. Scand. **109**, 21–27 (2004)
2. Olesen, J., Gustavsson, A., Svensson, M., Wittchen, H.-U., et al.: The economic cost of brain disorders in Europe. Eur. J. Neurol. **19**, 74–80 (2003)
3. Waterhouse, E.: New horizons in ambulatory electroencephalography. Eng. Med. Biol. Mag. **22**, 74–80 (2003)
4. Brown, L., Van de Molengraft, J., Yazicioglu, R.F., Torfs, T., et al.: A low-power, wireless, 8-channel EEG monitoring headset. In: 32nd Annual International Conference of the IEEE EMBS Engineering in Medicine and Biology Society (EMBC), pp. 4197–4200 (2010)
5. Carmo, J.P., Dias, N.S., Silva, H.R., Mendes, P.M., et al.: A 2.4-GHz low-power/low-voltage wireless plug-and-play module for EEG applications. Sens. J. **7**, 1524–1531 (2007)
6. Lin, C.-T., Ko, L.-W., Chiou, J.-C., Duann, J.-R., Huang, R.-S., Liang, S.-F., Chiu, T.-W., Jung, T.-P.: Noninvasive neural prostheses using mobile and wireless EEG. Proc. IEEE **96**, 1167–1183 (2008)
7. Emotiv (2013). http://emotiv.com
8. Quasar (2013). http://www.quasarusa.com
9. NeuroSky (2009). http://neurosky.com
10. Patki, S., Grundlehner, B., Verwegen, A., Mitra, S., Xu, J., Matsumoto, A., Yazicioglu, R.F., Penders, J.: Wireless EEG system with real time impedance monitoring and active electrodes. Biomedical Circuits and Systems Conference (BioCAS), 2012 IEEE, pp. 108–111 (2012)

11. Mihajlovic, V., Grundlehner, B., Vullers, R., Penders, J.: Wearable, wireless EEG solutions in daily life applications: what are we missing? IEEE J. Biomed. Health Inf. **99**, 1 (2014)
12. Tachakra, S., Wang, X.H., Istepanian, R.S.H., Song, Y.H.: Mobile e-Health: the unwired evolution of telemedicine. Telemedicine J. e-Health **9**(3), 247–257 (2003)
13. Chan, S.R., Torous, J., Hinton, L., Yellowlees, P.: Mobile tele-mental health: increasing applications and a move to hybrid models of care. Healthcare **2**(2), 220–233 (2014)
14. Lupu, C., Cosmin-Constantin, M.: Actual portable devices as base for telemedicine and e-health: Research and case study application. In: E-Health and Bioengineering Conference (EHB), 2013, PP. 1–4 (2013)
15. El Khaddar, M.A., Harroud, H., Boulmalf, M., ElKoutbi, M., Habbani, A.: Emerging wireless technologies in e-health trends, challenges, and framework design issues. In: 2012 International Conference on Multimedia Computing and Systems (ICMCS), pp. 440–445 (2012)
16. Aaron Smith: Smartphone Ownership 2013. Pew Research Center (2013). http://www.pewinternet.org/2013/06/05/smartphone-ownership-2013/
17. Micromed: Brain Quick (2014). http://www.micromed.eu
18. Riera, A., Dunne, S., Cester, I., Ruffini, G.: STARFAST: a Wireless Wearable EEG/ECG Biometric System based on the ENOBIO Sensor. In: Proceedings of the International Workshop on Wearable Micro and Nanosystems for Personalised Health (pHealth08) (2008)

# Indirect Blood Pressure Evaluation by Means of Genetic Programming

Giovanna Sannino[✉], Ivanoe De Falco, and Giuseppe De Pietro

Institue of High Performance Computing and Networking, CNR,
Via Pietro Castellino 111, 80131 Napoli, Italy
{giovanna.sannino,ivanoe.defalco,giuseppe.depietro}@na.icar.cnr.it
http://ihealthlab.icar.cnr.it/

**Abstract.** This paper relies on the hypothesis of the existence of a nonlinear relationship between Electrocardiography (ECG) and Heart Related Variability (HRV) parameters, plethysmography (PPG), and blood pressure (BP) values. This hypothesis implies that, rather than continuously measuring the patient's BP, both their systolic and diastolic BP values can be indirectly measured as follows: a wearable wireless PPG sensor is applied to a patient's finger, an ECG sensor to their chest, HRV parameter values are computed, and regression is performed on the achieved values of these parameters. Genetic Programming (GP) is a Computational Intelligence paradigm that can at the same time automatically evolve the structure of a mathematical model and select from among a wide parameter set the most important parameters contained in the model. Consequently, it can carry out very well the task of regression. The scientific literature of this field reveals that nobody has ever used GP aiming at relating parameters derived from HRV analysis and PPG to BP values. Therefore, in this paper we have carried out preliminary experiments on the use of GP in facing this regression task. GP has been able to find a mathematical model expressing a nonlinear relationship between heart activity, and thus ECG and HRV parameters, PPG and BP values. The experimental results reveal that the approximation error involved by the use of this method is lower than 2 mmHg for both systolic and diastolic BP values.

**Keywords:** Blood pressure · Wearable sensors · Heart rate variability · Plethysmography · Regression · Genetic programming

## 1 Introduction

Continuous non-invasive arterial pressure (CNAP) methods can be used to continuously measure arterial blood pressure (BP) in real time and without any need for patient's body cannulation. Such methods show two positive features related to clinical "gold standards". The first feature is that, since the measurement takes place in the invasive arterial catheter system, BP can be measured in real time in a continuous way. The second feature is that, as it is for the

© Springer International Publishing Switzerland 2015
A. Fred et al. (Eds): BIOSTEC 2015, CCIS 574, pp. 75–92, 2015.
DOI: 10.1007/978-3-319-27707-3_6

standard procedure based on upper arm sphygmomanometer, this method too is non-invasive. In the recent years, results in this field have been promising in terms of accuracy, ease of use, and acceptance in the clinical domain.

As it can be imagined, currently there is a high request for accurate and easy-to-use CNAP-systems. Consequently, an increasing focus on these devices exists by researchers, practitioners and the industry of medical devices. The use of small yet powerful microcomputers, and that of digital signal processors as well, make the development of efficient BP measurement instruments easier. In fact, an easy processing of complex and computationally intensive mathematical functions is allowed by small, cheap devices of this kind. Invasive catheters are used to continuously measure BP in only a small fraction, between 15 % and 18 %, of inpatient surgeries, as it has been reported by several researchers as, e.g., [1, 2]. Instead, intermittent, non-invasive blood pressure monitoring constitutes the practical standard of care for all the remaining inpatient surgeries, and also for outpatient surgeries. Unluckily, this latter type of monitoring is discontinuous, and this feature implies the possibility that some dangerous hypotensive episodes could be missed. As an example, during the monitoring of women undergoing Caesarean section, CNAP methods were found able to detect hypotensive phases in 39 % of the cases, whereas intermittent non-invasive methods could to this only in 9 % of the cases. Moreover, when CNAP was used for the measurement of systolic BP values higher than 100mmHg, dangerous foetal acidosis did not occur, as it is reported in [3]. Furthermore, it is reported in [4] that more than 22 % of hypotensive episodes were missed by intermittent non-invasive methods, thus leading to delayed treatments or even no treatments at all.

The detection of in-artery pressure changes from outside the arteries themselves is, of course, of high difficulty, whereas the measurement of changes in artery volume and flow is quite easy. The use of e.g. echography, light, impedance, and so on allows fulfilling this task. The drawback in these approaches is the absence of linear correlation between these volume changes and the arterial BP, and this is true especially when the measurement is performed in the periphery, as it is usually done because it is easier to access the arteries there.

Non-invasive devices, consequently, must convert the volume signal measured at the periphery into another one expressing the arterial BP. Typical techniques to do this are those relying on vascular unloading, tonometry, pulse transit time (PTT). In particular, this latter technique is based on the fact that, every time a stroke volume is ejected by the heart towards arteries, the BP wave takes a given transit time to arrive at the periphery. PTT is indirectly dependent on BP, namely the higher the pressure, the faster PTT is. This is very useful to detect changes in BP in a non–invasive way [5]. To get absolute values, the method needs calibration. Techniques relying on the use of PTT are good examples of indirect ways to continuously measure blood pressure. These techniques continuously take measures of other parameters, and hypothesize a non-linear relationship.

In the knowledge discovery area a problem as the one described above is named as a *regression problem*. In each of these problems, a relationship is hypothesized between some variables, called independent, and another one, called dependent, and the goal consists in searching for the explicit form of the mathematical model that connects them at best. The expression "independent variables" is used just to mean that such variables are the input ones to the problem, and it does by no way mean that they are not correlated one another. When performing traditional regression analysis, a user has to specify the general structure of the mathematical model he is looking for. A procedure based on a trial–and–error approach to hypothesize or experimentally find a good model is very laborious and time-consuming, and difficulties can be experienced by human minds when trying to guess which the most important independent variables affecting the dependent one are, and which the best formula relating them is.

Genetic Programming (GP) is a Computational Intelligence paradigm that, unlike the above regression approaches, can at the same time automatically evolve the structure of a mathematical model and select from among a wide parameter set the most important parameters contained in the model. Consequently, it can carry out very well the task of regression.

Nowadays the availability of wearable sensors is increasing, while their cost is decreasing. Among them, there is a frequent use of chest sensors able to capture electrocardiographic (ECG) signals. Starting from such a signal, a wide set of parameters describing a patient's ECG activity can be easily extracted with Heart Rate Variability (HRV) analysis. Moreover, to compute plethysmography (PPG) values wearable wireless sensors can be applied to a patient's finger.

This paper relies on the hypothesis of the existence of a nonlinear relationship between PPG and heart activity (and thus ECG and HRV parameters), and blood pressure. This hypothesis implies that, rather than continuously measuring the patient's BP, both their systolic and diastolic BP values can be indirectly measured as follows: a wearable wireless PPG sensor is applied to a patient's finger, an ECG sensor to their chest, HRV parameter values are computed by performing HRV analysis, and regression is performed on the achieved values of these parameters, so that BP values can be indirectly measured.

An analysis of the scientific literature of this field reveals that nobody has ever used GP aiming at relating parameters derived from HRV analysis and PPG to BP values. Therefore, in this paper we have carried out preliminary experiments on the use of GP in facing this regression task.

## 2    Related Work

A quite small number of papers can be found in the scientific literature aiming at the investigation of the relationship between the BP and some other variables. This section provides a short description of some of the most relevant ones.

In [7] the BP estimation method relies on the hypothesis that a relationship exists between the pulse wave velocity (PWV) in the arteries and BP. Measuring PWV requires registering two time markers. The first marker depends on the detection of R peaks for ECG, whereas the second on the detection of the pulse wave in peripheral arteries. Their experimental device for BP monitoring is made of two analogue modules for signal acquisition, namely one for PPG signal and the other for ECG signal. Namely, the ECG electrodes are positioned on patient's wrists, while a pulse oximetry finger to register PPG is placed on a finger.

In [8] the aim was the evaluation of whether PWV can reliably predict the longitudinal changes in systolic BP (SBP), and the incident hypertension. The authors measured PWV at baseline in 449 volunteers, partly normotensive and partly untreated hypertensive. Their average age was $53 \pm 17$. BP measurements were repeatedly carried out during an average follow-up of $4.9 \pm 2.5$ years. By considering covariates such as body mass index, age, and mean arterial pressure, the authors applied linear mixed effects regression models, and their conclusion was that PWV can independently determine the longitudinal SBP increase.

In [9] an attempt was made to design and build a wearable sensor for BP measurement. This sensor should have the features of placing a lower burden on the examinees, of being less influenced by patient's physical movements, and of being usable to continuously measure BP. They modified the existing Moens-Korteweg BP equation by hypothesizing that the following relation exists: $P_s = b_1/T_{PTT}^2 + b_2$, where $T_{PTT}$ is the pulse-wave transit time, $P_s$ is the systolic BP, and coefficients $b_1$ and $b_2$ can be derived by using measured values of an examinee's BP and measured values of $T_{PTT}$. They implemented a new system for the calculation of patient's systolic BP, that used electrocardiography and ear-lobe pulse waves. Through this system they were able to estimate patients' BP, and to also directly measure patient's arterial pressure. They found that their methodology was able to correctly capture trends in the variations in BP.

In [10] the aim was the creation of a function able to link PWV and SBP, and the testing of its reliability in determining suitable absolute SBP values by using a non-linear algorithm and a one-point calibration. In order to induce BP increase, they asked 63 volunteers to exercise, and obtained this nonlinear function: $BP_{PTT} = P1*PWV*e^{(P3*PWV)}+P2*PWV^{P4}-(BP_{PTT,cal}-BP_{cal})$, with P1 = 700, P2 = 766,000, P3 = -1, and P4 = 9. $BP_{PTT,cal}$ is the BP value, computed from PTT, corresponding to the BP value measured by the reference method, while $BP_{cal}$ is the BP measured by the reference method (cuff) at experiment beginning. The authors made use of this non–linear function to compute BP values. The comparison between SBP values measured by cuff and those obtained by using the PTT-based method showed a significant correlation.

This short review shows that researchers are striving to find a suitable relationship between independent and dependent variables for the problem of indirectly measuring blood pressure, nonetheless this activity has a high cost in terms of labour, time, and experiments to find the values of the coefficients.

## 3    Signal Processing

The regulation of BP is usually described in terms of homeostasis [11]. This is regulated by the autonomic nervous system (ANS) due to two opposing divisions: the sympathetic division and the parasympathetic division. Heart rate variability (HRV) is a tool representing the balance between the sympathetic and parasympathetic branches of the ANS. HRV is known from many studies [12–14] to be one of the most investigated non-invasive biomarkers of ANS activities, and it can be extracted by means of the use of a wearable ECG sensor.

The plethysmography (PPG) is an easy–to–use and cheap optical technique that is useful in the detection of changes in blood volume in the micro-vascular bed of tissue, as shown in, e.g., [15]. It is a physiological waveform that is related to the activity of the sympathetic nervous system [16], and its monitoring can be performed through the use of a wearable non-invasive finger pulse oximeter.

We investigate here the associations among HRV and PPG measurements and the Systolic BP (SYS) and Diastolic BP (DIA) with the aim to propose a mathematical model that could allow the indirect calculation of the SYS and the DIA values through the use of just a wearable ECG sensor and a pulse oximeter.

The MIMIC database [17], available on physionet.org, was used in order to obtain the mathematical model. This database is composed by data recorded from more than 90 intensive care unit patients. The data include both signals and periodic measurements coming from a bedside monitor as well as clinical data acquired from the medical record of the patient. The files available for each patient include the following ones: qrs (ECG beat labels, all beats labelled normal), al (annotations for alarms related to changes in the patient's status), in (annotations related to changes in the functioning of the monitor), abp (arterial blood pressure), pap (pulmonary arterial pressure), cvp (central venous pressure), and ple (fingertip plethysmograph) annotations.

## 3.1   ElectroCardioGraphy

The pre–processing of the ECG signal takes place by using Kubios [18,19], a software package for event-related bio–signal analysis based on Matlab and developed at the University of Kuopio, Finland. Kubios is a state–of–the–art computer program often used to extract HRV–related parameters and analyse them.

Kubios carries out a standard linear HRV analysis complying to the guidelines established by the European Society of Cardiology and the North American Society of Pacing and Electrophysiology [13]. Moreover, it also computes a set of additional nonlinear features following the relevant literature [20,21]. Table 1 summarizes all the features computed by Kubios.

## 3.2   Plethysmography and Arterial Blood Pressure Pulse Waveform

The PPG signal and the Arterial Blood Pressure (ABP) pulse waveform are processed by a Matlab script developed by us to automatically compute the minimum and the maximum values of PPG in each pulse, and the systolic and the diastolic blood pressure values in each pulse from the ABP waveform. Figure 1 reports an example of the signals processed and of the results obtained.

## 3.3   The Database

By making use of the data contained the MIMIC database [17], a new dataset has been created by us. This latter database should be used to develop the mathematical model for the BP regression problem.

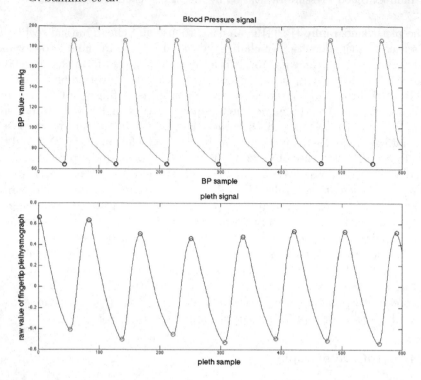

**Fig. 1.** An example of a record contains the ABP waveform and the PPG. Top pane: the purple circles indicate the systolic BP values, and the black circles the diastolic BP values. Bottom pane: the blue circles indicate the maximum PPG values, and the red circles the minimum PPG values (Color figure online).

The new database contains the measurements related to HRV, BP, and PPG. It contains 50 instances for each patient. The generic i–th instance in this new database is composed by the following information:

- $sub_{id}$: a numerical value identifying the patient;
- $SYS\_BP_i$: the average of the Systolic BP computed in the i-th 1-minute time slot;
- $DIA\_BP_i$: the average of the Diastolic BP computed in the i-th 1-minute time slot;
- $Pleth\_max_i$: the average of the maximum values of the PPG signal computed in the i-th 1-minute time slot;
- $Pleth\_min_i$: the average of the minimum values of the PPG signal computed in the i-th 1-minute time slot;
- f: a vector containing the values for the 35 HRV features reported in Table 1 computed in the i-th 1-minute time slot;

Consequently, the generic i–th database instance is composed as follows:

$$i = sub_{id}; SYS\_BP_i; DIA\_BP_i; Pleth\_max_i; Pleth\_min_i; f$$

For the experiments reported in this paper four patients have been considered, therefore our database is composed by a total number of 200 instances.

**Table 1.** Linear and non Linear HRV features.

| Measure | Description (Unit) |
|---|---|
| **Time Domain** | |
| Mean RR | The mean of RR intervals (ms) |
| STD RR | Standard deviation of RR intervals (ms) |
| Mean HR | The mean heart rate (1/min) |
| STD HR | Standard deviation of instantaneous heart rate value (1/min) |
| RMSS | Square root of the mean squared differences between successive RR intervals (ms) |
| NN50 | Number of successive RR interval pairs that differ more than 50 m (count) |
| pNN50 | NN50 divided by the total number of RR intervals (%) |
| RR tri index | The integral of the RR interval histogram divided by the height of the histogram |
| TINN | Baseline width of the RR interval histogram (ms) |
| **Frequency Domain** | |
| Peak freq. VLF | VLF band peak frequencies (Hz) |
| Peak freq. LF | LF band peak frequencies (Hz) |
| Peak freq. HF | HF band peak frequencies (Hz) |
| Absol. Pow. VLF | Absolute powers of VLF band (ms2) |
| Absol. Pow. LF | Absolute powers of LF band (ms2) |
| Absol. Pow. HF | Absolute powers of HF band (ms2) |
| Rel powers VLF | Relative powers of VLF bands (%) |
| Rel powers LF | Relative powers of LF bands (%) |
| Rel powers HF | Relative powers of HF bands (%) |
| Normalized powers LF | Powers of LF bands in normalized units |
| Normalized powers HF | Powers of HF bands in normalized units |
| Total power | Total Value for the spectral power (ms2) |
| LF/HF ratio | Ratio between LF and HF band powers |
| EDR | Electrocardiogram Derived Respiration (Hz) |
| **NonLinear Domain** | |
| SD1 | The standard deviation of the Poincaré plot perpendicular to the line of identity (ms) |
| SD2 | The standard deviation of the Poincaré plot along to the line of identity (ms) |
| ApEn | Approximate entropy |
| SampEn | Sample entropy |
| D2 | Correlation Dimension |
| $\alpha 1$ | Short-term fluctuation slope in Detrended Fluctuation Analysis |
| $\alpha 2$ | Long-term fluctuation slope in Detrended Fluctuation Analysis |
| Mean line length | Mean line length in RP (beats) |
| Max line length | Maximum line length in RP (beats) |
| REC | Recurrence Rate (%) |
| DET | Determinism (%) |
| ShEn | Shannon Entropy |

## 4   Genetic Programming

### 4.1   The General GP Framework

Genetic Programming (GP) [6] is a heuristic optimization technique relying on the imitation in a computer of mechanisms that are characteristic of the evolution in populations of individuals in nature. GP is based on a set, called population, of solutions to a problem that has to be faced. Individuals in the population represent programs that are encoded as tree structures, normally being different one another in terms of shape and number of composing nodes. In each individual the internal nodes stand for functions, whereas the leaf nodes denote terminals, i.e. both problem variables and constant values. Reading the tree in pre-order allows obtaining the program encoded in it. In the example tree in Fig.2 the in-order expression $7.9 + (3.2 * y) - x$ is encoded.

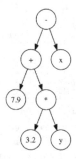

**Fig. 2.** An example of a tree in Genetic Programming.

One of the most important and delicate issues when using a GP is represented by the choice of a *fitness* function. This latter is a criterion that may represent in a quantitative way for any given solution the degree of goodness that solution has in solving the specific problem that is being faced. Obviously, the choice of a specific type of fitness function is dependent on the specific problem to be faced.

The following pseudo–code describes the typical search procedure of GP:

- input the data of the problem (e.g. the database for regression);
- generate randomly an initial population with a number of *Pop_size* individuals, each of which encodes a regression model;
- evaluate the quality of each individual in the current population through the use of a suitable *fitness function*;
- in each generation repeat the following steps until a new population is filled:
  - choose one operator from among *crossover*, *mutation*, and *copy*;
  - select within the current population a number of individuals suited to the operator chosen;
  - apply the operator chosen to the individuals selected in order to generate one offspring;

– insert the obtained offspring into the new population;

– evaluate the quality of the new offspring by means of the *fitness* function;

• repeat the above steps for each generation until a set maximum number of generations $Max\_gen$ is reached.

As the number of generations increases, better and better solutions in terms of better *fitness* values to the original problem will very likely be found.

*Selection* is a mechanism that chooses the individuals in the current population that will undergo the reproduction process. In it individuals with better fitness values should be preferred, i.e. chosen more frequently, yet also worse individuals should be allowed being selected, though with lower probability. The well–known tournament selection scheme has been used for the experiments described in this paper. In this scheme a number of $tourn\_size$ individuals contained in the current population is randomly chosen, and the best among them in terms of fitness function values will be selected.

The three genetic operators are shortly described in the following:

**Crossover.** Selection chooses two individuals, called *parents*, in the current population, and a subtree is randomly selected in each of these two trees. Then, the application of crossover means that these two selected subtrees are swapped from one parent individual to the other. This operator should check that the limit on the maximal depth allowed for a tree is not violated. If this violation takes place, then the offspring is discarded, one parent from among the two is randomly selected, and becomes the new offspring.

**Copy.** In this operator the selection mechanism randomly chooses one individual from the current population, and this individual is directly copied into the new population.

**Mutation.** In this case selection randomly chooses one parent individual in the current population, then mutation randomly selects a node in that tree, and generates a new subtree starting from it. The same check about the tree depth limit described above is performed: if the limit is violated, this new offspring is discarded and the parent individual is copied into the new generation.

Each time the choice of an operator must be made, it is performed by taking three probability values into account: $p\_cross$ is the probability value for crossover, $p\_copy$ is that for the direct copy, and $p\_mutate$ is the one for mutation. The sum of these three values must be equal to 1.0. For this choice, a random real value in $[0.0, 1.0]$ is generated and its value determines the specific operator that has to be used.

### 4.2   The General GP Framework

We have decided to avail ourselves of GP for the above described regression problem because it can automatically find a suitable model for the relationship between the independent variables and a dependent one, in this case one between the systolic and the diastolic BP values. Therefore, once chosen a suitable fitness function, facing this regression problem by GP means let it search,

**Table 2.** The set of the symbols representing the elementary functions, their description, and their arity.

| Symbol | Arity | Description |
|--------|-------|-------------|
| + | 2 | addition |
| - | 2 | subtraction |
| * | 2 | multiplication |
| / | 2 | protected division (returns 1 if the denominator is 0) |
| psqroot | 1 | protected square root (returns 0 for negative operands) |
| plog | 1 | Protected logarithm (rlog(0) is 0) |
| sqr | 1 | square |
| tanh | 1 | hyperbolic tangent |
| sin | 1 | sine |
| cos | 1 | cosine |
| exp | 1 | exponential |

from among the huge number of possible models, the one that best describes the fundamental features of this relationship. From a computational viewpoint it is obviously impracticable to perform an exhaustive search by considering the complete enumeration of all the possible models. Therefore we make reference here to GP: it is a heuristic method, so it cannot guarantee that it will achieve the globally optimum regression model, nonetheless it normally provides users with a suboptimal solution obtained in a reasonable computation time, and we hope the same will be true for this regression problem too.

The evolving population is made up by 'formulas', each encoding for one regression model. Specifically, these models are encoded as variable–depth trees, each of which is constituted by both elementary functions and terminals. The function set we have used for these experiments consists of 11 well-known elementary functions, that are listed in Table 2. In the table, the word 'Arity' represents for each function the number of arguments the function has.

The terminal set, on its turn, is made up by 37 symbols (the generic symbol $x_i$ stands for the $i$–th independent variable in the database), and by the *Const* symbol, that denotes a random constant value in a range suitable for the variable the constant refers to. The arity for all these terminal nodes is equal to 0.

With the aim at finding the (sub)–optimal model, the database data are suitably divided into three sets: the train, the test, and the validation sets. In the learning phase the approximation of the actual output values is performed on the items contained in the first set. The second set, instead, is used to evaluate the generalization ability of the model that has been trained over the first set. The third set, finally, is used to perform the real evaluation of the performance of the algorithm over previously unseen data.

If we let $S$ denote the model that is represented by a generic individual in the GP evolution, and let $f$ denote the function representing a regression model over a number of $n$ given instances, then the specific fitness function $\Phi$ used in this paper is the Root Mean Square Error (RMSE), i.e. in formulae:

$$\Phi = \frac{\sum_{i=1}^{n}(S(i) - f(i))^2}{n}$$

where $S(i)$ is the value forecasted by the model on the $i$–th problem instance. With this choice the regression problem becomes a minimization problem.

## 5   Experiments

For the experiments described in this paper GPTIPS [22], a GP tool working under MATLAB, has been used, and the GP parameters have been set at the following values: $Pop\_size = 500$, $Max\_gen = 200$, $tourn\_size = 7$, $p\_mutate = 0.10$, $p\_cross = 0.85$, and $p\_copy = 0.05$.

The database used in this paper, consisting of 50 instances for each of the four patients, has been divided into train, test, and validation sets as follows: for each patient, each item has been exclusively assigned in a random way to one of the three sets according to the following proportions: 44 % for the train set, 32 % for the test set, and 24 % for the validation set.

An important issue when using a GP is that this is a nondeterministic algorithm: it needs a random integer value to be assigned as seed to a random number generator. This value influences GP execution and the obtained results: different seeds can yield different results. To get rid of this, the GP algorithm has been run 25 times over the database. We consider as the best among those 25 runs the one which has obtained the lowest RMSE value over the validation. Actually, the model obtained in that execution shows the highest ability to correctly predict unknown data, hence it has the highest generalization capability.

The best run for the systolic blood pressure has achieved this formula:

$$SYS\_BP = 0.5064 \cdot cos(\frac{0.6095 MeanHR}{plog(MeanHR)})$$

$$- 0.5947 \cdot tanh(cos(\frac{Pleth\_max}{MeanHR}) - RRtriindex)$$

$$- 0.7316 \cdot e^{Pleth\_min} \cdot 0.7316 sin(Pleth\_min)$$

$$+ \frac{0.5857 \cdot sin(sin(sin(EDR)))}{e^{Pleth\_min} - sinPleth\_min)} + 1.112$$

Figure 3 shows how the values predicted by this formula fit the actual systolic BP values over the three sets: the top pane shows the behaviour over the train set, the middle pane that over the test set, and the bottom pane that over the validation set.

Over the validation set, i.e. over data never seen before by the GP algorithm, the results are very good, and the RMSE value obtained is 3.3679. This value means that over previously unseen data any actual systolic BP value and the

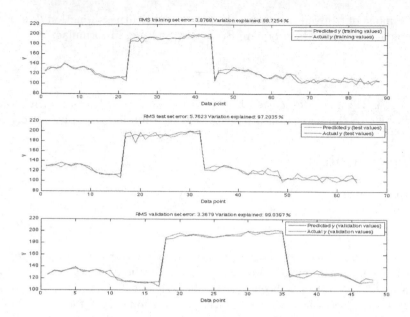

**Fig. 3.** Results for the systolic blood pressure.

corresponding computed one differ on average by $\sqrt{3.3679} = \pm 1.8352$ mmHg, which is a very good approximation.

In each set the data are loaded in the following sequential order: the data from patient #1, those from patient #2, those from patient #3 and finally those from patient #4. As it can be seen, the data from patient #2 are quite higher than those from patients #1, #2, and #4. Nonetheless, the formula obtained predicts very well the data from all the patients, including those related to patient #2.

Figure 4 shows useful information about the best run for the systolic BP.

Namely, the top left pane shows each item in each of the three sets as a dot with abscissa equal to the actual value, and ordinate equal to the predicted value. Moreover, a 45–degree line represents the ideal case of perfect prediction. Therefore, the closer a dot to this ideal line, the better the prediction for that item. This pane confirms that the approximation is good for all the four patients, including patient #2, over the three sets.

The top right pane, instead, shows the most frequently used variables in the best 5 % (in terms of fitness values) of the individuals in the final population of this run. Namely, the parameters $Pleth\_min$, $Pleth\_max$, $MeanRR$ and $MeanHR$ are contained in the 50 % of these individuals, whereas the parameters $RR\ tri\ index$ and $EDR$ are present in the 25 %.

The bottom left pane shows the so–called evolution for the best run, i.e. it reports at each generation the best fitness value contained in the population at that generation. As it can be seen, the evolution is very good during the first 30 generations or so, because for this minimization problem the fitness values quickly decrease during those generations. Then the evolution continues and the best fitness value continues to decrease, yet more slowly.

Finally, the bottom right pane represents each individual obtained during the best run as a dot with abscissa equal to the number of nodes composing it and with ordinate equal to its fitness value. The best individual is shown in a red circle. Most individuals have a number of nodes between 40 and 45.

The formula obtained shows that our methodology has allowed automatically selecting, out of the 37 in the database, the most important parameters that best help predict the systolic BP values.

These parameters are five, namely: *Pleth_min*, *Pleth_max*, *Mean HR*, *RR tri index*, and *EDR*. This selection task is hugely hard for a human being, however skilled he can be in the problem faced.

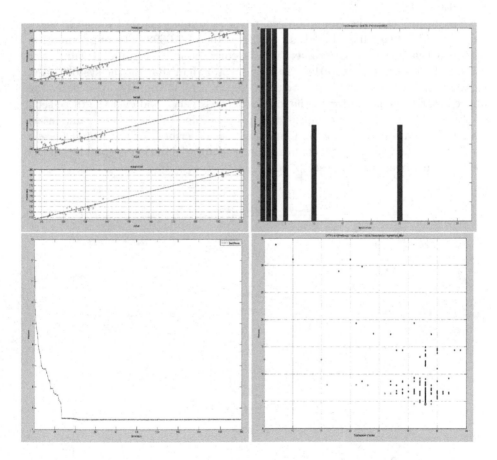

**Fig. 4.** Information about the best run for the systolic blood pressure. Top left: distance from the perfect forecasting in the three sets. Top right: the most used variables in the best 5 % of the final population. Bottom left: the evolution of the fitness. Bottom right: the distribution of the individuals in terms of number of nodes and fitness.

Similarly, the best run for the diastolic BP has achieved the following formula:

$$DIA\_BP = 0.3132 \cdot EDR - 2.142 \cdot Pleth\_min$$
$$+ 2.899 \cdot tanh(tanh(Pleth\_min + 0.6855))$$
$$+ 0.8554 \cdot plog(sin(sin(MeanHR))))$$
$$+ 1.829 \cdot tanh(e^{Pleth\_min}) - 1.386$$

Figure 5 shows how the values predicted by this formula fit the actual diastolic BP values over the three sets: the top pane shows the behaviour over the train set, the middle pane that over the test set, and the bottom pane that over the validation set.

The results over the validation set, never examined before by the GP algorithm, yield for the diastolic BP an RMSE value of 2.6692. The approximation of any actual diastolic BP value with its corresponding computed value over previously unseen data is in this case even better than that for the systolic case, as their difference is now equal to $\sqrt{2.6692} = \pm 1.6338$ mmHg. Also here, the approximation achieved for all the four patients is very good over the three sets.

Figure 6 shows in four panes for the diastolic blood pressure the same interesting information about the best run that was described for the systolic case.

Namely, the top left pane confirms that the approximation is good for all the four patients over the three sets.

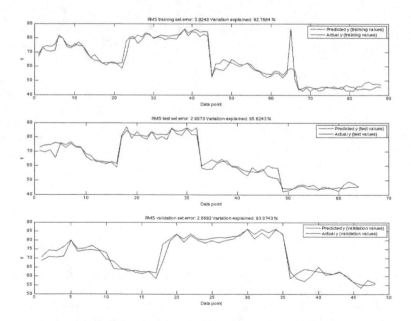

**Fig. 5.** Results for the diastolic blood pressure.

The top right pane, instead, says that the most frequently used variables in the best 5 % (in terms of fitness values) of the individuals in the final population of this run are the variables *Pleth_min*, *Pleth_max*, and *EDR*, which are contained in the 75 % of these individuals, followed by the parameters *StdRR* and *Mean HR*, which are present in the 50 % of these best individuals. Finally, the parameters *Peak Frequency VLF*, *Absolute Power VLF*, *Absolute Power LF*, and *ApEn* are contained in the 25 % of these best individuals.

The bottom left pane shows the evolution for the best run. In this case the evolution is even better than in the previous case, because fitness values decrease a lot during the first 70 generations or so. Then the best fitness value continues to decrease until the end of the run better than it does in the systolic case.

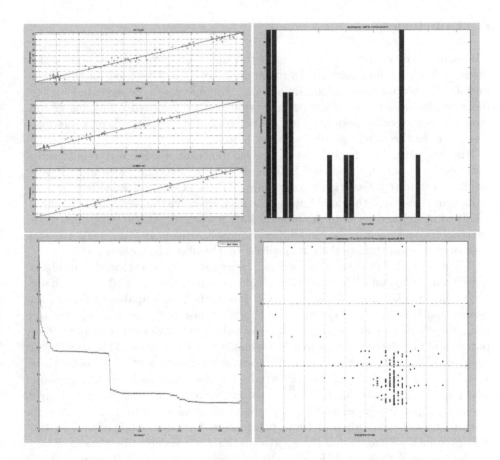

**Fig. 6.** Information about the best run for the diastolic blood pressure. Top left: distance from the perfect forecasting in the three sets. Top right: the most used variables in the best 5 % of the final population. Bottom left: the evolution of the fitness. Bottom right: the distribution of the individuals in terms of number of nodes and fitness.

Finally, the bottom right pane shows that in this case too most individuals have a number of nodes between 40 and 45. Also in this case the best individual is shown in a red circle.

Here the independent parameters automatically selected are three, namely $Pleth\_min$, $Mean\ HR$, and $EDR$.

A comparison of the two formulae shows that three parameters, i.e. $Pleth_{min}$, Mean HR, and EDR, are contained in both, therefore they strongly influence both BP values. Also two more parameters, i.e. $Pleth_{max}$ and RR tri index, are important, because they appear in one of the two rules. Consequently, a conclusion that can be drawn from these experiments is that both PPG values and ECG-related ones are of high value in the indirect estimation of BP values.

## 6    Conclusions

The issue of the continuous measurement of blood pressure is important in the medical domain. Obviously, a sphygmomanometer cannot be used for this task, and alternative ways should be found. One possible way is represented by the indirect measurement of blood pressure that can be achieved by means of the measurement of other patient's vital parameters, if the hypothesis of a relationship between the former and these latter holds true.

In this paper the hypothesis of the existence of a nonlinear relationship between heart activity, and thus ECG and HRV parameters, PPG and BP values has been tested. Genetic Programming (GP) is a Computational Intelligence paradigm that can at the same time automatically evolve the structure of a mathematical model and select from among a wide parameter set the most important parameters contained in the model. Consequently, it can carry out very well the task of regression such as the one involved by the above hypothesis.

To investigate the above hypothesis, experiments on a real-world database have been carried out. The numerical results obtained have proved that actually a non–linear relationship exists, and GP has turned out capable of finding an explicit mathematical model expressing this relationship. An interesting outcome from these experiments is that both PPG values and ECG-related ones are of paramount importance to indirectly estimate the BP values. This conclusion suggests that, instead of continuously measuring the patient's BP, both their systolic and diastolic BP values can be indirectly measured as follows: a wearable wireless PPG sensor is applied to the patient's finger, an ECG sensor to their chest, HRV parameter values are computed, and regression is performed through GP on the achieved values of these parameters. On the quantitative side, the experiments have shown that the approximation error involved by the use of this method is lower than 2 mmHg for both systolic and diastolic BP values. These results confirm also the significant association between HRV and BP that can be used in a large set of interesting health applications, for example to timely predict imminent falls [23].

In our future work we wish to carry out experiments by using 10–fold cross-validation. Moreover, we wish to investigate aiming at assessing the maximal number of items that we can profitably use in the database. In order to create this larger database, we will use cases from a higher number of patients. Finally, we wish to compare our model against other approaches from the literature.

# References

1. Maguire, S., Rinehart, J., Vakharia, S., Cannesson, M.: Technical communication: respiratory variation in pulse pressure and plethysmographic waveforms: intraoperative applicability in a North American academic center. Anesth. Analg. **112**(1), 94–96 (2011)
2. von Skerst, B.: Market survey, N=198 physicians in Germany and Austria. December 2007 - March 2008. InnoTech Consult GmbH, Germany (2008)
3. Ilies, C., Kiskalt, H., Siedenhans, D., Meybohm, P., Steinfath, M., Bein, B., Hanss, R.: Detection of hypotension during Caesarean section with continuous non-invasive arterial pressure device or intermittent oscillometric arterial pressure measurement. Br. J. Anaesth. **109**(3), 413–419 (2012)
4. Dueck, R., Jameson, L.C.: Reliability of hypotension detection with noninvasive radial artery beat-to-beat versus upper arm cuff BP monitoring. Anesth. Analg. **102**(Suppl), S10 (2006)
5. Sotera wireless. http://www.soterawireless.com
6. Koza, J.: Genetic Programming: On the Programming of Computers by Means of Natural Selection. MIT Press, Cambridge (1992)
7. Meigas, K., Lass, J., Karai, D., Kattai, R., Kaik, J.: Pulse Wave Velocity in Continuous Blood Pressure Measurements. In: IFMBE Proceedings, World Congress on Medical Physics and Biomedical Engineering 2006, Volume 14, pp 626–629. Springer (2007)
8. Najjar, S., Scuteri, A., Shetty, V., Wright, J.G., Muller, D.C., Fleg, J.L., Spurgeon, H.P., Ferrucci, L., Lakatta, E.G.: Pulse wave velocity is an independent predictor of the longitudinal increase in systolic blood pressure and of incident hypertension in the baltimore longitudinal study of aging. J. Am. Coll. Cardiol. **51**(14), 1377–1383 (2008)
9. Inajima, T., Imai, Y., Shuzo, M., Lopez, G., Yanagimoto, S., Iijima, K., Morita, H., Nagai, R., Yahagi, N., Yamada, I.: Relation between blood pressure estimated by pulse wave velocity and directly measured arterial pressure. J. Robot. Mechatron. **24**(5), 811–821 (2012)
10. Gesche, H., Grosskurth, D., Kuechler, G., Patzak, A.: Continuous blood pressure measurement by using the pulse transit time: comparison to a cuff-based method. Eur. J. Appl. Physiol. **112**, 309–315 (2012)
11. Vukovich, R., Knill, J.: Blood Pressure Homeostasis. In: Case, D., Sonnenblick, E., Laragh, J. (eds.) Captopril and Hypertension, pp. 3–13. Springer, Heidelberg (1980)
12. Berntson, G.G., Bigger Jr, J.T., Eckberg, D.L., Grossman, P., Kaufmann, P.G., Malik, M., Nagaraja, H.N., Porges, S.W., Saul, J.P., Stone, P.H., van der Molen, M.W.: Heart rate variability: origins, methods, and interpretive caveats. Psychophysiology **34**(6), 623–648 (1997)
13. Task Force of the European Society of Cardiology, the North American Society of Pacing and Electrophysiology: Heart Rate Variability: Standards of Measurement, Physiological Interpretation, and Clinical Use. Circulation **93**(5), 1043–1065 (1996)

14. Karapetian, G.K.: Heart Rate Variability as a Non-invasive Biomarker of Sympatho-vagal Interaction and Determinant of Physiologic Thresholds. Doctoral Thesis, Wayne State University (2008)
15. Golparvar, M., Naddafnia, H., Saghaei, M.: Evaluating the relationship between arterial blood pressure changes and indices of pulse oximetric plethysmography. Anesth. Analg. **95**(6), 1686–1690 (2002)
16. Allen, J.: Photoplethysmography and its application in clinical physiological measurement. Physiol. Meas. **28**(3), R1 (2007)
17. Goldberger, A.L., Amaral, L.A.N., Glass, L., Hausdorff, J.M., Ivanov, P.C., Mark, R.G., Mietus, J.E., Moody, G.B., Peng, C.K., Stanley, H.E.: PhysioBank, physiotoolkit, and physionet: components of a new research resource for complex physiologic signals. Circulation **101**(23), e215–e220 (2000)
18. Niskanen, J.P., Tarvainen, M.P., Ranta-Aho, P.O., Karjalainen, P.A.: Software for advanced HRV analysis. Comput. Methods Programs Biomed. **76**(1), 73–81 (2004)
19. Tarvainen, M.P., Ranta-Aho, P.O., Karjalainen, P.A.: An advanced detrending method with application to HRV analysis. IEEE Trans. Biomed. Eng. **49**(2), 172–175 (2002)
20. Melillo, P., Bracale, M., Pecchia, L.: Nonlinear heart rate variability features for real-life stress detection. Case study: students under stress due to university examination. Biomed. Eng. Online **10**, 96 (2011)
21. Rajendra Acharya, U., Paul Joseph, K., Kannathal, N., Lim, C.M., Suri, J.S.: Heart rate variability: a review. Med. Biol. Eng. Comput. **44**(12), 1031–1051 (2006)
22. Searson, D.: GPTIPS: Genetic Programming and Symbolic Regression for MATLAB (2009). http://gptips.sourceforge.net
23. Sannino, G., Melillo, P., De Pietro, G., Stranges, S., Pecchia, L.: To what extent it is possible to predict falls due to standing hypotension by using HRV and wearable devices?. In: Study Design and Preliminary Results from a Proof-of-Concept Study, pp. 167–170. Springer International Publishing, Ambient Assisted Living and Daily Activities (2014)

# Non-invasive Wireless Bio Sensing

Artur Arsenio[1(✉)], João Andrade[2], and Andreia Duarte[2]

[1] Universidade da Beira Interior, IST-ID and YDreams Robotics, Covilhã, Portugal
artur.arsenio@ist.utl.pt
[2] Instituto Superior Técnico, Universidade de Lisboa, Porto Salvo, Portugal
{joao.sousa.andrade,andreia.duarte}@ist.utl.pt

**Abstract.** Wireless sensing technologies are increasingly being employed on Health systems, aiming to improve the data communication flow between patients and clinical experts. This is especially important for patients located at remote locations or facing mobility constraints. In order to fully exploit the advantages of wireless communications, it is necessary biosensors that collect data about user's health, possibly integrated on a personal wireless sensor network. With this goal in mind, a wireless solution is described that presents an innovative wireless heart rate device, as well as user interface technologies for enabling real-time data visualization on mobile devices by patients and medical experts.

**Keywords:** Non-invasive · Bio-sensing · Wireless communications · Remote monitoring · Personal networks

## 1 Introduction

Health technologies are becoming increasingly less invasive. The initial trend of replacing treatments based on chirurgical procedures by others without such requirement, is now leading to other practices of collecting bio-data with a minimal impact on users' lives. This trend follows recent technology advances in sensing, computation, storage, and communications [1, 2]. Furthermore, the automatic collection of data enables to perform data analytics, both at user's level as well by processing data at community level, through the intelligent analysis of data collected by a large number of sensing devices [3]. This also enables the development of people-centric sensing applications and systems [2, 4], to facilitate both monitoring and sharing of automatically gathered health data [1, 5]. As most people possess sensing-enabled phones, the main obstacle for the widespread adoption of smart medical devices is not the lack of an infrastructure. Rather, the technical barriers are related to performing non-invasive signal acquisition (dealing eventually with a large set of heterogeneous equipment), and addressing data privacy and lack of connectivity issues, whereas supplying users and communities with useful feedback [2].

This chapter addresses these problems. Although the presented methodology and apparatus is generic to various health sensing devices (and can therefore be also applied for monitoring different user communities), the analysis will focus on a particular application, that of remotely sensing the health status of fetus during pregnancy. This work

© Springer International Publishing Switzerland 2015
A. Fred et al. (Eds): BIOSTEC 2015, CCIS 574, pp. 93–109, 2015.
DOI: 10.1007/978-3-319-27707-3_7

was developed under the scope of a HMSP collaborative project "Improving Perinatal Decision-Making: Development of Complexity-based Dynamical Measures and Novel Acquisition Systems". The work addresses the development of Remote Fetus Monitoring from biosensors, involving medical groups at Harvard and Portugal, two sensing device companies (Omniview Sisporto by Speculum in Portugal, and DynaDX in Taywan), and two groups focusing on applying technologies in biophysics and wireless communications. The further extension of the solution to other bio sensing devices, namely for elderly people monitoring, as well as sensing therapeutic patients to gather performance data, was motivated by recent requirements roused by the AHA-Augmented Human Assistance research project.

The chapter organization is as follows. Section 1 presents the motivation behind this work. Challenges faced by health sensing systems are discussed on Sect. 2. Section 3 describes the use-cases, and correspondent implementation, of the fetal heart rate wireless sensing device. The overall mobile application is described in Sect. 4, and the main conclusions discussed on Sect. 5.

## 1.1 Fetus Sensing

Throughout pregnancy, the placenta is responsible for supplying the fetus with oxygen and nutrients, as well as removing carbon dioxide and other waste gases from the fetal environment. Therefore, malfunction of the placenta may result in low oxygen delivery repercussions, in a condition known as fetal hypoxia. This condition is associated with as much as 10 % of perinatal deaths and 15 % of long-term damage cases, such as cerebral palsy [6].

Electronic fetal monitoring (EFM) may provide detection of fetal hypoxia on a early stage, thus enabling medical interventions before irreversible changes take place, which is why the relevance of continuous EFM in reducing neonatal mortality and morbidity has been acknowledged for some years [7–9]. In fact, nowadays EFM is used as standard care during pregnancy [9] and labor [10] in most developed countries. Such monitoring has special relevance in high-risk pregnancies, which include maternal hypertensive disorders and intrauterine growth restriction [11].

Current EFM methods have various contraindications, which include active genital Herpes infection, Hepatitis, HIV and lacked monitoring before 34 weeks of gestation. Cardiotocography (CTG, see Fig. 1), the most common EFM method, is associated with highly complex fetal heart rate (FHR) patterns, making standardization difficult [10]. Moreover, this method is not suitable for long term monitoring for a number of reasons, namely: it is active, restrictive and requires large power [10, 12].

On the other hand, the use of transabdominally recorded electrocardiogram (fECG) carries a number of advantages: it is passive, uses low-cost electronic components and standard ECG electrodes. fECG, is also suitable for long-term ambulatory recording, not relying on the presence of highly trained professionals [13–15]. Other benefits include the possibility of extracting beat-to-beat FHR data, along with averaged fECG waveforms, which are easier to opine on [12].

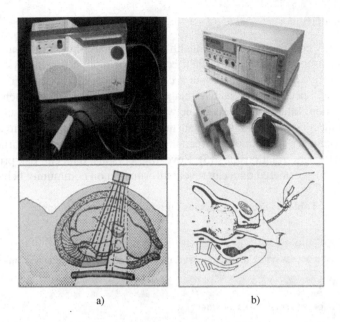

a)                                        b)

**Fig. 1.** Traditional methods for fetal heart rate measurements: auscultation, (a) doppler ultrasound (external), and (b) cardiotocography (internal).

Making fECG a reality will enrich the knowledge on FHR tracings and improve current protocols and signal interpretation [7, 16, 17]. Furthermore, computer analysis of fECG is 100 % reproducible and can include parameters difficult to evaluate visually, something especially important if we consider reports associating over 50 % of *intra-partum* deaths with CTG use and interpretation [18].

## 1.2 Devices' Heterogeneity

The aforementioned heterogeneity found on remote fetus monitoring poses several challenges (see Fig. 2). Besides diversity on sensing devices and technology, there is also significant heterogeneity on wireless technologies for sensors, from cellular communications on mobile phones, to WiFi or Bluetooth on mobile devices, or WiMax.

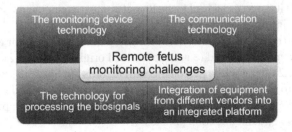

**Fig. 2.** Remote monitoring challenges.

## 1.3   Data Security

Relevant security concerns for remote health monitoring are shown in Table 1.

Respecting users' privacy is a critical concern for mobile sensing system [2, 3]. People are sensitive about how their data is captured and used, especially if it contains their location [2], speech [19], sensitive images [2], or personal records such as private health information. Interestingly, social network application's users may take privacy as a less relevant concern [4]. Collected data may inadvertently reveal information about people. For instance, a connection between mobile sensors and observed parties may be implicit in their users' relationships [5]. Revealing personal data can risk privacy, and sharing community gathered data can reveal information on community behaviors [3].

**Table 1.**   Security concerns for remote health monitoring.

| Privacy: Protection involves |
| --- |
| • identity (who access data), granularity (level of data revealed), time (retention time of data) |

| Authentication |
| --- |
| • Large number of users makes it harder |

| User control: Control over data sharing |
| --- |
| • keeping sensitive relations from being exposed, either by local filtering<br>• or by providing users with an interface to review data before it is released |

| Anonymization: not revealing the user identity |
| --- |
| • Personal sensing: processing data locally<br>• Community sensing: using the group membership approach, sensitive information is only shared within the groups in which users have existing trust relationships |

| Trust: Ensuring data sources (and info) are valid |
| --- |
| • In opportunistic sensing schemes user trust may become a barrier to wide-scale adoption<br>• k-anonymous tasking provides sensing device users with a notion of anonymity |

Countermeasures such as pausing the collection of sensor data, are not suitable as they may cause a noticeable gap in the sensing data stream [2]. Revealing too much context can potentially compromise anonymity and location privacy. The inability to associate data with its source can lead to the loss of context, reducing the system's ability to generate useful information [5].

Mobile health (mHealth) security has recently gathered significant attention. New attack and defense models surged (e.g. unauthorized origin crossing). Zapata et al. [20] analyzed a total of 24 free mobile personal health records applications, for Android and iOS, with respect to security issues. MedApp [21] explores privacy and security options for the accessibility of the medical data records in mHealth, enhancing privacy through the implementation of authentication policies.

## 2  Health Sensing

A low-cost, ambulatory device allows monitoring to be performed at home or in outpatient clinics [13–15]. In order to provide this kind of solution, efficient wireless transmission techniques were investigated to enhance the clinical utility of the signal processing technology, addressing several challenges (see Table 2). The portable monitoring device brings the additional requirement of Internet connectivity, i.e., transmitting bio signal information, and warnings in case of pathological dynamics detection, from the sensor device to a healthcare expert at a clinic.

**Table 2.**  Health Sensing Challenges.

**Safety due to proximity to the recording device**
- communications power should be kept to a minimum
- Information from the recording device to the mobile device employing low-power wireless technologies (Bluetooth) and USB cable.

**Mobile device acts as an edge gateway**
- Relays information to the healthcare provider
- through any of the available wireless technologies such as mobile cellular or WiFi

**Internet connectivity to a healthcare expert at a clinic**
- transmission of heart rate information.
- warnings in case of pathological dynamics detection
- deal with temporary lack of connectivity

**Low energy consumption**
- energy is a scarce resource on portable devices
- it can affect user's mobility.
- communications can consume a significant % of a portable device power.

Due to the patient proximity to the monitoring device (for instance, a fetus inside a pregnant woman), communications power should be kept to a minimum for safety reasons [22]. Information from the recording device will thus be first relayed through the mobile device employing low-power wireless technologies for body area networks like Bluetooth, or through USB cable. The mobile device therefore acts as an edge gateway, connecting to the healthcare provider through any of the available wireless technologies at a specific location, such as mobile cellular technologies (supporting various generation mobile technologies), as well as through WiFi (IEEE 802.11 technology).

In cases where the patient has internet connectivity available, our system also supports direct connection to the internet without bridging through the mobile device (these capabilities, although already existent in other applications and/or devices, were not available to the best of our knowledge at any other fECG sensing device).

Another challenge to be tackled is energy consumption, since communications can consume a significant percentage of a portable device power. Energy is a scarce resource on these portable devices and this constraint can directly affect a pregnant woman's mobility. Applications should offer good reliability and user experience without significantly altering the lifetime of the sensing devices [4]. Some sensors use a varying amount of power depending on external factors. Lack of sensor control limits the management of energy consumption [2]. A real time sensing system should usually supply sensor data at high rates. However, such an approach yields high-energy costs. Mobile data upload can consume a large amount of energy, especially when the sensing device is far from base stations [4].

## 3  Wireless Biosensors

The DynaDx device was used to acquire the fetus ECG. Initially the data was stored in the device's internal memory. After the acquisition period ended, it was then possible to extract this information at a medical provider. This approach is common in other marketplace solutions besides FHR.

To overcome this limitation, the device was integrated in a first version with a wireless communication (Bluetooth) *Arduino* module, allowing the wireless transmission of stored signals to a mobile phone or a server on a delivery room. A software program running on Arduino was responsible for reading data from the sensor and sending it, as well as for receiving reading commands and trigger the start of sensor readings. A second version of the device was created afterwards enabling the real-time communication of these signals (see Fig. 3) to a mobile device, which forwards such data to a backend infrastructure.

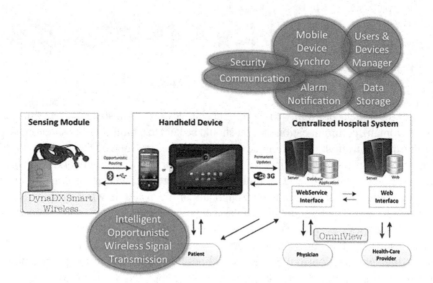

**Fig. 3.**  Solution architecture, consisting of a wireless biosensing device, mobile communication devices, user interfaces, and a backend infrastructure running support services for storing and processing data (as well as "intelligent" services based on machine learning techniques).

## 3.1  Wireless Communication Use-Cases

Current EFM methods still constrain the patient and thus are not suitable for long-term monitoring. Our solution allows continuous monitoring of the fetus throughout pregnancy and *antepartum* (the integration of other bio devices would allow as well monitoring other bio data). In fact, using a low-cost fECG sensing device (appropriately connected to a mobile phone or a tablet, or directly to the internet), all the data registered will be reliably transferred to a remote server.

We consider several usage scenarios. Figure 4 presents three of them:

(1) USB tablet/mobile phone: the sensing device is connected to a tablet or mobile phone using a USB cable;
(2) Bluetooth mobile phone: the sensing device transmits the data to a mobile phone or a tablet using Bluetooth technology. The later will then delivery the data to a backend using opportunistic routing through a wireless technology (WiFi or cellular such as GPRS);
(3) Direct Bluetooth: the sensing device transmits directly the data to a backend infrastructure for further processing, using Bluetooth interface.

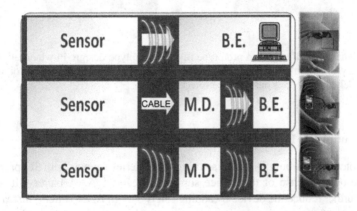

**Fig. 4.** Communication scenarios' use-cases with the sensing and mobile device components.

The major advantage of this solution is mobility. Featuring an always-on connection, the continuous monitored data will be available to the health care provider in quasi-real time, since we have to consider the raw data must be processed using CPU expensive data processing algorithms. Furthermore, the mobile device is provided with an interface that monitors GPS location of the patient and allows exchange of alerts between the patient and health care provider. Both can visualize the data collected by the sensing device using other software at a backend infrastructure (using for instance cloud resources).

## 3.2  Computer Assisted Design

With the goal of extending an existing device (by DynaDX), to a wireless communication device, it was designed (using Solidworks, a computer assisted design - CAD - tool), an additional case for supporting the overall electronics necessary to introduce the additional functionality. Figure 5a shows an exploded view of the first prototype design, in which both the electronics and batteries were placed outside DynaDX protective case. The system was later further compacted so that the electronics fits inside the DynaDX original device. Solely the batteries stayed outside, as shown in Fig. 5b. This final solution solely required an additional component part to secure the batteries to the original device.

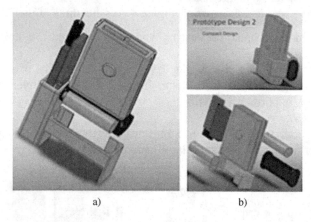

a)                                b)

**Fig. 5.** CAD design of the system prototype (a) first prototype design, in which all additional electronics stay outside the original protective case of the device; (b) final prototype design containing solely the batteries outside the protective case.

### 3.3  3D-Printing for Fast Prototyping

The final prototype was built by manufacturing the batteries support with 3D-printing (see Fig. 6). The original case of the device supports already 1 AAA battery. It would be possible to further reduce the device by adding solely 1 extra AAA battery, instead of two.

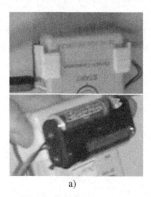

a)                                b)

**Fig. 6.** Building the final prototype: a) battery holder printed by b) an Air-Prusa2 3D-printer.

Hence, the current device's battery compartment could be extended in the future to take an additional battery, so that the overall system would fit inside the protective case.

### 3.4 Wireless Electronics Integration

The final integration of components consisted in connecting the DynaDx device to a MAX232 circuit board, which converted the RS232 signal to a Transistor-Transistor Logic (TTL) signal. This board was connected to an Arduino processing unit, which communicates through a Bluetooth shield, enabling the device to receive and transmit. The system is compact, portable, and solely requires an additional AAA battery, summing up to a total of two AAA batteries per device. All the electronics fits within the original sensor case.

The final system for wirelessly acquiring fECG signals is shown in Fig. 7. A sensing device carried by a pregnant woman acquires fetus biosignal data in real-time and transmits it to a custom hardware module. This module is capable of offloading data through a Bluetooth communication interface. Possible receivers are any such devices that can communicate in the aforementioned communication protocol. In the developed system these devices can either be a mobile device running an Android application or a machine running the Omniview-SisPorto software client. The mobile device is also capable of communicating the acquired data wirelessly in an opportunistic fashion to a back-end middleware, in order to save energy and to retransmit data after lacking connectivity.

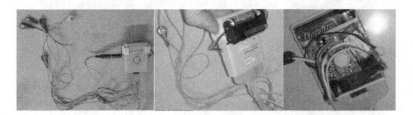

**Fig. 7.** The upgraded DynaDX ECG sensing device for continuous, wireless transmission.

## 4   Mobile Solution's Architecture

The architecture for remote fetus health monitoring comprises two main set of devices (as previously shown in Fig. 3): sensing modules and handheld devices, which can be mobile phones or tablets. The mobile device collects sensor data, being responsible for permanent updates to a centralized hospital system. The webserver receives and stores the raw data, processes it and makes it available for both the patient and health care provider sites, offering user interfaces to properly display the data. Furthermore, users (patients and health professionals) can also use the web interface at any given time to visualize the data, with no location restrictions.

This monitoring solution aims to target a large number of users (such as pregnant women), which is feasible given the conditions it offers, namely: low cost sensing device, handheld device and cellular technologies common in developed countries, low restrictions in mobility of specific users associated with the sensing device. Furthermore, a network comprising patients, physicians and other health care providers can be established on the solution social web application. All the gathered information is securely integrated and made available between strictly defined subjects.

### 4.1 Mobile Application

A mobile application (MA) was designed complying with the system requirements (see Fig. 8). The first step in using the system is registration. From the end-user point of view, only the unique pair username-password (login data) is relevant. However transparent to the user, there are two additional authentication levels, related to the handheld device in use and the communication protocol with the backend (BE) platform.

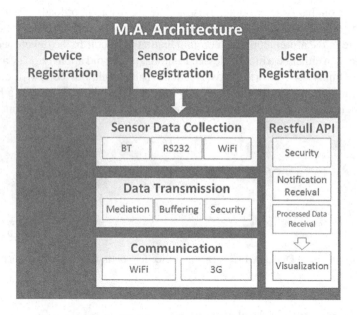

**Fig. 8.** Mobile application architecture.

Each handheld device (either a tablet or a mobile phone) has a unique identity, which is recognized by systems and allows improved security: only registered devices are able to exchange information with the BE platform. Upon enrolling in trial with this platform, the user is provided with a pair username-password and associated with the handheld device identity in the BE. The personal data included also refers name, age, gestation time and associated doctor, together with relevant characteristics to the health care provider.

The second level of authentication is transparent to the user and is renewed each time the application is started. The first interaction between the end-user and the MA is authentication using the login data. This pair is tested using a webservice and, in case of success, returns a cookie to be used throughout the session, i.e., until the program is closed and login verification is required again. This cookie is associated with each webservice based request used in the application, to ensure the identification of the request source by the BE.

The next step is the selection of the monitoring device. With this goal in mind, the mobile application was built in a modular manner, so that sensing devices can be exchanged. In fact, the temporary databases associate a parameter "type" with the data, which makes it easier to increment the available options. The possibilities are nearly endless. For instance, if the new version of the current sensing device includes uterine contractions measurement, the upgrade for this new feature will be trivial. On the other hand, the connection type between the sensing and the mobile devices can be extended as well, with limitations that can only include available hardware. For options besides RS232 and Bluetooth, exposure security cautions must be of course kept in mind [22].

Upon authentication, the device connection starts several actions besides opening the main menu. To simplify the user experience, all the collecting and uploading of data is transparent to the user, which means the user only has to start the connection to the device by clicking an option and the entire data gathering and upload processes are done in background services. The GPS monitoring is independent of the sensing device, since it is only based on the handheld device, and is transparent to the user as well. All the data exchanged between the mobile application and the webserver is encrypted using a security certificate created for this system, to ensure the privacy of all the participants.

The data collected from the sensing device and internal mobile phone GPS is temporarily saved in local databases. This method ensures that no information is lost. The MA is built to choose a WiFi connection if present, but use 3G if the first is not available. If there is no internet connection in any given moment, the data is buffered on the mobile application until a connection is available and uploads are re-established.

Besides security, the Restful API has two additional modules: notification and processed data reception, which in turn allow for data visualization. Notification reception refers to the exchange of private messages between the associated members of the network, for instance patient and physician. Whenever a new message is received, the mobile application generates a user alert, which remains visible until the message is read. This feature is relevant namely for recommendations from the physician or health care provider, as well as doubts from the patient.

Also concerning the Restful API is the visualization of data plots. This is relevant for evaluating the patient's condition at any given time. If the user is a physician or health care provider, one can choose the desired patient and data. There are relevant options included before visualizing the data plot, namely type of data and time span wanted. By default, current time is selected, which results in quasi-real time monitoring. The mobile application uploads raw data, which is then processed by the webserver and sent back to the MA. Although resulting in a short delay, it is tolerable by physicians (current CTG methods carry a delay of some seconds as well).

A MA was developed for Android mobile devices, including both tablets and mobile phones, to perform the assigned tasks. Keeping in mind always on monitoring and user friendliness, some user options were agreed upfront to be present in the main menu. The end-user might be a patient or a healthcare provider, and the menu will adapt itself accordingly to the user.

The snapshots in Fig. 9 show the main menu options for patients and physicians, such as login, verifying device and internet connectivity, visualizing plotted data, using the web interface to access a restricted social network, exchanging alerts and updated user information, or configuring the visualization of the plotted data.

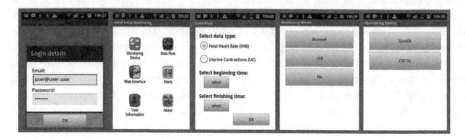

**Fig. 9.** Five snapshots of Android Mobile Application Graphical User Interface (GUI).

### 4.2   Backend Architecture

The backend architecture (as shown in Fig. 10) is constituted by the following modules: "Sensor Data Collection", "Transmission", "Display" modules, and RESTful API. The "Sensor Data Collection" module is responsible for acquiring system input. It supports different input sources, such as the DynaDx device. The communication of this input is performed with information security concerns: the confidentially of the transmitted data, supported by users authentication; the integrity of the backend data, made possible by the underlying authorization system that assigns a profile to a given user upon registration. It is also assured the mutual authentication of the backend and its users.

The "Transmission" module accounts for the communication of processed data (graphs and notifications) to the system's client (the MA). Notifications result from the complexity analysis of the pre-processed acquired signal that is performed by Omniview-SisPorto using MultiScale Entropy (MSE) technology developed by Harvard team. Graphs are built for the visualization of the time series that results from the acquisition

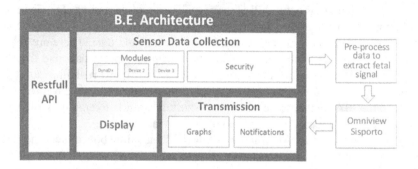

**Fig. 10.**   Backend architecture.

of the pre-processed sensed signal. Both these data types can be requested by a mobile device through the RESTful API.

The "Display" module is capable the different types of data that compose the system (namely the sensed signal graphs and associated notifications). The web application framework "Lift" is used to display this information. More specifically, Lift's Flot plugin is responsbile for displaying the graphs.

The "RESTful API" module establishes a uniform communication interface for the system. Communication is achieved through the use of a RESTful webservice-based interface. A webservice based approach tackles issues of system universality, as it allows different types of devices to access the system, providing flexibility in terms of programming languages [23]. This module defines a standard system API that is composed by a set of services that are made available.

Through this interface mobile clients can offload different kinds of data, such as historical traces of raw sensory data, the output of signal processing at the back-end, or notifications (see Fig. 11). Data is communicated asynchronously and must have an acquisition timestamp assigned by the client. In this approach, a publishing client on the sensing devices phone collects samples and uploads them using the web service interface, after applying data filters and according to network availability. A client (a patient, a doctor, a healthcare provider, a sensor, or a mobile device) must be authenticated in the system for information to be sent to the backend, so prior registration is required.

Fig. 11.   Snapshot of web interface for the backend platform and applications.

## 4.3   Integration with Other Systems

Under the scope of the HMSP project, our proposed solution was integrated with current commercial software: Omniview-SisPorto, which is employed for monitoring in-loco the patients. This tool (as shown in Fig. 12) uses a proprietary algorithm for detecting fetus abnormalities. A new methodology developed at Harvard, denoted multi-scale entropy, was also integrated into the tool.

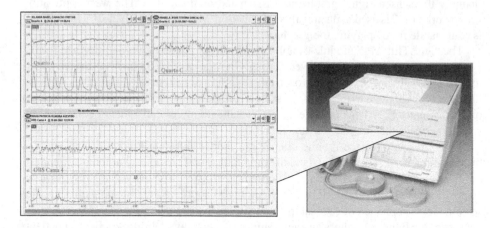

**Fig. 12.** Omniview-Sisporto® monitoring software.

Yet another goal was the upgrade of a portable ECG device, from DynaDx, to support wireless, continuous communications (previously, the device stored 24 h of data, to be posteriorly downloaded at a clinical facility). This portable device also uses its own proprietary technology for the discrimination of the fetus ECG signal from the mixed pregnant/fetus biosignal. Hence, a Javascript based software was created (see Fig. 13) to provide the user an Omniview-like User Interface on the mobile device. The main

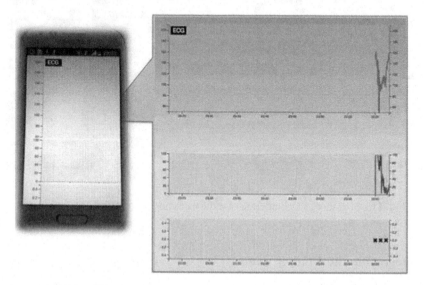

**Fig. 13.** Data Visualization on the smartphone according to current interface settings on non-mobile devices.

challenges addressed were the real-time refreshment of the graphics image given limited processing power, as well as complying with normalized display requirements. Indeed, scale units, plot size and data frequency, etc., need to comply with current non-mobile display systems.

## 5    Conclusions

An integrated sensing platform was presented. This solution aims at improving the current paradigm for remote patient monitoring. A solution was presented focused on pregnancy surveillance, which can further be applied on other scenarios such as elderly people monitoring. Mobile device and Web technologies were integrated into this paradigm, constituting a flexible modular platform that could be customized to the application domain's requirements. Opportunistic Routing on the mobile application adapts to varying wireless transmission conditions.

Multiple communication scenarios are supported for the flexible deployment of the solution according to specific needs. The final device increase of cost is not significant, and the overall system can be installed on a slightly larger protective case with an additional AAA battery.

The development of this solution had to address several challenges, since both DynaDX sensor and OmniView-SisPorto had already its own communication technologies implemented, and upgrading to a wireless protocol had an impact on these tools workflows. Additionally, fECG Data Visualization on the mobile devices had to comply with standardized rules, so it required a new approach for real-time visualization of remotely gathered data on the mobile devices.

Mobile Data security is currently an issue worth of further study. The presented architecture enables intelligent applications to run at the backend, and could be employed for processing big volume of data for a large number of biosensors, as suggested by [24] for epidemics estimation. Additionally, the current system should be applied in the future on other domains of bio sensing.

**Acknowledgements.** Part of this work was supported by Harvard Medical School Portugal Collaborative Research Award HMSP-CT/SAU-ICT/0064/2009: Improving perinatal decision-making: development of complexity-based dynamical measures and novel acquisition systems. Artur Arsenio has also been partially funded by CMU-Portuguese program through Fundação para Ciência e Tecnologia, AHA-Augmented Human Assistance project, AHA, CMUP-ERI/HCI/0046/2013.

## References

1. Campbell, A.T., Eisenman, S.B., Lane, N.D., Miluzzo, E., Peterson, R.A., Lu, H.L.H., Zheng, X.Z.X.: The rise of people-centric sensing. In: IEEE Internet Computing, pp. 12–21. IEEE Computer Society, doi:10.1109/MIC (2008)
2. Lane, N.D., Miluzzo, E., Lu, H.L.H., Peebles, D., Choudhury, T., Campbell, A.T.: A survey of mobile phone sensing. IEEE Commun. Mag. **48**, 140–150 (2010)

3. Zhang, D., Guo, B., Li, B., Yu, Z.: Extracting social and community intelligence from digital footprints: an emerging research area. In: Yu, Z., Liscano, R., Chen, G., Zhang, D., Zhou, X. (eds.) UIC 2010. LNCS, vol. 6406, pp. 4–18. Springer, Heidelberg (2010)

4. Miluzzo, E., Lane, N.D., Fodor, K., Peterson, R., Lu, H., Musolesi, M., Eisenman, S.B., Zheng, X., Campbell, A.T.: Sensing meets mobile social networks: the design, implementation and evaluation of the CenceMe application. Archit. Des. **10**, 337–350 (2008). doi:10.1145/1460412.1460445

5. Abdelzaher, T., Anokwa, Y., Boda, P., Burke, J., Estrin, D., Guibas, L., Kansal, A., Madden, S., Reich, J.: Mobiscopes for human spaces. IEEE Pervasive Comput. **6**(2), 20–29 (2007). doi:10.1109/MPRV.2007.38

6. Graham, E.M., Ruis, K., Hartman, A.L., Northington, F.J., Fox, H.E.: A systematic review of the role of intrapartum hypoxiaischemia in the causation of neonatal encephalopathy. Am. J. Obstet. Gynecol. **199**(6), 587–595 (2008)

7. Devoe, L.D.: Electronic fetal monitoring: does it really lead to better outcomes? Am. J. Obstet. Gynecol. **204**(6), 455–456 (2011)

8. Jenkins, H.: Technical progress in fetal electrocardiography - a review. J. Perinat. Med. **14**, 365–377 (1986)

9. Banta, D.H., Thacker, S.B.: Historical controversy in health technology assessment: the case of electronic fetal monitoring. Obstet. Gynecol. Surv. **56**(11), 707–719 (2001)

10. Alfirevic, Z., Devane, D., Gyte, G.M.L.: Continuous cardiotocography (CTG) as a form of electronic fetal monitoring (EFM) for fetal assessment during labour. Cochrane Database of Syst. Rev. **3**, CD006066 (2006)

11. American College of Obstetricians and Gynecologists: ACOG Practice Bulletin No. 106: Intrapartum fetal heart rate monitoring: nomenclature, interpretation, and general management principles. Obstet. Gynecol. **114**(1), 192–202 (2009)

12. Piéri, J.F., Crowe, J.A., Hayes-Gill, B.R., Spencer, C.J., Bhogal, K., James, D.K.: Compact long-term recorder for the transabdominal foetal and maternal electrocardiogram. Med. Biol. Eng. Comput. **39**(1), 118–125 (2001)

13. Crowe, J.A., Harrison, A., Hayes-Gill, B.R.: The feasibility of long-term fetal heart rate monitoring in the home environment using maternal abdominal electrodes. Physiol. Meas. **16**(3), 195–202 (1995)

14. Graatsma, E.M., Jacod, B.C., Van, Egmond L.A.J., Mulder, E.J.H., Visser, G.H.A.: Fetal electrocardiography: feasibility of long-term fetal heart rate recordings. BJOG: Int. J. Obstet. Gynaecol. **116**(2), 334–337 (2009). discussion, pp. 337–338

15. Karvounis, E.C., Tsipouras, M.G., Papaloukas, C., Tsalikakis, D.G., Naka, K.K., Fotiadis, D.I.: A non-invasive methodology for fetal monitoring during pregnancy. Methods Inf. Med. **49**(3), 238–253 (2010)

16. Taylor, M.J.O., Smith, M.J., Thomas, M., Green, A.R., Cheng, F., Oseku-Afful, S., Wee, L.Y., Fisk, N.M., Gardiner, H.M.: Non-invasive fetal electrocardiography in singleton and multiple pregnancies. BJOG: Int. J. Obstet. Gynaecol. **110**(7), 668–678 (2003)

17. Thomas, M.J., Cleal, J.K., Hanson, M.A., Green, L.R., Gardiner, H.M.: Non-invasive fetal electrocardiography: validation and interpretation. In: 4th IET International Conference on Advances in Medical, Signal and Information Processing MEDSIP, pp. 1–4 (2008)

18. CESDI 7th Annual Report - CTG Education Survey. Maternal and Child Health Research Consortium, London, Technical report (2000)

19. Lu, H., Pan, W., Lane, N.D., Choudhury, T., Campbell, A.T.: SoundSense: scalable sound sensing for people-centric applications on mobile phones. In: Architecture, pp. 165–178 (2009)

20. Zapata, B., Hernandez Ninirola, A., Fernandez-Aleman, J., Toval, A.: Assessing the privacy policies in mobile personal health records. In: 36th Annual International Conference of the IEEE Engineering in Medicine and Biology Socicty (EMBC), pp. 4956–4959 (2014)
21. Lomotey, R., Deters, R.: Mobile-based medical data accessibility in mHealth. In: 2nd IEEE International Conference on Mobile Cloud Computing, Services, and Engineering (MobileCloud), pp. 91–100 (2014)
22. Gandhi, O.P., Morgan, L.L., De Salles, A.A., Han, Y.-Y., Herberman, R.B., Davis, D.L.: Exposure limits: the underestimation of absorbed cell phone radiation, especially in children. Electromagn. Biol. Med. 31(1), 34–51 (2012)
23. Kansal, A., Goraczko, M., Zhao, F.: Building a sensor network of mobile phones. In: Proceedings of the 6th International Conference on Information Processing in Sensor Networks, IPSN 2007 (2007)
24. Andrade, J., Arsenio, A.: Epidemic estimation over social networks using large scale biosensors. Publication at Advanced Research on Hybrid Intelligent Techniques and Applications, pp. 287–320. IGI Global, Hershey (2015)

# Bioimaging

# Crutchfield Information Metric: A Valid Tool for Quality Control of Multiparametric MRI Data?

Jens Kleesiek[1,2,3](✉), Armin Biller[1,3], Andreas J. Bartsch[1], and Kai Ueltzhöffer[1]

[1] Division of Neuroradiology, Heidelberg University Hospital, Heidelberg, Germany
[2] HCI/IWR, Heidelberg University, Heidelberg, Germany
[3] Division of Radiology, German Cancer Research Center, Heidelberg, Germany
kleesiek@uni-heidelberg.de

**Abstract.** We propose an information theoretic framework to automatically infer the physical relationship and asses the quality of multiparametric MRI sequences. The method is based on the Crutchfield information metric. This distance measure can be computed solely based on the voxel intensities. In a series of experiments we proof its usefulness. First, we show that given multiparametric MRI data sets it is possible to discover the physical relationship w.r.t. the acquisition parameters of the individual sequences. Next, we demonstrate that this relationship can be employed to perform a quality check of a large ($N = 216$) data set by identifying faulty components, e.g. due to motion artifacts. Finally, we use a multidirectional diffusion weighted data set to confirm that the approach is fine grained enough to even detect small differences of diffusion vectors as well as the direction of the phase encoding of an echo planar imaging (EPI) sequence. Future work aims at transferring the preliminary results of these promising experiments into clinical routine and at standardizing MRI protocols for large scale clinical trials.

**Keywords:** Multiparametric MRI · Crutchfield information metric · MRI quality control · Multidirectional diffusion weighted imaging

## 1 Introduction

Multiparametric magnetic resonance imaging (MRI) is required for diagnosis and therapy monitoring of diseases. However, in general there are no standardized protocols specified [2]. In the clinical routine workup it is not feasible to acquire every available sequence deposited at the scanner. This would result in an unreasonable long scanning time, very likely making the patient feel uncomfortable. Secondly, the images acquired at the end of a long scanning session tend to suffer from motion artifacts which impair their diagnostic value. Needless to say that this approach is economically unacceptable.

To shed light on the zoo of available MRI sequences we establish a framework that is rooted in information theory and allows us to capture and quantify

© Springer International Publishing Switzerland 2015
A. Fred et al. (Eds): BIOSTEC 2015, CCIS 574, pp. 113–125, 2015.
DOI: 10.1007/978-3-319-27707-3_8

the relative information content between MRI sequences. In the current work
we present a proof of concept and show that the framework can be used for
MRI quality control. Further, we evaluate a multidirectional diffusion weighted
(MDDW) data set to demonstrate the fine grained nature of this approach.
Future experiments will address how it can be an aid for standardizing MRI
protocols and possibly also be used for optimizing MRI sequence parameters.

Historically, information theory investigated the transmission between a sen-
der and a receiver [18] but it has also been extended to theoretical measures
that capture the information integration [21] and information distances of information sources [13]. A not well known yet very important information distance
was introduced by Crutchfield [3]. He showed that a proper metric space (in
a mathematical sense) of information sources can be defined. Given physically
or functionally related sources, i.e. different MRI sequences, that are activated
by an identical localized stimulus, in our case a patient that is examined, the
information distance between those sources can be determined using this metric.
In turn, due to the fact that it is a proper metric we can exploit the discovered
relationship geometrically.

To the best of our knowledge no comparable approach has yet been explored
for multiparametric MRI data. We were inspired by robotic experiments where
the Crutchfield information metric has been used to determine the informational
topology of a set of robot sensors, that consecutively was exploited for simple
visually guided movements [14] or unsupervised activity classification [10].

## 2    Materials and Methods

### 2.1    Theory

Each MRI sequence can be interpreted as an information source with the voxel
intensities as the respective measurements. Given two different information
sources $X$ and $Y$, e.g. corresponding T1- and T2-weighted data sets, it is possible
to compute the *conditional entropy* of the two sources:

$$H(X|Y) = -\sum_x \sum_y p(x,y) \, \log_2 p(x|y) \ . \tag{1}$$

Consecutively, given $H(X|Y)$ and the entropy $H(X)$ allows to determine the
*joint entropy*:

$$H(X,Y) = H(X) + H(Y|X)$$
$$= \sum_x p(x) \, log_2 \, p(x) + H(Y|X) \ . \tag{2}$$

After calculating these quantities the distance between the two information
sources can be computed in the form of the Crutchfield information metric [3]:

$$d_C(X,Y) = \frac{H(Y|X) + H(X|Y)}{H(X,Y)} \ . \tag{3}$$

This metric is related to the mutual information (MI). However, MI measures what two random variables have in common, whereas the Crutchfield information metric quantifies what they do not have in common [14]. In addition, in contrast to the MI the Crutchfield distance is a proper metric fullfilling the properties of:

1. *symmetry*: $d_C(X, Y) = d_C(Y, X)$,
2. *equivalence*: $d_C(X, Y) = 0$ iff $X$ and $Y$ are recoding-equivalent (as defined by Crutchfield [3]), $d_C(X, Y) = 1$ states that the two sources are independent,
3. *triangle inequality*: $d_C(X, Z) \leq d_C(X, Y) + d_C(Y, Z)$.

Being a metric implies that the information space has a structure that can be exploited geometrically. The first experiment is a proof of concept. In the second experiment we computed the metric for all combinations ($D = 171$) of sequences within a private multiparametric MRI data set (see Sect. 2.2) and interpreted the resulting matrix as a distance (dis-similarity) matrix and as a graph adjacency matrix. In the former case we used non-linear dimensionality reduction methods like *Isomap* [20] and *local linear embedding* [17] to embed the data in a 2D geometric space, in the latter case we used Kruskal's algorithm [12] to obtain the minimum spanning tree (MST). In a third experiment we computed the distance matrices ($N = 216$, $D = 6$) for a publicly available multiparametric MRI data set (see Sect. 2.2) and used the low dimensional embedding to identify impaired images. The fourth experiment emphasizes the fine grained nature of the approach. It displays the Crutchfield distance as a function of the angle differences of the diffusion vectors of a multidirectional diffusion weighted volume with 160 directions (see Sect. 2.2).

## 2.2 Data

**Data Set 1.** In total $N = 17$ multiparametric MRI data sets with $C = 19$ channels each were acquired in a clinical routine workup of patients using two different 3 Tesla MR system from the same manufacturer (Magnetom Tim Trio ($N = 9$) and Magnetom Verio ($N = 8$), Siemens Healthcare, Erlangen, Germany).

All patients were suffering from a *glioblastoma multiforme* (WHO grade IV). The data was anonymized. The reasoning for taken data from diseased subjects is motivated by the fact that patients suffering from this disease are scanned with a more detailed protocol that comprises more MRI sequences.

The following MRI images were acquired: native (T1) and contrast enhanced (T1CE) T1-weighted images with $TE = 4.04$ ms and $TR = 1710$ ms; T2-weighted TSE imaging (T2) with $TE = 85$ ms and $TR = 5500$ ms; T2-weighted fluid attenuated inversion recovery images (FLAIR) with $TE = 135$ ms and $TR = 8500$ ms; diffusion-weighted images with $TE = 90$ ms and $TR = 5300$ ms comprising a $b = 0$ (DWI_b0), a $b = 1200$ (DWI_b1200t) as well as an apparent diffusion coefficient (ADC) map; native (SWI) and contrast enhanced (SWICE) susceptibility weighted images with $TE = 19.7$ ms and $TR = 27$ ms, this set of sequences also includes a magnitude (SWI[CE]_MAG) and phase (SWI[CE]_PHA) image as well as a minimum intensity projection

(SWI[CE]_MIP); dynamic susceptbility contrast perfusion images with $TE = 37$ ms and $TR = 2220$ ms yielding the relative cerebral blood flow (PWI_CBF) and volume (PWI_CBV), the mean transit time (PWI_MTT) and the time to peak (PWI_TTP).

**Data Set 2.** For the third experiment we use the publicly available BraTS 2014 training data set provided via the Virtual Skeleton Database (VSD) [11]. It comprises $N = 216$ co-registered native and contrast enhanced T1-weighted images, as well as T2-weighted and T2-FLAIR images ($C = 4$). The data was acquired with MR scanners of different vendors, at different field strengths and using non-uniform protocols (i.e. physical parameters). The images contain low grade as well as high grade tumors.

**Data Set 3.** For the multidirectional diffusion weighted imaging experiment we used a data set acquired with a multiplexed EPI sequence [4]. The 160 diffusion directions were selected by an electrostatic repulsion model akin to [9] and recorded twice, each time with an anterior-posterior (AP) as well as an posterior-anterior (PA) phase encoding direction.

### 2.3 Preprocessing

**Experiments 1 and 2.** All sequences of the multiparametric data set 1 were co-registered intra-individually to the respective native T1-weighted images. A rigid 6-DOF registration was preformed using the BRAINSFit [8] command line interface of 3D-Slicer (3D Slicer v4.3 [1]). The registration accuracy was confirmed by a board-certified neuroradiologist. In the next step we used FMRIB's brain extraction tool (BET) [19], which is part of FSL (FMRIB s Software Library FSL v5.0 [5]), for deskulling of the T1-weighted images. The obtained mask was applied to all channels. Finally, all data was rescaled to be in the range [0, 1024]. To estimate the probabilities we used a standard frequency count method, after we confirmed that histogram equalization methods do not alter the results.

**Experiment 3.** Data set 2 was rescaled to be in the range [0, 1024] and the same standard frequency count method as in the other experiments was applied.

**Experiment 4.** The MDDW data was preprocessed using FSL scripts (topup, eddy) [5], comprising least-squares restoration, distortion correction using the reversed phase encoding (AP, PA) acquisition and a correction for eddy currents. Again, the data was rescaled ([0, 1024]) and the standard frequency count method was applied.

### 2.4 Data Manipulation

For the first experiment we manipulated the data. We added increasing levels of noise to the images using a normal distribution $\mathcal{N}(\mu = 0, \sigma^2)$ centered at zero but with varying values of sigma. Further, we generated an artificial sequence $Z$ by combining two recorded sequences $X$ and $Y$ with a weighted sum:

$$Z = (1 - \alpha)X + \alpha Y. \tag{4}$$

Finally, we applied a 6-DOF rigid transformation $T$ that allows for separately rotating around an axis or translating along an axis.

$$T(x, y, z, \alpha, \beta, \gamma) =$$
$$\begin{pmatrix} \cos\alpha\cos\beta & \cos\alpha\sin\beta\sin\gamma - \sin\alpha\cos\gamma & \cos\alpha\sin\beta\cos\gamma + \sin\alpha\sin\gamma & x \\ \sin\alpha\cos\beta & \sin\alpha\sin\beta\sin\gamma + \cos\alpha\cos\gamma & \sin\alpha\sin\beta\cos\gamma - \cos\alpha\sin\gamma & y \\ -\sin\beta & \cos\beta\sin\gamma & \cos\beta\cos\gamma & z \\ 0 & 0 & 0 & 1 \end{pmatrix}.$$
$$\tag{5}$$

Manipulating $\alpha$, $\beta$ or $\gamma$ allows to yaw, pitch and roll respectively. Manipulating $x$, $y$ or $z$ results in a translation along the chosen axis.

## 2.5   Implementation

Except the tools (BRAINSFit [8] and FSL [5]) noted above, all algorithms were implemented in custom python scripts (Python v2.7.6 [16]). For analyzing the distance matrices we adapted functions implemented in scikit-learn [15] and NetworkX [7].

# 3   Results

## 3.1   Experiment 1 – Proof of Concept

In the first experiment (Fig. 1) we confirmed that the Crutchfield information distance is a valid metric for our purposes. On this account we manipulated the MRI image data (data set 1) in several ways. The average course as well as the standard deviation (gray shaded area) are plotted for all experiments.

Initially we added an increasing level of Gaussian noise to the data. This manipulation was repeated for each T1 sequence of MRI data set 1 ($N = 17$). In Fig. 1A left it can be seen that the Crutchfield distance increases monotonically with the amount of noise added. In Fig. 1A right an exemplary axial T1-weighted image is shown without noise and with a noise level corresponding to $\sigma = 30$.

Secondly, we generated artificial MRI data by blending the T1-weighted and T2-weighted images of the same, co-registered data set using varying weighting factors (Eq. 4). We then computed the Crutchfield distance between the T1 sequence and the artificial images. This was repeated for all images of MRI data set 1. In Fig. 1B left it can be seen that, as expected, with an increasing T2-fraction of the artificial image also the Crutchfield distance to the T1 image increases. In Fig. 1B right an exemplary axial T1 slice as well as a mixed T1- and T2-weighted image using a factor of $\alpha = 0.3$ are shown.

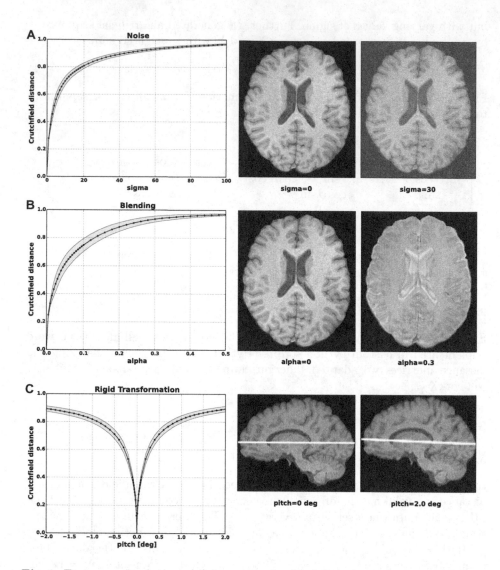

**Fig. 1. Data manipulation.** (*A*) Increasing levels of Gaussian noise were added to a T1-weighted image and the Crutchfield distance to the original sequence was determined (left). Exemplary axial T1 image with ($\sigma = 30$) and without noise (right). (*B*) Crutchfield distance between a T1-weighted image and artificially generated images (left). The artificial images were obtained by blending co-registered T1- and T2-weighted images. Exemplary axial T1 image as well as a mixed T1- and T2-weighted image using a factor of $\alpha = 0.3$ (right). (emphC) Crutchfield distance of an unmodified reference sequence to the identical, but in the interval of $[-2, 2]$ degrees rotated data set (left). Unrotated sagittal T1 image as well as an image pitched by 2 degrees (right). The average course for all data sets $N = 17$ as well as the standard deviation (gray shaded area) are plotted.

Next, we used a rigid transformation (Eq. 5) to selectively rotate a data set around an axis. We measured the Crutchfield distance of an unmodified reference sequence to the identical, but rotated data set in the interval of $[-2, 2]$ degrees. This is shown for the pitch movement in Fig. 1C left. It clearly can be seen that there is a well defined minimum at 0 degrees with an almost symmetrically increasing information distance in both rotation directions. As an example, an unrotated sagittal T1-weighted image as well as an image pitched by 2 degrees is presented in Fig. 1C right.

## 3.2    Experiment 2 – Application to Multiparametric MRI Data

Figure 2A shows the Crutchfield information distance for all combinations ($D = 171$) of sequences from data set 1. Note the symmetry of the matrix. We depicted the average of all $N = 17$ multiparametric MRI data sets. However, the structure that can be seen is also present at the individual level. This is also supported by the small standard deviation of the distances, which is on average 0.0035.

First, we interpreted the distance matrix as a dis-similarity matrix. We used the *Isomap* algorithm [20] to perform a 2D embedding of the data (Fig. 2B). Comparable results were obtained when we employed *local linear embedding* [17] to reduce the dimensionality of the data (not shown). It clearly can be seen that related sequences cluster in close proximity. In a second approach we used the distance matrix as an adjacency matrix of a fully connected graph and applied Kruskal's algorithm [12] to obtain the MST (Fig. 2C). Also this method allows to discover physically related sequences by grouping them at neighboring leaves in the tree.

## 3.3    Experiment 3 – Automatic Quality Control

To demonstrate a potential application of the proposed framework we compute the distance matrices for $N = 216$ data sets from the BraTS 2014 training data (data set 2) and use the *Isomap* algorithm [20] to perform a 2D embedding (Fig. 3A). Based on the distance to the cluster centers we were able to identify outliers. This is shown for four cases (numbered 1 to 4 in Fig. 3A). To confirm our hypothesis that the outliers correspond to impaired data sets that do not meet quality standards, we manually inspected them. Case 1 corresponds to data set *brats_tcia_pat313_1*. This data set contains no native T1 - instead the T2-FLAIR image was enclosed twice. Case 2 (*brats_tcia_pat216_1*) misses again a native T1 weighted image. Instead a contrast enhanced image with spherical artifacts was included (Fig. 3B left). Case 3 (*brats_tcia_pat230_2*) displays severe motion artifacts in T1 (Fig. 3B right). Case 4 corresponds to *brats_tcia_pat250_1* and does not contain a native T1, instead a T1CE was included twice.

Note, even if only one channel is corrupted this leads to changes of multiple entries in the distance matrix and thus can affect the position of (all) other channels in the low dimensional embedding.

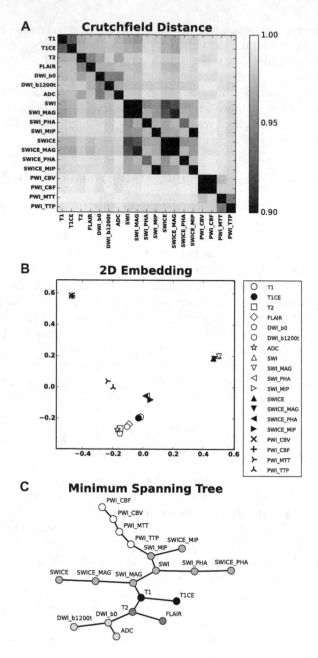

**Fig. 2. Crutchfield Matrix.** (A) Average Crutchfield information distance between all combinations of the multiparametric MRI data of data set 1. The scaling ([0.9, 1.0]) was chosen to better emphasize the structure. (B) 2D geometric embedding of the distance matrix depicted in A using the *Isomap* algorithm. (C) MST computed from the distance matrix depicted in A using Kruskal's algorithm. For the embedding we used the graphviz "spring model" layout (Graphviz [6]). For the abbreviations of the MRI sequences please refer to Sect. 2.2.

### 3.4   Experiment 4 – Multidirectional Diffusion Weighted Imaging

Yet another potential application is the evaluation of DTI data (Sect. 2.2). We compute the Crutchfield distances between an arbitrary reference diffusion direction (PA phase encoded) and all remaining diffusion directions of the data set. This distance is plotted as a function of the angle differences of the diffusion vectors (Fig. 4). As expected, the distance is largest when the two direction vectors are orthogonal to each other, i.e. the two volumes are most dissimilar. It

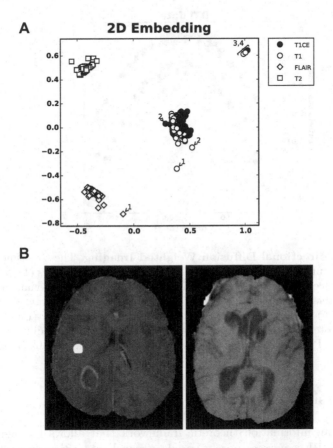

**Fig. 3. 2D Embedding of BraTS 2014 training data.** (*A*) Using the *Isomap* algorithm we embedded all $N = 216$ data sets in 2D. This allowed us to identify outliers (e.g. numbered 1 to 4) which indeed corresponded to impaired data sets. The scatter of the embedded points can be explained by the fact that the data was acquired with MRI scanners of different vendors as well as with different field strengths and protocols. For the abbreviations of the MRI sequences please refer to Sect. 2.2. (*B*) The left image corresponds to number 2 above and was labeled as a native T1 weighted image. Instead it is a contrast enhanced T1 image that contains multiple spherical hyperintense artifacts (a neuroradiologist confirmed that these do not correspond to hemorrhage). The image on the right side exhibits severe motion artifacts and corresponds to number 3 above.

decreases when the two directions get more parallel to each other again, resulting in an inverted u-shaped arc. Further, the approach is fine grained enough to distinguish between the two different phase encoding directions of the data set (PA and AP), resulting in two nested and well separable arcs. The upper arc (with larger distance values) corresponds to the AP phase encoding direction. This is most likely due to the fact that the reference sequence was chosen to be PA encoded.

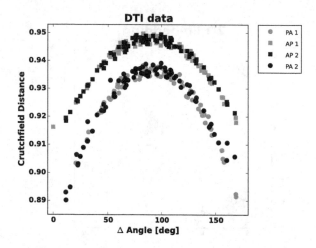

**Fig. 4. Multidirectional Diffusion Weighted Imaging.** The Crutchfield distance is plotted as a function of the angle differences of the diffusion vectors of a MDDW data set. The distance is largest when the two direction vectors are orthogonal, and decreases when the vectors get more parallel again. This highlights the fine grained resolution of the proposed framework. Circles correspond to a posterior-anterior (PA) and squares to anterior-posterior (AP) phase encoding direction. In total $4 \times 80$ diffusion directions were recorded (PA, AP, AP, PA) in two blocks, encoded in gray and black, respectively.

## 4    Discussion

We present an information theoretic framework that allows to infer the relationship of MRI sequences purely based on voxel intensities. It is shown that the Crutchfield information metric [3] is a suitable distance measure for MRI sequences and is able to capture the following relationship: the greater the (physical) distance between two MRI sequences, the less information they share. We manipulated images by adding noise (Fig. 1A), blending two MRI sequences (Fig. 1B) and purposefully applying a rigid transform to them (Fig. 1C). In all cases the Crutchfield distance increased monotonically with the amount of manipulation and showed only a small standard deviation across data set 1 ($N = 17$). If we measure the information distance between all combinations of sequences ($D = 171$) of data set 1, we can construct a distance matrix which already shows a structure that corresponds to the intrinsic physical relationship of the sequences (Fig. 2A). This relationship becomes more explicit if we perform a low dimensional (2D) embedding (Fig. 2B) or compute the MST (Fig. 2C).

Usually the physical relationship of the MRI sequences is known or can be obtained from the DICOM header. What is the benefit of the proposed method? For instance, consider data from a multicenter study which is designated for an automatic evaluation. Even if the data sets are acquired with similar parameters (e.g. TE and TR) they still originate from different scanners and thus might not be located in the same informational space. It also is very likely, as known from clinical routine, that some of the channels are affected by motion artifacts, which would also alter the informational structure. We demonstrated for a $N = 216$ data set that the proposed framework indeed can be used as an automated screening method for impaired images (Fig. 3). Employing the Crutchfield metric for quality control allows to identify data sets which are not located in the same informational space, e.g. are affected by motion artifacts (Fig. 3B right). Admittedly, so far this is a very coarse approach and it still has to be validated on a finer level with controlled experiments that determine sensitivity and specificity of the method.

To emphasize the sensitivity of the approach we compare the Crutchfield distance of 160 diffusion directions of a multidirectional diffusion weighted data set (Sect. 2.2) to an arbitrary chosen reference volume. The distance is largest when two direction vectors are orthogonal and decreases when the direction vectors get more parallel again (Fig. 4). In addition, the method allows to distinguish between PA and AP phase encoding. We propose that the shape of the informational structure, the inverted u-shaped arc, might be used to perform a quality assessment of the MDDW data set, e.g. to automatically check for motion artifacts or corrupted volumes. Further studies will have to establish how deviations of the crutchfield distance relate to tractography results.

Another potential application is to utilize this method for the assembly of standardized multiparametric MRI sequences. The information distance can be used as guideline for radiologists to select optimal subsets of the available sequences by e.g. pruning the MST to minimize the aquisition of redundant information. Further applications include MRI sequence optimization by choosing parameters of a set of sequences to maximize the coverage in information space, i.e. reducing redundancy within the sequences. Yet, this still requires a thorough study of the dependence of the Crutchfield distance on the differences in physical parameters of MRI sequences.

## 5    Conclusions

We demonstrated that the Crutchfield information metric in combination with methods for dimensionality reduction or from graph theory are suitable for discovering the physical relationship of various MRI sequences solely based on their voxel intensities. Initial experiments confirm that the proposed framework can be used for automatic MRI sequence quality control and is also fine grained enough to detect small changes in the input data. This has to be validated in future work.

**Acknowledgements.** Thanks to the anonymous reviewer who suggested the experiment with multiple diffusion directions. This work was supported by a postdoctoral fellowship from the Medical Faculty of the University of Heidelberg.

# References

1. 3D Slicer v4.3. http://www.slicer.org
2. Cornfeld, D., Sprenkle, P.: Multiparametric MRI: standardizations needed. Oncol. (Williston Park) **27**(4), 277–280 (2013)
3. Crutchfield, J.: Information and its metric. In: Lam, L., Morris, H.C. (eds.) Nonlinear Structures in Physical Systems. Woodward Conference. Springer, New York (1990)
4. Feinberg, D.A., Moeller, S., Smith, S.M., Auerbach, E., Ramanna, S., Gunther, M., Glasser, M.F., Miller, K.L., Ugurbil, K., Yacoub, E.: Multiplexed echo planar imaging for sub-second whole brain fmri and fast diffusion imaging. PLoS One **5**(12), e15710 (2010)
5. FMRIB's software Library FSL v5.0. http://fsl.fmrib.ox.ac.uk
6. Graphviz. http://www.graphviz.org
7. Hagberg, A.A., Schult, D.A., Swart, P.J.: Exploring network structure, dynamics, and function using NetworkX. In: Proceedings of the 7th Python in Science Conference (SciPy 2008), pp. 11–15. Pasadena, CA USA, August 2008
8. Johnson, H., Harris, G., Williams, K.: BRAINSFit: mutual information registrations of whole-brain 3D images, using the insight toolkit. The Insight J., October 2007. http://hdl.handle.net/1926/1291
9. Jones, D.K., Horsfield, M.A., Simmons, A.: Optimal strategies for measuring diffusion in anisotropic systems by magnetic resonance imaging. Magn. Reson. Med. **42**(3), 515–525 (1999)
10. Kaplan, F., Hafner, V.V.: Information-theoretic framework for unsupervised activity classification. Adv. Robot. **20**(10), 1087–1103 (2006). http://www.tandfonline.com/doi/abs/10.1163/156855306778522514
11. Kistler, M., Bonaretti, S., Pfahrer, M., Niklaus, R., Büchler, P.: The virtual skeleton database: an open access repository for biomedical research and collaboration. J Med. Internet Res. **15**(11), e245 (2013)
12. Kruskal, J.B.: On the shortest spanning subtree of a graph and the traveling salesman problem. Proc. Am. Math. Soci. **7**(1), 48–50 (1956). http://www.jstor.org/stable/2033241
13. Kullback, S.: Information Theory and Statistics. Dover, New York (1968)
14. Olsson, L.A., Nehaniv, C.L., Polani, D.: From unknown sensors and actuators to actions grounded in sensorimotor perceptions. Connect. Sci. **18**(2), 121–144 (2006). http://www.tandfonline.com/doi/abs/10.1080/09540090600768542
15. Pedregosa, F., Varoquaux, G., Gramfort, A., Michel, V., Thirion, B., Grisel, O., Blondel, M., Prettenhofer, P., Weiss, R., Dubourg, V., Vanderplas, J., Passos, A., Cournapeau, D., Brucher, M., Perrot, M., Duchesnay, E.: Scikit-learn: machine learning in python. J. Mach. Learn. Res. **12**, 2825–2830 (2011)
16. Python v2.7.6. http://www.python.org
17. Roweis, S.T., Saul, L.K.: Nonlinear dimensionality reduction by locally linear embedding. Sci. **290**(5500), 2323–2326 (2000)
18. Shannon, C., Weaver, W.: The Mathematical Theory of Communication. University of Illinois Press, Chicago (1949)

19. Smith, S.M.: Fast robust automated brain extraction. Hum. Brain Mapp. **17**(3), 143–155 (2002)
20. Tenenbaum, J.B., de Silva, V., Langford, J.C.: A global geometric framework for nonlinear dimensionality reduction. Sci. **290**(5500), 2319–2323 (2000)
21. Tononi, G., Edelman, G.M., Sporns, O.: Complexity and coherency: integrating information in the brain. Trends Cogn. Sci. **2**(12), 474–484 (1998)

# Medical Image Retrieval for Alzheimer's Disease Using Structural MRI Measures

Katarina Trojacanec[✉], Ivan Kitanovski, Ivica Dimitrovski,
and Suzana Loshkovska, for the Alzheimer's Disease Neuroimaging
Initiative*

Faculty of Computer Science and Engineering, Ss. Cyril and Methodius
University, Rugjer Boshkovik 16, PO Box 393, Skopje, Macedonia
{katarina.trojacanec,ivan.kitanovski,
ivica.dimitrovski,suzana.loshkovska}@finki.ukim.mk

**Abstract.** The aim of the paper is to study medical image retrieval for Alzheimer's Disease (AD) on the bases of structural MRI measures. The main goal of the strategy used in this paper is to improve the retrieval performance while using smaller number of features. The feature vector consists of the measurements of cortical and subcortical brain structures, including volumes of the brain structures and cortical thickness. The feature subset selection is additionally applied using the Correlation-based Feature Selection method to exclude irrelevant, redundant or possibly noisy data and to consider the most relevant and discriminative features. Six different scenarios for the image representation are studied: volumetric features, cortical thickness features, all imaging features, selected volumetric features, selected cortical thickness feature and selected imaging features. Euclidean distance is used as a similarity measurement. The dataset used for evaluation of the retrieval performance is provided by the Alzheimer's Disease Neuroimaging Initiative (ADNI). Experimental results show that the strategy used in this research outperforms the traditional one despite its simplicity and small number of features used for representation. Additionally, the performed analysis demonstrated that the selected features are highly stable through the leave-one-out strategy. Moreover, they are stressed in the literature as significant biomarkers for Alzheimer's Disease, which makes the strategy used in this research even more reasonable.

**Keywords:** CBIR · Alzheimer's Disease · VOI · Segmentation · Feature extraction · Feature selection · MRI · ADNI

*Data used in preparation of this article were obtained from the Alzheimer's Disease Neuroimaging Initiative (ADNI) database (adni.loni.usc.edu). As such, the investigators within the ADNI contributed to the design and implementation of ADNI and/or provided data but did not participate in analysis or writing of this report. A complete listing of ADNI investigators can be found at: http://adni.loni.usc.edu/wp-content/uploads/how_to_apply/ADNI_Acknowledgement_List.pdf

© Springer International Publishing Switzerland 2015
A. Fred et al. (Eds): BIOSTEC 2015, CCIS 574, pp. 126–141, 2015.
DOI: 10.1007/978-3-319-27707-3_9

# 1    Introduction

Alzheimer's Disease (AD) is one of the most common forms of neurodegenerative disorders and very progressive form of dementia for older adults. Finding relevant biomarkers, early diagnoses of the disease, monitoring the patient's condition changes, progression of the disease or reaction to the therapy, and identifying the patients who are most probable to ultimately develop AD, are considered as very important challenges for researchers in this domain [1].

Magnetic Resonance Imaging (MRI) as a powerful technique for diagnosis of AD and its prodromal stage, Mild Cognitive Impairment (MCI) provides rich information needed for detecting and understanding the disease pathology. This leads to enormously increased number of images stored in the medical databases. To improve and make easier the medical image analysis process, efficient organization and representation is needed in a way that will provide precise and clinically relevant retrieval and knowledge discovery. Being able to retrieve images from the database with similar Volume of Interest (VOI)/pathology/disease might be very useful in the clinical and research centres in two directions: (1) providing clinically relevant information to the physicians at right moment, thus supporting the diagnosis process and improving its quality and efficiency [2], and (2) for educational purposes [3].

The retrieval process involves generating descriptor that represents the image given as a query and then, comparing such descriptor to those already organized in the system [4]. The main concern is to find a good representation of the image content by using techniques for feature extraction that will properly represent relevant information. Taking into consideration the specific nature of the medical images and subtle changes that need to be detected and taken into consideration, the research is going towards its specialization in direction of particular diseases, such as interstitial lung diseases [5], AD [6–9] etc., body part such as brain [6–10], lung [5], or the medical imaging techniques used for acquisition (e.g. Magnetic Resonance Imaging (MRI) [10].

The focus of the paper is on medical image retrieval applied to AD. Most of the current techniques in this domain focus on visual information extraction following the standard procedures for feature extraction [6–9]. However, considering this approach in medical volumetric data context, the dimensionality and complexity become crucial problems. To overcome this and with the aim to provide and use relevant medical image representation, we base the retrieval process on the structural MRI measures closely related to AD [8]. Wide range of research is performed on their statistical dependence with respect to the disease [1, 11–13]. Some of them, including volume of the ventricular structures, hippocampal volume, amygdala volume or cortical thickness are used by the researchers to distinguish or automatically label patients as AD, MCI, or healthy controls [14, 15], or to generate high-level semantic words used subsequently for retrieval purposes [16].

In this paper, volumes of the brain structures and cortical thickness are used to generate the feature vector. Additionally, the advantage of using the feature subset selection to extract the most relevant information is also investigated. Six scenarios regarding the feature set used for image representation are tested: (1) all volumetric features, (2) all cortical thickness features, (3) all imaging features (volumes and cortical thickness), (4) selected volumetric features, (5) selected cortical thickness feature and (6) selected imaging features. All scenarios were applied on two subsets from ADNI database, baseline images obtained using 3T scanners, and screening images obtained using 1.5T scanners Performing the analysis based on baseline (screening) images only for evaluation of the retrieval capabilities using cross-sectional data is meaningful and significant in the case of the first visit for a patient in the clinical practice [17]. Experimental results show that the approach used in this study outperforms the traditional one despite its simplicity and small number of features needed for representation.

The paper is organized as follows. Section 2 provides the state of the art. The experimental setup is explained in Sect. 3, while the experimental results are given in Sect. 4. Section 5 provides concluding remarks and future directions.

## 2  State of the Art and Related Work

Medical image retrieval in the context of AD is subject of research in several studies [6–9]. Regarding the feature extraction in this context, one usual direction is following the procedure used in the traditional image retrieval taking into consideration features derived from the visual cues contained in the image. Intensity histograms, Local Binary Pattern (LBP) and Gradient Magnitude Histograms are applied to the middle slice for subsequent usage in automated diagnosis of AD [6]. Discrete Cosine Transform, Daubechie's Wavelet Transform and again LBP are used as descriptors in [8]. The feature extraction method in this case are also applied on 2D bases on a selected subset of slices by radiologists. Laguerre Circular Harmonic Functions expansions enabling capturing the local image patch structure directly are also used for feature extraction [9]. The Bag-of-Visual-Words approach is then applied on a specific region (hippocampus). Slice by slice analysis is performed in this research.

Two main critical aspects should be noticed in the performed research in these studies. Firstly, the feature extraction is performed only on one/several slice/s or slice by slice manner, meaning exclusion of possibly significant information that might be extracted from the volumetric data. The other critical point is the dimensionality of the feature vector which can lead to a high computational complexity. Examples include 256 features in each descriptor in [6], 13312 features for 3D Grey Level Co-occurrence Matrices, 1920 for 3D Wavelet Transforms, 9216 for Gabor Transforms and 11328 for 3D LBP per volume [18], all applied to brain MRI.

To overcome these disadvantages, this paper focuses on an alternative method for generating feature vector. Using the domain knowledge, it is based on structural changes considered as indicators for AD, including cortical thickness and volumes of the separate brain structures. While the traditional direction basically means extraction of the visual information itself, the alternative one utilizes the visual information to delineate the relevant brain structures on the bases of which the measurements are subsequently obtained. In the context of image retrieval, the first direction enables retrieval of images/Volumes of Interest (VOIs) with similar visual characteristics (not always semantically similar in the same time), while the second provides retrieval of images/VOIs with similar structural appearance with the query which in the context of AD is expected to lead to more relevant and semantically more precise results.

To enable this, the pipeline used in this study to construct the feature vector includes: (1) segmentation of the relevant brain structures/(VOIs), (2) calculating the measurements such as volume of the selected structures and cortical thickness.

There is plenty of research based on delineation and analysis of VOIs, relevant for detecting anatomical changes related to or imposed by AD, including hippocampus, amygdala, ventricular structures, and entorhinal cortex. For instance, methods for segmentation of thalamus, caudate, putamen, pallidum, hippocampus and amygdala are proposed in [12, 19–21]. Cortical segmentation and parcellation is studied in [22]. The anatomical segmentation of structural images of the human brain considering 83 regions is proposed in [23].

The segmentation and subsequent calculation of the measurements in this research are performed using the FreeSurfer software package [24], due to its powerful capabilities. Moreover, with the aim to improve the retrieval results and increase the efficiency, we apply the feature selection step using Correlation-based Feature Selection (CFS) [25]. As a result, better results are obtained in comparison to the reported results by the other authors while using very small number of features as image representation.

The contributions of our research using this approach are: (1) key information extracted from the medical volume itself, (2) possibility to be adapted to reflect the change of the patient condition/disease progress (for future use), and (3) efficiency.

## 3   Experimental Setup

### 3.1   Dataset

The images used for preparation of this paper were obtained from the Alzheimer's Disease Neuroimaging Initiative (ADNI) database (adni.loni.usc.edu). The ADNI was launched in 2003 by the National Institute on Aging (NIA), the National Institute of Biomedical Imaging and Bioengineering (NIBIB), the Food and Drug Administration (FDA), private pharmaceutical companies, and non-profit organizations as a $60 million, 5-year public–private partnership. ADNI aims to enable research on progression of mild cognitive impairment (MCI) and Alzheimer's Disease (AD) using serial magnetic resonance imaging (MRI), positron emission tomography (PET), other biological markers, such as cerebrospinal fluid (CSF) markers, APOE status and full-genome genotyping via blood sample, as well as clinical and neuropsychological assessments. Finding sensitive and specific markers of very early AD progression is aimed to support the development of new treatments, to improve the process of monitoring treatments effectiveness, and to reduce the time and cost of clinical trials.

The Principal Investigator of this initiative is Michael W. Weiner, MD, VA Medical Center and University of California – San Francisco. ADNI is a valuable product resulting from the efforts of many coinvestigators from a broad range of academic institutions and private corporations. Subjects have been recruited from over 50 sites across the U.S. and Canada. Starting with the initial goal to recruit 800 subjects, ADNI has been followed by ADNI-GO and ADNI-2, having recruited over 1500 adults, ages 55 to 90, to participate in the research. Cognitively normal individuals, adults with early or late MCI, and people with early AD can be distinguished in the dataset with different follow up duration of each group, specified in the protocols for ADNI-1, ADNI-2, and ADNI-GO. For up-to-date information, see http://www.adni-info.org.

In this study, two subsets of ADNI database: screening T1-weighted MRI obtained using 1.5T scanners, and baseline T1-weighted MRI acquired at 3T scanners. Both datasets have three categories of subjects: AD and MCI patients, and normal controls (NL). In the case of scans obtained by 1.5T scanners, six cases did not achieve successful FreeSurfer segmentation. These subjects were excluded from the retrieval process. The number of subjects included in this paper is given in Table 1. Demographic information about the subjects such as gender and age is represented on Fig. 1 (for the Screening 1.5T subset) and Fig. 2 (for the Baseline 3T subset). This information is not explicitly included in the performed research, hence it does not have direct influence on the retrieval performance.

**Table 1.** Subjects available for both data sets.

| Subset | Subjects available | | |
|---|---|---|---|
| Screening visits at 1.5T | Normal: 226 | MCI: 400 | AD: 185 |
| Baseline visits at 3T | Normal: 47 | MCI: 71 | AD: 33 |

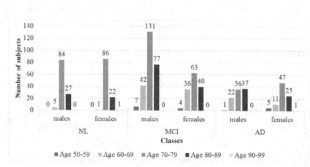

**Fig. 1.** Demographic information such as gender and age per class for the Screening 1.5T subset. The horizontal axis denotes gender per each class, while the vertical one – the number of subjects. In each class, the number of subjects grouped by age range and gender is represented.

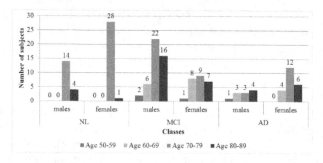

**Fig. 2.** Demographic information such as gender and age per class for the Baseline 3T subset. The horizontal axis denotes gender per each class, while the vertical one – the number of subjects. In each class, the number of subjects grouped by age range and gender is represented.

## 3.2  Segmentation and Quantitative Measurements

The images were processed using fully automated pipeline of the powerful software package FreeSurfer version 5.1.0. This way, the required measurements, volume of the brain structures and cortical thickness, were obtained. The main methods from the FreeSurfer pipeline are summarized in Table 2 [26]. Due to the lack of neuroradiology expert, the quality control on the FreeSurfer output is not addressed in this research.

**Table 2.** Methods of the FreeSurfer pipeline.

| Methods |
| --- |
| Motion correction and averaging |
| Removal of non-brain tissue using a hybrid watershed/surface deformation procedure |
| Automated Talairach transformation |
| Segmentation of the subcortical white matter and deep grey matter volumetric structures |
| Intensity normalization |
| Tessellation of the grey matter white matter boundary |
| Automated topology correction |
| Surface deformation following intensity gradients |
| Registration to a spherical atlas |
| Parcellation of the cerebral cortex into units based on gyral and sulcal structure |

## 3.3  Feature Representation

After the segmentation has been conducted and the measurements have been obtained, the feature vector is constructed using these measurements. For this purpose, 48 regional volumes of the separate brain structures are considered (the fifth ventricle, left and right white matter hypointensities, as well as left and right non-white matter hypointensities are excluded for further analysis because most subjects are characterized by zero values). Additionally, 68 cortical thickness measures (34 from each hemisphere) are used as features. Despite using the combination of all imaging features [27] resulting in 116 features in total in this research, two more scenarios are investigated, using volumetric features only, and using cortical thickness features only.

## 3.4  Feature Selection

For each feature subset, the feature subset selection method was applied. This step aims to improve the retrieval performance and further reduce the feature vector dimensionality. Correlation-based Feature Selection (CFS) method was applied. It evaluates subsets of features taking into account the usefulness of individual features for predicting the class along the degree of intercorrelation among them, meaning that good feature subsets contain features highly correlated with the class, yet uncorrelated to each other [25]. To get an unbiased results, the feature selection was performed independent of the query subject. Thus, the specific feature subset for each query subject was obtained. To be able to analyze the stability of the selected features, the inclusion rate for each feature in the experiments was calculated.

## 4   Experimental Results

Generated feature sets in all six scenarios, namely, volumetric features, cortical thickness features, all imaging features, selected volumetric features, selected cortical thickness features, and selected imaging features were evaluated. Taking into consideration the number of subjects which is not very large, leave-one-out strategy was performed. This means that each image was used as a query against all other images in the database. Euclidean distance was used as a similarity measurement. To evaluate the retrieval performance, mean average precision (MAP) was used. The retrieved image is considered as relevant if it belongs to the same class as the query (AD, NL, MCI). To be able to compare the results with those from the other researchers available for one of the subsets, the curves of average precision at the first N (up to N = 20) retrieved scans are provided. The retrieval performance was evaluated for both data sets in two cases: (1) considering only AD and NL subjects, and (2) considering all subjects (AD, NL and MCI).

### 4.1   Evaluation Performed on the 1.5T Screening MRI Dataset

The experimental results of the proposed strategy applied to the dataset containing the screening visits at 1.5T are presented in this subsection. Table 3 contains the inclusion rate for the features used in the experiments for the data set of 1.5T Screening MRIs in the case of two (NL and AD) and three classes (NL, AD and MCI). Only the features that were selected by the algorithm at least once are depicted. When using volumetric features only, left and right hippocampus and left amygdala were selected in all experiments (in the dataset with AD and NL subjects), and followed by the temporal horn of left lateral ventricle (Left-Inf-Lat-Vent) that was not selected only once. The other volumetric features were selected in dramatically less experiments (Table 3). Considering cortical thickness features, seven features were selected in all experiments and two more features were included in more than 98.5 % of the cases followed by two features involved in more than 75 %. When the feature selection is applied to the whole set of imaging features, the most selected features in the case of separated volumetric and cortical thickness features are again selected in all or most of the cases. The stability of the selected features in all three cases is very high.

In the case of all categories (AD, NL, MCI) included in the analyses, the situation with the feature selection is similar. Considering the volumetric features, the same four volumes as in the other dataset were selected in all experiments in this case. When cortical thickness features are used, seven of the features mostly selected in the other dataset (AD, NL only), are again selected in most of the cases, and the difference is in the

**Table 3.** Inclusion rate for the features used in the experiments 1.5T MRI Screening dataset.

| Features / Classes included | AD, NL | | | AD, NL, MCI | | |
|---|---|---|---|---|---|---|
| | Volumes | Cortical thickness | All imaging | Volumes | Cortical thickness | All imaging |
| Left-Inf-Lat-Vent | 99.76 | | 99.76 | 100 | | 100 |
| Left-Hippocampus | 100 | | 100 | 100 | | 100 |
| Left-Amygdala | 100 | | 100 | 100 | | 100 |
| Right-Inf-Lat-Vent | 16.06 | | 0.24 | 0.49 | | |
| Right-Putamen | 0.24 | | 0.49 | | | |
| Right-Hippocampus | 100 | | 100 | 100 | | 100 |
| Right-Amygdala | 15.82 | | 1.46 | 0.49 | | |
| Right-VentralDC | 0.49 | | | | | |
| CC_Central | 1.46 | | 2.19 | | | |
| lhCortexVol | | | | 0.49 | | |
| lh_bankssts_thickness | | 99.27 | 99.51 | | 99.63 | 97.66 |
| lh_entorhinal_thickness | | 100 | 100 | | 100 | 100 |
| lh_fusiform_thickness | | 5.35 | 0.49 | | 100 | 99.75 |
| lh_inferiorparietal_thickness | | 14.84 | 18.98 | | 99.01 | 99.88 |
| lh_inferiortemporal_thickness | | 100 | 100 | | 100 | 100 |
| lh_medialorbitofrontal_thickness | | 4.38 | 2.92 | | | 100 |
| lh_middletemporal_thickness | | 100 | 100 | | 100 | |
| lh_parahippocampal_thickness | | 100 | 100 | | 100 | 100 |
| lh_rostralanteriorcingulate_thickness | | 100 | 100 | | | |
| lh_insula_thickness | | 0.24 | 0.24 | | | |
| rh_bankssts_thickness | | 0.73 | | | | 1.23 |
| lh_precuneus_thickness | | | | | 0.37 | |
| rh_entorhinal_thickness | | 100 | 100 | | 100 | 100 |
| rh_fusiform_thickness | | 0.97 | | | | 0.12 |
| rh_inferiorparietal_thickness | | 76.89 | 27.98 | | 0.12 | |
| rh_inferiortemporal_thickness | | 77.86 | 78.83 | | 100 | 98.77 |
| rh_isthmuscingulate_thickness | | 0.24 | 0.24 | | | |
| rh_middletemporal_thickness | | 100 | 97.32 | | 100 | 100 |
| rh_parahippocampal_thickness | | 98.78 | 99.76 | | 100 | 100 |
| rh_posteriorcingulate_thickness | | 0.24 | 0.24 | | | |
| rh_precuneus_thickness | | 6.08 | 55.96 | | | |
| rh_insula_thickness | | 0 | 0.24 | | | |

other four. Except for one (lh_medialorbitofrontal_thickness vs. lh_middletemporal_thickness), all other features mostly selected in the case of volumetric and cortical thickness feature sets are also selected in the combined feature set in most of the cases.

The selected features are very stable in the case of the dataset with AD, NL, and MCI subjects, too. The selection is also very similar between the two datasets. Moreover, it should be noticed that most of the selected features are reported by the researchers as significant AD biomarkers.

Values of MAP for the Screening 1.5T MRI dataset are given in Table 4 for both datasets in all six scenarios regarding the feature sets. According to the obtained results, it should be noticed that in the case when only two classes are considered, using cortical thickness features only leads to the highest value of MAP of 0.662 without the feature selection step. However, after applying the CFS algorithm, the selected feature subset provides best overall MAP of 0.754 in the case of all imaging features. With this step, in most of the experiments (82.24 %) only 4 features were selected in the case of volumetric features. Considering the cortical thickness feature set, the length of the feature vector in 73.97 % was 11, while in the case of combined feature set, in most of the experimetns (75.18 %), 15 features were selected. By applying the feature selection algorithm, the retrieval performance is improved in all scenarios.

The situation is similar with the other dataset where all subjects categories are considered. Hoewver, the results are significantly lower values of MAP. This can be explained by the nature of the MCI condition, since it is usually a transitional condition that very often develops to AD and is difficult to automatically make a distinction between this condition and AD or NL. It is also recorded that the volume of hippocampus does not have as big discriminative power in distinguishing MCI and AD, as it has in distinguishing AD and NL [11]. As a result, additional research is needed in this case, and is part of our future work.

**Table 4.** Evaluation of the retrieval performance on the bases of MAP for 1.5T MRI Screening dataset (Classes: NL, AD and NL, AD, MCI) for all six feature sets.

| Dataset | Features | | | | | |
|---|---|---|---|---|---|---|
| | Volumes | Cortical thickness | All imaging | Volumes – selected | Cortical thickness - selected | All imaging - selected |
| 1.5T Screening (NL, AD) | 0.596 | 0.662 | 0.650 | 0.703 | 0.729 | 0.754 |
| 1.5T Screening (NL, AD, MCI) | 0.404 | 0.424 | 0.420 | 0.441 | 0.447 | 0.458 |

We additionally provide curves of average precision at the first N (up to N = 20) retrieved scans for the 1.5T Screening MRI dataset in the case when only NL and AD classes are considered and in the case where all three classes are considered (Fig. 3) for all scenarios: all imaging features [27], volumes only, cortical thickness, and all of these with the selected features. It is clearly noticeable that the feature selection improves the results. Additionally, when applied to the combined feature vector, it provides the best results for this dataset in both cases, with and without MCI class.

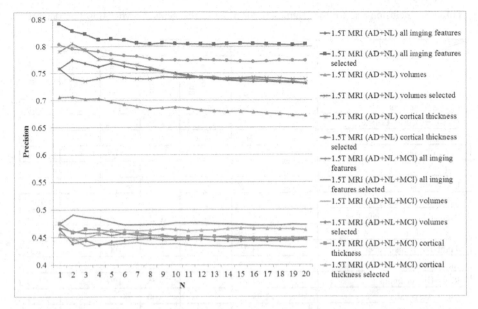

**Fig. 3.** Precision at N (up to N = 20) for the 1.5T Screening MRI subset of ADNI dataset, considering subjects with AD and NL (the upper group of curves), and AD, NL and MCI subjects (the upper group of curves). N denotes the number of retrieved scans. All scenarios for feature set are included: all imaging features, volumes only, cortical thickness, and all of these after performing the feature selection process.

## 4.2  Evaluation Performed on the 3T MRI Baseline Dataset

In this subsection, the results of the retrieval performance evaluation conducted on the 3T MRI baseline dataset are given. The retrieval performance was also evaluated for all six feature sets with application to 3T baseline images considering only AD and NL subjects, and in the case of all categories of subjects (AD, NL, MCI). CFS method was also applied in the same manner. Table 5 summarizes the inclusion rate for the features in the experiments for 3T dataset. Features not involved in any experiment are not included in the table. Considering the selected feature subsets, it should be noticed here again that most of them are found to be valuable indicators for AD in the literature, including volume of the hippocampus, amygdala, lateral ventricles, entorhinal thickness etc. Thus, using these feature subsets in the subsequent retrieval process is very reasonable. Particularly, in the case of the dataset where only AD and NL subjects are considered, regarding the volumetric features, mostly selected the temporal horn of left lateral ventricle, left and right hippocampus (the only feature selected in all experiments), right lateral ventricle, right thalamus proper, and right amygdala. Taking into consideration cortical thickness, five features are selected in all experiments and two in more than 95 %. There is small difference in the feature subset selected in the case of combined features. In the case of the dataset in which all three classes of subjects are considered, the selected volumetric features are slightly more stable in the scenario with volumetric features only and with the combined features.

**Table 5.** Inclusion rate for the features used in the experiments 3T MRI Baseline dataset.

| Classes included / Features | AD, NL | | | AD, NL, MCI | | |
|---|---|---|---|---|---|---|
| | Volumes | Cortical thickness | All imaging | Volumes | Cortical thickness | All imaging |
| Left-Inf-Lat-Vent | 92.5 | | | 100 | | 98.68 |
| Left-Cerebellum-Cortex | 1.25 | | 1.25 | | | |
| Left-Hippocampus | 100 | | 100 | 100 | | 100 |
| Left-Amygdala | | | | 92.05 | | 85.43 |
| CSF | 1.25 | | | | | |
| Left-VentralDC | 5 | | 1.25 | | | 0.66 |
| Left-choroid-plexus | | | | 1.99 | | 1.99 |
| Right-Lateral-Ventricle | 86.25 | | 1.25 | 100 | | 100 |
| Right-Inf-Lat-Vent | 8.75 | | | 100 | | 100 |
| Right-Thalamus-Proper | 93.75 | | 85 | 100 | | 100 |
| Right-Pallidum | 2.5 | | 1.25 | | | |
| Right-Hippocampus | 98.75 | | 73.75 | 100 | | 100 |
| Right-Amygdala | 98.75 | | 100 | 99.34 | | 99.34 |
| SubCortGrayVol | 22.5 | | 5 | 2.65 | | 2.65 |
| lh_bankssts_thickness | | 95 | 7.5 | | | |
| lh_entorhinal_thickness | | 100 | 100 | | 100 | 100 |
| lh_inferiorparietal_thickness | | 100 | | | 1.99 | 1.99 |
| lh_inferiortemporal_thickness | | | 100 | | 0.66 | 0.66 |
| lh_middletemporal_thickness | | 1.25 | | | | |
| lh_parahippocampal_thickness | | 21.25 | 40 | | 100 | 100 |
| lh_paracentral_thickness | | | | | 22.52 | 22.52 |
| lh_posteriorcingulate_thickness | | | | | 0.66 | 0.66 |
| lh_parsorbitalis_thickness | | 100 | 100 | | | |
| lh_superiorparietal_thickness | | | | | 0.66 | 0.66 |
| lh_superiorfrontal_thickness | | 1.25 | 1.25 | | | |
| rh_entorhinal_thickness | | 100 | 100 | | 100 | 99.34 |
| rh_inferiorparietal_thickness | | 98.75 | 97.5 | | | |
| rh_inferiortemporal_thickness | | | | | 100 | 100 |
| rh_parahippocampal_thickness | | 100 | 100 | | 5.3 | 5.3 |
| rh_paracentral_thickness | | 28.75 | 28.75 | | | |
| rh_rostralmiddlefrontal_thickness | | 1.25 | 1.25 | | | |
| rh_superiorparietal_thickness | | 2.5 | 3.75 | | 100 | |
| rh_temporalpole_thickness | | | | | | 100 |

The evaluation of the retrieval performance on the bases of MAP for each form of the feature vector considering the case with two and three classes for the Baseline 3T MRI dataset is summarized in Table 6. In the case of the whole feature vector used for image representation, without feature selection, the value of MAP is highest in the case of combined feature vector of all imaging features. While the situation remains the same for the dataset with two classes, this is not the case with the other dataset.

In all cases, feature selection comparing to the appropriate case without feature selection. For volumetric features six features were used in most experiments (62.5 %) in the case of two considered classes and eight in 90.73 % in the other case. Regarding cortical thickness features, eight or seven features were selected in 41.25 % and 2.5 % of the experiments respectively in the dataset with two classes, and five features for the other dataset. Most common numbers of features considering the combined imaging features, 9–11 features for the first and 13 features regarding the dataset with all three classes of subjects.

**Table 6.** Evaluation of the retrieval performance on the bases of MAP for 3T MRI Baseline dataset (Classes: NL, AD and NL, AD, MCI) for all six feature sets.

|  | Features | | | | | |
|---|---|---|---|---|---|---|
| Dataset | Volumes | Cortical thickness | All imaging | Volumes - selected | Cortical thickness - selected | All imaging - selected |
| 3T Baseline (NL, AD) | 0.639 | 0.619 | 0.649 | 0.711 | 0.657 | 0.713 |
| 3T Baseline (NL, AD, MCI) | 0.432 | 0.420 | 0.433 | 0.482 | 0.430 | 0.464 |

To be able to compare the results to the results of the research performed on the same subset of ADNI dataset in [9] (Baseline 3T dataset with AD and NL classes), we also provide the curves of average precision at the first N (up to N = 20) retrieved scans (Fig. 4). According to the results reported in [9] for this subset, the best average precision reaches 0.74 (at N = 1). In our case, better or comparable average precision is obtained regarding the feature set scenario, similarly as it was firstly noticed in our previous work [27]. This means that the method for using measures of the brain structures used in this paper reaches even higher average precision at N = 1 than [9] with reduced number of features. We additionally provide curves of average precision at the first N (up to N = 20) retrieved scans in the case when all three classes are considered (Fig. 4).

It should be emphasized that in all cases, the feature selection leads to improved retrieval performance, while decreasing the retrieval process complexity. The reason is that including this step leads to reduction of irrelevant, redundant or possibly noisy data. In fact, only the most relevant and discriminative features are considered. The feature vector dimension is significantly reduced with this step leading to feature vector constructed from only 3 features and not exceeding 18 features in some experiments. Most often this range is from 4 to 15 features. The selected features are highly stable. Moreover, features selected most of the time comply with the most significant AD markers listed in the literature (for example: hippocampus, amygdala, left and right lateral ventricles, entorhinal thickness etc.), which is very important looking from the perspective of the application domain. This gives additional impact and makes this step even more meaningful.

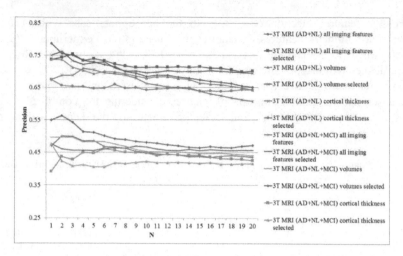

**Fig. 4.** Precision at N (up to N = 20) for the Baseline 3T subset of ADNI dataset, considered classes: AD and NL (the upper group of curves), and AD, NL and MCI (the lower group of curves). N denotes the number of retrieved scans. All scenarios for feature set are included: volumes only, cortical thickness, all imaging features, and all of these with included feature selection.

Without the feature selection step, cortical thickness feature set leads to the highest MAP for the 1.5T dataset, while in the case of 3T dataset - the combined feature vector of all imaging features. However, after the feature selection, the selected subset of all imaging features provides the best results (on the bases of MAP) in all cases except for the 3T Baseline MRI dataset when all three classes are considered. In this case, the selected volumes give highest value of MAP. In all cases the results significantly drop when the MCI class is included. Further analysis is needed in this case, which is a part of our future work.

In general, it can be concluded that the approach used in this paper provides very promising results. It gives better retrieval results than the research conducted on the same dataset (3T Baseline MRI dataset) with very small number of features. This is very important for the practical medical image retrieval system. Moreover, the performed analyses of different scenarios regarding the feature sets provides deeper understanding of the research problem and finding possible ways for improvement. Considering the retrieval problem by using baseline images is important for the perspective of enabling as precise and relevant information as possible during the first visit of the patient. Moreover, it gives a good bases for subsequent extension in the direction of longitudinal analysis, which is also a part of our future work.

## 5    Conclusions

Medical image retrieval strategy of based on the structural MRI measures characteristic for Alzheimer's Disease was researched in this paper. The alternative approach for feature extraction used for this research was performed to reflect the brain structural changes. Several scenarios for feature sets were then examined: volumetric features, cortical features, as well as a combination of all imaging features (volumes and cortical thickness). The feature subset selection using CFS method was also applied in each scenario. Stability of the selected features was also analyzed. The retrieval performance was evaluated on the Baseline 3T MRIs and Screening 1.5T MRIs from the ADNI database. The experiments were conducted in the case where only AD and NL subjects were taken into consideration, and in the case of all three categories, including MCI.

In this research, the results were significantly improved by involving feature selection procedure. Moreover, it should be emphasized that most of the features selected by the feature evaluator are highly stable and stressed in the literature as valuable indicators of AD. Comparing to the results obtained on the same subset (Baseline 3T MRI), the strategy used in this paper leads to better results with only 10 features selected in most of the experiments. This dimensionality is quite smaller than the traditional feature vector length. Considering the categories of subjects included in the research, the results of the retrieval process when the MCI group is excluded are significantly better. This is because of the nature of this condition and needs further research which is a part of our future work.

Regarding the feature vector scenarios, cortical thickness feature set leads to the highest MAP for the 1.5T dataset, while in the case of 3T dataset - the combined feature vector of all imaging features. After the feature selection is applied, the selected sub-set of all imaging features provides the best results in most of the cases.

The approach used in this research is very beneficial. It provides information extraction using the required volumetric data and efficient information representation. The usage of the measurements such as volumes and thickness of the brain structures as a medical volume representation in the CBIR system, enables answering the questions of type "find all subjects that have similar anatomical structure to the query one" utilizing the visual information rather than "find all subjects that have similar visual properties to the query image/VOI" (which is characteristic for the traditional approach). This is very important regarding the application domain. Moreover, the approach used in this research gives a good opportunity to extend this work with the aim to address the progression of the disease.

**Acknowledgements.** Data collection and sharing for this project was funded by the Alzheimer's Disease Neuroimaging Initiative (ADNI) (National Institutes of Health Grant U01 AG024904) and DOD ADNI (Department of Defense award number W81XWH-12-2-0012). The National Institute on Aging, the National Institute of Biomedical Imaging and Bioengineering, and through generous contributions from the following: Alzheimer's Association; Alzheimer's Drug Discovery Foundation; Araclon Biotech; BioClinica, Inc.; Biogen Idec Inc.; Bristol-Myers Squibb Company; Eisai Inc.; Elan Pharmaceuticals, Inc.; Eli Lilly and Company; EuroImmun; F. Hoffmann-La Roche Ltd and its affiliated company Genentech, Inc.; Fujirebio; GE Healthcare; IXICO Ltd.; Janssen Alzheimer Immunotherapy Research & Development, LLC.; Johnson & Johnson Pharmaceutical Research & Development LLC.; Medpace, Inc.; Merck & Co., Inc.; Meso Scale Diagnostics, LLC.; NeuroRx Research; Neurotrack Technologies; Novartis Pharmaceuticals Corporation; Pfizer Inc.; Piramal Imaging; Servier; Synarc Inc.; and Takeda Pharmaceutical Company are all funders of ADNI. ADNI clinical sites in Canada are supported and funded by the Canadian Institutes of Health Research. Private sector contributions are facilitated by the Foundation for the National Institutes of Health (www.fnih.org). The study is coordinated by the Alzheimer's Disease Cooperative Study at the University of California, San Diego, whereas the grantee organization for it, is the Northern California Institute for Research and Education. The Laboratory for Nero Imaging at the University of Southern California is dissemination the ADNI data.

Authors also acknowledge the support of the European Commission through the project MAESTRA - Learning from Massive, Incompletely annotated, and Structured Data (Grant number ICT-2013-612944).

# References

1. Nho, K., Risacher, L.S., Crane, P.K., DeCarli, C., Glymour, M.M., Habeck, C., Kim, S., et al.: Voxel and surface-based topography of memory and executive deficits in mild cognitive impairment and Alzheimer's disease. Brain Imaging Behav. **6**(4), 551–567 (2012)
2. Oliveira, M.C., Cirne, W., de Azevedo Marques, P.M.: Towards applying content-based image retrieval in the clinical routine. Future Gener. Comput. Syst. **23**(3), 466–474 (2007)
3. Rosset, A., Muller, H., Martins, M., Dfouni, N., Vallée, J.-P., Ratib, O.: Casimage project - a digital teaching files authoring environment. J. Thorac. Imaging **19**(2), 1–6 (2004)
4. Akgül, C.B., Rubin, D.L., Napel, S., Beaulieu, C.F., Greenspan, H., Acar, B.: Content-based image retrieval in radiology: current status and future directions. J. Digit. Imaging **24**(2), 208–222 (2011)
5. Depeursinge, A., Zrimec, T., Busayarat, S., Müller, H.: 3D lung image retrieval using localized features. SPIE Medical Imaging, pp. 79632E–79632E. International Society for Optics and Photonics, Bellingham (2011)
6. Akgül, C.B., Ünay, D., Ekin, A.: Automated diagnosis of Alzheimer's disease using image similarity and user feedback. In: Proceedings of the ACM International Conference on Image and Video Retrieval, p. 34 (2009)
7. Agarwal, M., Mostafa, J.: Image retrieval for Alzheimer's disease detection. In: Caputo, B., Müller, H., Syeda-Mahmood, T., Duncan, J.S., Wang, F., Kalpathy-Cramer, J. (eds.) MCBR-CDS 2009. LNCS, vol. 5853, pp. 49–60. Springer, Heidelberg (2010)
8. Agarwal, M., Mostafa, J.: Content-based image retrieval for Alzheimer's disease detection. In: 9th International Workshop on Content-based Multimedia Indexing (CBMI), pp. 13–18 (2011)
9. Mizotin, M., Benois-Pineau, J., Allard, M., Catheline, G.: Feature-based brain MRI retrieval for Alzheimer disease diagnosis. In: 19th IEEE International Conference on Image Processing (ICIP), pp. 1241–1244 (2012)
10. Simonyan, K., Modat, M., Ourselin, S., Cash, D., Criminisi, A., Zisserman, A.: Immediate ROI search for 3-D medical images. In: Greenspan, H., Müller, H., Syeda-Mahmood, T. (eds.) MCBR-CDS 2012. LNCS, vol. 7723, pp. 56–67. Springer, Heidelberg (2013)
11. Gerardin, E., Gaël, C., Marie, C., Rémi, C., Béatrice, D., Ho-Sung, K., Marc, N., et al.: Multidimensional classification of hippocampal shape features discriminates Alzheimer's disease and mild cognitive impairment from normal aging. Neuroimage **47**(4), 1476–1486 (2009)
12. Lötjönen, J., Robin, W., Juha, K., Valtteri, J., Lennart, T., Roger, L., Gunhild, W., Hilkka, S., Daniel, R.: Fast and robust extraction of hippocampus from MR images for diagnostics of Alzheimer's disease. Neuroimage **56**(1), 185–196 (2011)

13. Sabuncu, M.R., Desikan, R.S., Sepulcre, J., Yeo, B.T.T., Liu, H., Schmansky, N.J., Reuter, M., et al.: The dynamics of cortical and hippocampal atrophy in Alzheimer disease. Arch. Neurol. **68**(8), 1040–1048 (2011)

14. Cuingnet, R., Gerardin, E., Tessieras, J., Auzias, G., Lehéricy, S., Habert, M.O., Chupin, M., Benali, H., Colliot, O.: Automatic classification of patients with Alzheimer's disease from structural MRI: a comparison of ten methods using the ADNI database. Neuroimage **56**(2), 766–781 (2011)

15. Gray, K.R., Aljabar, P., Heckemann, R.A., Hammers, A., Rueckert, D.: Random forest-based similarity measures for multi-modal classification of Alzheimer's disease. Neuroimage **65**, 167–175 (2013)

16. Liu, S., Cai, W., Song, Y., Pujol, S., Kikinis, R., Feng, D.: A bag of semantic words model for medical content-based retrieval. In: MICCAI Workshop on Medical Content-based Retrieval for Clinical Decision Support (2013)

17. Eskildsen, S.F., Coupé, P., Fonov, V.S., Pruessner, J.C., Collins, D.L., and Alzheimer's Disease Neuroimaging Initiative: Structural imaging biomarkers of Alzheimer's disease: predicting disease progression. Neurobiology of aging **36**, S23-S31 (2015)

18. Qian, Y., Gao, X., Loomes, M., Comley, R., Barn, B., Hui, R., Tian, Z.: Content-based re-trieval of 3D medical images. In: eTELEMED 2011, 3rd International Conference on eHealth, Telemedicine, and Social Medicine, pp. 7–12 (2011)

19. Lötjönen, J.M., Wolz, R., Koikkalainen, J.R., Thurfjell, L., Waldemar, G., Soininen, H., Rueckert, D.: Fast and robust multi-atlas segmentation of brain magnetic resonance images. Neuroimage **49**(3), 2352–2365 (2010)

20. Chupin, M., Gérardin, E., Cuingnet, R., Boutet, C., Lemieux, L., Lehéricy, S., Benali, H., Garnero, L., Colliot, O.: Fully automatic hippocampus segmentation and classification in Alzheimer's disease and mild cognitive impairment applied on data from ADNI. Hippocampus **19**(6), 579–587 (2009)

21. Chupin, A., Hammer, A., Liu, R.S., Colliot, O., Burdett, J., Bardinet, E., Duncan, J.S., Garnero, L., Lemieux, L.: Automatic segmentation of the hippocampus and the amygdala driven by hybrid constraints: method and validation. Neuroimage **46**(3), 749–761 (2009)

22. Velayudhan, L., Proitsi, P., Westman, E., Muehlboeck, J.S., Mecocci, P., Vellas, B., et al.: Entorhinal cortex thickness predicts cognitive decline in Alzheimer's disease. J. Alzheimers Dis. **33**(3), 755–766 (2013)

23. Heckemann, R.A., Keihaninejad, S., Aljabar, P., Gray, K.R., Nielsen, C., Rueckert, D., Hajnal, J.V., Hammers, A.: Automatic morphometry in Alzheimer's disease and mild cognitive impairment. Neuroimage **56**(4), 024–2037 (2011)

24. FreeSurfer. https://surfer.nmr.mgh.harvard.edu

25. Hall, M.A., Holmes, G.: Benchmarking attribute selection techniques for discrete class data mining. IEEE Trans. Knowl. Data Eng. **15**(6), 1437–1447 (2003)

26. FreeSurfer methods. http://surfer.nmr.mgh.harvard.edu/fswiki/FreeSurferMethodsCitation

27. Trojacanec, K., Kitanovski, I., Dimitrovski, I., Loshkovska, S.: Content based retrieval of MRI based on brain structure changes in Alzheimer's disease. In: Proceedings of the International Conference on Bioimaging, pp. 13–22. doi:10.5220/0005182200130022 (2015)

# An Iterative Mesh Optimization Method for 3D Meristem Reconstruction at Cell Level

Guillaume Cerutti$^{(\boxtimes)}$ and Christophe Godin

Virtual Plants INRIA Team, INRIA, Montpellier, France
{guillaume.cerutti,christophe.godin}@inria.fr

**Abstract.** This paper focuses on the reconstruction of 3-dimensional multi-lay-ered triangular mesh representations of plant cell tissues, based on segmented images obtained from confocal microscopy of shoot apical meristems of model plant *Arabidopsis thaliana*. Obtaining good-quality meshes of cell interfaces in plant tissues is currently a missing step in the existing image analysis pipelines. We propose a method for optimizing the quality of such a mesh representation of the tissue simultaneously along several different citeria, starting from a low-quality mesh. An iterative process minimizes an energy functional defined over this discrete structure, by deforming its geometry and updating its connectivity at fixed complexity. This optimization results in a light discrete representation of the cell surfaces that enables fast visualization, and quantitative analysis, and gives way to *in silico* physical and mechanical simulations on real-world data. We also propose a complete quantitative evaluation scheme to measure the quality of the cell tissue reconstruction, that demonstrates the capacity of our method to fit multiple optimization criteria.

**Keywords:** Mesh optimization · Shoot apical meristem · Deformable models · Confocal microscopy · Cell reconstruction · Morphogenesis

## 1 Introduction

The spectacular development of 3-dimensional microscopy imaging techniques over the past few years has opened a brand new field of experimental investigation for developmental biology. The huge amounts of data produced require the development of complete software pipelines to process automatically the image sequences capturing the living tissues. This is particularly true in the context of developmental biology where the automatic segmentation of cells and the tracking of cell lineages over time allows to quantify tissue growth, cell deformation and gene expression patterns.

However, segmented images constitute massive objects, and lighter data structures representing the cells are preferable to the raw voxel information. Triangular meshes are a compressed representation that make visualization easier and various computations on surfaces and volumes much more efficient. It is also a necessary object for a great deal of physical simulations, notably those

A. Fred et al. (Eds): BIOSTEC 2015, CCIS 574, pp. 142–166, 2015.
DOI: 10.1007/978-3-319-27707-3_10

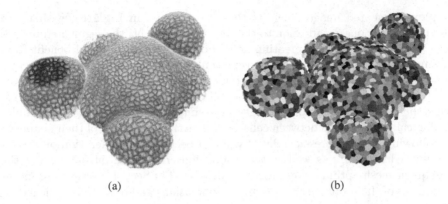

(a)                                             (b)

**Fig. 1.** Confocal laser scanning image of a shoot apical meristem of *Arabidopsis thaliana* (a) and result of the watershed segmentation obtained on the same image stack (b).

based on finite elements methods. We therefore consider here the problem of converting a 3-dimensional image stack of segmented cell tissue into a triangular mesh representing the surfaces the cells.

Our work is focused on images of shoot apical meristems (SAM) of *Arabidopsis thaliana* acquired by confocal laser scanning microscopy (CLSM) (Fig. 1(a)). SAMs constitute small niches of dividing cells where all the aerial organs of a plant (inflorescence, leaves or branches), originate from. The study of sequences of cell tissue on the SAM is therefore a key step for a better understanding of morphogenesis in plants, and the growth of organs over time. The identification of cells, the reconstruction of their lineages and extraction of shape and semantic features at a cellular level are necessary steps in this analysis process, as a way to unravel the dynamic processes linking individual cell behaviours and shape emergence, at the core of morphogenesis.

The novel possibility of monitoring growing tissues and organisms and modelling the processes at work at cell level is an unprecedented opportunity for developmental biology [19]. Many works aim at building a digital reconstruction of living cells, considering different subjects of study: capturing the dynamics of plant shoot apical meristems [5,11,28] or following animal embryos during different development stages [15,22,25,26] with an impressive increasing accuracy [1].

Concerning the development of SAM cells (mostly of *Arabidopsis thaliana*) the works initially focused on the reconstruction of the surface [20] to analyze the growth and division dynamics of the first layer of cells (L1) in the meristem [3]. More recently, complete reconstructions of the dynamic multi-layered tissue structure have emerged, based on CLSM images, using a watershed segmentation algorithm [11] or representing cells as truncated ellipsoids [6].

The method we use to segment the confocal image sequences is based on the MARS pipeline [11] that uses a seeded 3-dimensional watershed algorithm to extract the regions corresponding to each cell. The accurate identification of cells relies on the determination of seeds, but the method ensures that the regions converge towards each other, correctly defining interface surfaces. It results in complex segmented images such as the one depicted in Fig. 1(b), with up to 3000 labeled regions. In addition to their excessive weight, the segmented images inevitably present artifacts creating noisy boundaries, which could benefit from the smoothing induced by a coarser representation.

In this context, we propose a tool to reconstruct automatically the whole 3D structure of a SAM by a discrete representation of cell surfaces. This mesh transforms the voluminous complex images into simple 3D primitives, and includes topological relationships between cells as well as the estimation of their shape. In the following, we will present related work in Sect. 2. The generation of meshes is presented in Sect. 3 as well as the evaluation criteria we introduce, and the subsequent mesh optimization process is described in Sect. 4. Section 5 proposes an analysis of the results, as Sect. 6 draws conclusions and potential applications.

## 2   Related Works

The conversion of complex segmented SAM images into discrete representations is currently a missing step in cell reconstruction pipelines. Some methods already provide a lighter tissue reconstruction, under the form of a Voronoi-like anisotropic tessellation [5] or with tools allowing the definition of a triangular mesh on the surface of the meristem only [4]. Such a compact tissue reconstruction is a desirable output for fast quantification of cell properties, interactive visualization of large objects and models of tissue growth, but meshing the whole meristem tissue is still an open challenge.

A triangular mesh constitutes a discretization of a surface or a volume, and a generally more compressed view of an object, making it easier to manipulate. Different approaches have emerged to generate a mesh from a 3-dimensional image, most commonly using Marching Cubes [21] to produce a very high resolution mesh of the surfaces, or based on triangulated sample points [27] to represent a volume by tetrahedra. Compared to Marching Cubes [21] that generate a huge number of triangles, a triangulation offers the advantage of reducing drastically the volume of information, as it produces from the start a lighter object. The computation of object-filling tetrahedra instead of surface triangles also ensures that the topology of the mesh will be well-defined, which might not be the case for Marching Cubes, in particular for image cubes containing more that two different labels.

A common problem with mesh generation methods is the lack of control on the quality of the produced mesh. To obtain a satisfying result, some optimization is generally needed, either on the connectivity of the mesh elements (through local collapse, split or flip operations [8,16]) or on their shape [23]. To improve the shapes of the mesh elements, a common approach is to smooth the mesh by adjusting the positions of its vertices. The most widespread method is the Laplacian smoothing [12], where vertices are attracted by the barycenters of their neighbors, and that has been used widely in other mesh smoothing techniques [14,29].

Another method is to consider smoothing as an optimization problem, where the quality of the mesh can be estimated locally, and the location of any point updated to improve the quality of its surroundings [2,9,14]. In that case, the problem can be formulated as the minimization of an energy functional [16,17,31], a framework that can also be used for segmentation and tracking purposes [10]. Such approaches allow a precise definition of the properties that should be optimized by the smoothing, as well as the inclusion of external constraints introducing some prior knowledge, and are well suited for domain-specific applications such as ours.

## 3 Generating Tissue Mesh

The starting point of our mesh generation is a segmented image where voxels are labeled with a cell identifier, forming closed, neighboring 3-dimensional regions. These images are complex and heavy objects, as the typical number of regions in the image may reach several thousands, each of them containing roughly 20000 voxels.

### 3.1 Tetrahedral Mesh Generation

To represent the segmented cells by a mesh structure, our strategy consists in computing the mesh topology using a standard method, and then improve it relatively to the characteristics of our data. Among the various possibilities for generating a mesh topology from the segmented image data, we chose to use a tetrahedral discretization of the whole image domain, based on a Delaunay triangulation refinement algorithm [27].

Along with many other computational geometry algorithms, this tetrahedral mesh generation is implemented in the CGAL library [30], and we used this implementation to generate our mesh topology. The resulting object is a filling of the image by tetrahedra, where each tetrahedron is labeled with a cell identifier. We keep only the triangular faces that are common to two tetrahedra with different labels to produce the interface mesh. Each cell is then bounded by a set of triangles forming a closed surface. Each one of these triangles may be part of (at most) two cells, as the same interface triangle will be assigned to the boundary of the two cells it separates. The resulting mesh constitutes an approximation of the complete structure of the tissue, as illustrated in Fig. 2(a) and (b).

The precision of the image approximation provided by the mesh is controlled in the CGAL implementation by a distance parameter. This value sets the upper bound for the distance between the mesh triangles and the boundaries of their corresponding regions in the image. This parameter has also a strong influence on the complexity of the mesh, as more numerous and smaller triangles are necessary to fit the image boundaries with less error. The optimization performed by lowering this global distance constraint comes with an exponential rise in the number of primitives used to define the surfaces, and consequently does not operate on all the desirable properties of the mesh.

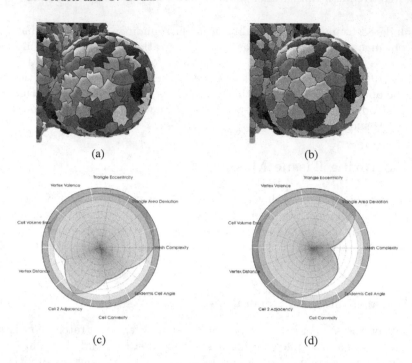

**Fig. 2.** Two meshes of the same image obtained with distance parameter values of 2.0 voxels (a) and 1.0 voxel (b) and comparison of their respective quality criteria measures (c) and (d).

## 3.2   Quality Criteria

The criteria we would like to see as optimal in the mesh are multiple and not necessarily compatible. The quality objective involves both geometrical and biological factors that should be met with a minimal complexity by the resulting mesh. It is then necessary to define accurately what is a good tissue mesh, and how it is possible to quantitatively evaluate its quality.

To perform this evaluation, we define **9 quality criteria** that account for the various objectives a cell tissue mesh should fulfill, measuring the precision with which the mesh fits the image, the consistency of the shapes it defines with those expected from SAM cells, the regularity of its triangles, and the number of geometric elements necessary to the representation. A thorough definition of these criteria can be found in Sect. 5.

- **Image Related Criteria:** Assessing that the mesh reconstructs faithfully the segmented voxel image $\mathcal{S}$, considered as the ground truth:
  - **Region Consistency:** global measure based on the comparison of cell volumes in the mesh and in the original image. (*Cell Volume Error*)
  - **Interest Point Preservation:** local measure based on the distance of identified cell vertices in the mesh with the ones in the image. (*Vertex Distance*)

- **Adjacency Preservation:** global measure of the similarity of neighborhood relationships between two cells in the image and the mesh. (*Cell 2 Adjacency*)
- **Shape Related Criteria:** Measuring the similarity of mesh cells with observed cell geometric properties for a biologically plausible meristem reconstruction:
  - **Cell Convexity:** global measure based on the convexity of the mesh cells, estimated as a volume ratio with their convex hull. (*Cell Convexity*)
  - **Surface Arrangement:** local measure based on the geometry of the surface cells, and the projected angles formed by adjacent cells. (*Epidermis Cell Angle*)
- **Triangle Related Criteria:** Quantifying the intrinsic quality of the triangles with the goal of homogeneous equilateral triangles, without small or flat elements:
  - **Triangle Quality:** global measure based on a triangle eccentricity estimation, measuring how far a triangle is from an equilateral one. (*Triangle Eccentricity*)
  - **Size Homogeneity:** global measure based on the standard deviation of the areas of triangles to estimate how regular the mesh is. (*Triangle Area Deviation*)
  - **Vertex Regularity:** global measure of how the valences of interface vertices deviate from the ideal target of 6 neighbors per vertex. (*Vertex Valence*)
- **Size Related Criteria:** Counting the mesh elements, with the idea that the aforementioned objectives should be reached with as few triangles as possible:
  - **Mesh Lightness:** global measure based on the number of triangles necessary to represent the surface of an individual cell. (*Mesh Complexity*)

A simultaneous view of all these quality criteria can be given by a spider chart, as soon as all measurement can be aligned on the same scale (by a normalization between 0 and 1 for instance). This results in a visualization such as those presented in Fig. 2(c) and (d), where the overall area of the domain delimited by the values of the different parameters has to be maximal. Any defect in one of the criteria will immediately be reflected in the visualization chart.

Applied to the surface meshes produced by the Delaunay refinement algorithm, such quality estimation highlights the complex optimization problem we are facing. The objective being to maximize all the criteria at the same time, improving the mesh by lowering the distance parameter is not satisfactory, given the drastic loss on the lightness criterion, as it appears strikingly in Fig. 2(d). An acceptable solution should be more of a compromise that tends to optimize all quality criteria for a fixed mesh complexity.

## 4 Mesh Optimization

The fulfilling of all the introduced criteria suggests a mesh optimization, that we choose to formulate as an energy minimization problem, in the same spirit as [16] or [17]. One objective being to limit the mesh complexity, this task can be performed on a light mesh without any change in the number of elements, by working only on deforming the mesh geometry and applying element-preserving topological operations.

## 4.1  Mesh Notations

The mesh object $\mathcal{M}$ we consider is a **boundary representation** (B-rep) of dimension 3, consisting of four sets of elements $\mathcal{W}_0, \mathcal{W}_1, \mathcal{W}_2, \mathcal{W}_3$ representing elements of dimensions 0 to 3, respectively vertices, edges, triangles and cells.

The elements themselves are defined by their boundaries in the lower dimension, which can be represented by a **boundary relationship** $\mathcal{B}_d$ at dimension $d$. For instance, an edge $e \in \mathcal{W}_1$ is defined by its two vertex extremities $\mathcal{B}_1(e) = \{v_1(e), v_2(e)\}, v_1 \neq v_2 \in \mathcal{W}_0$, a triangle $t \in \mathcal{W}_2$ by its three edges. A cell $c \in \mathcal{W}_3$ is the only type of element for which the number of boundary elements (triangles) is not fixed.

The boundary relation can be generalized by transition to consider even lower dimensions, for instance to retrieve the vertices of a triangle; this higher dimension boundary relation $\mathcal{B}_d^n$ links elements of dimension $d$ with their boundaries at dimension $d - n$, and can be defined recursively: $\forall w \in \mathcal{W}_d, \mathcal{B}_d^n(w) = \bigcup_{u \in \mathcal{B}_d(w)} \mathcal{B}_{d-1}^{n-1}(u)$ (with $\mathcal{B}_d^1 = \mathcal{B}_d$).

This first relationship gives birth to a converse **region relationship** $\mathcal{R}_d^n$ representing, for an element of dimension $d$, the elements of higher dimension for which it constitutes a boundary, for instance the triangles delimited by an edge:

$$\forall w \in \mathcal{W}_d, \mathcal{R}_d^n(w) = \{u \in \mathcal{W}_{d+n}(\mathcal{M}) \mid w \in \mathcal{B}_{d+n}^n(u)\}$$

Those two symmetrical relations define two possible **neighborhood relationships** $\mathcal{N}_d^+$ and $\mathcal{N}_d^-$ between elements of the same dimension, depending whether we consider that they are connected by a higher dimension region (like two vertices linked by an edge) or by a lower dimension boundary (like two cells sharing an interface triangle):

$$\mathcal{N}_d^+(w) = \{w' \in \mathcal{W}_d \mid \mathcal{R}_d(w) \cap \mathcal{R}_d(w') \neq \emptyset\}$$

$$\mathcal{N}_d^-(w) = \{w' \in \mathcal{W}_d \mid \mathcal{B}_d(w) \cap \mathcal{B}_d(w') \neq \emptyset\}$$

In addition to this topological aspect, the geometry of the mesh is defined by the association of each vertex $v \in \mathcal{W}_0$ with a point in $\mathbb{R}^3$ representing its **position** $\mathcal{P}(v)$ in the image referential. The definition of $\mathcal{P}$ determines the shape of every higher dimension element, and plays a predominant part in the quality of the mesh.

Using this representation, the optimization problem consists in searching the optimal positions $\mathcal{P}$ of the vertices, and the optimal definition of the mesh triangles contained in $\mathcal{B}_1$ and $\mathcal{B}_2$. This optimization can be performed while preserving the number of elements in the mesh, *i.e.* without affecting $\mathcal{W}_0$, $\mathcal{W}_1$, $\mathcal{W}_2$ nor $\mathcal{W}_3$, and with no cell-level topological changes, that would be induced by a modification of $\mathcal{B}_3$.

Provided an initial mesh $\mathcal{M}^{(0)} = \left( \mathcal{W}_0, \mathcal{P}^{(0)}, \mathcal{W}_1, \mathcal{B}_1^{(0)}, \mathcal{W}_2, \mathcal{B}_2^{(0)}, \mathcal{W}_3, \mathcal{B}_3 \right)$, obtained for instance by Delaunay refinement, the problem is to find the optimal mesh geometry $\mathcal{P}^*$ and connectivity $\{\mathcal{B}_1^*, \mathcal{B}_2^*\}$, as the ones that minimize an energy functional accounting for all of the introduced quality criteria.

## 4.2   Energy Formulation

Following the widespread approach concerning deformable models, from their first application to contour detection [18] to other commonly used models [7], the energy functional $E$ that we aim at minimizing is defined as a sum of energy terms accounting for the different criteria to be optimized. This general combination applied to 3-dimensional models [10] can be written as:

$$E(\mathcal{M}, \mathcal{S}) = E_{image}(\mathcal{M}, \mathcal{S}) + E_{prior}(\mathcal{M}) + E_{regularity}(\mathcal{M}) \tag{1}$$

There is an identity between the energy terms and the quality criteria we defined. The image, or data attachment, energy should be minimal when the region consistency and interest point preservation criteria are optimal, and the same goes for the prior energy and the cell convexity and surface arrangement criteria, and for the internal regularity energy and the triangle quality, size homogeneity and vertex regularity criteria. The mesh lightness and topology preservation criteria are out of reach given the frame of our optimization (no optimization of the mesh elements or of the cell triangles) and the initial mesh $\mathcal{M}^{(0)}$ is supposed to be produced with satisfactory values for these criteria.

Each one of the three terms of the energy is built on local energy potentials defined on the elements of $\mathcal{M}$ (and the geometry of their vertices) that estimate the validity of the local configuration regarding the corresponding criterion.

**Image Attachment Energy.** The surfaces represented by the mesh $\mathcal{M}$ are defined to fit the separations between cells in the segmented image $\mathcal{S}$, so that the regions delimiting cells in both representations superimpose as perfectly as possible.

More importance is given to special interest points which are cell corners. These points correspond to those where four cells intersect in $\mathcal{S}$ (or three cells at the surface of the tissue) and should particularly be preserved in the mesh for a realistic reconstruction. Such points are very easily accessible in the topology of $\mathcal{M}$ as the vertices $v$ for which $\mathcal{R}^3{}_0(v)$ contains four elements (respectively three).

Both the region consistency and the interest point preservation criteria refer to a notion of gradient magnitude in the segmented image $\mathcal{S}$, as separation interfaces are transitions leading to non-zero values of gradient, and cell corners, where many transitions occur, would correspond to local maximal values. However, unlike a color or intensity image, the segmentation $\mathcal{S}$ is a label image where the actual values do not matter, only the transitions are to be considered.

Therefore, we approximate the gradient magnitude $\|\nabla \mathcal{S}(\mathcal{P})\|$ by considering a spherical neighborhood of radius $\sigma$ around each voxel point $\mathcal{P}$ and count the number of different labels it encloses. To give more weight to exterior object boundaries, the number is doubled when the neighborhood intersects the exterior. This function is then smoothed by a gaussian filter of standard deviation $\sigma$ to obtain a more continuous field.

The energy $E_{gradient}$ associated with this boundary information is defined on the positions $\mathcal{P}$ of the vertices of the mesh, and should be minimal when a vertex fits well on a cell interface. We define it simply as the opposite of the gradient magnitude, so that high values of the gradient constitute optimal positions for the vertices of the mesh, weighted by a coefficient $\omega_{gradient}$:

$$E_{image}(\mathcal{M}, \mathcal{S}) = \sum_{v \in \mathcal{W}_0} \omega_{gradient} E_{gradient}(v, \mathcal{S}) = \sum_{v \in \mathcal{W}_0} -\omega_{gradient} \|\nabla \mathcal{S}(\mathcal{P}(v))\| \tag{2}$$

**Shape Prior Energy.** The second energy term gives the possibility of including prior knowledge on the geometry of cells or additional external constraints. The most simple characterization for a SAM cell, and the minimal constraint it should satisfy, is that its surface should form a convex solid with convex planar polygonal facets.

From biological expertise, we expect cells to present in most cases planar interfaces, and linear edges that form convex faces, properties that should fulfill the two cell shape criteria. They are translated into two energy potentials defined on the vertices of the mesh:

$$E_{prior}(\mathcal{M}) = \sum_{v \in \mathcal{W}_0} \omega_{plan} E_{plan}(v) + \omega_{line} E_{line}(v) \tag{3}$$

The planarity energy potential $E_{plan}(v)$ simply sums the distances of $\mathcal{P}(v)$ to the average planes of all the cell interfaces it belongs to. Those planes are estimated by averaging the normals of the triangles composing the interface, and the distance can be computed as a dot product.

The energy $E_{line}(v)$ is defined on the vertices belonging to at least two cell interfaces, thus part of the contour of a cell interface. To regularize those contours, we chose to sum the laplacian energies of all the interface contours going through $v$, computed at $\mathcal{P}(v)$.

**Triangle Regularity Energy.** The internal energy term accounts for the regularity of the mesh, and for the optimization of the triangle quality, size homogeneity and vertex regularity criteria. This leads to a definition of three energy potentials, this time defined on triangles, but that can commonly be restricted to the vertices of each triangle:

$$E_{regularity}(\mathcal{M}) = \sum_{v \in \mathcal{W}_0} \left( \frac{1}{3} \sum_{t \in \mathcal{R}^2_0(v)} \omega_{qual} E_{qual}(t) + \omega_{size} E_{size}(t) + \omega_{val} E_{val}(t) \right) \tag{4}$$

It is commonly accepted that a regular triangular surface mesh is composed of vertices of degree 6, connecting 6 even equilateral triangles. The quality energy potential $E_{qual}(t)$ defined on a triangle's geometry $\{\mathcal{P}(v) \mid v \in \mathcal{B}_2^2(t)\}$ should measure its eccentricity, its deviation from an equilateral configuration. Among the various possible triangle eccentricity measures [13], we chose the sum of the sinuses, that involves no maximum operator, and has good derivability properties, while being optimal in the equilateral case. We note $\theta_t(v)$ the angle of the triangle $t$ at vertex $v$:

$$E_{qual}(t) = 1 - \frac{2}{3\sqrt{3}} \sum_{v \in \mathcal{B}_2^2(t)} \sin\left(\theta_t(v)\right) \tag{5}$$

Concerning the size homogeneity potential $E_{size}(t)$, we simply use the squared distance of the triangle area $A$ to the average area $\overline{A}(\mathcal{W}_2)$ over all the mesh. Defined this way, this energy is of course minimal when all the triangles have the same size.

$$E_{size}(t) = \frac{1}{\overline{A}(\mathcal{W}_2)}\left(A\left(\{\mathcal{P}(v), v \in \mathcal{B}_2^2(t)\}\right) - \overline{A}(\mathcal{W}_2)\right)^2 \tag{6}$$

Finally, the vertex regularity energy measures how regular the arrangement of the triangles is, by looking at the gap between the numbers of neighbors of the triangle vertices and what these numbers would be in a regular mesh. In a regularly triangulated surface, each vertex should be connected to 6 equilateral triangles, assuming the surface is planar. This is true inside the cell interfaces, but at the junction between three cells, or on the exterior edge between two cells, this optimal number of neighbors becomes 9, assuming the vertex is on a straight edge (both assumptions are consistent with the prior on cell shapes).

We define a target number of neighbors $\mathbf{N}_0^*(v)$ that depends on the number of cells the vertex $v$ is in contact with, namely $|\mathcal{R}_0^3(v)|$, and on whether it is located on the exterior surface or not, the latter being true if at least one of its neighboring triangles is the border of only one cell. This gives rise to the following definition of the vertex regularity energy potential:

$$E_{val}(t) = \sum_{v \in \mathcal{B}_2^2(t)} \left(|\mathcal{N}_0^+(v)| - \mathbf{N}_0^*(v)\right)^2$$

$$\mathbf{N}_0^*(v) = \begin{cases} 3\left(|\mathcal{R}_0^3(v)| + 1\right) & \text{if} \quad \exists t \in \mathcal{R}_0^2(v), |\mathcal{R}_2(t)| = 1 \\ 3\left(|\mathcal{R}_0^3(v)|\right) & \text{otherwise} \end{cases} \tag{7}$$

## 4.3   Minimization and Evolution

The minimization of the energy functional $E$ is performed by local optimization of the vertex positions $\mathcal{P}$ and triangle connectivity $\{\mathcal{B}_1^*, \mathcal{B}_2^*\}$. These two entities belonging respectively to the geometrical and topological domain, their optimization involves different deformations, and has to be carried out separately, in a two-fold iterative optimization process. The process alternates two phases:

1. Vertex Shifting: small moves of the vertices decreasing the local value of the energy function (affecting $\mathcal{P}$),
2. Edge Flips: inversions of triangle edges inducing a decrease in the local energy (affecting $\mathcal{B}_1^*$ and $\mathcal{B}_2^*$).

At each iteration $i$, the vertex positions $\mathcal{P}$ are first updated, based on decisions made locally on the energy potentials. To add more control on this operation, we force the updated position $\mathcal{P}^{(i+1)}(v)$ of each vertex to remain within a sphere of radius $\sigma_v$ around its current position $\mathcal{P}^{(i)}(v)$. This limitation ensures that the mesh deforms smoothly and limits the risk of fold-overs and intersections between triangles.

Then, a succession of flipping operations is performed, only on edges delimiting exactly two triangles. The edge flip consists in replacing an edge by the edge linking the two other vertices of the triangles it is separating, thus converting those two triangles into two new ones. In the mesh representation we use, this operation has the advantage of not impacting the mesh elements $\mathcal{W}_1$ and $\mathcal{W}_2$ but only modifying the boundary relations $\mathcal{B}_1$ and $\mathcal{B}_2$ as shown in Fig. 3. Only energy-decreasing flips are retained in this second phase of the iteration.

The motion of the vertices in the first phase can be determined by a gradient descent of the local energy potential, following the opposite direction of the local gradient of $E$, $i.e.$ the first order derivative of the energy with respect to $\mathcal{P}$, while staying inside the sphere of radius $\sigma_v$ around $\mathcal{P}(v)$:

$$\mathcal{P}^{(i+1)}(v) = \mathcal{P}^{(i)}(v) - \min\left(1, \frac{\sigma_v}{\left\|\frac{\partial E}{\partial \mathcal{P}}\big|_v\right\|}\right) \frac{\partial E}{\partial \mathcal{P}}\bigg|_v \tag{8}$$

The only exception concerns cell corners, as a matching can directly be performed between those interest points in the mesh and in the image. Their optimal position can therefore be known in advance, and the iterative shifting for the vertices $v$ of $\mathcal{M}$ that could be matched to a cell corner in $\mathcal{S}$ will simply consist in smoothly traveling along the trajectory between their initial position $\mathcal{P}^{(0)}(v)$ and their corresponding image target point $\mathcal{P}^{(\mathcal{S})}(v)$.

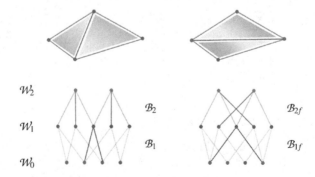

**Fig. 3.** Example of an edge-flip operation between two triangles and its translation in simple topological operations on the boundary representation.

For the remaining vertices of $\mathcal{M}$, the computation of the functional derivative of the energy can be done using the calculus of variations, and relies on the computation of the derivatives of each term of the overall energy.

**Energy Gradient Computation.** The formulation of the different energy terms as a sum of local potentials defined on the vertices of the mesh, makes it easy to compute a good approximation of the energy gradient at $\mathcal{P}(v)$ as the derivative of the concerned potential in the sum, considering only the vertex $v$.

The image attachment energy certainly has the simplest formulation regarding this computation, as it is possible to compute the gradient of the approximated gradient magnitude $\|\nabla \mathcal{S}\|$ in any point of the image. Consequently, the local energy gradient becomes:

$$\left. \frac{\partial E_{gradient}}{\partial \mathcal{P}} \right|_v = -\nabla \left\| \nabla \mathcal{S} \left( \mathcal{P}(v) \right) \right\| \tag{9}$$

Concerning the shape prior energies, the interface planarity potential is a sum of distances to planes. If we consider that the interface planes remain unchanged by shifting one vertex, the derivative of this distance is simply the unitary projective vector of $\mathcal{P}(v)$ to the plane, the direction of which is given by the plane's normal. The laplacian formulation of the energy defined for interface contours leads to an easy differentiation. It corresponds to the one involved in the laplacian smoothing operation, where vertices are attracted by the barycenters of their neighbors.

Finally, the gradients of the triangle-based regularity energy potentials, are estimated by considering the derivatives of the potential in one single triangle relatively to the position of one of its vertices. Note here that the derivative of the vertex valence energy potential is trivially zero, since it is not affected by the position of the vertices, only by their number of neighbors. The resulting energy gradient for the other energies at one vertex is simply obtained by summing the derivatives of the potentials at all neighboring triangles, for instance:

$$\left. \frac{\partial E_{qual}}{\partial \mathcal{P}} \right|_v = \sum_{t \in \mathcal{R}^2{}_0(v)} \frac{1}{3} \left. \frac{\partial E_{qual}(t)}{\partial \mathcal{P}} \right|_v \tag{10}$$

All the triangle energy potentials can be computed using the three lengths of the edges forming the boundary of the triangle, and their derivative can be decomposed over the two lengths $l_1$ and $l_2$ involving the considered vertex, as depicted in Fig. 4. The decomposition base consists then in two unitary vectors, located in the triangle's plane, and following the direction of the edges, giving the following equation:

$$\left. \frac{\partial E_{qual}(t)}{\partial \mathcal{P}} \right|_v = \frac{\partial E_{qual}(t)}{\partial l_1} \left. \frac{\partial l_1}{\partial \mathcal{P}} \right|_v + \frac{\partial E_{qual}(t)}{\partial l_2} \left. \frac{\partial l_2}{\partial \mathcal{P}} \right|_v \tag{11}$$

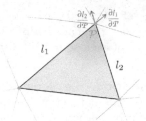

**Fig. 4.** One mesh triangle and the length derivatives used to compute the local energy gradient at point $\mathcal{P}$.

**Optimal Edge Flipping.** After a phase of vertex shifting, consisting in a sequence of $n_v$ motions following the local energy gradient, we expect the vertices to have reached a locally optimal position. This optimum however is very often a compromise, not optimal in regard of each individual energy component, especially in places where the different energies play an antagonist role. Most of all, it is often the triangle regularity energy that acts as a constraint impeding the other forces to express fully.

In the example of Fig. 5, the regularity energy acts as a force avoiding that the triangles become flat, thus preventing the boundary edge between two cells from getting straight, which would be the expected optimal outcome of the cell shape prior energy. In this case it is indeed the topology of the mesh, and the fact that a triangle links three boundary vertices, that impedes that this optimal solution could be reached.

**Fig. 5.** A case where suboptimality comes from topology and how edge flipping can help reaching an optimal solution.

To avoid such topologically induced local minima, we perform a phase of edge flipping operations with the aim of adjusting the local topology so that the next vertex shifting phase reaches a better minimum of the energy.

We only consider edges $e$ such that $\|R_1(e)\| = 2$, *i.e.* that separate exactly two triangles, or where the surface is locally manifold. The flip of the edge $e$ in a mesh $\mathcal{M}$ would produce a mesh $\mathcal{M}_f(e)$ where the boundary relations $\mathcal{B}_1$ and $\mathcal{B}_2$ around $e$ are changed into $\mathcal{B}_{1f}$ and $\mathcal{B}_{2f}$ as in Fig. 3. We compute the energy variation associated with the flip of each candidate edge, which sums up, as the image and shape prior energies are not affected by the $\mathcal{B}$ relations, as the variation of the regularity energy. As the energy is defined solely on triangles, this can be computed as the variation in the sum of the energy potentials of the two triangles linked to $e$ in each configuration (respectively $R_1(e)$ and $R_{1f}(e)$):

$$
\begin{aligned}
\Delta E_f(e) &= E_{regularity}\left(\mathcal{M}_f(e)\right) - E_{regularity}\left(\mathcal{M}\right) \\
&= \frac{1}{3}\Big( \sum_{t \in R_{1f}(e)} \omega_{qual}E_{qual}(t) \ + \ \omega_{size}E_{size}(t) \ + \omega_{val}E_{val}(t) \\
&\quad - \sum_{t \in R_1(e)} \omega_{qual}E_{qual}(t) \ + \ \omega_{size}E_{size}(t) \ + \omega_{val}E_{val}(t)\Big)
\end{aligned}
\tag{12}
$$

Once computed all the energy variations, the edges are ranked in the ascending order of $\Delta E_f$, and if the variation is negative, flipped. Each time an edge $e$ is flipped, its region neighbors $\mathcal{N}_1^+(e)$ are removed from the stack. This process is iterated until no change in the mesh connectivity is performed, and a new phase of vertex shifting, constrained by the newly defined triangles can begin.

**Energy Balancing.** As often in energy-based optimization methods, a lot of the algorithm's efficiency relies on the right combination of the weights associated with each energy term to produce a balanced behavior.

In our case, the energies taken separately have very different effects, with almost contradictory results. For instance, considered alone, the image energy will push the mesh vertices towards the local maxima of the approximated gradient $\|\nabla\mathcal{S}\|$ in a converging way: all the vertices in the neighborhood of one maximum will be attracted by it. This generates small and possibly irregular triangles as it is visible in Fig. 6(b), in complete contradiction with the regularity energy.

The same goes for planarity and linearity energies, as projections on a plane for example might create flat triangles depending on its original orientation, as in Fig. 6(c). The regularization energy will then generally have an antagonist effect on the behavior of the other forces. It acts as a rigidity constraint trying to preserve the good properties of the mesh triangles, through the global displacement induced by the remaining energies.

Acting alone however, it fails to produce realistic cell shapes, leading to noisy edges and irregular cell facets (as the shape of triangles is optimizes without regard on the consequences on the surface geometry). This is clear on the very regular mesh of Fig. 6(d) where almost equilateral triangles generate very irregular borders.

It is then very important to balance well the weight of the regularity term relatively to the shape an image terms to have an evolution where cell shapes improve (the rigidity of the mesh should not prevent it from deforming) while the regularity of the triangles is preserved (and not crushed by the destructive effects of the other energies).

To find the weights offering the best compromise between the intrinsic quality of the mesh and the its consistency with the data, we used the 9 quality criteria

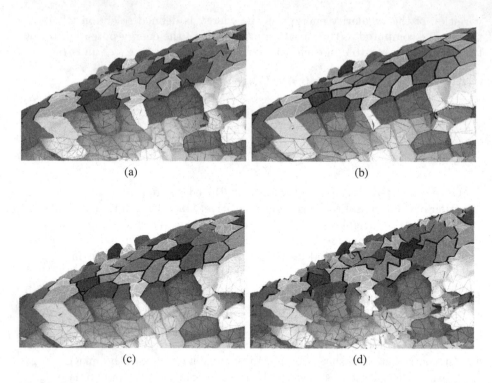

(a)               (b)

(c)               (d)

**Fig. 6.** Effects of the optimization of the different energy terms on the mesh: initial mesh (a), image energy (b) shape prior energy (c) and regularity energy (d).

our process is implicitly designed to maximize. By exploring the value space for the set of parameters $\omega$, and calculating the correlation of the quality estimators with the different values, we estimated a best joint configuration of the energy weights:

- $\omega_{gradient} = 0.17$
- $\omega_{line} = 1.3$
- $\omega_{size} = 0.005$
- $\sigma_v = 1.0$
- $\omega_{plan} = 0.47$
- $\omega_{qual} = 2.0$
- $\omega_{val} = 0.5$

**Optimization Results.** We applied our mesh optimization method to tissue meshes of different shoot apical meristems obtained form the Delaunay tetrahedral mesh generation. The positions cell corners are optimized with their extracted image position, and the rest of the mesh vertices follow a shifting by energy gradient descent. The edge flipping phases are repeated at every iteration of the vertex shifting to cope with the topology induced local minima.

The termination criterion is set to be a fixed number of iterations rather than a measure of deformation between two iterations. One of the reasons for this choice is that the energy minimum for the data attachment energy would be a mesh where vertices concentrate at local gradient maxima. Though the regularization energy works against this tendency, it is not guaranteed to find a globally stable configuration avoiding this problem, and it was sensible to stop the evolution as soon as the energies have all had the time to play their role, with 20 iterations.

The result of the optimization is a mesh with the same number of elements as the initial one, and the same adjacency relationships between cells, but with a configuration of the triangles (in both connectivity and vertex position) that makes it much more consistent with the cell regions in the image and with their biological reality at the same time. This better consistency visible in Fig. 7 is achieved without any unnecessary complication of the mesh, which remains constant all along the process.

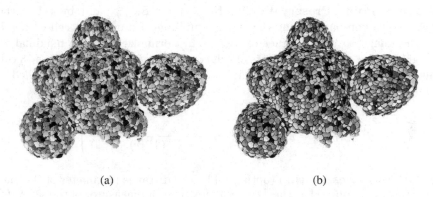

(a)                                   (b)

**Fig. 7.** Example of the application of our optimization process on a shoot apical meristem mesh: mesh generated by Delaunay refinement (a) and optimized mesh (b).

### 4.4   Implementation Details

The algorithms detailed above were implemented in Python 2.7, using the standard NumPy and SciPy libraries, and wrappers for the C++ CGAL library [30]. Data structures and visualization use the PlantGL library [24] of the OpenAlea platform.

An implementation constraint comes from the fact that 16-bit labels are not handled by CGAL the mesh generation function. Consequently, the processing of images with more that 255 cells requires a re-labeling step before the mesh generation, and the re-affectation of original labels to the connected components in the resulting tetrahedral complex.

Concerning execution times, the complexity proves to be linear with respect to the number of cells in the image, as the number of triangles (and therefore edges) varies almost proportionally with it. The typical computation times for a 2000 cell image is of 125 s for the mesh generation and 130 s for its optimization without edge flips, provided the costly filtering of the image necessary for the gradient and the cell corner extraction (depending on the image size) has been computed beforehand.

## 5    Evaluation

To assess the quality of the reconstructed tissue meshes, we generated meshes over several segmented SAM images, and computed the estimators for the nine quality criteria. We give here a formal definition of these normalized estimators, using the notations of our mesh representation:

– **Region Consistency** (optimized by $E_{image}$) computed using the average volume error, $V_{\mathcal{M}}(c)$ representing the volume of cell $c$ in the mesh and $V_{\mathcal{S}}(c)$ in the image:

$$q_{region} = 1 - \frac{1}{|\mathcal{W}_3|} \sum_{c \in \mathcal{W}_3} \frac{|V_{\mathcal{M}}(c) - V_{\mathcal{S}}(c)|}{V_{\mathcal{S}}(c)}$$

– **Interest Point Preservation** (optimized by $E_{image}$ and by cell corner extraction) computed using the average minimal distance of a cell corner $\mathcal{P}'$ in the image to a cell corner of the mesh, normalized by the maximal 26-neighborhood distance. The sets of cell corners (points where 4 cells or 3 cells and the exterior meet) for the image and the mesh are respectively called $\Gamma_{\mathcal{S}}$ and $\Gamma_{\mathcal{M}}$:

$$q_{point} = \min\left(1, \frac{\sqrt{3}}{\frac{1}{|\Gamma_{\mathcal{S}}|} \sum_{\mathcal{P}' \in \Gamma_{\mathcal{S}}} \min_{v \in \Gamma_{\mathcal{M}}} (\|\mathcal{P}(v) - \mathcal{P}'\|)}\right)$$

– **Adjacency Preservation** (optimized by the distance parameter of the mesh generation) computed as the Jaccard index (overlap measure) of the set $\mathcal{N}_3(\mathcal{S})$ of adjacencies between two cells in the image and the same set in the mesh $\mathcal{N}_3(\mathcal{M}) = \{(c_1; c_2) \mid c_2 \in \mathcal{N}_3^-(c_1), c_1 < c_2 \in \mathcal{W}_3\}$:

$$q_{adjacency} = jaccard(\mathcal{N}_3(\mathcal{M}), \mathcal{N}_3(\mathcal{S})) = \frac{|\mathcal{N}_3(\mathcal{M}) \cap \mathcal{N}_3(\mathcal{S})|}{|\mathcal{N}_3(\mathcal{M}) \cup \mathcal{N}_3(\mathcal{S})|}$$

– **Cell Convexity** (optimized by $E_{prior}$) computed using the difference between the volume $V_{\mathcal{M}}(c)$ of the cell $c$ and the volume of the convex hull $\mathcal{H}$ of its vertices:

$$q_{convexity} = \frac{1}{|\mathcal{W}_3|} \frac{\sum_{c \in \mathcal{W}_3} V\left(\mathcal{H}\left(\{\mathcal{P}(v) \mid v \in \mathcal{B}_3^3(c)\}\right)\right) - V_{\mathcal{M}}(c)}{\frac{1}{|\mathcal{W}_3|} \sum_{c \in \mathcal{W}_3} V_{\mathcal{M}}(c)}$$

– **Surface Arrangement** (implicitly optimized) computed by counting the cell angles that are outside the regular configuration (between 90 and 180), with a smooth gaussian weight at the borders. The projected surface cell angle of cell $c$ at vertex $v$ is written $\theta_c(v)$ ($\theta_{min} = \frac{\pi}{2} + \frac{\sigma_\theta}{2}$, $\theta_{max} = \pi - \frac{\sigma_\theta}{2}$, $\sigma_\theta = \frac{\pi}{9}$):

$$q_{angle} = 1 - \frac{1}{|\Gamma_{\mathcal{M}}|} \sum_{\Gamma_{\mathcal{M}}(v)} \frac{1}{|\mathcal{R}_0^3(v)|} \sum_{c \in \mathcal{R}_0^3(v)} \left(e^{-\left(\frac{\max(\theta_{min} - \theta_c(v), 0)}{\sigma_\theta}\right)^2} + e^{-\left(\frac{\max(\theta_c(v) - \theta_{max}, 0)}{\sigma_\theta}\right)^2}\right)$$

- **Triangle Quality** (optimized by $E_{regularity}$) computed using the average eccentricity of all the mesh triangles, estimated using the sum of sinuses:

$$q_{triangle} = \frac{1}{|W_2|} \sum_{t \in W_2} \left( \frac{2}{3\sqrt{3}} \sum_{v \in \mathcal{B}_2^2(t)} \sin(\theta_t(v)) \right)$$

- **Size Homogeneity** (optimized by $E_{regularity}$) computed using the standard deviation of the areas $A$ of all the mesh triangles, normalized by their average area $\bar{A}$:

$$q_{area} = \min \left( 1, \sqrt{2} - \frac{\sqrt{\sum_{t \in W_2} \left( A(t) - \bar{A} \right)^2}}{\sqrt{2} \sum_{t \in W_2} A(t)} \right)$$

- **Vertex Regularity** (optimized by $E_{regularity}$) computed using the error between the valences of the mesh and their optimal valence $\mathbf{N}_0^*$:

$$q_{valence} = 1 - \frac{1}{|W_0|} \sum_{v \in W_0} \frac{||\mathcal{N}_0^+(v)| - \mathbf{N}_0^*(v)|}{6}$$

- **Mesh Lightness** (optimized by the distance parameter of the mesh generation) computed using the average number of triangle necessary to represent one cell, normalized by the number $n_{oct} = 152$ obtained for a good triangulation of a space-filling truncated octahedron:

$$q_{lightness} = \min \left( 1, \frac{n_{oct}}{\frac{1}{|W_3|} \sum_{c \in W_3} |\mathcal{B}_3(c)|} \right)$$

(an empirical value of $\frac{1}{3\pi} \sqrt[3]{\frac{1}{|W_3|} \sum_{c \in W_3} V_S(c)}$ for the distance parameter proved to give values of $q_{lightness}$ between 0.8 and 1)

Such an evaluation framework opens the way to a rather exhaustive quantification of the quality of a SAM tissue mesh, both from an intrinsic point of view and from the comparison with real-life data. One problem issues from the necessity of normalizing all the quality estimators on a same scale, which has been done by setting evaluation parameters to reflect quality on reference high and low quality meshes. The values taken by the estimators are such that a range of acceptability can be defined between the values 0.8 and 1 for each criterion. As such, it constitutes an original tool to assess whether a mesh representing a complex plant cell tissue satisfies all the necessary computational and visual properties.

This evaluation framework is also flexible as it can be adapted to different types of *a priori* constraints in the definition of the quality criteria, for instance if one deals with other types of tissues. In such a case, it is possible to select and quantify additional criteria for mesh quality and integrate them in the energy functional as well as on the evaluation spider representation.

We applied this evaluation to a set of segmented meristem images containing 1000 to 3000 cells, on the initial meshes, and on the optimized meshes with

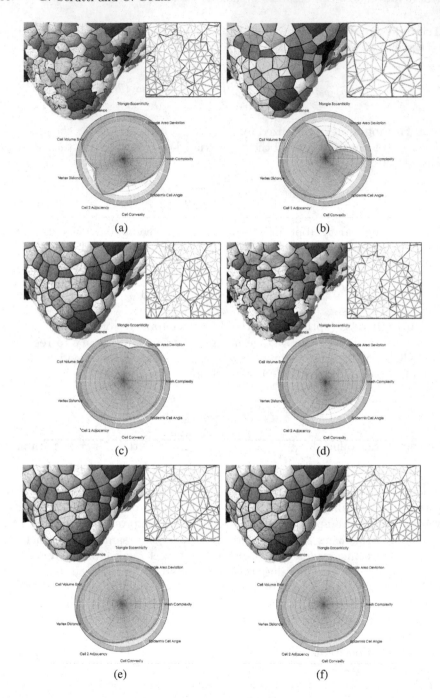

**Fig. 8.** Validation of the SAM mesh optimization results through the comparison of average mesh quality estimators on inital meshes (a) and on the corresponding optimized meshes using image attachment energy only (b), using shape prior energy only (c), using regularity energy only (d), with the complete balanced energy (e), and adding edge flips (f).

each energy separately, with the complete energy, and with or without the edge flipping phase. The results, presented in Fig. 8, are averaged over the different images, with the mean and standard deviation of the criteria displayed on the diagrams. They underline the contributions of the various components of our optimization process, that are analyzed in the following.

- **Image Attachment Energy** (Fig. 8(b)): the minimization of $E_{image}$ only leads to a valuable improvement of cell shapes while providing an excellent approximation of the image data. However the geometry of the mesh triangles is destroyed, which is clearly visible in the performance over regularity criteria. The apparition of degenerated triangles penalizes this scores and gives locally wrong configurations, around cell corners notably.
- **Shape Prior Energy** (Fig. 8(c)): using the cell interface planarization and cell edge fairing energies, we obtain very satisfactory cell shapes, with biologically plausible angles between them. The forces however fail to preserve the quality of the triangles at the cell boundaries, which leads primarily to a visibly low performance on the triangle eccentricity criterion, as the triangles get compressed or stretched at the cell edges.
- **Regularity Energy** (Fig. 8(d)): the optimization based on the minimization of $E_{regularity}$ (along with the optimal cell corner shifting) appears as an effective way of preserving, if not improving, the overall quality of the mesh triangles, throughout the deformation induced by the motion of the vertices. Still, the visual impression produced by this optimization appears very poor, which is confirmed by the mitigated performance reached on the cell shape quality criteria.
- **Complete Energy** (Fig. 8(e)): it appears then clearly that the addition of the other energy terms is essential to obtain a globally satisfying result. The quantitative measures show that regularization forces and shape/image forces have antagonist effects, the latter tending to improve the cell shape related criteria with a strong counterpart on the triangle quality criteria. Their simultaneous application constitutes an improvement on both sides, as the rigidity provided by the regularity term compensates the natural tendency of the other terms to flatten the triangles.
- **Complete Energy with Edge Flips** (Fig. 8(f)): the optimization of the mesh geometry only still shows some problem in areas where the opposite effects of forces lead unsatisfying local minima (noisy cell edges, small triangles around high valence vertices). The application of topological operations based on the same energy proves to solve some of those problems and lead to a global improvement of the mesh quality on all criteria (see Table 1). Following this two-fold optimization process, the cell reconstructions we obtain prove reach satisfactory values on all the objective criteria, as well as a visually convincing representation of the tissue.

To provide a more thorough validation of the proposed approach, we compared its results on the introduced quality criteria with the ones of other commonly used mesh geometry optimization methods. These methods are generally designed to reduce high-frequency noise on a surface mesh while preserving its global shape and if possible improving the quality of its triangles. In our case where the surfaces are actually nested and triangles connected to possibly more than one neighbor, we tried to implement the surface smoothing as if it was performed cell by cell to get more realistic methods. We chose three very common and competitive iterative mesh smoothing methods:

- **Laplacian Smoothing** [12]: at each iteration, the vertices are attracted towards the isobarycenter of their neighbors. In our case the neighbors are evaluated cell by cell so that the smoothing reflects a cell-wise smoothing, and the cell corners follow their optimal trajectory, as in our method.
- **Taubin Smoothing** [29]: each iteration consists of two phase of gaussian smoothing (vertices attracted by a gaussiang weighted barycenter of their neighbors) one with a positive coefficient, and one with a slightly greater negative coefficient. We used 0.33 and $-0.34$, as in the original paper, and used the same neighbor computing and optimal cell corner trajectory as in the Laplacian case.
- **Taubin Smoothing with Edge Flips**: same algorithm as the previous one, with the improvement of performing optimal edge flips between two phases of smoothing, with the decision based only on the $E_{val}$ energy. The cell corners still follow their optimal trajectory.

**Table 1.** Average quality estimator measures on meshes optimized with different methods: different energies of the proposed method, Laplacian smoothing and Taubin smoothing; the proposed method appears as the one providing a higher value for the minimum of all estimators.

| Method | $q_{area}$ | $q_{triangle}$ | $q_{valence}$ | $q_{region}$ | $q_{point}$ | $q_{convexity}$ | $q_{angle}$ | Mean $q$ | Min $q$ |
|---|---|---|---|---|---|---|---|---|---|
| Initial Mesh | 1.0 | 0.92 | 0.888 | 0.947 | 0.636 | 0.52 | 0.666 | 0.813 | 0.52 |
| Optimized (Image) | *0.256* | *0.333* | 0.888 | 0.865 | 1.0 | 0.813 | 0.576 | 0.719 | 0.24 |
| Optimized (Shape) | 0.948 | *0.693* | 0.888 | 0.873 | 1.0 | **0.82** | **0.835** | 0.866 | 0.693 |
| Optimized (Regularity) | 1.0 | **0.962** | 0.888 | 0.917 | 1.0 | *0.51* | *0.695* | 0.857 | 0.51 |
| Optimized (Complete) | 0.991 | 0.874 | 0.888 | 0.886 | 1.0 | 0.767 | 0.809 | **0.884** | 0.767 |
| Optimized (Flip) | 0.997 | 0.888 | **0.898** | 0.884 | 1.0 | 0.783 | 0.825 | **0.89** | **0.783** |
| Laplacian | 1.0 | 0.828 | 0.888 | *0.743* | 1.0 | *0.726* | 0.822 | 0.861 | 0.713 |
| Taubin | 1.0 | 0.896 | 0.888 | 0.932 | 1.0 | *0.693* | 0.812 | **0.885** | 0.693 |
| Taubin (Flip) | 1.0 | 0.891 | **0.904** | 0.932 | 1.0 | *0.715* | 0.832 | **0.89** | 0.715 |

The results obtained on all the different methods applied to the same set of segmented images are given in Table 1. The Laplacian smoothing performs quite similarly to the shape energy, but with a better preservation of the triangle quality. Its main problem comes from its well-known drawback of shrinking the surfaces it is applied on. This induces a wrong estimation of the cell regions that will generally end up a little reduced, as well as the appearance of concave parts especially on the outer surfaces whose corners don't shrink since they follow an optimal trajectory. Both phenomena explain of the rather low image consistency and convexity scores.

On the other hand, the Taubin smoothing is designed specifically to avoid the shrinkage of surfaces and performs much better on this aspect, with a notably high score on the image region preservation measure. The overall results are very good, but they still display noisy cell edges, which explain the low convexity score. The edge flips allow to correct this problem to some extent as shown in Fig. 9, and makes the method reach a top performance globally. But if we look at the optimization goal to make all the criteria reach satisfying values simultaneously, the Taubin smoothing fails to show the flexibility of the proposed method, as its "worst" quality score lies significantly lower than the one of our method.

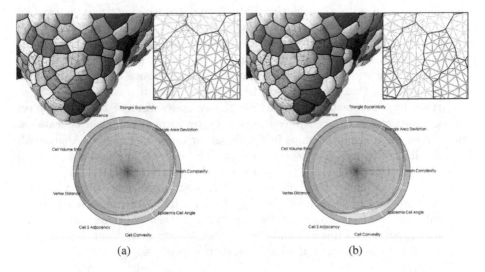

(a)                                                        (b)

**Fig. 9.** Comparison of the results of the proposed mesh optimization method (a) with the Taubin smoothing method with edge flips (b).

The optimization through the balanced combination of the different energies appears then as the only way to obtain an overall optimal compromise that does not neglect any aspect of the quality criteria. The modification of the mesh topology proves to add a significant contribution by eliminating some problematic configurations and reach a better minimum of the energy. The result is a mesh that models in a visually convincing way the shape and disposition of the cells in the meristem, with a high regularity of the triangles, and a strong consistency with the original image, while preserving an overall complexity low enough to make it exploitable for visualization and simulation uses.

## 6   Conclusions

The method we presented to reconstruct a complex triangular mesh of a shoot apical meristem cell tissue actually bridges a gap between experimental data, and higher-level computational simulations. The mesh representation we obtain constitutes a ready-to-process object, including local shape information as well as tissue-scale topological relations. It opens the way to a great deal of potential applications, from fast shape feature extraction using discrete geometry, to statistical computations (average shapes, extended to average tissue) or physical and mechanical growth simulations. Using growing real-world examples rather than hand-built model structures would constitute a major step for the validation of such biophysical development models.

All of this of course holds only if the provided reconstruction presents the properties needed to make these applications possible and sensible. Our optimization process guarantees that the produced mesh reconstructs faithfully the experimental data, with a complexity that will make processing times reasonable, and by geometrical elements regular enough to expect correct simulations. The quantitative quality evaluation framework we designed ensures that the compromise between these hardly consonant aspects fulfills the necessary criteria. It additionally provides an objective and complete measure of the quality of a SAM tissue mesh, that could be used to compare different methods.

Further improvements might include the consideration of the higher-level topology, optimizing the cell-to-cell adjacency by local topological operations, either based on the relationships in the image or on cell geometry. Along with the mesh lightness, it is the aspect on which the process has no influence and having a way to modify the neighborhood of the cells might constitute a valuable improvement. The other axis of research is the inclusion of the temporal dimension, working on sequences following the same individual meristem in time. The goal would be the reconstruct a continuous 4D mesh structure based on the sequence of segmented images, and the knowledge extracted from the neighbor frames may help to optimize meshes towards this purpose.

# References

1. Amat, F., Lemon, W., Mossing, D.P., McDole, K., Wan, Y., Branson, K., Myers, E.W., Keller, P.J.: Fast, accurate reconstruction of cell lineages from large-scale fluorescence microscopy data. Nat. Methods **11**(9), 951–958 (2014)
2. Amenta, N., Bern, M.W., Eppstein, D.: Optimal point placement for mesh smoothing. In: Proceedings of the ACM-SIAM Symposium on Discrete Algorithms, pp. 528–537 (1997)
3. Barbier de Reuille, P., Bohn-Courseau, I., Godin, C., Traas, J.: A protocol to analyse cellular dynamics during plant development. The Plant J. **44**(6), 1045–1053 (2005)
4. de Reuille, P.B., Robinson, S., Smith, R.S.: Quantifying cell shape and gene expression in the shoot apical meristem using MorphoGraphX. In: Žárský, V., Cvrčková, F. (eds.) Plant Cell Morphogenesis. Methods in Molecular Biology, pp. 121–134. Springer, New York (2014)
5. Chakraborty, A., Perales, M.M., Reddy, G.V., Roy Chowdhury, A.K.: Adaptive geometric tessellation for 3d reconstruction of anisotropically developing cells in multilayer tissues from sparse volumetric microscopy images. PLoS One **8**(8), e67202 (2013)

6. Chakraborty, A., Yadav, R., Reddy, G.V., Roy Chowdhury, A.K.: Cell resolution 3d reconstruction of developing multilayer tissues from sparsely sampled volumetric microscopy images. In: BIBM, pp. 378–383 (2011)
7. Chan, T., Vese, L.: Active contours without edges. IEEE Trans. Image Process. 10(2), 266–277 (2001)
8. Clark, B., Ray, N., Jiao, X.: Surface mesh optimization, adaption, and untangling with high-order accuracy. In: Proceedings of the 21st International Meshing Roundtable, IMR 2012, pp. 385–402 (2012)
9. Desbrun, M., Meyer, M., Schröder, P., Barr, A.H.: Implicit fairing of irregular meshes using diffusion and curvature flow. In: Proceedings of the 26th Annual Conference on Computer Graphics and Interactive Techniques, SIGGRAPH 1999, pp. 317–324 (1999)
10. Dufour, A., Thibeaux, R., Labruyere, E., Guillen, N., Olivo-Marin, J.-C.: 3-D active meshes: Fast discrete deformable models for cell tracking in 3-D time-lapse microscopy. IEEE Trans. Image Process. 20(7), 1925–1937 (2011)
11. Fernandez, R., Das, P., Mirabet, V., Moscardi, E., Traas, J., Verdeil, J.-L., Malandain, G., Godin, C.: Imaging plant growth in 4D : robust tissue reconstruction and lineaging at cell resolution. Nat. Methods 7, 547–553 (2010)
12. Field, D.A.: Laplacian smoothing and delaunay triangulations. Commun. Appl. Numer. Methods 4(6), 709–712 (1988)
13. Field, D.A.: Qualitative measures for initial meshes. Int. J. Numer. Methods Eng. 47(4), 887–906 (2000)
14. Freitag, L.A.: On combining laplacian and optimization-based mesh smoothing techniques. In: Trends in Unstructured Mesh Generation, pp. 37–43 (1997)
15. Guignard, L., Godin, C., Fiuza, U.-M., Hufnagel, L., Lemaire, P., Malandain, G.: Spatio-temporal registration of embryo images. In: IEEE International Symposium on Biomedical Imaging (2014)
16. Hoppe, H., DeRose, T., Duchamp, T., McDonald, J., Stuetzle, W.: Mesh optimization. In: Proceedings of the 20th Annual Conference on Computer Graphics and Interactive Techniques, SIGGRAPH 1993, pp. 19–26 (1993)
17. Jiao, X., Wang, D., Zha, H.: Simple and effective variational optimization of surface and volume triangulations. Eng. Comput. (Lond.) 27(1), 81–94 (2011)
18. Kass, M., Witkin, A., Terzopoulos, D.: Snakes: active contour models. Int. J. Comput. Vis. 1(4), 321–331 (1988)
19. Keller, P.J.: Imaging morphogenesis: technological advances and biological insights. Science 340(6137) (2013)
20. Kwiatkowska, D.: Surface growth at the reproductive shoot apex of Arabidopsis thaliana pin-formed 1 and wild type. J. Exp. Bot. 55(399), 1021–1032 (2004)
21. Lorensen, W.E., Cline, H.E.: Marching cubes: a high resolution 3d surface construction algorithm. In: Proceedings of the 14th Annual Conference on Computer Graphics and Interactive Techniques, SIGGRAPH 1987, pp. 163–169 (1987)
22. Michelin, G., Guignard, L., Fiuza, U.-M., Malandain, G., et al.: Embryo cell membranes reconstruction by tensor voting. In: IEEE International Symposium on Biomedical Imaging (2014)
23. Owen, S.J.: A survey of unstructured mesh generation technology. In: International Meshing Roundtable, pp. 239–267 (1998)
24. Pradal, C., Boudon, F., Nouguier, C., Chopard, J., Godin, C.: Plantgl: A python-based geometric library for 3D plant modelling at different scales. Graph. Models 71(1), 1–21 (2009)
25. Rizzi, B., Peyrieras, N.: Towards 3D in silico modeling of the sea urchin embryonic development. J. Chem. Biol. 7(1), 17–28 (2014)

26. Robin, F.B., Dauga, D., Tassy, O., Sobral, D., Daian, F., Lemaire, P.: Time-lapse imaging of live Phallusia embryos for creating 3D digital replicas. Cold Spring Harb. Protoc. **10**, 1244–1246 (2011)
27. Shewchuk, J.R.: Tetrahedral mesh generation by Delaunay refinement. In: Proceedings of the Symposium on Computational Geometry, SCG 1998, pp. 86–95 (1998)
28. Tataw, O.M., Reddy, G.V., Keogh, E.J., Roy Chowdhury, A.K.: Quantitative analysis of live-cell growth at the shoot apex of arabidopsis thaliana: algorithms for feature measurement and temporal alignment. IEEE/ACM Trans. Comput. Biol. Bioinform. **10**(5), 1150–1161 (2013)
29. Taubin, G.: Curve and surface smoothing without shrinkage. In: Proceedings of the Fifth International Conference on Computer Vision (ICCV 1995), pp. 852–857 (1995)
30. Cgal. Cgal, Computational Geometry Algorithms Library (1996). http://www.cgal.org
31. Vidal, V., Wolf, C., Dupont, F.: Combinatorial mesh optimization. The Vis. Comput. **28**(5), 511–525 (2012)

# Bioinformatics Models, Methods and Algorithms

# Latent Forests to Model Genetical Data for the Purpose of Multilocus Genome-Wide Association Studies. Which Clustering Should Be Chosen?

Duc-Thanh Phan[1], Philippe Leray[1], and Christine Sinoquet[2]($\boxtimes$)

[1] LINA, UMR CNRS 6241, POLYTECH, University of Nantes, Rue Christian Pauc, BP 50609, 44306 Nantes, France
duc-thanh.phan@etu.univ-nantes.fr, philippe.leray@univ-nantes.fr
[2] LINA, UMR CNRS 6241, Faculty of Sciences, University of Nantes,
2 Rue de la Houssinière, 44322 Nantes, France
christine.sinoquet@univ-nantes.fr

**Abstract.** The aim of genetic association studies, and in particular genome-wide association studies (GWASs), is to unravel the genetics of complex diseases. In this domain, machine learning offers an attractive alternative to classical statistical approaches. The seminal works of Mourad et al. [1] have led to the proposal of a novel class of probabilistic graphical models, the forest of latent trees (FLTM). The design of this model was motivated by the necessity to model genetical data at the genome scale, prior to a multilocus GWAS. A multilocus GWAS fully exploits information about the complex dependences existing within genetical data, to help detect the loci associated with the studied pathology. The FLTM framework also allows data dimension reduction. The FLTM model is a hierarchical Bayesian network with latent variables. Central to the FLTM construction is the recursive clustering of variables, in a bottom up subsuming process. This article focuses on the analysis of the impact of the choice of the clustering method used in the FLTM learning algorithm, in a GWAS context. We rely on a real GWAS data set describing 41400 variables for each of 3004 controls and 2005 cases affected by Crohn's disease, and compare the impact of three clustering methods. We compare statistics about data dimension reduction as well as trends concerning the ability to split or group putative causal SNPs in agreement with the underlying biological reality. To assess the risk of missing significant association results due to subsumption, we also compare the clustering methods through the corresponding FLTM-based GWASs. In the GWAS context and in this framework, the choice of the clustering method does not influence the satisfying performance of the GWAS.

**Keywords:** Genome-wide association study · Multilocus association study · Linkage disequilibrium · Probabilistic graphical model · Latent variable · Data dimension reduction · Bayesian network

© Springer International Publishing Switzerland 2015
A. Fred et al. (Eds): BIOSTEC 2015, CCIS 574, pp. 169–189, 2015.
DOI: 10.1007/978-3-319-27707-3_11

# 1  Introduction

With the finalization of the Human Genome Project in 2003, it was confirmed that any two individuals share, on average, 99.9 % of their genome with one another. It is then the sole 0.1 % genetic variations that may explain why individuals are physically different or should inherit a greater risk of contracting genetic disorders, such as coronary heart disease, diabetes, autism, some cancers. A heavy burden as regards public health, complex genetic diseases also generate a considerable socio-economic impact. For instance, in France, in 1960, 2001 and 2009, public health expenditure represented respectively 4.2 %, 8.7 % and 11.9 % of the gross domestic product. Various predictions estimate that the health spending weight will fall in the range [7.3 % − 22.3 %] with respect to the gross domestic product, at the horizon 2050. According to [2], in the next twenty years, in advanced countries, these percentages will continue to rise by 3 % of the gross domestic product on the average. In Europe, this increase could reach 2 % on average, and more than 3 % in seven other countries. A growth of 5 % is estimated for the USA. For a main part, this ever increasing share of public health expenditure is to be related to the gain in longevity, which favours the emergence of chronic diseases by elderly subjects. Such diseases include old-age onset genetic diseases.

Identifying the genetic factors underlying these diseases potentially plays a crucial role in prediction, monitoring subjects with risks, as well as developing new treatments. The medicine of the future, or personalized medicine, intends to target the therapy best adapted to the patient's genotypic background; early gene susceptibility detection aims at a better prevention or surveillance. Thus, deciphering the putative causes of complex genetic diseases has been one of the main focuses of human genetics research during the last thirty years [3]. Among different approaches that have been proposed, association studies stand out as one of the most successful paths, even though their potential is yet to be fully tapped.

Thanks to new advances in techniques for genotyping and sequencing genomes, researchers started to work on seeking genetic variations potentially associated to common diseases throughout the entire genome. In the following years, the HapMap Project [4] and its successor, the 1000 Genomes Project [5], were launched with the hope to establish a catalogue of human genome regions in which people of different populations have differences.

When no clue is available about the genome regions likely to contain one of the putative causes for a studied disease, geneticists are compelled to resort to genome-wide association studies (GWASs). Genetic markers are used for this purpose, as well as a population of affected and unaffected individuals. Genetic markers represent as many DNA sequences, spraid over the whole genome, with a known location, where the DNA variations within a given population may be observed. In a nutshell, GWASs seek to identify genetic markers whose variants vary systematically between affected (cases) and unaffected (controls) individuals [6]. The standard GWAS consists in comparing variant frequencies in cases and controls, on massive genotypic datasets (tens of thousands of individuals each described by hundreds of thousands up to a few millions of genetic markers). The goal is to identify the loci on the genome for which the distributions of variants are significantly different between cases and controls, using dependence - namely *association* - tests (e.g. the $Chi^2$ test). The unit variants, called single nucleotide polymorphisms (SNPs), which refer to single base pair changes in the DNA sequence represent the most abundant type of variants in human; they are very often used as markers in GWASs.

The key to GWASs lies in this interesting phenomenon known as "linkage disequilibrium" (LD) where variants for different SNPs tend to co-occur non-randomly [7] (the corresponding SNPs are said to be in LD). The case would be exceptional if a genetic marker, which is observed in the population, coincided with a genetic causal factor. Nevertheless, thanks to LD, a dependence exists between the non observed causal factor and a genetic marker nearby the former. On the other hand, by definition, a dependence exists between the causal factor and the disease of interest. Therefore, it is likely that a dependence will be detected between the nearby genetic marker and the disease.

In the human genome, the HapMap project confirmed evidence of the linkage disequilibrium, this latent structure organized in the so-called "haplotype blocks" [8]. Therein, regions showing high dependences between contiguous markers (blocks) alternate with shorter regions characterized by low statistical dependences. In general, LD exhibited among physically close loci is stronger than LD between SNPs that are farther apart. In other words, LD decays with distance.

However, standard GWASs do not fully exploit LD. Some authors proposed to test combinations of SNPs - haplotype blocks - against the disease, rather than merely each SNP against the disease: this is the principle of multilocus approaches. First, if the causal SNP has low frequency and is not in high LD with any one of the genotyped SNPs, then the multilocus test will tend to be more powerful. Besides, the advantage to the GWAS is that the LD is likely to reveal an excess of haplotype sharing around a causative locus, amongst cases. Third, testing haplotypes instead of SNPs is a way to implement data dimension reduction. In this context, fine LD modeling at genome scale is required.

Few works have focused on LD modeling at genome scale, which is a challenging task. The proposals of [9,10] both rely on the use of Markov random fields, a popular kind of probabilistic graphical models. Two scalable models designed for the specific purpose of multilocus GWASs have been described by [1,11]. The approach in [11] relies on a variable length Markov chain (VLMC), a Markov model where the size of the memory conditioning the prediction of the variant at a given location is flexible. The model obtained is a graph where each path from the root to a given node represents an haplotype shared by individuals in the population. In this graph, edges converging in the same child node delineate a cluster of haplotypes relevant for association testing. In contrast with this block-based method, the works in [1] seek to subsume clusters of SNPs through latent variables. SNPs within the same cluster are not necessarily contiguous. Such latent variables are intented to be tested against the disease. Both methods account for the fuzzy nature of LD since block boundaries are not accurately defined over the genome. However, being blocked-based, the method in [11] cannot take into account long-range dependences. Moreover, LD is intrinsically hierarchical, with clusters of SNPs recursively structured in clusters of lower and lower correlated SNPs. To attempt a faithful representation of LD upstream of a GWAS, hierarchical clustering is one of the key ingredients of the learning algorithm of the Bayesian model used in [1]. Since clustering is central to learning the model in [1], namely the forest of latent tree models (FLTM), this paper analyses the impact of the choice of the clustering method in a GWAS context.

## 2    Objectives and Structure of the Paper

In the remainder of this paper, data partitioning - or clustering - denotes the generation of a set of non overlapping clusters. Such a task is NP-complete [12]. Though, choosing a clustering method to learn an FLTM must comply with the scalability goal. This paper compares the native clustering method used in [1] ($CAST_{bin}$) with a relaxed version ($CAST_{real}$) and another clustering method (DBSCAN). In this framework, two aims of the paper are to evaluate whether FLTM learning is robust to the choice of the clustering method and how close a clustering method approximates the underlying biological reality. However, there exists no generic method dedicated to the comparison of the two partitions yielded by any two clustering methods [13]. To fulfill the first goal, a protocol is used that relies on assessing how much two partitions agree. The second objective is met by applying the previous protocol to compare each clustering method to a reference partition supposed to be close to biological reality. The Haploview software program is the tool chosen to derive such a reference partition. Focusing on the data dimension reduction aspect, a third objective of the paper is to analyze the impact of the choice of the clustering method on data subsumption quality. By construction, an FLTM-based GWAS processes data subsumed through latent variables, to hopefully pinpoint the interesting regions of a genome without testing each SNP for association. Thus, the third objective of this paper is to assess whether the choice of the clustering method impacts the risk of missing significant association results through subsumption. FLTM-driven GWASs are run to study this impact. In this paper, we will focus on the GWAS data set relative to the Crohn's disease.

The remainder of the paper is organized in five sections. Section 3 first offers a brief introduction to Bayesian networks, the kind of probabilistic graphical models FLTM is based upon. Then Sect. 3 provides a broad brush description of the FLTM learning algorithm together with a sketch of a GWAS strategy based on FLTM. Section 4 describes the native clustering method used in FLTM ($CAST_{bin}$) and its relaxed version ($CAST_{real}$); it then motivates the choice of the alternative clustering method (DBSCAN) plugged in the FLTM learning algorithm and depicts the principle of the DBSCAN method. Then, Sect. 5 explains the design of the protocols and methods used in our work. First, we discuss the protocol used to assess how much two partitions agree. Second, we justify the

use of the Haploview software program to derive the reference partition, suppos-
edly the closest representation of the underlying reality. In Sect. 6, we describe
the Crohn's disease GWAS data used in our study. Section 7 is devoted to the
presentation and discussion of the results observed.

# 3    Framework and FLTM Model

The FLTM model is a tree-structured Bayesian network (BN). Therefore this
section first briefly introduces Bayesian networks, to further focus on the FLTM
model. The principle of the FLTM learning algorithm is then presented. Finally,
the principle for a multilocus GWAS based on the FLTM model is sketched.

## 3.1    A Brief Reminder About Probabilistic Graphical Models

When probabilistic graphical models are learnt from scratch, one has to learn
their two fundamental components from a data matrix. In this matrix, the lines
correspond to the observations and the columns correspond to the variables $X_i$
($1 \leq i \leq n$). For example, in the case of genetical data, the observations are the
individuals (cases and controls) in the population studied, and the variables are
the SNPs. The qualitative component of a BN is a graph where the variables are
represented as nodes. The connections between the nodes represent the direct
dependences between the variables. More specifically, the qualitative component
of a BN is a directed acyclic graph. The quantitative component of a BN is
a collection of probability distributions, denoted as "the parameters $\theta$". If the
variable $X_i$ has no parent in the graph, then $\theta_i$ is merely an *a priori* distribution ($\theta_i = \mathbb{P}(X_i)$). If the variable $X_i$ has a set of parents $Pa_{X_i}$, then $\theta_i$ is the
conditional distribution $\theta_i = \mathbb{P}(X_i \mid Pa_{X_i})$. In particular, Bayesian networks
offer a practicable framework: exploiting the network structure, this framework
allows to compute the joint probability of the variables, $\mathbb{P}(X)$, as a product of
low-dimensional functions.

It may happen that the data observed is thought to embed a latent structure,
depicted through latent variables and their connections in the learnt BN. In this
case, learning the Bayesian network encompasses the task of inferring the latent
variables, and their connections within the BN.

The FLTM model is a forest of latent tree models (LTMs). Figure 1(a) shows
that an LTM is characterized by a hierarchical structure organized in layers. The
first layer is composed of the observed variables. The other layers are composed
of latent variables. The learning algorithm of the FLTM (see Fig. 1(b)) relies on
the simplest LTM that may be described, the latent class model (LCM). A latent
class model connects a single latent variable to child variables; no connections
are allowed between the latter (see Fig. 1(c)).

## 3.2    Outline of the FLTM Learning Algorithm

Learning a BN is a hard task that consists in inferring both the graph structure and the parameters. For example, there exists respectively 25, 29 × 10³

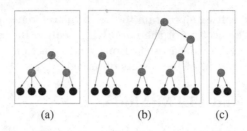

(a)                    (b)                    (c)

**Fig. 1.** (a) Latent tree model (LTM). (b) Forest of latent tree models (FLTM). (c) Latent class model. Observed and latent variables are represented in dark and light color shades, respectively.

and $4.2 \times 10^{18}$ possible structures for a BN with 3, 5 and 10 observed variables [14]. Learning a BN with latent variables is far more complicated. First, one does not even know how many latent variables have to be inferred; it has been shown that the number of different possible LTMs that may be inferred from $n$ observed variables is upper bounded by $2^{3n^2}$ [15]. Second, in a BN without latent variables, the parameters are estimated to maximize the likelihood, that is the probability of the (observed) data given the parameters. In contrast to this rapid algorithm, a slow procedure has to be employed for BNs with latent variables, the expectation-maximization algorithm dedicated to learn parameters in the case of missing data. Prior knowledge (the hierarchical LD structure) is used by the specific procedure described in [1], to provide a scalable learning algorithm. Figure 2 depicts the principle of this iterative algorithm, based on an ascending hierarchical clustering procedure.

In the case of LD modeling, the observed variables are the SNPs. The cardinality of these observed variables is equal to 3, which codes for minor homozygosity, heterozygosity and major homozygosity. The first iteration starts with the partitioning of the observed variables into non overlapping clusters of pairwise highly dependent SNPs. No two variables are allowed in the same cluster if their physical distance on the genome is above a given threshold, $\delta$. For each cluster, an LCM is constructed whose child variables are the variables in the cluster and whose latent variable is created. The approach of [1] considers discrete latent variables whose cardinalities may be different from one another; a heuristic is used to determine the specific cardinality of each latent variable. This specific cardinality is computed as an affine function of the number of variables in the cluster. Then, the parameters of each LCM are estimated through the standard expectation-maximization procedure. Knowing the parameters of each LCM further allows to impute the data corresponding to its latent variable. For each observation $x^j$ (i.e. the $j^{th}$ individual), the value for the latent variable $L$ is drawn from the probabilistic distribution $\mathbb{P}(L \mid \mathbf{x}^j)$ derived from the LCM's parameters. Such imputed data are used by a validation step; this step relies on a normalized mutual information criterion to examine whether each novel latent variable is sufficiently informative to subsume its child variables. The data is updated with the validated latent variables replacing their child variables.

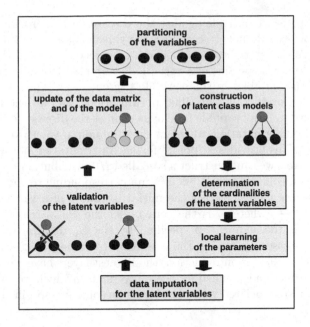

**Fig. 2.** Sketch of the iterative FLTM learning algorithm.

The validated latent variables can thus be considered as observed variables and an iteration begins anew.

It is now easy to see how an iterative ascending procedure constructs the whole forest: after iteration $i$ has inferred and imputed novel latent variables, iteration $i + 1$ partitions into clusters the set composed of these newly created variables and of the remaining variables. These remaining variables are the previous validated latent variables and the observed variables that could never be incorporated in a valid cluster so far. This ascending process is iterated until no valid cluster can be identified, or until a single cluster of maximal size is obtained. Among other parameters of the learning procedure, the clustering procedure plugged in the algorithm is likely to impact the quality of the LD modeling, and therefore the quality of the GWAS performed downstream.

### 3.3  Performing a GWAS Guided by the FLTM Model

Central to the use of the FLTM model for a GWAS purpose are the request for data dimension reduction and the motivation for a multilocus strategy. In the study described here, we have implemented a multilocus GWAS strategy as follows: in the lowest layers, we traverse the forest top down, following a best-first search strategy which only tests all child nodes for the nodes whose association significances (i.e. p-values) are below a threshold. The way to compute this threshold is specific to the layer the variable belongs to. These child nodes are selected in turn with respect to the appropriate threshold. The standard Chi$^2$

test is applied to test a variable against the disease and to provide a pointwise p-value. We now explain how to compute the thresholds for the variables in the lowest layers. To test statistical significance, we consider a global threshold, say $\alpha = 5\%$. The pointwise p-value of a variable is corrected based on permutations applied on all the variables in this variable's latent layer. Thus, the correction is layer specific. Considering, say, 1000 permutations, and all the layers in the FLTM model, we test any variable in a layer against the disease, for each permutation. For a given permutation, we select the minimum p-value over all variables in a given layer. This provides a "point" for each permutation, for a given layer. Thus, for this layer, we can construct a so-called $H0$ distribution (no association with the disease), collecting the 1000 minima corresponding to the 1000 permutations. Finally, we merely correct the initial p-value $p$ of a variable in layer $\ell$ into $p_{corrected} = \frac{a_p}{a_t}$, where $a_t$ is the aera under the $H_0$ distribution curve for layer $\ell$, and $a_p$ is the aera $x \leq p$ specified by the $H_0$ distribution. Statistical significance is assessed by testing the condition (p-value$_{corrected} \leq \alpha$). On the other hand, due to dimension reduction, the highest layers have a low number of variables. No correction is applied for these layers, from which we systematically select the top most associated variables, based on some pre-specified $\beta$ threshold (e.g. $\beta = 10\%$).

# 4 The CAST and DBSCAN Partitioning Methods

Performing an optimal clustering is NP-hard [12]. Therefore, heuristics must be designed instead. In this paper, we focus on two partitioning methods, CAST and DBSCAN, to study how they impact LD modeling and a further downstream GWAS analysis.

Since we address high-dimensional data, we could not envisage the use of ascending hierarchical clustering (AHC), whose complexity scales in $O(n^3)$ where $n$ is the number of objects to be asssigned to clusters. CAST and DBSCAN are well known partitioning methods whose complexity is lower than AHC's. Moreover, in contrast to AHC and k-means, another well known partitioning method, CAST and DBSCAN do not request the tuning of the number of clusters.

Partitioning objects into clusters relies on pairwise distances (alternatively pairwise similarities). Storing a pairwise similarity matrix at the genome scale is intractable. Thus, following [1], we acknowledge a physical constraint, $\delta$, expressed in $kbp$ (kilobase pairs), in both implementations of the CAST and DBSCAN methods. This constraint $\delta$ represents the physical distance on the genome beyond which two objects (in our case two variables) are not allowed in the same cluster. Additional calculus is required to estimate the distance between two variables one of which at least is a latent variable.

## 4.1 The CAST Partitioning Method

The CAST (Cluster Affinity Search Technique) algorithm was proposed in [16] and is depicted in [17]. Its theoretical runtime complexity scales in

$O(n^2 \ (log(n))^c)$, and its empirical complexity allows to handle high-dimensional data. CAST is the native clustering method used in the FLTM learning algorithm depicted in [1].

To decide cluster membership, CAST relies on an affinity measure. The affinity measure of an object is defined as the average of the similarity measures between objects in a given cluster and this former object outside the cluster. To grow a cluster, the CAST algorithm successively adds the object with greatest affinity to this cluster as long as this maximum affinity satisfies a threshold constraint (see Algorithm 1, lines 15 to 24). When the maximum affinity drops

---

**Algorithm 1.** CAST clustering algorithm $(M, \ n, \ \tau)$.

---

**INPUT:** M, an n-by-n similarity matrix, pairwise-comparing $n$ objects
$\quad \quad \tau$, an affinity threshold.

**OUTPUT:** $\mathcal{C}$, a partition of the $n$ objects into clusters.

1: /* initialization */
2: $\mathcal{C} \leftarrow \emptyset$ /* collection of closed clusters */
3: $\mathcal{U} \leftarrow \{1, \ 2, \cdots n\}$ /* objects not yet assigned to a closed cluster */

4: **while** $(\mathcal{U} \neq \emptyset)$
5: $\quad C \leftarrow \emptyset$ /* start a new cluster */
6: $\quad$ reset all affinity measures $aff$ to 0
7: $\quad change\_occurs \leftarrow yes$
8: $\quad$ **while** $(change\_occurs)$
9: $\quad \quad change\_occurs \leftarrow no$
10: $\quad \quad add(C, change\_occurs)$
11: $\quad \quad remove(C, change\_occurs)$
12: $\quad$ **end while**
13: $\quad \mathcal{C} \leftarrow \mathcal{C} \cup \{C\}$ /* close the current cluster */
14: **end while**

15: **procedure add**$(C, change\_occurs)$
16: $\quad$ **while** $(\max\limits_{x \in \mathcal{U}} (aff(x)) \geq \tau \mid C \mid)$
17: $\quad \quad o \leftarrow \underset{o \in \mathcal{U}}{argmax}(aff(o))$ /* select an object with maximum affinity */
18: $\quad \quad C \leftarrow C \cup \{o\}; \ \mathcal{U} \leftarrow \mathcal{U} \setminus \{o\}; \ change\_occurs \leftarrow yes$

19: $\quad \quad$ /* update affinity */
20: $\quad \quad$ **for each** $x \in \mathcal{U} \cup C$
21: $\quad \quad \quad aff(x) \leftarrow aff(x) + M(x, o)$
22: $\quad \quad$ **end for**
23: $\quad$ **end while**
24: **end procedure**

25: **procedure remove**$(C, change\_occurs)$
26: $\quad o \leftarrow \underset{o \in \mathcal{U}}{argmin}(aff(o))$ /* select an object with minimum affinity */
27: $\quad C \leftarrow C \setminus \{o\}; \ \mathcal{U} \leftarrow \mathcal{U} \cup \{o\}; \ change\_occurs \leftarrow yes$

28: $\quad$ /* update affinity */
29: $\quad$ **for each** $x \in \mathcal{U} \cup C$
30: $\quad \quad aff(x) \leftarrow aff(x) - M(x, o)$
31: $\quad$ **end for**
32: **end procedure**

---

below the threshold, CAST removes the object with the minimum affinity with respect to the cluster (see lines 25 to 32). Additions and removals are operated as long as the current cluster undergoes modifications (see lines 10 and 11). Finally, the cluster is closed (see line 13).

In the implementation of CAST adapted to FLTM learning, the binary similarity measure is assessed as the thresholded mutual information (MI). The mutual information between variables $X_1$ and $X_2$ is computed as: $MI(X_1, X_2) = H(X_1) + H(X_2) - H(X_1, X_2)$ where the entropy for the discrete variable $X$ defined on domain $Dom(X)$ (e.g. $\{0, 1, 2, 3, 4\}$) is $H(X) = -\sum_{c \in Dom(X)} \mathbb{P}(X = c) \log_2 \mathbb{P}(X = c)$, and the joint entropy of two variables $X_1$ and $X_2$ is $H(X_1, X_2) = -\sum_{c_1 \in Dom(X_1), c_2 \in Dom(X_2)} \mathbb{P}(X_1 = c_1, X_2 = c_2) \log_2 \mathbb{P}(X_1 = c_1, X_2 = c_2)$. A parameter $q_{pairwise}$ (e.g. 50 %) allows to compute the MI quantile (e.g. median) over the pairs of variables whose physical distance is below $\delta$. This quantile allows to assign a binary similarity (0/1), as in the native FLTM learning algorithm. We also consider the unthresholded version. These two CAST versions are denoted $CAST_{bin}$ and $CAST_{real}$.

## 4.2   The DBSCAN Partitioning Method

The DBSCAN (Density-Based Spatial Clustering of Applications with Noise) algorithm was proposed in [18]. Its theoretical runtime complexity is $O(n^2)$, where $n$ is the number of objects to be assigned to clusters. However, its empirical complexity is known to be lower. The DBSCAN principle lies in constructing clusters from the estimated density distribution of the objects to be clustered. This method requires two parameters: $R$, the maximum radius of the neighborhood to be considered, and $N_{min}$, the minimum number of neighbors needed for a cluster. DBSCAN exploits the fact that an object in a cluster also has its R-neighborhood in this cluster. The R-neighborhood of an object $o$ is merely the set of objects whose distance from $o$ is less than or equal to $R$ (see Algorithm 2, line 4). If this R-neighborhood is dense (i.e. its size is greater than or equal to $N_{min}$), then, $o$'s R-neighborhood is grown through the addition of the proper R-neighborhoods of $o$'neighbors, provided that these neighborhoods are themselves dense (see lines 7 and 16). In the end, the grown neighborhood augmented with $o$ represents a new cluster. If the R-neighborhood of an object is not sufficiently dense, this object is labeled as noise (see line 6). This object might later be found in the sufficiently dense R-neighborhood of a further visited object, and hence be assigned to the cluster constructed from this latter point (see line 16). Then, a new unvisited point is retrieved and processed, leading to the discovery of a further cluster or to the labeling as noise.

We have chosen DBSCAN as it is resistant to noise and can handle clusters of different shapes and sizes. Notably, DBSCAN is known to be able to identify a cluster embedded in another cluster. On the genome line, long-range LD corresponds to this situation.

---

**Algorithm 2.** DBSCAN clustering algorithm $(X, R, N_{min})$.

---

**INPUT: X**, the data set consisting in objects
    $R$, the maximum radius of the neighborhood to be considered,
    $N_{min}$, the minimum number of neighbors needed for a cluster.

**OUTPUT:** $\mathcal{C}$, a partition of the $n$ objects into clusters.

1: $\mathcal{C} \leftarrow \emptyset$ /* collection of clusters */

2: **for each** *unvisited object* $o \in X$
3:    $mark(o) \leftarrow visited$
4:    $Neigh_o \leftarrow neighborhood(o, R)$
5:   **if** $(\mid Neigh_o \mid < N_{min})$
6:       **then** $mark(o) \leftarrow noise$
7:       **else** $C \leftarrow expandCluster(o,\ Neigh_o,\ R,\ N_{min});\ \mathcal{C} \leftarrow \mathcal{C} \cup C$
8:   **end if**
9: **end for**

10:  **function expandCluster**$(o,\ Neigh_o,\ R,\ N_{min})$
11:  $C \leftarrow \{o\}$
12:  **for each** *object* $x \in Neigh_o$
13:     **if** $(x\ is\ unvisited)$
14:         $mark(x) \leftarrow visited$
15:         $Neigh_x \leftarrow neighborhood(x, R)$
16:         **if** $(\mid Neigh_x \mid \geq N_{min})\ Neigh_o \leftarrow Neigh_o \cup Neigh_x$ **end if**
17:     **end if**

18:     **if** $(x\ belongs\ to\ no\ cluster)$
19:         $C \leftarrow C \cup \{x\}$
20:     **end if**
21:  **end for**

22:   **return** $C$
23: **end function**

---

# 5   Methods

In this section, we first present the protocol used to evaluate how much two partitions agree when focusing on the top most associated SNPs found by a GWAS. Then, we motivate how we derived the so-called reference partition (to be further defined). This methodological section ends with the presentation of the protocol used to compare the impact of the choice of the partitioning method on the subsequent GWAS. For this purpose, we rely on GWAS results published in the literature.

## 5.1   Comparing Two Partitions

To cope with the genome scale, we were compelled to select a simple method: we focused on the comparison of the partitions respectively obtained for the first layer (SNPs) by two partitioning methods, and we examined how the top most associated SNPs identified by a GWAS are distributed among the clusters.

    The methods dedicated to the comparison of two partitions may be categorized into three main groups [19]. Two groups attempt to map a partition onto the other, either from set matching functions, or from information theory-centered methods. The third category relies on counting for how many pairs of elements two partitions agree or disagree. The FLTM-driven GWAS strategy is a multilocus strategy by definition. In this multilocus GWAS framework, it is relevant to analyze pair agreement between two partitioning methods, for a selection of top most associated SNPs. A counting method fits well this purpose of focusing on a subset of SNPs.

Given two partitions over the same set of objects, and a pair of objects (in our case, a pair of variables), an agreement means that the two partitions both group the two variables in a cluster or both assign two different clusters to the two variables. Consequently, a disagreement means that the variables belong to a cluster according to one partition, but belong to two different clusters according to the other partition. Given two partitions $P_1$ and $P_2$, let

- $N_{11}$, the number of pairs both partitions assign to one cluster,
- $N_{00}$, the number of pairs both partitions assign to different clusters,
- $N_{10}$, the number of pairs kept in the same cluster by $P_1$ but splitted by $P_2$,
- $N_{01}$, the symmetric case of the latter.

From here, a large set of comparison measures is available. We selected three measures to perform the following comparisons: $CAST_{bin}$ versus $CAST_{real}$, $CAST_{bin}$ versus DBSCAN, $CAST_{real}$ versus DBSCAN, and each of the three methods $CAST_{bin}$, $CAST_{real}$ and DBSCAN versus the reference partition (to be defined in Sect. 5.2). The comparison measures selected are:

- the Rand index [20]:

$$RI = \frac{N_{11} + N_{00}}{N_{11} + N_{00} + N_{10} + N_{01}} \tag{1}$$

for which we used instead an adjusted corrected-for-chance version ($ARI = \frac{RI - expected\ RI}{maximum\ RI - expected\ RI}$) (for the detailed description, see [13]);
- the Mirkin distance [21]:

$$MI = \frac{S_{P_1} + S_{P_2} - 2S_{P_1 P_2}}{n^2} \tag{2}$$

with

$$S_{P_j} = \sum_{cluster_i \in P_j} |\ cluster_i\ |^2, j = 1, 2$$

$$S_{P_1 P_2} = \sum_{cluster_i \in P_1,\ cluster_j \in P_2} |\ cluster_i\ |\ |\ cluster_j\ |,$$

and $n$ the number of objects to be assigned to clusters and $|\ S\ |$ the size of set $S$;
- the Fowlkes-Mallows index [22]:

$$FM = \sqrt{\frac{N_{11}}{N_{11} + N_{10}} \cdot \frac{N_{11}}{N_{11} + N_{01}}}. \tag{3}$$

Our concern is to study the relevance of the partitioning method used in the FLTM model to represent the LD, and further allow an efficient multilocus GWAS. For this purpose, not all pairs of SNPs are interesting to assess the agreement between two partitioning methods: we are only interested in pairs of SNPs selected among the top SNPs found most associated with the studied disease. As using the top SNPs selected by an FLTM-based GWAS would introduce a bias in our comparisons, the standard tool PLINK was used to identify these top SNPs [23] (http://pngu.mgh.harvard.edu/purcell/plink/).

## 5.2 Deriving the Reference Partition

The reference partition intends to be the closest representation of the underlying reality, that is the haplotype blocks. We used the Haploview software program [24] for this purpose. This application allows to select commonly used block definitions to partition the genome into regions of strong LD [24,25]. As this block generation is dedicated to handle genetical data, Haploview can only be used for the first layer (observed variables). This reason explains why the partitioning method of the Haploview application has not been plugged in the FLTM learning algorithm.

## 5.3 GWAS-oriented Analysis of a Partition

We have performed three FLTM-driven GWASs, using the method described in Sect. 3.3, and respectively relying on the three partitioning methods: $CAST_{bin}$, $CAST_{real}$ and DBSCAN. To pursue our investigations relative to the impact of the choice of the partitioning method, we have analyzed how published associated SNPs are distributed within the clusters of the first layer, according to the three partitioning methods used. For this purpose, it was necessary to consider a disease for which published causal SNPs are available. In this study, we relied on the results published on the Crohn's disease, one of the most studied genetic diseases.

# 6  Crohn'S Disease GWAS Data

The Crohn's disease data set we used is made available by the WTCCC Consortium (http://www.wtccc.org.uk/); it consists of 5009 individuals genotyped using the Affymetrix GeneChip 500 K Mapping Array Set (3004 controls, 2005 cases). We performed the same data quality control as the WTCCC. We excluded individuals, using exactly the same criteria as the WTCCC ([26], page 26) (e.g. individuals with more than 3 % missing data across all SNPs; individuals sharing more than 86 % of identity with other ones). The rules to exclude SNPs were also modelled after those of the WTCCC (e.g. missing rate over 5 %; if MAF (minor allele frequency) under 5 %, missing rate threshold decreased to 1 %) ([26], page 27).

In this paper, we focus on chromosome 2, known to harbour SNPs with susceptibility towards Crohn's disease. The initial WTCCC data set describes 41400 SNPs. After the quality control step, our data consisted of 38730 SNPs.

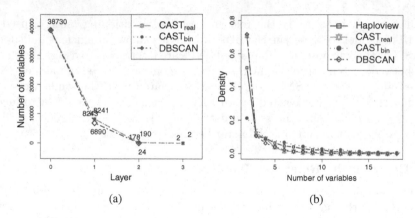

**Fig. 3.** Impact of the choice of the partitioning methods $CAST_{bin}$, $CAST_{real}$ and DBSCAN on the structure of the FLTM model. (a) Impact on the number of variables per layer. (b) Impact on the sizes of the clusters for the first layer (observed layer).

## 7    Results and Discussion

The parameter $t_{CAST}$ (see details in [17]) specific to the CAST method, whatever the version (*bin* or *real*), was empirically set to 0.50. The parameter $q_{pairwise}$ specific to the $CAST_{bin}$ clustering method was empirically chosen to be 50%. The $N_{min}$ and $R$ parameters specific to DBSCAN were tuned to 2 and 0.2 respectively. The FLTM learning algorithm requires the setting of six parameters. We systematically evaluated the coefficients of the affine function used to determine the cardinality of each latent variable, $\ell_1$ and $\ell_2$, in $[0.2, 0.3, 0.4, 0.5] \times [1, 2]$. We observed no differences between the eight settings, with regard to the sizes and contents of the clusters. Thus, $\ell_1$ and $\ell_2$ were set 0.5 and 1. Following [1], we fixed the maximum cardinality as 20, the physical distance constraint $\delta$ as 45 *kbp* and the number of restarts of the stochastic expectation-maximization procedure as 10. The threshold for the quality control of the candidate latent variables was set to a low value, 0.01. The GWAS thresholds $\alpha$ and $\beta$ were fixed to 5% and 10%. The study was conducted using a 3.3 GHz processor. We had to adapt the generic versions of the CAST and DBSCAN algorithms, to store a sparse similarity matrix instead of a pairwise similarity matrix (see Sect. 4).

### 7.1    FLTM Architectures

On average, the running time observed for FLTM learning with each clustering method is in the order of 60 hours. A closer examination shows that clustering and other operations only required at most 1 min for each layer, and that practically all the running time was spent in the expectation-maximization procedure (see Sect. 3.2). Moreover, it is likely that the presence of a few clusters of large size (size up to 50) severely increases running times for the expectation-maximization procedure.

We first analyze the impact of the partitioning method on the structure of the FLTM model constructed prior to a GWAS. Figure 3(a) compares the impacts of the three partitioning methods on the data dimension reduction. We observe that for any layer, the total number of latent variables created using $CAST_{real}$ is always greater than that created using DBSCAN. Moreover, layer 3 does not exist for DBSCAN whereas it exists for $CAST_{bin}$ and $CAST_{real}$. Indeed, for DBSCAN, no more variables can be grouped in layer 2: all candidate clusters are singletons. The numbers of variables in layers 1 and 3 are either very close or similar between $CAST_{bin}$ and $CAST_{real}$. Again, among the three methods, the numbers of variables in layer 2 are the closest for $CAST_{bin}$ and $CAST_{real}$.

Figure 3(b) provides the histogram for the sizes of the clusters in the first (observed) layer, for each of the three partitioning methods, together with the histogram of the reference Haploview partitioning. It has to be mentioned that, for reasons of presentation, the histograms have been truncated. Very few clusters of large sizes are observed: the maximum sizes observed are 18, 45 and 50 for $CAST_{real}$, DBSCAN and $CAST_{bin}$, respectively. Such clusters would normally appear far apart on the right section of Fig. 3 (b).

First, we observe that from size 3, the $CAST_{bin}$ curve is slightly above the $CAST_{real}$ and DBSCAN curves. Besides, the latter curves are nearly superimposed. Finally, we note that from size 3, the curve relative to the reference partitioning is located slightly below that of $CAST_{bin}$, on the one hand, and slightly above the quasi superimposed curves of $CAST_{real}$ and DBSCAN, on the other hand.

Therefore, the general conclusion to draw for this section is the propensity for DBSCAN to produce a lower number of variables than $CAST_{bin}$ and $CAST_{real}$, but with no clear impact on the differences between the cluster size histograms.

## 7.2   Comparison of the Partitioning Methods in a GWAS Context

In a GWAS context, we wish to focus in priority on pairs of SNPs selected among the top SNPs found most associated with the studied disease. The standard tool PLINK was used to identify these top SNPs [23] (http://pngu.mgh.harvard.edu/purcell/plink/). Relying on PLINK, we performed a single-SNP GWAS on the WTCCC data set relative to chromosome 2. The association test used was the $Chi^2$. We have extended the agreement analysis of two partitions to embedded sets of associated SNPs, increasing the size of the set of top associated SNPs up to 1000.

Figure 4(a) and (b) compare the partioning methods $CAST_{bin}$, $CAST_{real}$ and DBSCAN together with Haploview, following two of the three comparison criteria described in Sect. 5.1.

The adjusted Rand index is all the higher as the agreement between two partitioning methods is high. Thus, we observe that $CAST_{bin}$ does not agree with the reference (Haploview) partitioning as well as $CAST_{real}$ and DBSCAN. This specificity of $CAST_{bin}$ is explained by the conversion of real mutual information values into binary values (see the role of parameter $q_{pairwise}$ in Sect. 4). This discretization therefore entails slightly larger cluster sizes for $CAST_{bin}$, as seen in Sect. 7.1.

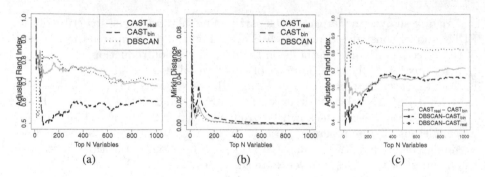

**Fig. 4.** Agreement of two partitioning methods, in a GWAS context. (a) and (b) Agreement of a partitioning method with the reference block partitioning method used by Haploview. Comparison for the partitioning methods $CAST_{bin}$, $CAST_{real}$ and DBSCAN. Impact of the number of top SNPs considered on the agreement. The top SNPs considered are those found most significantly associated by a standard single-SNP GWAS. (a) Adjusted Rand index. (b) Mirkin distance. (c) Pairwise comparison of the partitioning methods $CAST_{bin}$, $CAST_{real}$ and DBSCAN. Impact of the number of top SNPs considered on the agreement. Adjusted Rand index.

On the left section of Fig. 4 (a), the index is computed from few top SNPs. We observe that $CAST_{bin}$ and $CAST_{real}$ show a high Rand index in contrast to DBSCAN. However, in a GWAS context, we do not wish to examine only, say, the 20 top significantly associated SNPs. Thus, the most relevant section to focus on is around 50-100 top SNPs. In this latter section of Fig. 4 (a), we observe that the $CAST_{real}$ and DBSCAN curves are comparatively close and clearly located higher than the $CAST_{bin}$ curve. This trend is observed up to the 1000 top most associated SNPs.

In Fig. 4(b), a low Mirkin distance indicates a high agreement between two partitioning methods. The observations in Fig. 4(b) confirm that $CAST_{bin}$'s agreement with Haploview clustering is always worse than the other two methods'. We have not shown the results for the Fowlkes-Mallows index as the curves obtained are quasi superimposable with those plotted for the adjusted Rand index.

The first general conclusion to draw from this first series of agreement comparisons on the Crohn's disease data set is that DBSCAN and $CAST_{real}$ show a high level agreement with Haploview partitioning, both being quite clearly better than $CAST_{bin}$.

Figure 4(c) displays the results for pairwise comparisons: $CAST_{real}$ versus $CAST_{bin}$, DBSCAN versus $CAST_{bin}$ and DBSCAN versus $CAST_{real}$. According to the adjusted Rand index, DBSCAN and $CAST_{real}$ show a high agreement. Given our previous observations, we expected that $CAST_{bin}$ and $CAST_{real}$ would show a low level agreement, which is confirmed. DBSCAN and $CAST_{bin}$ yield partitions that almost always disagree more than for the two former couples of partitioning methods. This trend is confirmed with the Mirkin distance and the Fowlkes-Mallows index (results not shown).

As a second general conclusion of this section, we cross-confirm one of our previous observations: DBSCAN and $CAST_{real}$ each show a high agreement with Haploview. This fact is therefore also reflected by a high agreement between DBSCAN and $CAST_{real}$.

## 7.3   FLTM-driven GWASs

In Fig. 5, the comparison of plots (a) to (c) and plot (d) shows how the dimension reduction allows to pinpoint the potentially most interesting regions on the genome. Thus, "sparse" association profiles are produced, as opposed to the dense output of the standard single-SNP GWAS.

The two putative causal SNPs located on chromosome 2 respectively reported in the WTCCC study [26] and in [27] are identified by the three FLTM-driven GWASs. Given that we used the same data set as in [26], one of the two results was expected. However, this result was not guaranteed, because of the data dimension reduction and of the subsumption involved in an FLTM-driven GWAS. Besides, it must be highlighted that the study in [27] analyzed 8059 individuals (3230 cases and 4829 controls), whereas the WTCCC data set describes a population of size 5009. Table 1 shows that $CAST_{bin}$ and $CAST_{real}$ capture exactly the same four highly associated SNPs through the latent variables $L_1$ and $L_2$, belonging to layer 1. These variables are the right-most latent variables in layer 1, on the plots (a) and (b) of Fig. 5. The virtual location of a latent variable is computed as the average of the locations of its child variables.

**Table 1.** Analysis of the latent variables in layer 1 found significantly associated with Crohn's disease, by the three FLTM-driven GWASs with plug-in $CAST_{bin}$, $CAST_{real}$ and DBSCAN, respectively. For each clustering method, the latent variable is described on the first line. On the following lines, the highly associated SNPs subsumed by this latent variable are depicted. The identifier of each SNP is provided (rsXXXXXXX). The • character highlights the SNPs which are common children of latent variables $L_1$ (or $L_2$) and $L_3$. * Note that the association tests used may differ between studies.

| Clustering method | Variable | Location | p-value | p-value reported in another study* |
|---|---|---|---|---|
| $CAST_{bin}$ | latent $L_1$ | 233837691    (1) | $5.86 \times 10^{-14}$ | |
| | rs6752107 | 233826187 •   (2) | $9.65 \times 10^{-14}$   (3) | |
| | rs6431654 | 233826508 •   (2) | $9.96 \times 10^{-14}$   (4) | |
| | **rs3828309** | **233845149 •**   (2) | $\mathbf{2.30 \times 10^{-13}}$   (5) | $\mathbf{2 \times 10^{-32}}$ [27] |
| | rs3792106 | 233855479   (2) | $3.70 \times 10^{-12}$ | |
| $CAST_{real}$ | latent $L_2$ | see (1) | $5.52 \times 10^{-14}$ | |
| | | see (2) | | |
| DBSCAN | latent $L_3$ | 233830355 | $6.58 \times 10^{-14}$ | |
| | **rs10210302** | **233823578** | $\mathbf{4.60 \times 10^{-14}}$ | $\mathbf{7 \times 10^{-14}}$ [26] |
| | rs6752107 | 233826187 • | see (3) | |
| | rs6431654 | 233826508 • | see (4) | |
| | **rs3828309** | **233845149 •** | see (5) | $\mathbf{2 \times 10^{-32}}$ [27] |

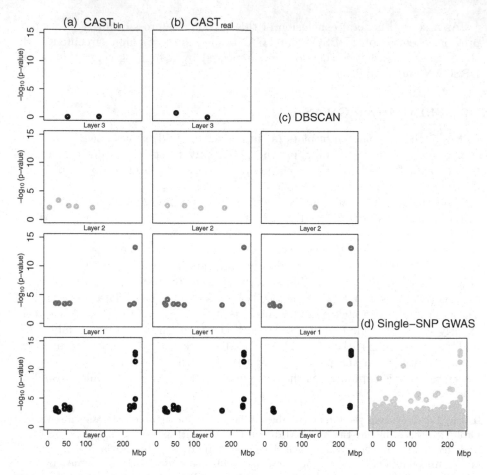

**Fig. 5.** Impact of the choice of the partitioning method on the multilocus GWAS results. For the FLTM-based GWASs ((a) to (c)), one "sparse" association profile is displayed for each layer, as not all variables in a layer are examined. The single-SNP GWAS in (d) was performed using the gold standard PLINK [23]. Its output only deals with variables in layer 0 (observed variables). All plots show initial (i.e. non corrected) p-values.

Thus, the location of $L_1$ (or $L_2$) is 233837691 bp. The p-values computed for $L_1$ and $L_2$ differ since the data imputed for these latent variables differ. For either $CAST_{bin}$ or $CAST_{real}$, the SNP published in [26] is not grouped with other SNPs into a cluster, in contrast to DBSCAN. Table 1 shows that for DBSCAN, the latent variable $L_3$ subsumes SNPs among which are the two already published putative causal SNPs. $L_1$ and $L_3$ share three highly associated SNPs, including the putative causal SNP published in [27]. The virtual location of $L_3$ is 233830355 bp. We can see that $L_1$ captures LD on a slightly wider range than $L_3$, since the regions encompassed by the former and the latter variables spread over 29292 and 21571 bp, respectively.

A more thorough analysis of the Affymetrix array indicates that the region encompassed by $L_1$, [233826187, 233855479], contains four highly associated SNPs, interspersed with three non associated SNPs. Similarly, the interval covered by $L_3$, [233823578, 233845149], contains eight SNPs, including four non associated SNPs. Clearly, among the four highly associated SNPs pinpointed by each of $L_1$ and $L_3$, respectively three and two SNPs are highly associated with the disease because they are in LD with a putative causal SNP (see Table 1). However, not every SNP close to a putative causal SNP has been incorporated in the cluster subsumed by $L_1$, $L_2$ or $L_3$. To confirm the relevance of the clustering performed, an in-depth examination shows that these former close SNPs that are not in LD with putative causal SNPs are found poorly associated with the disease (in the order of $10^{-1}$). Importantly, even the SNP flanking on the left the causal putative SNP published in [27] and having a p-value equal to $1.32 \times 10^{-5}$, was not retained in $L_1$ or $L_3$'s cluster. This observation shows that a fine-grain clustering is achieved for each of the three partitioning methods.

Therefore, a first remarkable result is that the subsumption process does not hinder the informativeness of $L_1$, $L_2$ and $L_3$: $L_1$, $L_2$ and $L_3$ are still found highly associated with the disease ($5.86 \times 10^{-14}$, $5.52 \times 10^{-14}$, $6.58 \times 10^{-14}$ respectively).

Moreover, a second remarkable result is obtained. The standard GWAS (Fig. 5(d)) identifies two SNPs with a high statistical significance (rs13394205, located at around 18 Mbp (17849508), and rs11887827, located at around 81 Mbp (81519665)). The p-values of these two SNPs are respectively $2.28 \times 10^{-9}$ and $1.81 \times 10^{-11}$. None of these SNPs were reported in former studies [26,27], which identified them as false positives. In the layers 0 of the plots (a) to (c) of Fig. 5, none of these two SNPs either appears. The reason lies in that during the top down traversal of the FLTM, the parents of these SNPs are detected as not significantly associated with the studied disease. Consequently, the descendants of these latent variables are not examined (and not displayed in the sparse outputs). Therefore, the FLTM-driven GWAS strategy exerts an efficient control of the number of false positives. Furthermore, all layers potentially allow to exert such a control, with a pruning effect on the forest structure guiding the GWAS.

In the context of this study, the general conclusion to draw from this section is that the three FLTM-driven GWASs capture the SNPs reported associated by two other studies and correctly detect false positive associations. Second, the differences reported in Sects. 7.1 and 7.2 between $CAST_{bin}$ and the two other clustering methods do not impact the quality of the corresponding FLTM-driven GWAS.

# 8    Conclusion and Future Work

In this paper, we have analyzed the influence of the choice of the clustering method to be plugged in the FLTM learning algorithm, for the purpose of a GWAS application. We have started examining this impact focusing on two scalable clustering methods, adding a relaxed variant of one of them. For this purpose, a methodological framework has been designed, which allows to compare the three clustering methods according to the following viewpoints: (1) effective ability to split or group the top associated SNPs, according to the underlying linkage disequilibrium structure; (2) data dimension reduction and associated risk of missing significant results through subsumption; (3) relevance of the partitioning method to guide an FLTM-based GWAS pinpointing regions with significantly associated SNPs. The $CAST_{bin}$ clustering method was shown slightly different from $CAST_{real}$ and DBSCAN, from the clustering viewpoint. However, this discrepancy was not reflected by a difference in GWASs' performances. Therefore, to the initial question "Which clustering method should be chosen", the answer for the Crohn's disease WTCCC data set relative to chromosome 2 would rather prioritize easiness in tuning parameters. In our experiments so far, the FLTM learning algorithm seems robust to the choice of the clustering method, provided that the intrinsic parameters of the latter are appropriately set. Further works include extending the current analysis to other chromosomes, for the WTCCC data set, as well as to other diseases, and extending our analysis to other clustering methods.

It was the first time that the FLTM learning algorithm was run on real GWAS data. It is questionable whether the present study should be complemented by intensive experiments run on simulated GWAS data sets. Given the high processing times required as soon as GWASs are addressed, and the recurring question of generating sufficiently realistic GWAS data, a less systematic approach, encompassing more diseases, seems wholly relevant.

Finally, to return to the multilocus aspect of the type of GWAS addressed here, one of our next tasks is to compare the FLTM-based GWAS strategy with the few other scalable multilocus approaches existing, including BEAGLE [11].

# References

1. Mourad, R., Sinoquet, C., Leray, P.: A hierarchical bayesian network approach for linkage disequilibrium modeling and data-dimensionality reduction prior to genome-wide association studies. BMC Bioinformatics **12**, 16 (2011)
2. International Monetary Fund: Macro-fiscal Implications of Health Care Reform in Advanced and Emerging Economies. IMF Policy Paper, Washington (2010)
3. Hechter, E.: On Genetic Variants Underlying Common Disease. Ph.D. thesis, University of Oxford (2011)
4. Gibbs, R.A., Belmont, J.W., Hardenbol, P., et al.: The international hapmap project. Nature **426**(6968), 789–796 (2003)
5. The 1000 Genomes Project Consortium: A map of human genome variation from population-scale sequencing. Nature 467, 7319, 1061–1073 (2010)
6. Balding, D.J.: A tutorial on statistical methods for population association studies. Nat. Rev. Genet. **7**(10), 781–791 (2006)
7. Pritchard, J.K., Przeworski, M.: Linkage disequilibrium in humans: models and data. Am. J. Hum. Genet. **69**(1), 1–14 (2001)
8. Patil, N., Berno, A.J., Hinds, D.A., Barrett, W.A., Doshi, J.M., Hacker, C.R., Kautzer, C.R., Lee, D.H., Marjoribanks, C., McDonough, D.P., et al.: Blocks of limited haplotype diversity revealed by high-resolution scanning of human chromosome 21. Science **294**(5547), 1719–1723 (2001)

9. Abel, H.J., Thomas, A.: Accuracy and computational efficiency of a graphical modeling approach to linkage disequilibrium estimation. Stat. Appl. Genet. Mol. Biol. 10(1), Article 5 (2011)
10. Verzilli, C.J., Stallard, N., Whittaker, J.C.: Bayesian graphical models for genome-wide association studies. Am. J. Hum. Genet. **79**, 100–112 (2006)
11. Browning, B.L., Browning, S.R.: Efficient multilocus association testing for whole genome association studies using localized haplotype clustering. Genet. Epidemiol. **31**, 365–375 (2007)
12. Ackerman, M., Ben-David, S.: Clusterability: A theoretical study. In: 12th International Conference on Artificial Intelligence and Statistics, vol. 5, pp. 1–8 (2009). J. Mach. Learn. Res
13. Hubert, L., Arabie, P.: Comparing partitions. J. Classif. **2**(1), 193–218 (1985)
14. Robinson, R.W.: Counting unlabeled acyclic digraphs. In: Little, C.H.C. (ed.) Combinatorial Mathematics V. Lecture Notes in Mathematics, vol. 622, pp. 28–43. Springer, New York (1977)
15. Zhang, N.L.: Hierarchical latent class models for cluster analysis. J. Mach. Learn. Res. **5**(6), 697–723 (2004)
16. Ben-Dor, A., Shamir, R., Yakhini, Z.: Clustering gene expression patterns. In: 3rd Annual International Conference on Computational Molecular Biology, pp. 33–42 (1999)
17. Cahill, J.: Error-Tolerant Clustering of Gene Microarray Data. Bachelor's Honors thesis, Boston College, Massachusetts (2002)
18. Ester, M., Kriegel, H.-P., Sander, J., Xu, X.: A density-based algorithm for discovering clusters in large spatial databases with noise. In: 2nd International Conference on Knowledge Discovery and Data Mining, pp. 226–231. AAAI Press (1996)
19. Meila, M. Comparing clusterings: an axiomatic view. In: 22nd International Conference on Machine learning, pp. 577–584 (2005)
20. Rand, W.M.: Objective criteria for the evaluation of clustering methods. J. Am. Stat. Assoc. **66**(336), 846–850 (1971)
21. Mirkin, B.: Mathematical classification and clustering: from how to what and why. J. Classifi. **2**(1), 193–218 (1998)
22. Fowlkes, E.B., Mallows, C.L.: A method for comparing two hierarchical clusterings. J. Am. Stat. Assoc. **78**(383), 553–569 (1983)
23. Purcell, S., Neale, B., Todd-Brown, K., et al.: PLINK: a toolset for whole-genome association and population-based linkage analysis. Am. J. Hum. Genet. **81**(3), 559–575 (2007)
24. Gabriel, S.B., Schaffner, S.F., Moore, J.M., Roy, J., Blumenstiel, B., Higgins, J., DeFelice, M., Lochner, A., Faggart, M., Liu-Cordero, S.N., Rotimi, C., Adeyemo, A., Cooper, R., Ward, R., Lander, E.S., Daly, M.J., Altshuler, D.: The structure of haplotype blocks in the human genome. Science **296**(5576), 2225–2229 (2002)
25. Wang, N., Akey, J.M., Zhang, K., Chakraborty, R., Jin, L.: Distribution of recombination crossovers and the origin of haplotype blocks: the interplay of population history, recombination, and mutation. Am. J. Hum. Genet. **71**(5), 1227–1234 (2002)
26. Wellcome trust case control consortium: genome-wide association study of 14,000 cases of seven common diseases and 3,000 shared controls. Nature **447**(7145), 661–678 (2007)
27. Barrett, J.C., Hansoul, S., Nicolae, D.L., et al.: Genome-wide association defines more than 30 Distinct susceptibility loci for crohn's disease. Nat. Genet. **40**(8), 955–962 (2008)

# Crosstalk Network Biomarkers
# of a Pathogen-Host Interaction Difference
# Network from Innate to Adaptive Immunity

Chia-Chou Wu and Bor-Sen Chen[✉]

Control and Systems Biology Laboratory, Department of Electrical Engineering,
National Tsing Hua University, Hsinchu, Taiwan
d9761820@oz.nthu.edu.tw, bschen@ee.nthu.edu.tw
http://www.ee.nthu.edu.tw/bschen

**Abstract.** Crosstalks between host and pathogen are crucial in the infection process. To obtain insight into the defense mechanisms of the host and the pathogenic mechanisms of the pathogen, pathogen-host interactions in the infection process have become a novel and promising research subject in the field of infectious disease. In this study, two pathogen-host dynamic crosstalk networks were constructed to investigate the transition of pathogenic and defensive mechanisms from the innate to adaptive immune system in the entire infection process based on two-sided time course microarray data of *C. albicans*-zebrafish infection model and database mining. Potential crosstalk network biomarkers for the transition from innate to adaptive immunity were identified based on proteins with larger interaction variations inside the host and pathogen, and at the interface between the host and pathogen. The crosstalk network biomarkers consist of proteins with larger interaction variation scores in the pathogen-host interaction difference network. From the crosstalk network biomarkers, the molecular mechanisms of innate and adaptive immunity were successfully investigated from a systems biology perspective. In view of these results, the proposed crosstalk network biomarkers may serve as potential therapeutic targets of infectious diseases.

**Keywords:** Pathogen-host interaction · Dynamic crosstalk network · Crosstalk network biomarkers · Interaction difference network · Interaction variation score

## 1 Introduction

Our immune system protects us from deadly threats from pathogens. To fulfil the requirements, the immune system has to detect the invasion of exogenous pathogens, watch for the pathogenic conversion of endogenous microbes, communicate the threats to the other systems in our bodies, e.g., the nervous [1–4] and digestive system [5–7], and then coordinate the systems to evade the threats. Obviously, the immune system cannot function alone. In the past, the studies

© Springer International Publishing Switzerland 2015
A. Fred et al. (Eds): BIOSTEC 2015, CCIS 574, pp. 190–205, 2015.
DOI: 10.1007/978-3-319-27707-3_12

regarding the immune system [8–10] focused on the molecular functions and cellular constitution of the immune system itself, and on the physiological effects of immune-related molecules and cells. However, the immune system is one part of a biological organism. Hence, from a systematic perspective, we should consider all systems as a whole, and not view the immune system in isolation.

Immune-related molecules (e.g., chemokines, cytokines, interferons, etc.) and cell types (e.g., lymphocytes, monocytes, mast cells, etc.) are commonly studied with respect to the molecular functions and cellular constitution of the immune system. After activating the first line of the host defense mechanisms (i.e., innate immunity), several cell types (e.g., macrophages, dendritic cells, natural killer cells, etc.) are recruited to protect the host from pathogen invasion and eliminate the threats from pathogens. The recognition of pathogen-associated molecular patterns (PAMPs) and/or damage-associated molecular patterns (DAMPs) by pattern recognition receptors (PRRs) (e.g., toll-like receptors, C-type lectin receptors, etc.) [10,11] can be viewed as a starting point in a series of the following complex mechanisms. The PRRs initiate downstream pathways that promote the activation of other parts of the innate immune system and the clearance of pathogens (e.g., production and secretion of cytokines, chemokines, and chemotactic cues to recruit more leukocytes). Meanwhile, the macrophages and dendritic cells are responsible for presenting antigens to induce the synthesis of the antibodies specific to the presented antigens if it is the first exposure of the host to the pathogen (Fig. 1A). If it was not the first exposure of the host to the pathogen, existing immunological memory cells proliferate and induce the synthesis of antibodies (Fig. 1A). In short, the interplays between T cells, B cells, macrophages, dendritic cells, etc. have been elaborated in detail at the physiological level. For the treatment of infectious diseases, current drug targets focus on some key molecules rather than the cellular level. Therefore, investigation of the systematic offensive and defensive mechanisms at a molecular level is the most important topic from a drug discovery and design perspective.

Compared with the host immune system, pathogenic mechanisms, not to mention interspecies protein-protein interactions (PPIs) between the host and pathogen, have attracted less attention. The battle is a two-sided affair, that is, the interplays between the host and pathogen shape the whole infection process, from the first exposure to the pathogen to the final outcomes of the infection [12]. Therefore, about a decade ago, the traditional viewpoint to treat the host and pathogen separately shifted to a more holistic viewpoint that included both players in the infection process. This viewpoint transition resulted from (i) the realization of the indispensability of pathogen-host interactions (PHIs) in infectious diseases and (ii) the advent of omics biotechnology to quantify genes, transcripts, and proteins at whole cell/organism levels [13]. This permitted a comprehensive interrogation of both the pathogen and host at the whole-genome, transcriptome, and proteome levels. Despite tremendous advances in understanding pathogenic mechanisms and the subsequent triumphs in drug development [14], the remaining issues (e.g., drug resistance) of infectious diseases have become more troublesome. The dynamic and complex interactions between

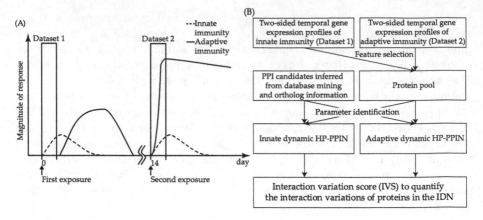

**Fig. 1.** Study design and the flowchart of PH-PPIN construction. (A) The first and second exposure induced the innate and adaptive immune responses, respectively, and the two-sided temporal gene expression profiles were recorded by microarray experiments (the rectangles are the observation windows of microarray experiments). (B) The flowchart delineates the procedures used in this study. Selected proteins of interest based on the microarray data formed a protein pool. The PPI candidates collected from the database mining and ortholog information were further pruned into the innate and adaptive dynamic PH-PPINs by the dynamic interaction model, system order detection method, and microarray data. Finally, the interaction variation scores were used to evaluate the significance of proteins in the interaction difference network, which was derived from the two constructed PH-PPINs.

the host and pathogen may partially explain why certain drugs are often not effective *in vivo* [15]. Hence, to investigate infection processes from a systematic perspective, in this study we constructed dynamic pathogen-host PPI networks (PH-PPINs) of innate and adaptive immunity.

To obtain systematic molecular interaction networks for targeted therapy, we utilized the *C. albicans*-zebrafish infection model [16]. We measured the temporal gene expression profiles of *C. albicans* and zebrafish during the infection process, constructed the interspecies PPIs using a dynamic interaction model, and identified the crosstalk network biomarkers with proposed interaction variation scores (Fig. 1B). Given the success of the *C. albicans*-zebrafish infection model [16] as well as its amenability to genetic manipulation [17], the zebrafish is a novel and potential model organism to study immunity. Furthermore, the zebrafish and human immune systems are remarkably similar and more than 75 % of human genes implicated in diseases have counterparts in zebrafish [18]. This provides a strong connection between the zebrafish and humans with respect to pathogenic mechanisms as well as immune responses, which are important for biomedical applications. The immune system of zebrafish as well as other vertebrates can be divided into two subsystems: i.e., innate and adaptive immunity [11]. The first dataset (GSE32119, [19]) we used to construct a dynamic PH-PPIN measured the two-sided gene expression profiles during the first 18 hours after zebrafish was first exposed to *C. albicans* to induce primary responses.

The second dataset (GSE51603, [20]) measured the two-sided gene expression profiles during the first 42 hours after zebrafish was secondarily exposed to *C. albicans* to induce secondary responses. To extract the interaction information from the time course microarray data, two dynamic PH-PPINs were constructed for innate and adaptive immunity in the infection process. By evaluating interaction variations based on the corresponding interaction variation scores, critical proteins and crosstalk network biomarkers of larger interaction variations in the infection process were identified. These crosstalk network biomarkers suggest the strategies taken by the host and pathogen during the infection process. Thus, these crosstalk network biomarkers could be potential drug targets when battling infectious diseases [13].

## 2    Material and Methods

### 2.1    Overview of Microarray Data

In this study, we used two microarray datasets: one was the two-sided temporal gene expression profiles of the host (zebrafish) and pathogen (*C. albicans*) in the period after first exposure (GSE32119, [19]), which were used to record the pathogen-host interaction information of innate immunity; the other was the two-sided temporal expression profiles of the host and pathogen in the period after secondary exposure (GSE51603, [20]), which were used to record the pathogen-host interaction information of adaptive immunity. For the first dataset, an experiment was performed to simultaneously profile the genome-wide gene expressions of innate immunity in both *C. albicans* and zebrafish during the infection process. *C. albicans* (SC5314 strain) was intraperitoneally injected into Adult AB strain zebrafish. The second dataset measured the genome-wide gene expressions of adaptive immunity in both *C. albicans* and zebrafish after the second exposure to *C. albicans*, fourteen days after the first exposure. Then, a two-step homogenization/mRNA extraction procedure was performed using the whole zebrafish infected with *C. albicans*. This approach can provide separate pools of gene transcripts from both the host and the pathogen, which provides individual estimates of specific gene expression profiles in either the host or pathogen using sequence-targeted probes derived from the individual genomes. Agilent *in situ* oligonucleotide microarrays, which cover 6,202 and 26,206 genes for *C. albicans* and zebrafish respectively, were used to profile temporal gene expressions; the first dataset consisted of three replicates of each organism measured at 9 time points (0.5, 1, 2, 4, 6, 8, 12, 16, and 18 h post-injection), and the second dataset consisted of two replicates of each organism measured at 8 time points (2, 6, 12, 18, 24, 30, 36, and 42 h post-re-injection). Both datasets were prepared under similar experimental conditions.

### 2.2    Protein Pool Selection and Database Integration

There are two steps that need to be completed before a dynamic protein-protein interaction (PPI) network with a dynamic interaction model can be constructed.

The first step is to have a protein pool from which the nodes in the resultant networks will be selected, and the second step is to obtain all possible PPIs among the proteins in the protein pool by integrating the interaction information from database mining. Here, our protein pool consisted of the union of the differentially expressed genes in the first and second datasets and the differentially expressed genes between the first and second microarray datasets. The criterion used to select the differentially expressed genes in the first and second microarray datasets was to compute the $p$-value of the ANOVA test to determine whether the average expression levels differed over time (i.e., for the first dataset, the null hypothesis was $\mu_1 = \cdots = \mu_9$, and the average expression levels were the same for all 9 time points; for the second dataset, the null hypothesis was $\mu_1 = \cdots = \mu_8$, and the average expression levels were the same for all 8 time points), and then to select those proteins with a Bonferroni corrected $p$-value$< 0.05$ for inclusion in the protein pool. In addition, the genes in the top 5 % of the expression difference between the first and second datasets were selected for the protein pool. Next, to know all possible interactions between the proteins in the protein pool, interaction information for the zebrafish-zebrafish, *C. albicans-C. albicans*, and zebrafish-*C. albicans* pairs are needed. However, the lack of information about these three kinds of interactions makes it difficult to collect all possible interactions. In addition, it is impossible to consider all interactions between the proteins in the protein pool. To overcome this issue, interaction information from human and yeast was used because of their similarity to our study subjects (zebrafish and *C. albicans*) and data availability. To infer the possible interactions of the study subjects (zebrafish and *C. albicans*), the ortholog information in the Inparanoid database [21] was used to convert the interactions of human and yeast [19,22] into the interactions of zebrafish and *C. albicans*. It should be noted that the interactions inferred from the ortholog-based method were derived under different experimental conditions. Consequently, the data do not accurately reflect the actual biological condition of the pathogen-host interactions during the *C. albicans* infection process; that is, false positive interactions exist in the complete set of inferred possible interactions of zebrafish and *C. albicans*, and these false positive interactions need to be validated and removed using real microarray data. Therefore, the false positive interactions were deleted from the candidate PPIs and realistic pathogen-host crosstalk PPI networks in innate and adaptive immunity were constructed using the two-sided microarray data and the dynamic model of PPI interaction in the following section.

### 2.3   Pathogen-Host Protein-Protein Interaction Network (PH-PPIN) Construction

To construct the interspecies PPI network from the protein pool and candidate PPIs, the dynamic protein-protein interaction model was used to determine the realistic PH-PPIN using individual proteins in succession. Given that the total numbers of the host and pathogen proteins are $N$ and $M$, respectively, then for a host target protein $i$ in the PH-PPIN, the dynamic interaction model is as follows [26]:

$$p_i^{(h)}[k+1] = \sigma_i^{(h)}p_i^{(h)}[k] + \sum_{n=1}^{N}\alpha_{in}^{(h)}p_n^{(h)}[k] + \sum_{m=1}^{M}\gamma_{im}p_m^{(p)}[k] + \beta_i^{(h)} + \epsilon_i^{(h)}[k+1]$$

$$(1)$$

where $p_i^{(h)}[k]$ denotes the protein level of the host target protein $i$ at time $k$, $\epsilon_i^{(h)}[k]$ denotes the environmental noise at time $k$, $\sigma_i^{(h)}$ denotes the self-regulation ability of the host target protein $i$, $\alpha_{in}^{(h)}$ denotes the interaction strength between the host protein $n$ and the host target protein $i$, $\gamma_{im}$ denotes the interaction strength between the pathogen protein $m$ and the host target protein $i$, and $\beta_i^{(h)}$ denotes the basal level of the host target protein $i$. Similarly, the dynamic interaction model of a pathogen target protein $j$ can be written as follows:

$$p_j^{(p)}[k+1] = \sigma_j^{(p)}p_j^{(p)}[k] + \sum_{m=1}^{M}\alpha_{jm}^{(p)}p_m^{(p)}[k] + \sum_{n=1}^{N}\gamma_{jn}p_n^{(h)}[k] + \beta_j^{(p)} + \epsilon_j^{(p)}[k+1] \quad (2)$$

The biological significance of this formulation is that the protein level of the host (pathogen) target protein $i$ ($j$) in the future (at time $k+1$) is determined by its current protein level (at time $k$) with self-regulation ability $\sigma_i^{(h)}$ ($\sigma_j^{(p)}$), the interaction strength between the host (pathogen) target protein $i$ ($j$) and the proteins of the same species $\alpha_{in}^{(h)}$ ($\alpha_{jm}^{(p)}$) and the other species $\gamma_{im}$ ($\gamma_{jn}$), the basal level $\beta_i^{(h)}$ ($\beta_j^{(p)}$), and the environmental noise $\epsilon_i^{(h)}$ ($\epsilon_j^{(p)}$) in the future. Due to the unavailability of proteomic data, the expression levels measured by the two-sided microarray experiments were used to represent the protein levels in the dynamic interaction model. The dynamic interaction model for the host target protein $i$ can be further rewritten into a concise form as follows:

$$\mathbf{p}_i^{(h)} = \Phi_i\theta_i^{(h)} + \epsilon_i^{(h)} \tag{3}$$

where $\mathbf{p}_i^{(h)} = \left[p_i^{(h)}[1] \cdots p_i^{(h)}[K]\right]^T$, $\theta_i = \left[\alpha_{i1}^{(h)} \cdots \alpha_{iN}^{(h)} \gamma_{i1} \cdots \gamma_{iM} \sigma_i^{(h)} \beta_i^{(h)}\right]^T$, $\epsilon_i^{(h)} = \left[\epsilon_i^{(h)}[1] \cdots \epsilon_i^{(h)}[K]\right]^T$, and

$$\Phi_i = \begin{bmatrix} p_1^{(h)}[0] & \cdots & p_N^{(h)}[0] & p_1^{(p)}[0] & \cdots & p_M^{(p)}[0] & p_i^{(h)}[0] & 1 \\ \vdots & \ddots & \vdots & \vdots & \ddots & \vdots & \vdots & \vdots \\ p_1^{(h)}[K-1] & \cdots & p_N^{(h)}[K-1] & p_1^{(p)}[K-1] & \cdots & p_M^{(p)}[K-1] & p_i^{(h)}[K-1] & 1 \end{bmatrix}.$$

The dynamic model for the pathogen can also be rewritten into a similar form. The only unknown parameter $\theta_i^{(h)}$ can then be estimated using parameter estimation methods, such as the least-squares estimation. However, due to the lack of large-scale measurements of host and pathogen protein levels, we used the temporal gene expression profiles as a substitute of protein activities to identify the parameter $\theta_i^{(h)}$ in the model. Furthermore, to ensure the model was not unnecessarily complex, the Akaike information criterion (AIC) was introduced to detect the true model order (the number of interactions). The true model order with minimum AIC was considered as the criterion to delete false positive interactions in the candidate PH-PPINs. Hence, the final dynamic PH-PPINs encompass the dynamic interaction model of each protein with the minimum AIC value to remove the false positive PPIs. Finally, after identifying the parameters for each protein in the protein pools, the identified interactions parameters $\left(\alpha_{in}^{(h)}, \alpha_{jm}^{(p)}, \gamma_{im}, \text{ and } \gamma_{jn}\right)$ formed the final dynamic PH-PPIN.

## 2.4   Interaction Variation Score (IVS) Calculation

To target the network biomarkers in the PH-PPINs, the IVSs were calculated for proteins to correlate proteins with the transition of the pathogen-host interactions from innate to adaptive immunity. The proteins in the PH-PPINs with the largest PPI variations from innate to adaptive immunity can be considered as crosstalk network biomarkers in the entire infection process and are considered as significant drug targets. Therefore, we investigated these crosstalk network biomarkers as follows. The IVS is a measurement of the variation of the interaction strength under a biological condition transition. According to the dynamic interaction models, the constructed PH-PPIN under a specific condition (innate or adaptive) can be written as follows:

$$
\begin{bmatrix} p_1^{(h)}[k+1] \\ \vdots \\ p_N^{(h)}[k+1] \\ p_1^{(p)}[k+1] \\ \vdots \\ p_M^{(p)}[k+1] \end{bmatrix} = \begin{bmatrix} \sigma_1^{(h)} & \cdots & \alpha_{1N}^{(h)} & \gamma_{11} & \cdots & \gamma_{1M} \\ \vdots & \ddots & \vdots & \vdots & \ddots & \vdots \\ \alpha_{N1}^{(h)} & \cdots & \sigma_N^{(h)} & \gamma_{N1} & \cdots & \gamma_{NM} \\ \gamma_{11} & \cdots & \gamma_{1N} & \sigma_1^{(p)} & \cdots & \alpha_{1M}^{(p)} \\ \vdots & \ddots & \vdots & \vdots & \ddots & \vdots \\ \gamma_{M1} & \cdots & \gamma_{MN} & \alpha_{M1}^{(p)} & \cdots & \sigma_M^{(p)} \end{bmatrix} \begin{bmatrix} p_1^{(h)}[k] \\ \vdots \\ p_N^{(h)}[k] \\ p_1^{(p)}[k] \\ \vdots \\ p_M^{(p)}[k] \end{bmatrix} + \begin{bmatrix} \beta_1^{(h)} \\ \vdots \\ \beta_N^{(h)} \\ \beta_1^{(p)} \\ \vdots \\ \beta_M^{(p)} \end{bmatrix} + \begin{bmatrix} \epsilon_1^{(h)}[k+1] \\ \vdots \\ \epsilon_N^{(h)}[k+1] \\ \epsilon_1^{(p)}[k+1] \\ \vdots \\ \epsilon_M^{(p)}[k+1] \end{bmatrix}
\tag{4}
$$

where the notations are the same as in the dynamic interaction models. The above equation can be written in a more concise form:

$$
\mathbf{p}[k+1] = A\mathbf{p}[k] + \beta + \epsilon[k+1]
\tag{5}
$$

where $A$ is a systematic interaction matrix of the PH-PPIN constructed under a specific condition. The interaction difference of two PH-PPINs between innate and adaptive immunity can be expressed in the following interaction difference matrix form:

$$
D_{\text{adaptive-innate}} = A_{\text{adaptive}} - A_{\text{innate}}.
\tag{6}
$$

If the variation of the interaction strength of a protein is larger during a biological condition transition (innate→adaptive immunity in this study), this may imply the protein has a more important role in the transition from innate to adaptive immunity. Therefore, the IVS used to evaluate the interaction variability of a protein in the transition from innate to adaptive immunity can be defined as follows:

$$
IVS_p = \frac{\sum_{q=1}^{Q} |d_{pq}|}{\text{Degree of protein } p}
\tag{7}
$$

where $d_{pq}$ is the $pq$-entry of $D_{\text{adaptive−innate}}$, that is, the average interaction variation of the protein $p$ in the transition from innate to adaptive immunity. The degree of protein $p$ is the number of non-zero elements in the $p$th row of the interaction difference matrix $D_{\text{adaptive−innate}}$. Those proteins with larger IVSs are considered as significant proteins that play an important role in the transition from innate to adaptive immunity in the infection process.

## 3  Results

### 3.1  The Pathogen-Host Protein-Protein Interaction Networks (PH-PPINs) of Innate and Adaptive Immunity

In this study, we aimed to investigate the systematic offensive and defensive mechanisms of pathogen and host at the molecular level. In particular, we aimed to understand the roles of pathogen-host interactions (PHIs) in innate and adaptive immunity from a systems biology perspective. The outcomes of interactions between the host and pathogen were recorded based on the two-sided temporal gene expression profiles of *C. albicans* and zebrafish that were simultaneously measured during the primary and secondary response periods in the infection process. During the two periods (the rectangles in Fig. 1A), the observed variations in the gene expression levels were mainly due to innate and adaptive immunity, respectively. We further selected 1620 proteins of interest for the protein pool, including those with differentially expressed features and the top 5 % of the expression level difference between the two datasets. The comparison of their temporal profiles (Fig. 2A) implied that their expression patterns changed: the activation of a group of pathogens genes was delayed and the repression of a group of hosts' genes was advanced. The changes in the gene expression patterns implied the PHIs in these two periods should have corresponding variations. To determine the variations of the underlying PHIs, 26060 PPI candidates inferred from the database mining and ortholog information were further pruned using the dynamic interaction models, model order detection method, and two-sided microarray data (Fig. 1B) and then the innate and adaptive dynamic PH-PPINs were formed (Fig. 2B). In particular, the two constructed PH-PPINs were the underlying mechanisms used to explain the observed changes in the gene expression patterns in the infection process.

The resultant PH-PPINs consisted of 1512 proteins (1431 *C. albicans* proteins; 81 zebrafish proteins) and 5721 PPIs (5510 intracellular interactions inside *C. albicans*; 145 interspecies interactions; 66 intracellular interactions inside zebrafish) for innate immunity, and 1578 proteins (1480 *C. albicans* proteins; 98 zebrafish proteins) and 3755 PPIs (3577 intracellular interactions inside *C. albicans*; 96 interspecies interactions; 82 intracellular interactions inside zebrafish) for adaptive immunity. Looking at the amount of variation in the nodes and edges of the pathogen, although most of the pathogenic nodes are shared between innate and adaptive immunity, the number of edges changed from 5511 to 3577: that is, only 1203 edges are shared. This implies that the pathogen may use almost the same set of proteins (85 %) but with different links to interact with the host and to regulate functions within the pathogen itself in response to various challenges from innate and adaptive immunity. In contrast, the host may use a different strategy since a different distribution of node and edge numbers was found compared with the pathogen. In the zebrafish, there are three more

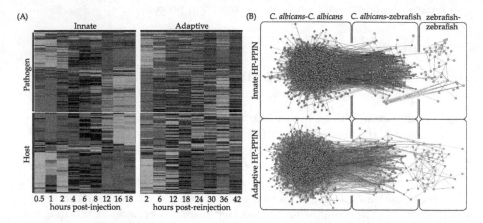

**Fig. 2.** Temporal gene expression profiles of the proteins of interest and the constructed innate and adaptive dynamic PH-PPINs. (A) The horizontal axis indicates the sampling time points in the microarray experiments. The vertical axis shows the genes clustered according to their expression patterns in innate immunity. (B) The innate and adaptive PH-PPINs consist of PPIs in three domains: pathogen-pathogen, pathogen-host, and host-host.

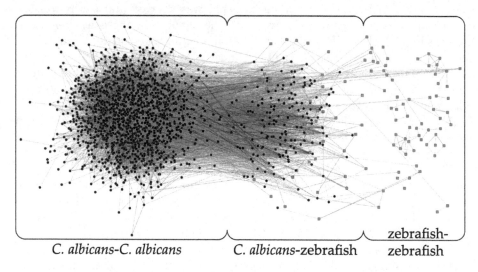

**Fig. 3.** The interaction difference network between innate and adaptive immunity. The IDN between innate and adaptive immunity consists of interactions in the three domains: pathogen-pathogen, pathogen-host, and host-host domains. The round and square nodes indicate the pathogen and host proteins, respectively.

significantly enriched functions (angiogenesis, coagulation, and circadian clock) in the adaptive PH-PPIN compared with the innate PH-PPINs (metabolic processes, immune responses, and apoptosis). In addition, in *C. albicans*, there are two more significantly enriched functions (circadian clock and filament growth) compared with the innate PH-PPINs (response to stimulus, redox status, and budding). The new functions in the adaptive PH-PPIN compared with the innate PH-PPIN indicated changes in the response strategies of the host and pathogen. To efficiently identify and evaluate the significance of proteins in the innate and adaptive dynamic PH-PPINs, we differentiated the two PH-PPINs into an interaction difference network (IDN) (Fig. 3), i.e., the matrix $D$ in Eq. (6), and then used interaction variation scores (IVSs) to evaluate the interaction variations of proteins in the IDN.

## 3.2  Identifying Crosstalk Network Biomarkers in the IDN Using IVS

Cell signaling depends on dynamic PPINs [23]. Hence, the interaction variation in the PPINs indicates the change in cell signaling and the corresponding consequences in the cellular functions. To illustrate the variation of PPINs, we adopted the notations of node color, edge color, and edge line style as shown in Fig. 3 to illustrate the existence of proteins and their interactions, and the variation of interactions from innate to adaptive dynamic PH-PPINs. Further, to focus on the proteins with significant variations, the IVS stated in Eq. (7) was used to evaluate the average interaction variation of a protein: that is, the ratio of the total interaction variation of a protein to the number of links possessed by the protein. Hence, the IVS can quantify the extent of the interaction variations, which may signify the importance of the proteins in the transition from innate to adaptive immunity, i.e., the IDN between innate and adaptive PH-PPINs (Fig. 3). In the following, we focused on the proteins with the ten highest IVSs in the three domains, that is, the host-host (zebrafish-zebrafish), pathogen-pathogen (*C. albicans-C. albicans*), and pathogen-host (*C. albicans-zebrafish*) domain, and determined the crosstalk network biomarkers in these domains.

**The Crosstalk Network Biomarkers in the Host-Host Domain.** In the host-host domain of the IDN, the ten proteins with the highest IVSs show close relationships with innate and adaptive immune responses. Extracting the ten proteins and their first neighbors from the IDN, there are five components in the host-host domain (Fig. 4A). The largest component consists of *f2*, LOC798231, LOC793315, *ace2*, *gnai1*, and their first neighbors (the left part of Fig. 4A). *gnat2*, a host G-protein that formed one end of the interspecies interaction, has connections with chemokine-related proteins (*ccl-c5a* and si:dkey-269d20.3) and chemotaxis-related proteins (ENSDARP00000105159 and ENSDARP0000111107). The angiogenesis- and coagulation-related proteins (*agt*, *ace2*, *f2*, and ENSDARP00000098661) are connected to these chemokine-related proteins. This component also consists of two other proteins: i.e., serine proteinase inhibitor (*serpinc1*) and prokineticin (ENSDARP00000109666). The roles of angiogenesis, coagulation, and chemokines are manifested in innate and adaptive immunity in this component. The second component mainly consists of complements (*c7b, c8g, c8a, c8b,* and *c9*) and vitronectins (*vtna* and *vtnb*). Given the well-known roles of the complements system in immunity, vitronectins have recently attracted much attention in the field of immunity [24]. The *cd36* and apolipoproteins (*apob1, apoba,* and *apobb*) form the third component (the lower right part of Fig. 4A). CD36 plays a pivotal role in macrophage foam-cell formation and atherogenesis, which is reduced by apolipoproteins. Although the last two components are much less documented, the roles of versican (*vcanb*) and tank in inflammation have been reported [25].

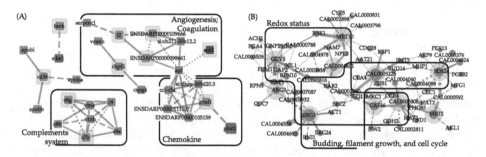

**Fig. 4.** The crosstalk network biomarkers in the host-host and pathogen-pathogen domains. (A) Chemokines, the complements system, and angiogenesis and coagulation are the three major crosstalk network biomarkers in the host-host domain of the IDN owing to the higher IVSs of their members. (B) Redox status and pathogen expansion are the two major crosstalk network biomarkers in the pathogen-pathogen domain of the IDN owing to the higher IVSs of their members. The shadowed nodes represent the proteins with the ten highest IVSs in their domains.

**The Crosstalk Network Biomarkers in the Pathogen-Pathogen Domain.** In the pathogen-pathogen domain, the ten proteins with the highest IVSs and their first neighbors form a single component (see Fig. 4B). In this component, the importance of redox status in the innate and adaptive immune responses is re-emphasized [26]. ERG1, CAL0005908, MET10, and GCV3 are all related to the redox status of *C. albicans*. In addition, CAL0005225, ERG1, and SDS24 are related to the expansion of *C. albicans* due to their functions in budding, filament growth, and the cell cycle, respectively. In particular, MET10 is related to the response to stress from the host and environment. Another major function in this component is transferase activity. MET2 is a homoserine acetyltransferase that can transform homoserine, a toxin for *C. albicans*, and is important for *C. albicans* survival. ARG3 facilitates the production of citrulline, which can induce pseudohyphal morphogenesis. The morphological transformation of *C. albicans* has been proven to be important in its pathogenesis. The hydrolase CAF16 exerts its influence on RNA polymerase II although the specific targeted genes are still unknown.

**The Crosstalk Network Biomarkers in the Pathogen-Host Domain.** In the pathogen-host domain, we also selected ten proteins from both the host and pathogen. These interspecies proteins form crosstalks that are more complicated than those in the pathogen-pathogen domain (Fig. 5). A possible mechanism for the correlation between redox status in the host and pathogen is shown in the pathogen-host domain, i.e., the interaction between thioredoxin (*txn*) and ribonucleotide reductase 1 (RNR1). In addition to its role in redox status, RNR1 also influences the iron utility, filament growth, and cell cycle of *C. albicans*. This implies that the effect of redox status on the pathogen is multifaceted. Compared with chemokine-related functions in the host-host domain, the role of chemokine-related functions in the PHIs of the pathogen-host domain are more interesting. CAG1, one protein involved in how chemokine-related functions affect the pathogen, is related to the hyphal growth, mating, and biofilm formation of the pathogen, which are all important in pathogenesis. In contrast to the appearance of redox status and chemokines in the pathogen-pathogen and host-host domains, respectively, gene transcription and the circadian clock can only be seen in the pathogen-host domain. Interactions between TAF60,

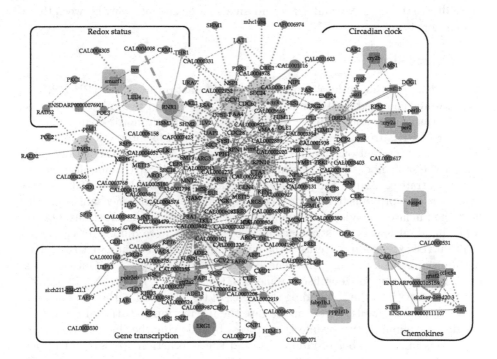

**Fig. 5.** The crosstalk network biomarkers in the pathogen-host domain. Redox status, circadian rhythm, gene transcriptions, and chemokines are the four major crosstalk network biomarkers in the pathogen-host domain. The shadowed nodes represent the proteins with the ten highest IVSs in the pathogen-host domain. The round and square nodes indicate the pathogen and host proteins, respectively.

*gtf2a2*, and *polr2e* emerged in adaptive immunity. TAF60, a transcription factor, is responsible for the drug responses in the pathogen, and *gtf2a2* and *polr2e* are related to gene transcription in the host. Their interactions indicate a possible mechanism as to how the PHIs affect the gene expression level. In addition to gene transcription, the circadian clock has an interesting function in the host and pathogen. The circadian clock-related proteins of the host (*cry2a*, *cry2b*, and *per2*) and pathogen (HRR25) form a sub-network in the host-pathogen domain. The circadian rhythms in the host and pathogen are correlated and numerous functions of the pathogen (yeast-hyphal switch, gene transcription, pathogenesis, etc.) are affected through HRR25.

In summary, we found that the proteins of the most variable interactions in the IDN are the elements related to chemokines, angiogenesis, coagulation, redox status, pathogen expansion, gene transcription, and circadian clock functions: i.e., the so-called crosstalk network biomarkers. Thus, these crosstalk network biomarkers change considerably in the transition from innate to adaptive immunity in the infection process and are potential targets for treatment and vaccination. To further evaluate the plausibility of the crosstalk network biomarkers, we selected angiogenesis, coagulation, redox status, and the circadian clock due to their systemic influence and investigated the interplay between these biomarkers based on the IDN.

## 4    Discussion

We presented pathogen-host protein-protein interaction networks (PH-PPINs), which were generated by dynamic interaction models and two-sided microarray data during innate and adaptive response periods. The dramatic changes in the number of PHIs from innate to adaptive immunity (145 interactions in innate immunity, 96 interactions in adaptive immunity, and 36 shared interactions) and almost the same nodes appearing in both innate and adaptive immunity suggest that the strategy used by the pathogen are characterized by the use of almost the same subset of proteins to respond to the two different defense mechanisms of the host (i.e., innate and adaptive immunity) but with different interactions. On the other hand, although the strategies used by the host were quite subtle, we can tell that the expression patterns of the host genes change from innate to adaptive immunity based on the temporal gene expression profiles (Fig. 2A). Once the lack of PPI information for zebrafish is remedied, the resultant PH-PPIN can provide further insight into the responding strategies of the host and pathogen. For now, we were able to quantitatively investigate the interaction variations from innate to adaptive immunity in the infection process by following the clues regarding the variations in the number of interaction and the interaction strengths identified in the dynamic interaction models.

To focus our investigation over a smaller and more meaningful range, we utilized IVS to evaluate the average interaction variation of a protein in the transition from innate to adaptive immunity in the infection process. The IVS rules out the possibility of a large IVS being caused by many small interaction variations, a weakness of the carcinogenesis relevance value (CRV) of [28].

Hence, the IVS could better focus on proteins with large interaction variations. Further, we visualized the interaction difference matrix from innate to adaptive immunity as an IDN (Fig. 3B), which can be divided into three domains according to the types of interactions involved. For the three domains, we focused on the proteins with the ten highest IVSs to investigate their interaction variations and determined the crosstalk network biomarkers. Not surprisingly, several immune-related and pathogenic crosstalk network biomarkers emerged: chemokines, cytokines, the complement system, pathogen expansion, and redox status. Nevertheless, three additional crosstalk network biomarkers—circadian clock, angiogenesis, and coagulation—were found for the larger interaction variations of their components. Although these functions are not totally new in immunity research, crosstalk among these crosstalk network biomarkers is a novel contribution of this study. In particular, the influences of circadian clock, redox status, angiogenesis, and coagulation are systemic. The samples and sampling time points of the microarray data provided us an opportunity to gain insight into the mechanisms of how these systemic crosstalk network biomarkers interact. The whole fish body samples provided a holistic view of the systemic variations of the transcriptomes from innate to adaptive immune response. The observation windows of the microarray experiments (Fig. 1A) revealed the involvement of the circadian clock in innate and adaptive immunity, which may be concealed if there are not enough sampling points over several days. Thus, we identified several significant proteins and crosstalk network biomarkers in the three domains based on their larger interaction variations from innate to adaptive immunity and then explored them by taking a closer look at the IDN.

In summary, our findings underpin the criticality of the circadian clock crosstalk network biomarker in terms of the type of immune response generated by an organism [27] and further show how the circadian clock, redox status, angiogenesis, and coagulation crosstalk network biomarkers are tightly coupled with pathogenesis and the host immune systems. This provides an opportunity to design new and efficient therapeutic guidelines for drug targets and the time window for treatments.

# References

1. Ren, K., Dubner, R.: Interactions between the Immune and nervous systems in pain. Nat. Med. **16**, 1267–1276 (2010)
2. Capuron, L., Miller, A.H.: Immune system to brain signaling: Neuropsychopharmacological implications. Pharmacol. Ther. **130**, 226–238 (2011)
3. Ransohoff, R.M., Brown, M.A.: Innate immunity in the central nervous system. J. Clin. Invest. **122**, 1164–1171 (2012)
4. Nataf, S.: The sensory immune system: a neural twist to the antigenic discontinuity theory. Nat. Rev. Immunol. **4**, 280 (2014)
5. Sekirov, I., Russell, S.L., Antunes, L.C.M., Finlay, B.B.: Gut Microbiota in health and disease. Physiol. Rev. **90**, 859–904 (2010)

6. Esplugues, E., Huber, S., Gagliani, N., Hauser, A.E., Town, T., Wan, Y.S.Y., O'Connor, W., Rongvaux, A., Van Rooijen, N., Haberman, A.M., Iwakura, Y., Kuchroo, V.K., Kolls, J.K., Bluestone, J.A., Herold, K.C., Flavell, R.A.: Control of $T_H17$ cells occurs in the small intestine. Nature **475**, 514–518 (2011)

7. Lee, W.J., Hase, K.: Gut Microbiota-generated metabolites in animal health and disease. Nat. Chem. Biol. **10**, 416–424 (2014)

8. Sorokin, L.: The impact of the extracellular matrix on inflammation. Nat. Rev. Immunol. **10**, 712–723 (2010)

9. Pulendran, B., Ahmed, R.: Immunological mechanisms of vaccination. Nat. Immunol. **12**, 509–517 (2011)

10. Romani, L.: Immunity to fungal infections. Nat. Rev. Immunol. **11**, 275–288 (2011)

11. Trede, N.S., Langenau, D.M., Traver, D., Look, A.T., Zon, L.I.: The use of zebrafish to understand immunity. Immunity **20**, 367–379 (2004)

12. Tierney, L., Kuchler, K., Rizzetto, L., Cavalieri, D.: Systems biology of host-fungus interactions: turning complexity into simplicity. Curr. Opin. Microbiol. **15**, 440–446 (2012)

13. Schmidt, F., Volker, U.: Proteome analysis of host-pathogen interactions: investigation of pathogen responses to the host cell environment. Proteomics **11**, 3203–3211 (2011)

14. Arnold, R., Boonen, K., Sun, M.G.F., Kim, P.M.: Computational analysis of interactomes: current and future perspectives for bioinformatics approaches to model the host-pathogen interaction space. Methods **57**, 508–518 (2012)

15. Meijer, A.H., Spaink, H.P.: Host-pathogen interactions made transparent with the zebrafish model. Curr. Drug Targets **12**, 1000–1017 (2011)

16. Chao, C.C., Hsu, P.C., Jen, C.F., Chen, I.H., Wang, C.H., Chan, H.C., Tsai, P.W., Tung, K.C., Wang, C.H., Lan, C.Y., Chuang, Y.J.: Zebrafish as a model host for Candida albicans infection. Infect. Immun. **78**, 2512–2521 (2010)

17. Gratacap, R.L., Wheeler, R.T.: Utilization of zebrafish for intravital study of eukaryotic pathogen-host interactions. Dev. Comp. Immunol. **46**, 108–115 (2014)

18. Schier, A.F.: Genomics: zebrafish earns its stripes. Nature **496**, 443–444 (2013)

19. Chen, Y.Y., Chao, C.C., Liu, F.C., Hsu, P.C., Chen, H.F., Peng, S.C., Chuang, Y.J., Lan, C.Y., Hsieh, W.P., Wong, D.S.H.: Dynamic transcript profiling of Candida albicans Infection in zebrafish: a pathogen-host interaction study. PLoS One **8**, e72483 (2013)

20. Lin, C., Lin, C.N., Wang, Y.C., Liu, F.Y., Chuang, Y.J., Lan, C.Y., Hsieh, W.P., Chen, B.S.: The role of TGF-$\beta$ signaling and apoptosis in innate and adaptive immunity in zebrafish: a systems biology approach. BMC Syst. Biol. **8**, 116 (2014)

21. Ostlund, G., Schmitt, T., Forslund, K., Kostler, T., Messina, D.N., Roopra, S., Frings, O., Sonnhammer, E.L.L.: InParanoid 7: new algorithms and tools for eukaryotic orthology analysis. Nucleic Acids Res. **38**, D196–203 (2010)

22. Croft, D., Mundo, A.F., Haw, R., Milacic, M., Weiser, J., Wu, G.M., Caudy, M., Garapati, P., Gillespie, M., Kamdar, M.R., Jassal, B., Jupe, S., Matthews, L., May, B., Palatnik, S., Rothfels, K., Shamovsky, V., Song, H., Williams, M., Birney, E., Hermjakob, H., Stein, L., D'Eustachio, P.: The Reactome pathway knowledgebase. Nucleic Acids Res. **42**, D472–477 (2014)

23. Yasui, N., Findlay, G.M., Gish, G.D., Hsiung, M.S., Huang, J., Tucholska, M., Taylor, L., Smith, L., Boldridge, W.C., Koide, A., Pawson, T., Koide, S.: Directed network wiring identifies a key protein interaction in embryonic stem cell differentiation. Mol. Cell. **54**, 1034–1041 (2014)

24. Gerold, G., Ajaj, K.A., Bienert, M., Laws, H.J., Zychlinsky, A., de Diego, J.L.: A toll-like receptor 2-integrin$\beta_3$ complex senses bacterial lipopeptides via vitronectin. Nat. Immunol. **9**, 761–768 (2008)
25. Wight, T.N., Kang, I., Merrilees, M.J.: Versican and the control of inflammation. Matrix Biol. **35**, 152–161 (2014)
26. Wang, Y.C., Lin, C., Chuang, M.T., Hsieh, W.P., Lan, C.Y., Chuang, Y.J., Chen, B.S.: Interspecies protein-protein interaction network construction for characterization of host-pathogen interactions: a Candida albicans-zebrafish interaction study. BMC Syst. Biol. **7**, 79 (2013)
27. Curtis, A.M., Bellet, M.M., Sassone-Corsi, P., O'Neill, L.A.J.: Circadian clock proteins and immunity. Immunity **40**, 178–186 (2014)
28. Wang, Y.C., Chen, B.S.: A network-based biomarker approach for molecular investigation and diagnosis of lung cancer. BMC Med. Genomics **4**, 2 (2011)

# Template Scoring Methods for Protein Torsion Angle Prediction

Zafer Aydin[1]([✉]), David Baker[2], and William Stafford Noble[3]

[1] Department of Computer Engineering, Abdullah Gul University,
38080 Kayseri, Turkey
zafer.aydin@agu.edu.tr
[2] Department of Biochemistry, University of Washington, Seattle, WA 98195, USA
[3] Department of Genome Sciences, Department of Computer Science
and Engineering, University of Washington, Seattle, WA 98195, USA

**Abstract.** Prediction of backbone torsion angles provides important constraints about the 3D structure of a protein and is receiving a growing interest in the structure prediction community. In this paper, we introduce a three-stage machine learning classifier to predict the 7-state torsion angles of a protein. The first two stages employ dynamic Bayesian and neural networks to produce an ab-initio prediction of torsion angle states starting from sequence profiles. The third stage is a committee classifier, which combines the ab-initio prediction with a structural frequency profile derived from templates obtained by HHsearch. We develop several structural profile models and obtain significant improvements over the Laplacian scoring technique through: (1) scaling templates by integer powers of sequence identity score, (2) incorporating other alignment scores as multiplicative factors (3) adjusting or optimizing parameters of the profile models with respect to the similarity interval of the target. We also demonstrate that the torsion angle prediction accuracy improves at all levels of target-template similarity even when templates are distant from the target. The improvement is at significantly higher rates as template structures gradually get closer to target.

## 1 Introduction

Protein 3D structure prediction benefits greatly from prediction of various 1D and 2D structural attributes such as secondary structure, backbone torsion (dihedral) angles, solvent accessibility, disordered regions, and contact maps [6]. Methods that predict structural properties of proteins typically employ sequence-based frequency profiles in their feature sets to utilize information in similar proteins. These profiles can be in the form of position specific scoring matrices (PSSM) or hidden Markov models (HMM) and can be derived by aligning the amino acid sequence of the query with sequences in a large protein database using an efficient algorithm such as PSI-BLAST [1] or HHBlits [16]. Despite the many efforts for improving the quality of sequence-based alignments and their profiles, the accuracy of 1D and 2D predictions has come to saturation due to the difficulty of eliminating false positives especially when the query sequence diverges from those in the protein database considerably. Recently, there has been a growing interest in using

A. Fred et al. (Eds): BIOSTEC 2015, CCIS 574, pp. 206–223, 2015.
DOI: 10.1007/978-3-319-27707-3_13

structural profiles as input features for predicting various structural characteristics of proteins. A structural frequency profile is a position specific scoring matrix (PSSM) that is constructed from the structural labels of templates (*i.e.*, hit proteins) obtained by aligning the target (*i.e.*, query) against a set of proteins. To date structural profiles have been derived mainly for protein secondary structure [7,12]; backbone structural motifs, solvent accessibility, contact density [13]; and shape strings [23].

Among the various structural properties of proteins, predicted secondary structure has been widely employed by many structure prediction methods to reduce the conformational space that must be explored. One limitation of using secondary structure is the inability to impose constraints for loop (or coil) segments, which do not have a well-defined structure [30]. On the other hand, predictions of backbone torsion angles can provide powerful and more fine-grained restraints as compared to secondary structure. It is anticipated that accurately predicted torsion angles will one day replace the dominant role played by secondary structure in tertiary structure prediction [15].

Despite the variety of methods proposed for predicting backbone torsion angles of proteins [4,8,17,18,22,27], less effort has been made to systematically incorporate structurally related templates into torsion angle predictions [13]. Furthermore, to the best of our knowledge, there is no work in the literature that inspects the accuracy of torsion angle predictions at all levels of target-template similarity (*i.e.*, from easy to difficult targets).

One approach for incorporating template information is to construct a frequency profile matrix, in which the occurrence frequencies of template residues are accumulated followed by a normalization step. This matrix is later included to the feature set of a machine learning classifier, which predicts the structural class labels of the amino acids. Methods that have been developed for profile matrices mainly use Laplacian counts, which is a technique that gives equal weights to templates [12]. As an alternative to the Laplacian count method, a new scoring technique has been proposed which scale the templates by the third power of the sequence identity score and the structural quality factor [14,24].

In this paper, we propose new structural profile methods that incorporate various score terms of HHsearch alignments [19] to predict torsion angles of proteins. We combine structural profiles with ab-initio predictions obtained from a two-stage classifier using a linear committee approach. Our method is able to generate specific and effective predictions for targets at all difficulty levels. We achieve this by optimizing certain parameters of profile models with respect to the similarity interval of the target.

## 2    Methods

### 2.1    Backbone Torsion Angles

Each amino acid residue has three associated backbone torsion angles: $\phi$, $\psi$, and $\omega$. The angle $\phi$ denotes rotation about the $C_\alpha$-N bond of the residue, $\psi$ denotes rotation about the bond linking $C_\alpha$ and the carbonyl carbon, and $\omega$

denotes rotation about the bond between the carbonyl carbon of the current residue and the nitrogen of the next residue. We compute $\phi$, $\psi$, and $\omega$ from the 3-D coordinate information in Protein Data Bank (PDB), which is the database of solved protein structures. Each of these angles is constrained to the range $[-180, 180]$.

Following [5], we first subdivided residues into five torsion angle classes, which represent the major clusters observed in PDB. However, to reduce the imbalance in the sizes of these classes, we further subdivided the two most common labels (A and B) according to whether the secondary structure class is loop or not. The resulting seven labels are described in Table 1.

**Table 1. 7-state torsion angle labels.** The table lists the seven torsion angle classes, their definitions, and the percentage of residues assigned to each class in the PDB-PC90 data set. ss denotes the secondary structure label of the amino acid residue.

| Label | Definition | Percent |
|---|---|---|
| L | $|\omega| \geq 90$, $\phi < 0$, $-125 < \psi \leq 50$, ss $=$ loop | 11.94 |
| A | $|\omega| \geq 90$, $\phi < 0$, $-125 < \psi \leq 50$, ss $\neq$ loop | 38.21 |
| M | $|\omega| \geq 90$, $\phi < 0$, $\psi \leq -125$ OR $\psi > 50$, ss $=$ loop | 20.08 |
| B | $|\omega| \geq 90$, $\phi < 0$, $\psi \leq -125$ OR $\psi > 50$, ss $\neq$ loop | 22.27 |
| E | $|\omega| \geq 90$, $\phi \geq 0$, $|\psi| > 100$ | 1.92 |
| G | $|\omega| \geq 90$, $\phi \geq 0$, $|\psi| \leq 100$ | 4.73 |
| O | $|\omega| < 90$ | 0.84 |

## 2.2   Torsion Angle Class Prediction

Based on the definition given in Table 1, the 7-state torsion angle prediction problem can be stated as follows. For a given protein, the goal is to assign to each amino acid a torsion angle label from the alphabet $\{L, A, M, B, E, G, O\}$ as shown in Fig. 1.

LWGLVKQGLKCEDCGMNVHHKCREKVANLC
MMELMGLBBBBLLLGMBBMAAAALLMMLMO

**Fig. 1. 7-state torsion angle class prediction problem.** The first row shows the amino acid sequence of the target and the second row is the sequence of 7-state torsion angle labels, which are defined according to Table 1.

## 2.3   Alignment Methods

**Deriving Templates for Structural Profiles.** In this paper, we used the HHsearch method [19] to detect the templates that are similar to a given target.

HHsearch first derives an HMM-profile for the target and aligns it against a database of HMM-profiles [19]. At the end of the alignment, it ranks the templates (*i.e.* hits) according to a probability score ranging from 0 % to 100 % and reports the ones that score above a threshold. An example alignment is shown in Fig. 2. We used the following commandline to compute HMM-HMM alignments for each target: ./hhsearch -i protein.hhm -d hhm3 -o protein.hhr -cpu 2 -mact 0.05 -ssw 0.11 -atab protein. start.tab -realign -E 100 -cov 20 -b 20. We then selected the HMM-HMM alignments that score above the given threshold as the templates. Note that, HHsearch uses predicted secondary structure to be able to compute sensitive HMM-HMM alignments. We used the PSIPRED version 2.61 [11] to predict secondary structures. All these alignments were generated in 2011. Further details on HHsearch and the HMM-HMM alignments can be found in the corresponding documentation [20,21].

```
No 3
>3fy3A
Probab=100.00  E-value=3.6e-41  Score=236.01  Aligned_columns=201  Identities=22%

Q ss_pred            CCEEECCC---EEE--ECCCCCEEEECCCC---EEEE-CHHCCCCCCCEEEEC-----------------CCCHHHHH
Q ss_conf           99896176---479--87898379944863---4985-133664889879976------------------88256532
Q 2odlA.fasta      4 GMDVVHGT---ATM--QVDGNKTIIRNSVD---AIIN-WKQFNIDQNEMVQFL------------------QENNNSAV  55
(372)
Q Consensus        4 g~~v~~g------i----------i~q~s--------n-w-sFnIg~~~~v~f~-----------------q~~~a~vi  55
(372)
                     .+.|+.+.   -.+  ..++.+.|+..+|.   ..||  |++|||++.+.+.+|               ...+|++|
T Consensus        1 ~gIv~~~~~~~~~v~~~~nG~~vinI~~Pn~~GiS~N~y~~FnV~~~G~vlNNs~~~~~t~l~G~i~~NpnL~~~~A~~I  80
(234)
T 3fy3A           1 NGIVPDAGHQGPDVSAVNGGTQVINIVTPNNEGISHNQYQDFNVGKPGAVFNNALEAGQSQLAGHLNANSNLNGQAASLI  80
(234)
T ss_dssp           CCCEECSSTTCCEEEEETTTEEEEEBCCCCCTTSEEEEEEEECCBCTTCEEEEECSSCEEETTTEEECCCTTSSSCCCSEE
T ss_pred           CCEEEECCCCCCEEEEECCCCCEEEEEBCCCCCCEEEEEEEECCCCCEEEECCCCCCCCCCCEECCCCCCCCCCCCCEE
T ss_conf           956637989886777559993289866788885345320121039997599657554530002022157533787642 89
```

**Fig. 2.** An example HMM-HMM alignment obtained by HHsearch.

**Generating Position-Specific Scoring Matrices for The Ab-Initio Method.** We employed PSSMs generated by the PSI-BLAST [1] and HHMAKE [19] algorithms as input features. We used BLAST version 2.2.20 and the NCBI's non-redundant (NR) database dated June 2011 to generate PSI-BLAST PSSMs. We generated the HMM-profiles by HHsearch version 1.5.1 [19]. Note that in deriving the HHMAKE PSSMs we did not perform any HMM-HMM alignments. After deriving PSSMs we scaled them to the interval $[0,1]$ by applying a sigmoidal transformation. Detailed descriptions of the PSSMs and the sigmoidal transformation can be found in [2].

### 2.4   Structural Frequency Profiles for 7-State Torsion Angles

Our structural profile is a $7 \times N$ matrix where rows represent the torsion angle classes and columns denote the amino acids of the target. An example structural profile is illustrated in Fig. 3.

The entries of a structural frequency profile matrix represent the propensity of having a particular torsion angle class at a given amino acid residue of the target protein. It therefore acts as a signature summarizing the 7-state torsion label expectancy of the target residues. The entries of this matrix are normalized

|   | 1 | 2 | ... | N |
|---|------|------|-----|------|
| L | 0.40 | 0.20 | ... | 0.18 |
| A | 0.17 | 0.30 | ... | 0.02 |
| M | 0.01 | 0.13 | ... | 0.08 |
| B | 0.03 | 0.07 | ... | 0.12 |
| E | 0.19 | 0.11 | ... | 0.04 |
| G | 0.16 | 0.09 | ... | 0.06 |
| O | 0.04 | 0.10 | ... | 0.50 |

**Fig. 3.** A structural frequency profile for 7-state torsion angle representation. Rows represent the torsion angle classes and columns denote the amino acids of the target. Each column sums to 1.

to the interval [0,1] and each column sums to 1 similar to a marginal *a posteriori* probability distribution. For this reason, we denote a structural profile by $P_s(t_j|y)$, where $t_j$ is the torsion angle class of the $j^{th}$ residue and $y$ is the amino acid sequence of the target.

A structural profile can be obtained by collecting the occurrence frequencies of the structure labels of the template proteins. There could be various approaches for computing a structural frequency profile. The following sections explain the methods that have been implemented in this paper.

**Laplacian Counts.** The most straightforward approach for deriving a structural frequency profile is to count the occurrence frequencies of torsion angle labels that match to a given position of the target, which is followed by a normalization step. The Laplacian count method is employed by most of the approaches that have been proposed for computing structural profiles to date [12]. First a count matrix $C$ is obtained that contains the occurrence frequencies of the structure labels of the templates, which is formulated as follows

$$C(i, j) = \sum_{A(j,k)} \delta(T(j, k), i) \tag{1}$$

where $C(i, j)$ is the $(i, j)^{th}$ entry of the count matrix such that $i \in \{L, A, M, B, E, G, O\}$ is the torsion angle class, $j$ is the residue position of the target, $A(j, k)$ is the residue of the $k^{th}$ database protein aligned to the $j^{th}$ position, $T(j, k)$ is the corresponding torsion angle class of the template, and $\delta(T(j, k), i)$ is the Kronecker delta function defined as

$$\delta(t, i) = \begin{cases} 1 \text{ if } t = i \\ 0 \text{ otherwise} \end{cases} \tag{2}$$

In other words, each torsion angle label that is aligned to the $j^{th}$ position of the target contributes by a count of 1, which is also known as the Laplacian count method. Once the count matrix is obtained it is normalized so that each column sums to 1. This is formulated as

$$M_a(i,j) = \begin{cases} \frac{C(i,j)}{\sum_i C(i,j)} & \text{if } |A(j)| > 0 \\ 0 & \text{otherwise} \end{cases} \tag{3}$$

where $A(j)$ is the set of all residues aligned to the $j^{th}$ residue of the target, $|A(j)|$ is the number of residues in $A(j)$, and $M_a$ is the normalized count matrix. In this formulation, the residues of the target are divided into two categories. In the first group, we have "aligned" positions (represented by the condition $|A(j)| > 0$) where at least one residue is aligned from a database protein and in the second group there is the set of "unaligned" positions (*i.e.*, the case where $|A(j)| = 0$) for which no residues are aligned from any hits. The second condition is realized for positions that correspond to gapped regions and for positions that are left out of the aligned regions when a local alignment algorithm is employed.

After the normalization step, the structural profile matrix can be computed as

$$M(i,j) = \begin{cases} M_a(i,j) & \text{if } |A(j)| > 0 \\ M_b(i,j) & \text{otherwise} \end{cases} \tag{4}$$

where $M_b(i,j)$ is the background probability of aligning a template residue with torsion class $i$ to the $j^{th}$ residue of the target. In this paper, we use predictions from the ab-initio classifier for the background distribution of torsion angle labels.

**Weighing Hits by Integer Powers of Sequence Identity Scores.** A second method for computing structural profiles weights templates by integer powers of the sequence identity score. In HHsearch, this score is computed for each target template pair and is represented by the "Identities" field as shown in Fig. 2. We first divide this score by 100 and convert it to a weight value. We then compute an integer power of this weight, which is used to scale templates that contribute to the structural profile. This is expressed in the equations below

$$C(i,j) = \sum_{A(j,k)} \theta(T(j,k), i) \tag{5}$$

$$\theta(t,i) = \begin{cases} I^a & \text{if } t = i \\ 0 & \text{otherwise} \end{cases} \tag{6}$$

where $C$ is the count matrix, $\theta$ is the new occurrence count function replacing the Kronecker delta in (2), $I$ is the sequence identity score of the $k^{th}$ template and $a$ is an integer that represents the strength of the amplification one wishes to impose on the structurally similar templates. The remaining terms are the same as their counterparts in (1) and (2). This type of template scaling has two benefits. The first one is related to scaling templates by sequence identity scores, which increases the contribution of structurally closer templates while reducing the votes of distant ones. The second benefit comes by taking integer powers of $I$, which manages the situation where a handful of structurally similar templates are followed by many less similar or distant templates. In such a scenario, if we use the Laplacian counts as in Sect. 2.4 or weigh templates by sequence identity scores only (*i.e.*, $a = 1$) the contribution of the similar templates would be suppressed by many structurally less similar candidates. To further amplify the effect of structurally similar templates and to reduce the contribution of false positives (*i.e.*, noise) it is useful to take integer powers of the sequence identity scores as formulated in (6). Once we compute the count matrix, we normalize it as in (3). All the other steps in deriving the structural profile are the same as in Sect. 2.4.

**Incorporating Quality of Templates.** In addition to taking integer powers of the sequence identity score, it is also possible to include other weight factors to the score function in (6). One such measure assesses the experimental quality of the templates and is proposed in [14,24]. When we employ this approach to score the templates, (6) takes the following form:

$$\theta(t,i) = \begin{cases} \frac{I^a}{q} & \text{if } t = i \\ 0 & \text{otherwise} \end{cases} \tag{7}$$

where $q$ is the quality of the template computed as X-ray resolution + R-factor / 20 as proposed in [10]. According to this measure, a template with a higher experimental quality has a lower $q$ parameter. Since this measure requires the X-ray resolution of the template, we apply it to those templates that have been solved by the X-ray method only ignoring the remaining templates for the target.

**Incorporating Other Alignment Scores.** When two proteins are aligned to each other, typically several score terms are calculated for assessing the statistical significance including e-value, raw similarity score, and percentage of sequence identity. Employing these terms in scaling the templates could also be useful in constructing a structural profile. With this motivation, we first incorporated the e-value score into the occurrence count function by converting it to a multiplicative weight factor as in [28]. This is formulated as

$$\theta(t,i) = \begin{cases} w_e \, I^a & \text{if } t = i \\ 0 & \text{otherwise} \end{cases} \tag{8}$$

where $w_e$ is the e-value weight defined as

$$w_e = \begin{cases} 1 & \text{if } E < 10^{-10} \\ -0.05 \log_{10}(E) + 0.5 & \text{if } 10^{-10} \leq E < 10^{10} \\ 0 & \text{otherwise} \end{cases} \tag{9}$$

such that $E$ is the E-value of the alignment. Note that we dropped the quality term as it did not bring any significant benefits, which is verified by our simulations. According to (9), $w_e$ is set to 1 when the target-template similarity is above a certain threshold ($E - value < 10^{-10}$) and decreases linearly as the E-value of the target-template alignment is greater than $10^{-10}$ until it becomes considerably high (*i.e.*, $10^{10}$), in which case it is set to zero.

In addition to the E-value, we also considered incorporating the overall raw similarity score of the alignment into our structural profiles. For this purpose, we modified the occurrence count function as

$$\theta(t, i) = \begin{cases} s\, w_e\, I^a & \text{if } t = i \\ 0 & \text{otherwise} \end{cases} \tag{10}$$

where $s$ is the raw score of the alignment. For HHsearch, this is the overall similarity score obtained at the end of the HMM-HMM alignment, which is computed as the sum of the similarities of the aligned profile columns minus the gap penalties [19]. A slight variation of this approach normalizes the raw score with the length of the aligned region as

$$\theta(t, i) = \begin{cases} \frac{s}{L}\, w_e\, I^a & \text{if } t = i \\ 0 & \text{otherwise} \end{cases} \tag{11}$$

such that $L$ is the length of the aligned region and is given as the field denoted as "Aligned_columns" in HHsearch's output (see Fig. 2).

**Scaling Columns of the Alignment.** Up to this point, we scaled the templates uniformly throughout the aligned positions without discriminating the individual columns of the alignment. In this section, we explain an approach for amplifying local regions within an alignment that could potentially contribute more accurate torsion label information and suppressing those that could be locally more distant. For this purpose, we include the similarity score between the aligned residues from a BLOSUM matrix into the occurrence count function as formulated below

$$\theta(t, i) = \begin{cases} \frac{s}{L}\, w_e\, I^a\, e^b & \text{if } t = i \\ 0 & \text{otherwise} \end{cases} \tag{12}$$

where $b$ is the similarity score such as BLOSUM matrix score between the $j^{th}$ residue of the target and the residue in the template that is aligned to the target residue. When the two residues are biologically similar to each other we expect this score term to be larger than the term obtained from dissimilar pairings. This approach has the potential to amplify local matches between motifs that are common both in the target and the template. Note that normalizing the sequence alignment score with $L$ is optional. For instance, a slightly modified version of (12) does not perform such type of normalization:

$$\theta(t, i) = \begin{cases} s\, w_e\, I^a\, e^b & \text{if } t = i \\ 0 & \text{otherwise} \end{cases} \tag{13}$$

**Selecting Templates Within a Score Window.** To improve the speed of predictions, it is possible to select those alignments that score above a given

HHsearch probability threshold denoted as $s_T$. Let $s_B$ represent the probability score of the top scoring alignment. Then, $s_T$ is computed as

$$s_T = s_B - W \tag{14}$$

where $W$ is the probability score window, which is a hyper-parameter of the template based predictor.

## 2.5  Prediction Model

**Ab-Initio Predictor.** Our *ab-initio* torsion class predictor is a hybrid architecture, in which four dynamic Bayesian network (DBN) models are combined with a neural network. Two of the DBNs use PSSMs derived by PSI-BLAST [1] and the other two use PSSMs from the HHMAKE module of the HHsearch method [19]. Details of the ab-initio predictor can be found in [2,3]. For simplicity we treat the output signal of the neural network as a probability distribution due to the constraints it satisfies (*i.e.*, it sums to 1 and takes values from 0 to 1). This distribution is denoted as $P_a(t_j|x)$, which represents the *ab-initio* likelihood of the $j^{th}$ residue to have $t_j$ as the torsion angle label given $x$, the set of input features around position $j$. Hence our neural network predicts the torsion angle label of the residue at the center of the feature window by selecting the particular label with the maximum discriminant score at the output layer. This is formulated as

$$t_j^* = \arg \max_{t_j} P_a(t_j|x). \tag{15}$$

**Committee Predictor.** The committee predictor combines the ab-initio predictions of torsion angle classes with the structural frequency profile according to the following equation:

$$P_c(t_j|x,y) = \begin{cases} (1-\lambda)P_a(t_j|x) + \lambda P_s(t_j|y) & \text{if aligned} \\ P_a(t_j|x) & \text{if unaligned} \end{cases} \tag{16}$$

where $P_c(t_j|x,y)$ is the combined likelihood of having torsion angle label $t_j$ for the residue at position $j$, $\lambda$ is the weight assigned to the structural profile, $x$ is the feature set of the ab-initio predictor described in Sect. 2.5, $y$ is the amino acid sequence of the target, $P_a(t_j|x)$ is the distribution of torsion angle classes obtained from the ab-initio predictor, and $P_s(t_j|y)$ is the structural profile computed from the templates. According to this equation, the combined likelihood is the weighted average of the ab-initio predictions and the structural profile for positions that are aligned to at least one template residue and it becomes equal to the likelihood of the ab-initio predictor only for the remaining positions.

After computing $P_c(t_j|x, y)$, we predict the torsion angle class of the $j^{th}$ residue as the particular label that maximizes $P_c(t_j|x, y)$ as

$$t_j^* = \arg \max_{t_j} P_c(t_j|x, y), \qquad (17)$$

where $t_j^*$ is the final prediction of the committee method.

## 2.6 Optimizing Model Parameters

We optimized the weight $\lambda$ in (16) that is used to combine the structural profiles with the ab-initio predictor, the probability score window $W$ in (14) that is used to select the HHsearch templates within a window and the power parameter $a$ in (6). For this purpose, we followed an iterative strategy in which we select the optimums that yield the best overall accuracy. In the first iteration, we optimized $\lambda$, followed by $W$ and $a$. In the second and third iterations, we refined our estimates by repeating the optimization of $\lambda$, $W$ and $a$ in sequence. At each optimization step, we set the remaining two parameters to their optimums obtained up to that point. We performed this optimization on the validation set described in Sect. 2.7 using the structural profile model in (6) together with the windowing approach in (14). Once we optimize these parameters, we trained the ab-initio predictor model on the set of 4205 proteins (i.e. the training set in Sect. 2.7) and computed predictions using the optimums.

## 2.7 Datasets

**PDB-PC90 Dataset.** To obtain the PDB-PC90 dataset, we used the PISCES server [25,26] with the following set of criteria: percent identity threshold of 90 %, resolution cutoff of 2.5 Å, and R-value cutoff of 1.0. We also used PISCES to filter out non-X-ray and $C_\alpha$-only structures and to remove short ($< 30$ amino acids) and long ($> 10000$ amino acids) chains. This dataset contained 17056 chains.

**Training, Validation and Test Sets.** We randomly selected 5161 proteins from the PDB-PC90 dataset. Among those, we randomly selected a set of 994 proteins for the first test set. From the set of 5161 proteins, we then removed those proteins that are similar to the test set using a 10 % sequence identity threshold. The remaining set contained 4205 chains, which is used to train our ab-initio prediction method. We computed the HHsearch alignments for the set of 994 proteins, which are used for computing the structural profiles of torsion angle classes and for predicting the torsion angle classes.

We further split the test set into two each containing 497 proteins. The first half is used as a validation set to optimize selected parameters of the model and the second is used to evaluate the prediction accuracy when the optimum parameters are employed during profile construction.

**Similarity Intervals and Subsets of the Test Set.** To distinguish easy targets from difficult ones, we defined similarity intervals using the HHsearch alignments from half of the proteins in our test set (see Sect. 2.7). For each target, we first selected the maximum sequence identity score from the set of target-template alignments. Then we ranked those scores and defined percentile intervals of sequence identity with increments of 5 %. This initially produced a total of 20 intervals. We combined the eighth and ninth intervals as the maximum sequence identity scores for those targets were very close to each other. We also combined the tenth up to the twentieth intervals since the maximum sequence identity score was 100 % for all the targets in those bins. This procedure resulted in a total of 9 sequence identity intervals. In the last step, we further reduced the number of intervals to 5 by combining the 2nd and 3rd, 4th and 5th, and 7th up to 9th intervals. The resulting intervals are tabulated below

**Table 2. Intervals of sequence identity scores.** The intervals are defined by selecting the target-template alignments with maximum sequence identity scores followed by sorting these scores in ascending order. Percentile increments of 5 % results in a total of 20 bins, which are further reduced to 5 intervals.

| Interval | Percentiles (%) | Max Identity (%) |
|---|---|---|
| Low | 0–5 | 0.0–26.0 |
| Medium-Low | 5–15 | 26.0–35.0 |
| Medium | 15–25 | 35.0–50.0 |
| Medium-High | 25–30 | 50.0–80.0 |
| High | 30–100 | 80.0–100.0 |

According to Table 2, the first interval represents targets with the most distant templates (*i.e.*, those with the maximum sequence identity score of 26 % or less) and the last interval represents targets that contain highly similar templates (*i.e.*, those with the maximum sequence identity score greater than 80 %). Based on this binning, we further divided our test set of 994 proteins (see the previous section) into 5 subsets such that each contained those targets that fall into one and only one of the intervals defined in Table 2. The number of proteins and amino acids in each of these subsets are summarized in Table 3.

Note that the number of proteins in each subset is not uniformly the same (especially true for the last set that contains targets from the "High" category) mainly because datasets have been constructed by random sampling from PDB without enforcing specific constraints for having equal number of samples in each interval. Nonetheless we have enough samples in each subset mainly because the torsion angle class prediction is performed on each residue separately. Furthermore the proportion of target positions that are aligned to at least one template residue is considerably high. This shows that a structural profile column is computed using the aligned templates for most of the target residues.

**Table 3. The five subsets of the test set with 994 proteins.** The number of proteins, the total number of amino acids and the number of amino acids that are aligned to at least one template residue are shown for each subset. The subsets are derived based on the intervals defined in Table 2.

| Subset | # proteins | # residues | # aligned res |
|---|---|---|---|
| Low | 56 | 12903 | 12665 |
| Medium-Low | 99 | 21792 | 21682 |
| Medium | 95 | 22596 | 22561 |
| Medium-High | 62 | 15326 | 15295 |
| High | 682 | 160037 | 159993 |
| Total | 994 | 232654 | 232196 |

We did a similar partitioning for the validation and the second test set where each subset contained proteins from the corresponding similarity interval. As a result of this procedure the validation subsets contained 25, 51, 52, 24, and 345 proteins and the second test set contained 31, 48, 43, 38, and 337 proteins respectively starting with the lowest similarity up to the highest. To reduce the computation time of optimizations in the high similarity interval we only used 35 proteins instead of 345.

# 3 Results

## 3.1 Accuracy of Structural Profiles only

We first compare the torsion angle label accuracy of the structural profiles on positions that are aligned to templates only. For this purpose, we implemented the profile methods summarized in Table 4.

**Table 4. The implemented structural profile methods and their descriptions.** Further details can be found in Sect. 2.4.

| Structural profile and description |
|---|
| SP1: Eq. (1), Laplacian |
| SP2: Eq. (6), $a = 1$ |
| SP3: Eq. (6), $a = 3$ |
| SP4: Eq. (6), $a$ varies wrt similarity interval |
| SP5: Eq. (7), quality, $a = 3$, (Pollastri et al.) |
| SP6: Eq. (8), E-value, $a = 3$ |
| SP7: Eq. (10), E-value, align. score, $a = 3$ |
| SP8: Eq. (11), E-value, norm. align. score, $a = 3$ |
| SP9: Eq. (12), E-value, norm. align. score, BLOSUM62, $a = 3$ |
| SP10: Eq. (13), E-value, align. score, $a$ varies, BLOSUM62 scores if target is in High interval only |
| SP11: SP10 with windowing in (14) |

In SP4, we modify the power of the sequence identity score term (*i.e.*, *a*) according to the similarity interval the target belongs to. For this purpose, we use the following mapping to define *a*:

$$a = \begin{cases} 1 \text{ if target in Low Interval} \\ 3 \text{ if target in Medium-Low Interval} \\ 5 \text{ if target in Medium Interval} \\ 7 \text{ if target in Medium-High Interval} \\ 9 \text{ if target in High Interval} \end{cases} \tag{18}$$

where the interval of the target is defined according to Table 2. In SP9 and SP10 we use the BLOSUM62 matrix to scale individual columns of the alignment as formulated in (12) and (13). In SP9, we employed the BLOSUM scores uniformly for all columns of the alignment whereas in SP10, we utilized the BLOSUM scores for targets in the "High" interval only. If the target belongs to one of the remaining four intervals then we turn off this score term in (13). In that case the equation takes the following form

$$\theta(t, i) = \begin{cases} s \, w_e \, I^a \text{ if } t = i \\ 0 \quad \text{otherwise} \end{cases} \tag{19}$$

Once we compute a structural profile, we predict the torsion angle class of the aligned target residues by selecting the particular label that yields the maximum value in the corresponding column of the profile.

Following these definitions, the torsion angle class prediction accuracy of the structural profiles listed in Table 4 is summarized in Table 5 below. In this table, Overall 1 is the number of correctly predicted amino acids divided by the total number of amino acids for which a structural profile column is computed (*i.e.*, those that are aligned to at least one template). The second up to the sixth columns show accuracies for the five similarity intervals and are computed on each subset of the test set. Overall 2 is the average of the five accuracies obtained for Low to High intervals. It estimates the accuracy we would obtain had we used equal number of amino acids for each of the five intervals. Based on these results, the most accurate structural profile method is SP10 though other methods such as SP7, SP4 and SP8 are also quite effective. SP10 outperforms SP1, (the Laplacian count method), by 14.53 %, which is a statistically significant improvement. It is also better than SP5 by 1.2 %, which was proposed in [14, 24]. This improvement is also statistically significant (with a p-value < 0.0001 from a two-tailed Z-test at a significance level of 0.01).

## 3.2   Combining Structural Profiles with Ab-Initio Predictions

After establishing that the new structural profile methods contain more accurate torsion angle information than the approaches proposed in the literature, we evaluated the accuracy when the structural profiles are combined with an ab-initio predictor. For this purpose, we trained our ab-initio method using the training set described in Sect. 2.7 and computed 7-state torsion angle predictions

**Table 5. 7-state torsion angle prediction accuracy of structural profiles.** L: Low, ML: Medium-Low, M: Medium, MH: Medium-High, H: High. Only target residues that are aligned to at least one template are considered.

| Profile | L | ML | M | MH | H | Overall 1 | Overall 2 |
|---|---|---|---|---|---|---|---|
| SP1 | 66.49 | 69.83 | 71.94 | 74.46 | 75.66 | 74.17 | 71.68 |
| SP2 | 66.85 | 70.71 | 73.93 | 80.07 | 81.96 | 79.18 | 74.70 |
| SP3 | 66.68 | 72.20 | 77.82 | 86.64 | 92.87 | 87.64 | 79.24 |
| SP4 | 66.85 | 72.20 | 79.40 | 87.92 | 93.47 | 88.30 | 79.97 |
| SP5 | 66.53 | 72.13 | 77.28 | 87.07 | 92.73 | 87.50 | 79.15 |
| SP6 | 67.15 | 73.26 | 78.89 | 87.13 | 93.06 | 88.03 | 79.90 |
| SP7 | 66.48 | 74.11 | 80.23 | 87.66 | 93.30 | 88.40 | 80.36 |
| SP8 | 66.59 | 73.91 | 80.13 | 87.31 | 93.25 | 88.33 | 80.24 |
| SP9 | 65.05 | 72.19 | 78.30 | 86.39 | 93.68 | 88.14 | 79.12 |
| SP10 | 66.55 | 74.11 | 80.71 | 87.36 | 93.69 | 88.70 | 80.48 |

on all the amino acids of the test set. We then combined those predictions with a structural profile as in (16). For the target residues that were not aligned to any template, we simply took predictions from the ab-initio method. Regarding the $\lambda$ parameter (*i.e.*, the weight of the structural profile) we considered two possibilities. The first approach sets $\lambda$ to 0.5 and the second one modifies it according to the similarity interval of the target according to the following function

$$\lambda = \begin{cases} 0.5 \text{ if target in Low Interval} \\ 0.6 \text{ if target in Medium-Low Interval} \\ 0.7 \text{ if target in Medium Interval} \\ 0.8 \text{ if target in Medium-High Interval} \\ 0.9 \text{ if target in High Interval} \end{cases} \tag{20}$$

In this equation, $\lambda$ is gradually increased as the similarity interval of the target approaches to the "High" interval thereby giving more weight to the structural profile than the ab-initio predictor. Table 6 summarizes the accuracy of committee predictors that combine the ab-initio method with various structural profiles. In addition to the overall accuracy measure, we also included the segment overlap (SOV) measure that is used in 1D structure prediction to assess the accuracy at the segmental level [29]. The SOV measure depicts how well the predicted torsion label segments match the true segments and is biologically more meaningful than the residue level accuracy.

According to this table, the ab-initio+SP10 method (with variable $\lambda$ parameter) is better than the ab-initio+SP5 method in all categories. The improvements are 2.78 % in Overall 1, 3.40 % in Overall 2, 3.57 % in SOV, 1.65 % in Low interval, 2.74 % in Medium-Low interval, 4.90 % in Medium interval, 4.82 % in Medium-High interval and 2.86 % in High interval. When ab-initio+SP10 is compared with ab-initio+SP1 (*i.e.*, the Laplacian method) for $\lambda = 0.5$,

**Table 6. 7-state torsion angle prediction accuracy of methods that incorporate structural profiles with ab-initio predictions.** The accuracy measures are computed on the first test set with 994 proteins. L: Low, ML: Medium-Low, M: Medium, MH: Medium-High, H: High. All target residues in the test set are considered.

| Method | L | ML | M | MH | H | Overall 1 | Overall 2 | SOV |
|---|---|---|---|---|---|---|---|---|
| Ab-initio | 72.36 | 73.96 | 73.58 | 73.35 | 74.01 | 73.83 | 73.45 | 71.33 |
| Ab-initio + SP1, $\lambda = 0.5$ | 74.47 | 75.42 | 76.10 | 76.94 | 77.96 | 77.28 | 76.18 | 74.49 |
| Ab-initio + SP5, $\lambda = 0.5$ | 72.39 | 74.19 | 74.19 | 75.53 | 78.28 | 76.99 | 74.92 | 74.52 |
| Ab-initio + SP10, $\lambda = 0.5$ | 74.59 | 78.01 | 80.69 | 87.37 | 93.62 | 89.10 | 82.86 | 88.11 |
| Ab-initio + SP5, $\lambda$ as in (20) | 72.94 | 75.21 | 77.22 | 83.86 | 90.99 | 86.70 | 80.04 | 85.19 |
| Ab-initio + SP10, $\lambda$ as in (20) | 74.59 | 77.95 | 82.12 | 88.68 | 93.85 | 89.48 | 83.44 | 88.76 |

the improvements are 12.20 % in Overall 1, 6.68 % in Overall 2, 13.62 % in SOV, 0.12 % in Low interval, 2.59 % for Medium-Low interval, 4.59 % in Medium interval, 10.43 % in Medium-High interval and 15.66 % in High interval. Adjusting the $\lambda$ parameter with respect to the similarity interval was particularly useful for the ab-initio+SP5 method. In other words, when $\lambda$ is set to 0.5 uniformly for all similarity intervals, the accuracy of ab-initio+SP5 dropped significantly higher than the ab-initio+SP10. This shows that the proposed structural profile SP10 is more useful than SP5 when combined with the ab-initio method. This is because torsion label errors of SP10 and the ab-initio method overlap less as compared to SP5 and therefore SP10 provides a better complement to the ab-initio predictor. Another observation one can make is the improvement over the ab-initio method when structural profiles are incorporated. This is true even for the Low interval (an improvement of 2.23 %) and is partly because of the sensitive nature of HMM-HMM profile alignments and also because HHsearch uses predicted secondary structure from PSIPRED [11] to align a pair of HMMs.

## 3.3    Optimizing Parameters

We optimized the parameters $a$, $W$ and $\lambda$ on the validation set as explained in Sect. 2.6. We observed that taking different values for the window parameter $W$ did not alter the accuracy of predictions considerably. Therefore this parameter is selected as nearly constant (Table 7).

Table 8 compares the accuracy of 7-state torsion angle prediction of various methods on the second test set of 497 proteins. In this table, except for the last row all the hits that score above the threshold are used during profile construction (i.e. no windowing is applied). Based on these results, optimizing the parameters of the profile models improves the accuracy of predictions as compared to the case where $a$ and $\lambda$ are fixed. However selecting the optimums (see the sixth row of Table 8) was not considerably better than the results in row five. The reason is because the cost curve that we are trying to optimize is not strongly convex (i.e. nearly flat around the maxima) and these are merely the variations caused by using different datasets for optimization and accuracy estimation.

**Table 7. Optimum values for $\lambda$, $W$, and $a$.** The values that give the best amino acid level accuracy on the validation set are selected. The optimization is performed separately for each HHsearch similarity interval. The improvements are computed with respect to the accuracy of *ab-initio* predictions.

| Interval | L | ML | M | MH | H |
|----------|-----|-----|-----|-----|-----|
| $\lambda^*$ | 0.5 | 0.6 | 0.6 | 0.6 | 0.7 |
| $a^*$ | 2 | 3 | 4 | 7 | 7 |
| $W^*$ | 20.0 | 10.0 | 10.0 | 10.0 | 10.0 |

**Table 8. 7-state torsion angle prediction accuracy of methods that incorporate structural profiles with ab-initio predictions.** The accuracy measures are computed on the second test set with 497 proteins. L: Low, ML: Medium-Low, M: Medium, MH: Medium-High, H: High. All target residues in the test set are considered.

| Method | L | ML | M | MH | H | Overall 1 | Overall 2 | SOV |
|--------|------|------|------|------|------|-----------|-----------|------|
| Ab-initio | 71.77 | 74.67 | 73.04 | 74.05 | 74.18 | 73.97 | 73.54 | 71.49 |
| Ab-initio + SP1, $\lambda = 0.5$ | 73.29 | 76.08 | 75.29 | 77.70 | 78.19 | 77.41 | 76.27 | 74.46 |
| Ab-initio + SP10, $a = 3$, $\lambda = 0.5$ | 73.29 | 77.87 | 80.82 | 87.24 | 93.24 | 89.11 | 82.49 | 87.97 |
| Ab-initio + SP10, $a$ as (18), $\lambda = 0.5$ | 73.44 | 77.87 | 81.15 | 87.99 | 93.61 | 89.20 | 82.81 | 88.09 |
| Ab-initio + SP10, $a$ as (18), $\lambda$ as (20) | 73.44 | 77.74 | 82.52 | 89.03 | 93.89 | 89.60 | 83.32 | 88.77 |
| Ab-initio + SP11, $W$, $\lambda$, $a$ optimized | 73.31 | 77.58 | 82.23 | 89.02 | 93.97 | 89.33 | 83.22 | 88.34 |

## 3.4   Other Approaches

In addition to the structural profile methods described above, we also considered three other approaches. The first one incorporates the probability score of HMM-profile alignments into (13) as a multiplicative factor to globally scale the templates and the second method incorporates the column score of HHsearch alignments to amplify local regions that are well conserved (e.g. motifs). These two approaches did not bring any reasonable change in the accuracy measures (result not shown). As a third approach we considered employing Henikoff weights [9] to scale the count information of templates before constructing the structural profiles. We applied this weighting procedure for the following three scenarios: (1) weights based on matched residues only, (2) weights based on torsion angle labels only, (3) weights based on residue and torsion angle tuples. Unfortunately, in all three cases, torsion angle prediction accuracy was significantly lower than the level achieved by other scaling methods considered in this paper (result not shown). Note that we did not consider utilizing a background distribution in (4) for torsion angle labels mainly because we use predictions from our ab-initio method, which would eventually contain a more accurate torsion angle representation than a simple background distribution.

Finally, we would like to state that we are unable to compare our torsion angle class predictor with the literature mainly because there is no other work that performs 7-state torsion angle prediction on the same set of alphabet. However we had shown in an earlier paper that a 5-state version of our predictor provides results comparable to the state-of-the-art in the ab-initio setting [3].

## 4    Conclusion

In this paper, we propose novel methods for scaling templates to construct structural profiles of torsion angle states. Though we use the score terms of the HHsearch method, our approach is generic and most of the structural profile methods proposed in this work can also be implemented using other alignment methods including PSI-BLAST. Second, the scaling techniques can be applied in many other tasks such as secondary structure prediction, solvent accessibility prediction, contact map prediction, and 3D structure prediction. Third, they can easily be incorporated into other methods that have been developed for deriving structural profiles from templates.

The proposed methods can be improved in several ways. First of all, the templates can be scaled in a position-specific manner using the confidence scores, which are now available in HHBlits (the new version of HHsearch). Second, instead of using a linear model, the ab-initio predictions can be combined with structural profiles using more advanced models such as neural networks. We expect that all these approaches will potentially improve the accuracy of structure prediction tasks further.

## References

1. Altschul, S.F., Madden, T.L., Schaffer, A.A., Zhang, J., Zhang, Z., Miller, W., Lipman, D.J.: Gapped BLAST and PSI-BLAST: a new generation of protein database search programs. Nucleic Acids Res. **25**, 3389–3402 (1997)
2. Aydin, Z., Singh, A., Bilmes, J., Noble, W.S.: Learning sparse models for a dynamic Bayesian network classifier of protein secondary structure. BMC Bioinform. **12**, 154 (2011)
3. Aydin, Z., Thompson, J., Bilmes, J., Baker, D., Noble, W. S.: Protein torsion angle class prediction by a hybrid architecture of bayesian and neural networks. In: 13th International Conference on Bioinformatics and Computational Biology (2012)
4. Berjanskii, M.V., Neal, S., Wishart, D.S.: PREDITOR: a web server for predicting protein torsion angle restraints. Nucleic Acids Res. **34**, W63–W69 (2006). (Web Server Issue)
5. Blum, B., Jordan, M., Kim, D., Das, R., Bradley, P., Baker, D.: Feature selection methods for improving protein structure prediction with Rosetta. In: Platt, J., Koller, D., Singer, Y., Roweis, S. (eds.) Advances in Neural Information Processing Systems 20, pp. 137–144. MIT Press, Cambridge (2008)
6. Cheng, J., Tegge, A.N., Baldi, P.: Machine learning methods for protein structure prediction. IEEE Rev. Biomed. Eng. **1**, 41–49 (2008)
7. Cong, P., Li, D., Wang, Z., Tang, S., Li, T.: Spssm8: an accurate approach for predicting eight-state secondary structures of proteins. Biochimie **95**(12), 2460–2464 (2013)
8. Faraggi, E., Zhang, T., Yang, Y., Kurgan, L., Zhou, Y.: SPINE X: improving protein secondary structure prediction by multistep learning coupled with prediction of solvent accessible surface area and backbone torsion angles. PLoS One **7**(2), e30361 (2012)
9. Henikoff, S., Henikoff, J.G.: Position-based sequence weights. J. Mol. Biol. **243**, 574–578 (1994)

10. Hobohm, U., Sander, C.: Enlarged representative set of protein structures. Protein Sci. **3**, 522–524 (1994)
11. Jones, D.T.: Protein secondary structure prediction based on position-specific scoring matrices. J. Mol. Biol. **292**, 195–202 (1999)
12. Li, D., Li, T., Cong, P., Xong, W., Sun, J.: A novel structural position-specific scoring matrix for the prediction of protein secondary structures. Bioinformatics **28**(1), 32–39 (2012)
13. Mooney, C., Pollastri, G.: Beyond the twilight zone: automated prediction of structural properties of proteins by recursive neural networks and remote homology information. Proteins Struct. Funct. Bioinform. **77**, 181–190 (2009)
14. Pollastri, G., Martin, A.J.M., Mooney, C., Vullo, A.: Accurate prediction of protein secondary structure and solvent accessibility by consensus combiners of sequence and structure information. BMC Bioinform. **8**, 201 (2007)
15. Rangwala, H., Karypis, G.: Introduction to Protein Structure Prediction: Methods and Algorithms. Wiley, Hoboken (2011)
16. Remmert, M., Biegert, A., Hauser, A., Soding, J.: Hhblits: lightning-fast iterative protein sequence searching by hmm-hmm alignment. Nat. Meth. **9**(2), 173–175 (2011)
17. Shen, Y., Delaglio, F., Cornilescu, G., Bax, A.: TALOS+: a hybrid method for predicting protein backbone torsion angles from nmr chemical shifts. J. Biomol. NMR **44**(4), 213–223 (2009)
18. Singh, H., Singh, S., Raghava, G.P.S.: Evaluation of protein dihedral angle prediction methods. PLoS One **9**(8), e105667 (2014)
19. Soding, J.: Protein homology detection by HMM-HMM comparison. Bioinformatics **21**, 951–960 (2005)
20. Soding, J.: Quick guide to HHsearch (2006). ftp://toolkit.genzentrum.lmu.de/pub/HHsearch/old/HHsearch/HHsearch1.5.1/HHsearch-guide.pdf
21. Soding, J., Remmert, M., Hauser, A.: HH-suite for sensitive sequence searching based on HMM-HMM alignment (2012). ftp://toolkit.genzentrum.lmu.de/pub/HH-suite/hhsuite-userguide.pdf
22. Song, J., Tan, H., Wang, M., Webb, G.I., Akutsu, T.: TANGLE: two-level support vector regression approach for protein backbone torsion angle prediction from primary sequences. PLoS One **7**(2), e30361 (2012)
23. Sun, J., Tang, S., Xiong, W., Cong, P., Li, T.: Dsp: a protein shape string and its profile prediction server. Nucleic Acids Res. **40**(W1), W298–W302 (2012)
24. Walsh, I., Bau, D., Martin, A.J.M., Mooney, C., Vullo, A., Pollastri, G.: Ab initio and template-based prediction of multi-class distance maps by two-dimensional recursive neural networks. BMC Struct. Biol. **9**, 5 (2009)
25. Wang, G., Dunbrack Jr., R.L.: PISCES: a protein sequence culling server. Bioinformatics **19**, 1589–1591 (2003). http://dunbrack.fccc.edu/PISCES.php
26. Wang, G., Dunbrack Jr., R.L.: PISCES: recent improvements to a pdb sequence culling server. Nucleic Acids Res. **33**, W94–W98 (2005)
27. Wu, S., Zhang, Y.: ANGLOR: A composite machine-learning algorithm for protein backbone torsion angle prediction. PLoS One **3**(10), e3400 (2008)
28. Wu, S., Zhang, Y.: MUSTER: improving protein sequence profile-profile alignments by using multiple sources of structure information. Proteins Struct. Funct. Bioinform. **72**(2), 547–556 (2008)
29. Zemla, A., Venclovas, C., Fidelis, K., Rost, B.: A modified definition of Sov, a segment-based measure for protein secondary structure prediction assessment. Proteins **34**, 220–223 (1999)
30. Zhou, Y., Duan, Y., Yang, Y., Faraggi, E., Lei, H.: Trends in template/fragment-free protein structure prediction. Theo. Chem. Acc. **128**, 3–16 (2011)

# Evaluating the Robustness of Correlation Network Analysis in the Aging Mouse Hypothalamus

Kathryn M. Cooper[1], Stephen Bonasera[2], and Hesham Ali[1(✉)]

[1] College of Information Science and Technology,
University of Nebraska Omaha, Omaha, USA
hali@unomaha.edu
[2] Division of Geriatrics, Department of Internal Medicine,
University of Nebraska Medical Center, Omaha, USA

**Abstract.** Volumes of high-throughput assays been made publicly available. These massive repositories of biological data provide a wealth of information that can harnessed to investigate pressing questions regarding aging and disease. However, there is a distinct imbalance between available data generation techniques and data analysis methodology development. Similar to the four "V's" of big data, biological data has volume, velocity, heterogeneity, and is prone to error, and as a result methods for analysis of this "biomedical big data" have developed at a slower rate. One promising solution to this multi-dimensional issue are network models, which have emerged as effective tools for analysis as they are capable of representing biological relationships *en masse*. Here we examine the need for development of standards and workflows in the usage of the correlation network model, where nodes and edges represent correlation between expression pattern in genes. One structure identified as biologically relevant in a correlation network, the gateway node, represents genes that change in co-expression between two different states. In this research, we manipulate parameters used to identify the gateway nodes within a given dataset to determine the consistency of results among network building and clustering approaches. This proof-of-concept is extremely important to investigate as there is a growing pool of methods used for various steps in our network analysis workflow, causing a lack of robustness, consistency, and reproducibility. This research compares the original gateway nodes analysis approach with manipulation in (1) network creation and (2) clustering analysis to test the consistency of structural results in the correlation network. To truly be able to trust these approaches, it must be addressed that even minor changes in approach can have sweeping effects on results. The results of this study allow the authors to call for stronger studies in benchmarking and reproducibility in biomedical "big" data analyses.

**Keywords:** Gateway nodes · Correlation networks · Aging · SPICi · Robustness

© Springer International Publishing Switzerland 2015
A. Fred et al. (Eds): BIOSTEC 2015, CCIS 574, pp. 224–238, 2015.
DOI: 10.1007/978-3-319-27707-3_14

# 1 Introduction

Recently, network analysis methods have been developed to analyze and draw signal from large, high-throughput datasets. These methods include the use of correlation networks, protein-protein interaction networks, genetic interaction networks, metabolic networks, and more. Commonly used to describe networks of co-expression, the correlation network model uses nodes to represent genetic probes and edges to represent a correlated pattern of gene expression between samples, defined by condition, time, or other environmentally quantifiable criterion. This technique has been proposed for identifying differentially expressed genes where traditional methods (such as Gene Set Enrichment Analysis) do not always return desirable results [1–3]. As such, correlation networks also serve as a valid supplement (note, not as a replacement) to traditional approaches.

While typically used for studying one particular state individually, the correlation network can also be used for comparison of states. A recent study by Dempsey and Ali [4] uses clustering in correlation networks, particularly clustering that identifies small, densely connected groups of genes, to compare datasets from the same cell lines under different conditions. This analysis revealed that clusters between states typically do not overlap except for at a limited number of genes. These genes that connect differentially to two different states are termed "gateway nodes." It has been proposed that these gateway nodes, which are thought to represent genes that are co-expressed with two different sets of genes at different states, can reveal a small, finite set of genes related to the phenotype under scrutiny, making this approach appealing when using high-throughput data – typically, in studies comparing 10,000 to 40,000 genes in two or more different states, typically only 20 to 100 gateway nodes result from analysis, depending on parameterization. Further studies on clusters in correlation networks have found that almost all clusters contain predicted and actual transcription factor binding sites for common regulatory elements [4]. This indicates that potentially, gene co-expression and even possibly co-regulated could be mined from this type of network, if such a relationship exists.

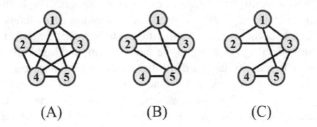

(A)          (B)          (C)

**Fig. 1.** An example of three different clusters.

Since 1999, the network model that is representative of biological data has found structure and function to be related [5], particularly when the network is built using clean data. In the protein-protein interaction networks, high degree or hub nodes typically are more likely to be lethal [6–8] clusters in these same networks represent proteins that complex together for functional purpose [9]. In a genetic interaction network, which represents the relationships between genes when both are simultaneously knocked out or down, the relationship represents some measure of how beneficial (or, more likely, detrimental) the duel silencing is on the organism [10]. Structures identified in these networks can lead to identifying of genes with common pathways. The correlation network is also known for these structure function relationships – hubs, while not as obviously lethal, can be enhanced to reveal lethal properties [11], clusters have been found to represent real sets of functionally related genes [3], and gateway nodes give insights into which genes play a pivotal role in the changes in expressions from one environment to another.

To investigate the novelty of gateway nodes in a number of datasets with mediocre differential gene analysis results, three datasets from varying brain tissues of mice at 2 to 3 ages were analyzed using the gateway nodes approach. It can be speculated that cluster density has an impact on biological function in correlation networks. The nature of correlation network construction suggests that in a network where genes are nodes and edges are correlated patterns of expression, a clique (a network where all nodes are connected to all other nodes) is theoretically a more reliable or likely representation of co-expression than a less connected cluster (also known as a semi-clique). Consider two "clusters" of 5 genes each, one where all 5 nodes are completely connected (10 edges) and another where the cluster is only semi-complete (say, having 7 edges or 70 % edge density). In the example 1 below, clusters A, B, and C all contain 7 edges – in example B, it seems likely that edge 4—5 is incorrect, and 1-2-3-4 are likely co-expressed. In C, it seems likely by examining $K_3$s that 1-2-3, 1-3-5, and 3-4-5 are all highly correlated, but if that were truly the case, it would stand to reason that 2-5 and 2-4 should also be connected.

The best evidence without examining cluster substructure is example A in this case, the densest. To investigate the influence of density related cutoffs on the gateway node, clusters were analyzed density filters of 50–100 % (50 %), 65–100 % (65 %), 75–100 % (75 %) and 85–100 % (85 %). The goal of this initial study was to analyze aging in the brain and possibly identify the pathway players with roles in neuronal growth and differentiation. Gateway node analysis was again used for its application to aging and for its design for identifying temporal expression changes. The beauty of this and other case studies is that they satisfy a need for application of methods to real world data and testing of hypotheses. The results of this study reveal a number of genes that are known players of change in aging in the mouse brain, and highlights how gateway nodes can be used to identify targets of further study in similar cases.

To probe the robustness and consistency of our results, we implemented two manipulations of our workflow to how implementation affects results. This was achieved by making slight changes to network creation (using a different method for building correlation network model) and changing the clustering approach (to another method that is fast and designed for identifying dense complexes in protein-protein interaction networks). Extensive research performed by Dempsey et al. also found links between structure and function in networks built from correlation [12–17]. In other words, there is a growing body of research that indicates the correlation network is a valid method for identifying co-expression and potentially co-regulation from high-throughput gene expression data. However, despite these studies, the issue of noise remains in correlation networks. Correlation does not imply causation, and using only one method of quantification does not necessarily capture all relationships between expression patterns. While studies have found that this level of noise diminishes as the number of samples used increases [16], costs and experimental constraints often times limits the number of samples available per high throughput experiment. In the National Center for Biotechnology's (NCBI) Gene Expression Omnibus (GEO) library, the majority of experiments have sample counts of 20 or less [18]. Therefore, the core of the manipulated network creation approach lies in removing samples to determine the variability of the correlation itself, and the dependency of the edge on each individual sample.

For example, if one were to compare two gene probes A and B, each with 10 samples, perhaps the Pearson Correlation Coefficient (PCC) between those two probes is around 50 %. After removing sample 1, the correlation could hypothetically fall to 48 %. After removing sample 2 but adding sample 1 back into the probes, the correlation may rise to 55 %. The total deviation when samples are removed could inform the confidence in a given edge. This method can be used to create a correlation network that necessarily tests for sample removal. Between this Iterative Removal approach and traditional Pearson Correlation approach described in the Methods, both result in networks where gene probes represent genes and their edges represent correlation. In the Iterative Removal approach code, both positive and negative correlations are considered, whereas in the Pearson Correlation network approach, only positive correlations are considered.

# 2 Methods

The network models used were created using data prepared and analyzed with pairwise Pearson Correlation (see *Network Creation and Enrichment Analysis*) and was then clustered and gateway nodes were identified (see *Gateway Node Analysis*). Targets were then identified via model creation and gateway analysis.

## 2.1 Network Creation and Enrichment Analysis

Data was drawn from three microarray expression datasets for this analysis; three were prepared in total: (1) Cerebellum from Balb/C mice at three time points (Young, Middle-aged, and Aged), (2) Striatum from Balb/C mice at three time points (Young, Middle-aged, and Aged), and (3) Hypothalamus from male C57 mice at two time points (Young and Middle-aged). Correlation networks were generated using probes and expression values from using pairwise computation of the Pearson Correlation Coefficient ($\rho$) and correlation threshold of $0.85 \leq \rho \leq 1.00$.

The traditional Pearson Correlation network creation method used is described as such: For each pairwise correlation computation, the Pearson Correlation coefficient (Eq. 1) was determined and hypothesis testing using the Student's T-test was performed; only significant correlations within the threshold of $0.85 \leq \rho \leq 1.00$ (P-value <0.0005) were kept. For some vector $X = <x_1, x_2, \ldots, x_s>$ and some vector $Y = <y_1, y_2, \ldots, y_s>$ where sample size is equal to $s$, $corr(X,Y)$ is equal to Pearson

Correlation ($\rho$). This measures the linear relationship between two vectors and always falls within the range to $-1.00$ to $1.00$.

$$\rho_{x,y} = \frac{\sum_{i=1}^{s} (x_i - \mu_x) * (y_i - \mu_y)}{\sqrt{\sum_{i-1}^{s} x_i^2 * \sum_{i-1}^{n} y_i^2}}, \tag{1}$$

Gene Set Enrichment Analysis was performed using the GeneTrail Analysis tool (http://genetrail.bioinf.uni-sb.de/) [19]. Parameters for each analysis were set as follows:

- Organism: *Mus musculus*
- Analysis type: KEGG, Gene Ontology (manually curated only)
- P-value adjustment: FDR Adjustment
- P-value threshold: 0.05
- Minimum # categories: 2

The Iterative Removal approach is modified from Dempsey and Ali [12] and described as such: In the Iterative Removal approach, each probe pair is once again examined. To determine correlation between edges, the Pearson Correlation Coefficient (Eq. 1) with a threshold of $\pm 0.85 \leq \rho \leq \pm 1.00$ using the Student's T-test (P-value <0.0005). If the correlation for a pair passes both threshold and hypothesis testing, the original PCC is added to a blank correlation array, and the algorithm iteratively removes each sample from both probes. The new correlation with the sample value missing from both vectors is calculated. If the new correlation is above correlation and significance thresholds, it is added to a correlation array for the current probe pair. The removed sample is then added back into the probe vectors (so probes do not get iteratively smaller, but are always their original length minus1 in the iterative removal) and the next sample is removed until all samples have been iteratively removed. Significance thresholds are updated to reflect the new degrees of freedom during iterative sample removal, and correlation arrays where the final size is 1 are not reported (i.e. only the original correlation was above thresholds, but none of the sample removal correlations passed correlation and significance thresholds). Standard deviation was calculated based on the resulting correlation arrays, and a standard deviation threshold of 0.50 was used to determine which edges to keep and which to reject.

## 2.2   Gateway Node Analysis

In brief, gateway nodes are identified by first clustering networks, then networks are overlapped and nodes that have edges in networks from both conditions are iteratively identified. To perform clustering initially, AllegroMCODE [9] was used on each network under the following parameters: Degree cutoff: 10, Node Score: 0.2, K-Core: 10, MaxDepth: 10. Clustering time ranged from 89.436 s (Male C57 young network) to 29,499.495 s (Cerebellum Balb/C young network). Clustering correlation networks is known to improve the lethality enrichment of high degree nodes, largely because important hub nodes in correlation networks are understood to be contained within clusters. While the choice of clustering method may vary, the lethality enrichment

findings. were conducted using AllegroMCODE which identifies many small, dense clusters. As such, this work also includes a cursory review of how clustering density impacts the robustness of the gateway node. After clustering was performed, clusters were filtered to three different arbitrarily chosen density thresholds: clusters at or above 65 % density, at or above 75 % density, and at or above 85 % density. Density is defined as the number of total edges in the network divided by the number of possible edges – in a network with $N$ nodes and no duplicate edges or self-loops, the number of possible edges is equal to $[N*(N-1)]/2$. As the density threshold changes, the number of gateway nodes present within the overlaid network changes, and as such, it is important to consider numerous thresholds to see if a gateway exists as an artifact of clustering or it exists as a true gateway node, or gene that is co-expressed with a unique group of genes in two or more different states.

In the following studies manipulating the original approach, the "Speed and Performance in Clustering" or SPICi algorithm was used to perform clustering. SPICi was chosen for its speed and design for use in biological protein-protein interaction networks to identify clusters [20]. Indeed, where AllegroMCODE could take many hours to run on one network, SPICi is able to cluster networks in this research in under one minute on average. SPICi was run using default parameters except to set a minimum cluster size of 10 (-s 10) and to turn on the "large sparse graph" option (-m 2), in addition to manipulating cluster density. Cluster density filters used for SPICi were 50–100 % (heretofore referred to as 50 %), 65–100 % (65 %), 75–100 % (75 %), and 85–100 % (85 %).

After the clustering step, networks are overlaid on top of one another to identify gateway nodes. The process used to identify these nodes in an automatic way is extensively described in Dempsey and Ali 2014. Briefly, for each node in the clustered, overlaid network, each node is first classified as having edges in one or both networks. If the node has edges in both networks, it is technically considered a gateway. Scoring is then performed by examining the number of edges per gateway per cluster overlap versus the total number of gateway edges (excluding intra-gateway edges). This ratio is the gatewayness score, and reflects the "responsibility" of each gateway in terms of how many edges pass through that particular node from one stage to another. An example of the difference is shown in Fig. 1. Gateway nodes were identified at each density threshold, heretofore referenced as 50 % (at or above 50 % cluster density), 65 % (at or above 65 % cluster density), 75 % (at or above 75 % cluster density), and 85 % (at or above 85 % cluster density).

In the initial analysis, after determination of gateway nodes at each density threshold and Gene Ontology (GO) enrichment of each gateway-connected cluster [21], a model was drawn to connect genes based on shared processes in which the gateway nodes are involved, if known. This model was manually curated using the following resources: Literature via PubMed search and review, KEGG pathway database [22], and NCBI, and included regulatory relationships, inhibitory relationships, binding relationships, etc.

## 3  Results

Before clustering, network sizes ranged from 38k–41k nodes and 312k–7,600k edges. After clustering, node counts ranged from 30–8k and edges from 300–86k. Thus, network sizes changed depending on state and tissue. As described in Table 1, the Male c57 Hypothalamus dataset contained the fewest gateway nodes (3), with only one gene (*Stk30*) found to be robust to changes at 65 %, 75 %, and 85 % cluster densities. The other two gateway nodes were only found at 65 % density (*Tmem204*, *Msx2*). All three gateway nodes in this case had scores of 100 % gatewayness. The Balb/c cerebellar dataset contained 7 gateway nodes, none of which were robust to all three thresholds. Three were robust to two thresholds, but only one of these are non-RIKEN probes (*Extl1*). Two gateways in this set that did not have 100 % gatewayness were found only at 65 % cluster density and were shared between two clusters – *Gm8221* at 48.63 % gatewayness and *Apol7c* at 51.37 % gatewayness. The Balb/c striatum dataset contained the most gateway nodes at 67; however, 19 of these were RIKEN or unknown/unnamed genes. All the gateway nodes in this set were not robust past 65 % cluster density.

The top gateway nodes identified from Table 1 that were robust to density changes were *Stk30*, two RiKEN hypothetical genes, and *Extl11*. Gene Ontology enrichment analysis was performed on all three datasets; the enrichment data was not used particularly for gateway analysis but for consistency and integrity of analysis to ensure biological functions were found, indicating cluster relevance. Results for GO enrichment on Male C57/Bl/6 mice clusters are shown in Table 2 (results for Balb/c datasets not shown). Based on the gateway node analysis, *Stk30* (coding for the RAGE protein) and *Extl1* are the only gateways that are non-RIKEN genes that are robust to multi-clustering thresholds. Based on literature collection and model curation, *Stk30* (aka RAGE) is the most upstream target that interacts with *reactive oxygen species* and is also upstream of *NF-kB*. Gateway node *Msx2* is also upstream of the NF-kB pathway acting as an inhibitor of *Tax* gene which induces NF-kB enhancing transcription factors. *Myb* is a downstream target gateway that has ties to the apoptotic pathway and the NF-kB pathway. *Plcg1* is acted upon by multiple proteins and goes on to influence DAG and IP3, both (way) upstream of NF-kB. Upstream of the same route to NF-kB as *Plcg1,* the gateway node *Grin2b* is influenced by glucose (Fig. 2).

In the second half of our study, we examine how changes to (1) network creation methods and (2) clustering method affect our results. Using the same three datasets, correlation networks were created with the Pearson Correlation (PC) approach and the Iterative Removal (IR) approach. The network sizes for these (Table 3) were generally similar with a few outliers - the closest in size by edge count being Striatum – Aged dataset where there was only a 3,403 edge count difference between PC and IR approaches, and the largest being Hypothalamus – Aged dataset where over a 60 million edge count difference. Already evidence points towards potential for huge inconsistencies between relatively similar approaches (Fig. 3).

Subsequently, every network developed using the PC and IR approach was clustered using the SPICi clustering standalone code assuming a large sparse network and a minimum cluster size of 10. SPICi reports clusters and their respective nodes, average

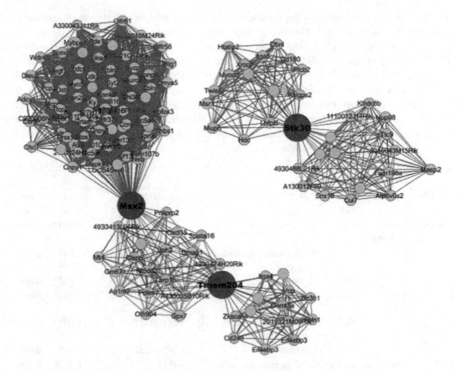

**Fig. 2.** The 65 % gateway clusters from Male c57 Hypothalamus networks. Gateway nodes are in red. These are the clusters examined using Gene Ontology in Table 2 (Color figure online).

**Table 1.** Gateways by dataset. Gateways not present at 1 density only not shown. Column 1: Dataset in which gateway was found. Column 2: Array ID for the gateway. Column 3: Gene Symbol for the gateway. Column 4: Edges running through the gateway. Column 5: Total edges running through gateways connecting the two clusters. Column 6: Gatewayness score. Column 7: If the gateway was found using 65 % edge density clustering, the box is marked. Column 8: If the gateway was found using 75 % edge density clustering, the box is marked. Column 9: If the gateway was found using 85 % edge density clustering, the box is marked.

| Dataset | Array ID | Gene symbol | Edges | Total edges | Gateway score | 65 % | 75 % | 85 % |
|---------|----------|-------------|-------|-------------|---------------|------|------|------|
| Male c57 Hypothalamus | A_51_P493919 | Stk30 | 31 | 31 | 100.00 % | X | X | X |
| | A_51_P478132 | 2210019G11Rik | 205 | 205 | 100.00 % | X | X | |
| | A_52_P78684 | D330040H18Rik | 174 | 174 | 100.00 % | X | X | |
| Balb/c Cerebellum | A_51_P346893 | Extl1 | 182 | 182 | 100.00 % | X | X | |

cluster size, and total number of clusters identified as a part of standard I/O. We used SPICi for its design in biological network application, but also for its speed. It is much more likely that a research scientist working on a large network will choose a command-line option (such as SPICi) over a GUI-based plug-in (such as

**Table 2.** Cluster Gene Ontology Set Enrichment Analysis for Male c57 Hypothalamus dataset. Column 1: The gateway name. Column 2: The cluster connecting that gateway  young or mid (not both combined) and edge color. Gene Ontology enrichment (GO) or KEGG enrichment (KEGG). Column 3: GO/KEGG annotation or pathway name. Column 4: Annotation/pathway ID. Column 5: The number of genes in that annotation/pathway name. Column 6: The p-value associated with that enrichment. Column 7: If down, the cluster has fewer genes in that annotation/pathway than expected for random. If up, the cluster has more genes in that annotation/pathway than expected for random. *FDR Adjustment was used, but if a * is included in the P-value column, this indicates the annotation did not survive P-value adjustment and the noted P-value is the unadjusted value.

| Gateway | Cluster desc. | Source | Category | ID | # Genes | P-value | Enrich. |
|---|---|---|---|---|---|---|---|
| Tmem204 | Aged-Blue | G.O. | Cell | GO:0005623 | 4 | 0.04* | down |
| | | G.O. | Cell part | GO:0044464 | 4 | 0.04* | down |
| Tmem204 | Yng Green | G.O. | Membrane | GO:0016020 | 2 | 0.044* | down |
| | | G.O. | Multicellular organismal process | GO:0032501 | 4 | 0.045* | down |
| | | G.O. | Cytoplasm | GO:0005737 | 5 | 0.048* | down |
| Msx2 | Yng Green | G.O. | Membrane | GO:0016020 | 2 | 0.044* | down |
| | | G.O. | Multicellular organismal process | GO:0032501 | 4 | 0.045* | down |
| | | G.O. | Cytoplasm | GO:0005737 | 5 | 0.048* | down |
| Msx2 | Aged-Blue | KEGG | ECM-receptor interaction | 4512 | 2 | 0.013* | down |
| | | KEGG | Malaria | 5144 | 2 | 0.013* | down |
| | | KEGG | Olfactory transduction | 4740 | 3 | 0.015* | down |
| | | G.O. | Biological regulation | GO:0065007 | 22 | 0.009* | down |
| | | G.O. | Cell | GO:0005623 | 24 | 0.012* | down |
| | | G.O. | Cell part | GO:0044464 | 24 | 0.012* | down |
| | | G.O. | Regulation of biological quality | GO:0065008 | 4 | 0.012* | down |
| | | G.O. | Multicellular organismal process | GO:0032501 | 19 | 0.012* | down |
| | | G.O. | Cellular process | GO:0009987 | 26 | 0.017* | Down |
| | | G.O. | Membrane part | GO:0044425 | 7 | 0.024* | Down |
| | | G.O. | Regulation of biological process | GO:0050789 | 21 | 0.026* | Down |
| | | G.O. | Non-membrane-bounded organelle | GO:0043228 | 6 | 0.034* | Down |
| | | G.O. | Intracellular non-membrane-bounded organelle | GO:0043232 | 6 | 0.034* | Down |
| | | G.O. | Regulation of localization | GO:0032879 | 4 | 0.035* | Down |
| | | G.O. | Cellular component assembly | GO:0022607 | 7 | 0.035* | Down |
| | | G.O. | Cellular component biogenesis | GO:0044085 | 7 | 0.035* | Down |
| | | G.O. | Negative regulation of biological process | GO:0048519 | 7 | 0.041* | Down |
| | | G.O. | Membrane | GO:0016020 | 10 | 0.044* | Down |
| | | G.O. | System process | GO:0003008 | 12 | 0.045* | Down |

*(Continued)*

**Table 2.** (*Continued*)

| Gateway | Cluster desc. | Source | Category | ID | # Genes | P-value | Enrich. |
|---------|---------------|--------|----------|-----|---------|---------|---------|
| | | G.O. | Regulation of cellular process | GO:0050794 | 16 | 0.046* | Down |
| | | G.O. | Molecular_function | GO:0003674 | 51 | 0.046* | Up |
| Stk30 | Aged-Blue | KEGG | Phagosome | 4145 | 2 | 0.038* | Up |
| | | G.O. | Binding | GO:0005488 | 8 | 0.021* | Up |
| | | G.O. | Plasma membrane | GO:0005886 | 4 | 0.031* | Up |
| | | G.O. | Membrane | GO:0016020 | 4 | 0.031* | Up |
| | | G.O. | Cytosol | GO:0005829 | 3 | 0.035* | Up |
| | | G.O. | Regulation of localization | GO:0032879 | 2 | 0.038* | Up |
| Stk30 | Yng Green | G.O. | Organelle | GO:0043226 | 2 | 0.042 | Up |
| | | G.O. | Membrane-bounded organelle | GO:0043227 | 2 | 0.042 | Up |
| | | G.O. | Intracellular organelle | GO:0043229 | 2 | 0.042 | Up |
| | | G.O. | Intracellular membrane-bounded organelle | GO:0043231 | 2 | 0.042 | Up |
| | | G.O. | Intracellular part | GO:0044424 | 2 | 0.042 | Up |

**Table 3.** Network sizes for the PC and IR approach.

| Dataset | Network | # Samples | Pearson correlation edge count | Iterative removal edge count |
|---------|---------|-----------|-------------------------------|------------------------------|
| Cerebellum | Young | 6 | 882,618 | 1,919,975 |
| | Middle-aged | 7 | 6,048,899 | 12,544,493 |
| | Aged | 4 | 708,989 | 1,774,660 |
| Hypothalamus | Young | 5 | 402,084 | 682,679 |
| | Aged | 5 | 399,620 | 60,666,219 |
| Striatum | Young | 3 | 439,701 | 141 |
| | Middle-aged | 5 | 873,089 | 1,451,391 |
| | Aged | 6 | 3,145,291 | 3,141,888 |

AllegroMCODE) to avoid runtime bottlenecks. As such, we also report the real runtime needed to cluster using SPICi in Table 4.

Table 4 shows a high degree of consistency in terms of average cluster sizes (even despite the difference in network size) but does not tell much about cluster content except in the Striatum – Young dataset. These results lead to the important question - are the clusters found in the PC approach similar to, or the same as, any of the clusters found in the IR approach? We cannot rely on cluster size or count alone to determine biological similarity of the findings. We also find that runtime is *very* fast compared to AllegroMCODE and with regard to runtime overall, while the IR approach seems to take slightly longer on average than the PC approach, the length of runtime in any dataset never reaches over 6 min – an exceedingly fast algorithm for these models.

Basic gateway analysis was performed on both the PC and IR SPICi-clustered networks. Table 5 shows not only the number gateways for each intra-dataset network

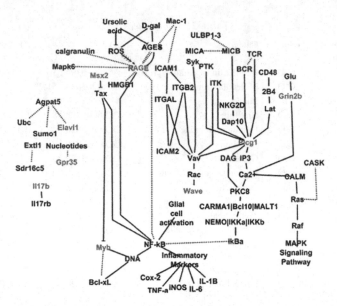

**Fig. 3.** The curated gateway nodes model. Genes/proteins in red are gateway nodes as listed above. Not all nodes listed above are in the model if they do not fit or pathway information is not available. This model is not comprehensive (Color figure online).

comparison (Young vs. Middle-aged, Middle-Aged vs. Aged, or Young vs. Aged) but also shows the overlap between gateways for IR and PC approaches. There is a very high number of gateways when using SPICi clustering (always ~ 10,000 gateways) when compared to AllegroMCODE (in the hundreds), and comparing the overlap between PC and IR network creation approaches shows that not many of the gateways identified by the approaches overlap. The percent overlap described in Table 5 is calculated as *Percent overlap = Number of gateways in both IR and PC approach/Total number of gateways*. As such, if the gateways were completely the same, the percent overlap would be at or near 100 %, and if they were totally discrete, would be at or near 0 %. With percent overlap scores for all comparisons below 50 %, this shows that consistency when manipulating results should be performed very carefully with a critical eye. To avoid the common "garbage-in, garbage-out" occurrence when manipulating biomedical big data, one should become intimately acquainted with their approaches instead of blindly running them as described.

## 4   Discussion

Based on the gateway node analysis, *Stk30* (coding for the RAGE protein) and *Extl1* are the only gateways that are non-RIKEN genes that are also robust to multi-clustering thresholds. The literature collected and resulting model reveal that *Stk30* aka RAGE is the most upstream target that interacts with *reactive oxygen species* and is also upstream of *NF-kB*. In entirety, the model proposed above points to activation of

**Table 4.** Clustering runtimes, counts, and average sizes for both PC and IR approaches.

| Dataset | Age | Density | Pearson Correlation | | | Iterative Removal | | |
|---------|-----|---------|---------|------------|---------------------|---------|------------|---------------------|
| | | | Runtime | # Clusters | Avg Cluster Size | Runtime | # Clusters | Avg Cluster Size |
| Cereb. | Young | 50 | 0m5.116s | 768 | 27.7565 | 0m9.635s | 987 | 36.9757 |
| | | 65 | 0m5.083s | 647 | 22.3754 | 0m10.956s | 1269 | 27.6493 |
| | | 75 | 0m5.015s | 806 | 20.1402 | 0m9.791s | 1349 | 24.0667 |
| | | 85 | 0m5.030s | 735 | 17.8503 | 0m11.401s | 1379 | 20.8463 |
| | Middle-aged | 50 | 0m32.577s | 513 | 35.5088 | 1m1.222s | 729 | 48.3141 |
| | | 65 | 0m33.223s | 521 | 33.6468 | 1m0.870s | 847 | 39.1133 |
| | | 75 | 0m33.520s | 513 | 31.0643 | 1m0.691s | 874 | 35.3856 |
| | | 85 | 0m34.027s | 438 | 30.726 | 1m0.490s | 823 | 33.4569 |
| | Aged | 50 | 0m4.056s | 664 | 32.1235 | 0m7.980s | 804 | 48.2786 |
| | | 65 | 0m4.037s | 751 | 27.1132 | 0m8.058s | 998 | 38.485 |
| | | 75 | 0m3.995s | 795 | 24.0943 | 0m7.921s | 1160 | 32.4216 |
| | | 85 | 0m4.046s | 794 | 21.4509 | 0m8.031s | 1281 | 27.751 |
| Hypo | Young | 50 | 0m2.413s | 1051 | 18.607 | 0m3.566s | 1340 | 28.8694 |
| | | 65 | 0m2.413s | 1051 | 18.607 | 0m3.572s | 1756 | 19.3166 |
| | | 75 | 0m2.474s | 1088 | 15.8879 | 0m3.639s | 1886 | 15.851 |
| | | 85 | 0m2.460s | 877 | 14.2896 | 0m3.615s | 1551 | 13.5145 |
| | Aged | 50 | 0m2.384s | 905 | 23.3238 | 5m8.276s | 36 | 1139.81 |
| | | 65 | 0m2.388s | 1041 | 18.4352 | 5m10.709s | 52 | 788.86 |
| | | 75 | 0m2.447s | 1044 | 15.9866 | 5m17.714s | 68 | 602.779 |
| | | 85 | 0m2.412s | 857 | 14.3349 | 5m3.965s | 92 | 444.837 |
| Striatum | Young | 50 | 0m2.592s | 825 | 29.297 | 0m0.099s | 0 | 0 |
| | | 65 | 0m2.670s | 892 | 26.6715 | 0m0.099s | 0 | 0 |
| | | 75 | 0m2.620s | 966 | 23.8747 | 0m0.099s | 0 | 0 |
| | | 85 | 0m2.542s | 1033 | 20.5653 | 0m0.099s | 0 | 0 |
| | Middle-aged | 50 | 0m5.082s | 619 | 33.538 | 0m7.279s | 989 | 35.6087 |
| | | 65 | 0m5.118s | 734 | 26.5763 | 0m7.025s | 1232 | 26.9854 |
| | | 75 | 0m4.960s | 777 | 22.9897 | 0m7.052s | 1314 | 22.968 |
| | | 85 | 0m4.927s | 735 | 20.8517 | 0m7.037s | 1289 | 19.5896 |
| | Aged | 50 | 0m17.676s | 951 | 38.4175 | | 949 | 38.4826 |
| | | 65 | 0m17.309s | 1190 | 29.3639 | 0m15.323s | 1190 | 29.3723 |
| | | 75 | 0m17.339s | 1235 | 25.9717 | 0m15.478s | 1227 | 26.1565 |
| | | 85 | 0m17.697s | 1218 | 23.2537 | 0m15.190s | 1220 | 23.2852 |

inflammation via NF-kB and RAGE as a map for aging in normal Balb/C and C57 mouse brain. A 2009 review by Kriete and Mayo confirms a link between NF-kB activation and aging, but calls for further investigation of the role of NF-kB outside its well-studied role in the innate immune system [23]. In our model, gateway node *Msx2* is also upstream of the NF-kB pathway acting as an inhibitor of *Tax* gene which induces NF-kB enhancing transcription factors. *Myb* is a downstream target gateway that has ties to the apoptotic pathway and the NF-kB pathway. *Plcg1* is acted upon by multiple proteins and goes on to influence DAG and IP3, both (way) upstream of NF-kB. Upstream of the same route to NF-kB as *Plcg1* is *Grin2b*, influenced by glucose. All of these genes have potential as effectors for change in the NF-kB pathway, either upstream or downstream, but perhaps the most important element in the model due to gateway robustness is the RAGE protein, encoded by gateway node *Stk30*. A 2003 study by Deane et al. revealed that RAGE is a mediator of

**Table 5.** Gateway node counts and percent overlaps for both PC and IR network creation approaches using SPICi clustering.

| Dataset | Density | Gateway Comparison | PC Gateway Count | IR Gateway Count | % Identity |
|---|---|---|---|---|---|
| Cerebellum | 50 | Yng vs Mid | 16,051 | 31,521 | 44.55% |
| | 65 | Yng vs Mid | 13,188 | 28,738 | 40.28% |
| | 75 | Yng vs Mid | 10,680 | 25,221 | 36.38% |
| | 85 | Yng vs Mid | 7,730 | 20,436 | 31.72% |
| | 50 | Mid vs Aged | 16,682 | 33,252 | 45.19% |
| | 65 | Mid vs Aged | 14,386 | 30,971 | 41.79% |
| | 75 | Mid vs Aged | 12,597 | 28,391 | 38.96% |
| | 85 | Mid vs Aged | 9,684 | 24,076 | 34.48% |
| Hypothalamus | 50 | Yng vs Aged | 17,167 | 35,890 | 41.32% |
| | 65 | Yng vs Aged | 14,735 | 33,082 | 35.36% |
| | 75 | Yng vs Aged | 11,469 | 29,779 | 28.60% |
| | 85 | Yng vs Aged | 6,359 | 20,856 | 17.76% |
| Striatum | 50 | Yng vs Mid | 19,109 | N/a | 0.00% |
| | 65 | Yng vs Mid | 17,826 | N/a | 0.00% |
| | 75 | Yng vs Mid | 15,940 | N/a | 0.00% |
| | 85 | Yng vs Mid | 12,863 | N/a | 0.00% |
| | 50 | Mid vs Aged | 18,542 | 31,433 | 48.53% |
| | 65 | Mid vs Aged | 16,688 | 28,498 | 45.76% |
| | 75 | Mid vs Aged | 14,212 | 24,082 | 42.82% |
| | 85 | Mid vs Aged | 10,974 | 18,164 | 0.00% |

disease-causing amyloid-beta proteins into the central nervous system, and even suggests it as a target for potential future therapies for Alzheimer's disease [24]. RAGE has been found to be up-regulated in Alzheimer's patients [25]. A 2004 study of transgenic mice with manipulated RAGE (mAPP⁻/RAGE⁻) expression by Arancio et al. found that pups displayed issues with spatial memory and the NF-kB pathway is activated, and again find it a potential target for Alzheimer's intervention [26]. Multiple other evidences exist to substantiate the speculation that RAGE plays a role in normal aging; an October 2013 PubMed search of "RAGE" + "Aging" reveals over 100 articles relating RAGE and aging dating back to 1999.

Application of the gateway nodes approach allows for the utilization of the gateway nodes approach to determine better targets for study in the aging mouse brain. The accompanying model provides a roadmap that points us toward RAGE, Msx2, and Plcg1 as upstream targets for manipulation of expression in the mouse brain. These genes all have indirect roles in the NF-kB pathway; it has recently been shown that inhibition of NF-kB in the mouse hypothalamus resulted in a 20 % increase in lifespan, improved cognition, and levels of muscle, bone, and skin tissue typically observed in younger mice. This suggests that the gateway nodes approach is able to identify genes with major roles in aging, particularly using a robust approach. This method is able to take sets of 30,000+ genes or gene probes and narrow it down to only a few targets of interest, and their potential relationships based on network modeling of expression correlation and integration of publicly available databases. Particularly in areas of research where little is understood, funding is not readily available, or resources are tight, the gateway nodes approach can provide a robust, reproducible, and reliable way to identify targets of interest in further research.

Certainly, current methods for analyzing gene expression capture just a snapshot of cellular activity at a given time, not a dynamic process. However, the minimal overlap of co-expression relationships in the network form confirm that the cellular environment is dynamic and spontaneous. This begs the question – does a snapshot of the cell, even in multiple replicates – accurately capture the goings-on of cellular activity? If we were able to understand how we got from point A to point B, we would better understand how these gateway nodes came about. Surely on a short-term basis gateway nodes could arise form differential regulation of expression, but in the long term, the question is whether the clusters captured are a result of a short-term cellular change or a compensatory effect of loss of previous gene function. To improve the dimensionality of these analyses without vastly increasing the data load, one might consider modifying their gene expression research design to include 3 or more time points and to include a high number of replicates for each time point (ideally, 5 or more). While this is certainly not always feasible due to cost, labor, or difficulty in sample preparation, it could be considered to help understanding of cellular dynamics using a network model.

This research has also demonstrated the vulnerability of network based approaches to inconsistencies in results based on workflow manipulation. No doubt that while some of the gateway nodes identified in all three approaches described above do overlap, the amount of noise generated by switching from one method to the next – be that model creation, clustering, or analysis – might yield vastly different results. We have shown that scientists need to guard against this inconsistency and place value on creating workflows that are reproducible and robust to change- however possible this may be.

# References

1. Benson, M., Breitling, R.: Network theory to understand microarray studies of complex diseases. Curr. Mol. Med. **6**(6), 695–701 (2006)
2. Reverter, A., Chan, E.K.: Combining partial correlation and an information theory approach to the reversed engineering of gene co-expression networks. Bioinformatics **24**(21), 2491–2497 (2008). doi:10.1093/bioinformatics/btn482
3. Horvath, S., Dong, J.: Geometric interpretation of gene coexpression network analysis. PLoS Comput. Biol. **4**(8), e1000117 (2008). doi:10.1371/journal.pcbi.1000117
4. Dempsey, K.M., Ali, H.H.: Identifying aging-related genes in mouse hippocampus using gateway nodes. BMC Syst. Biol. **8**, 62 (2014). doi:10.1186/1752-0509-8-62
5. Barabasi, A.L., Albert, R.: Emergence of scaling in random networks. Science **286**(5439), 509–512 (1999). doi:7898. [pii]
6. Jeong, H., Mason, S.P., Barabasi, A.L., Oltvai, Z.N.: Lethality and centrality in protein networks. Nature **411**(6833), 41–42 (2001). doi:10.1038/35075138
7. Barabasi, A.L., Oltvai, Z.N.: Network biology: understanding the cell's functional organization. Nat. Rev. Genet. **5**(2), 101–113 (2004). doi:10.1038/nrg1272
8. Albert, R.: Scale-free networks in cell biology. J. Cell Sci. **118**(Pt 21), 4947–4957 (2005). doi:118/21/4947. [pii]

9. Bader, G.D., Hogue, C.W.: An automated method for finding molecular complexes in large protein interaction networks. BMC Bioinform. **4**, 2 (2003)

10. Michaut, M., Baryshnikova, A., Costanzo, M., et al.: Protein complexes are central in the yeast genetic landscape. PLoS Comput. Biol. **7**(2), e1001092 (2011). doi:10.1371/journal.pcbi.1001092

11. Dempsey, K., Thapa, I., Bastola, D., Ali, H.: Functional identification in correlation networks using gene ontology edge annotation. Int. J. Comput. Biol. Drug Des. **5**(3–4), 222–244 (2012). doi:10.1504/IJCBDD.2012.049206

12. Dempsey, K., Ali, H.: On the robustness of the biological correlation network model. In: International Conference on Bioinformatics Models, Methods and Algorithms (BIOINFORMATICS 2014), pp. 186–195 (2014)

13. Dempsey, K., Ali, H.: On the discovery of cellular subsystems in gene correlation networks using measures of centrality. Curr. Bioinform. **8**(3), 305–314 (2013)

14. Dempsey, K., Bhowmick, S., Ali, H.: Function-preserving filters for sampling in biological networks. Procedia Comput. Sci. **9**, 587–595 (2012). doi:10.1016/j.procs.2012.04.063

15. Dempsey, K., Duraisamy, K., Bhowmick, S., Ali, H.: The development of parallel adaptive sampling algorithms for analyzing biological networks. In: 2013 IEEE International Symposium on Parallel and Distributed Processing, Workshops and PhD Forum, pp. 725–734. doi:10.1109/IPDPSW.2012.90 (2012)

16. Dempsey, K., Thapa, I., Cortes, C., Eriksen, Z., Bastola, D.K., Ali, H.: On mining biological signals using correlation networks. In: 2013 IEEE 13th International Conference on Data Mining Workshops, pp. 327–334. doi:10.1109/ICDMW.201 (2013)

17. Khazanchi, R., Dempsey, K., Thapa, I., Ali, H.: On identifying and analyzing significant nodes in protein-protein interaction networks. In: 2013 IEEE 13th International Conference on Data Mining Workshops, pp. 343–348. doi:10.1109/ICD (2013)

18. Barrett, T., Wilhite, S.E., Ledoux, P., et al.: NCBI GEO: archive for functional genomics data sets–update. Nucleic Acids Res. **41**(Database issue), D991–D9915 (2013). doi:10.1093/nar/gks1193

19. Backes, C., Keller, A., Kuentzer, J., et al.: GeneTrail–advanced gene set enrichment analysis. Nucleic Acids Res. **35**(Web Server issue), W186–W192 (2007). doi:10.1093/nar/gks1193

20. Jiang, P., Singh, M.: SPICi: a fast clustering algorithm for large biological networks. Bioinformatics **26**(8), 1105–1111 (2010). doi:10.1093/bioinformatics/btq078

21. Ashburner, M., Ball, C.A., Blake, J.A., et al.: Gene ontology: tool for the unification of biology. the gene ontology consortium. Nat. Genet. **25**(1), 25–29 (2000). doi:10.1038/75556

22. Aoki, K.F., Kanehisa, M.: Using the KEGG database resource. Curr. Protoc. Bioinform. Chapter 1: Unit 1.12. 10.1002/0471250953.bi0112s11 (2005)

23. Kriete, A., Mayo, K.L.: Atypical pathways of NF-kappaB activation and aging. Exp. Gerontol. **44**(4), 250–255 (2009). doi:10.1016/j.exger.2008.12.005

24. Deane, R., Du Yan, S., Submamaryan, R.K., et al.: RAGE mediates amyloid-beta peptide transport across the blood-brain barrier and accumulation in brain. Nat. Med. **9**(7), 907–913 (2003). doi:10.1038/nm890

25. Leclerc, E., Sturchler, E., Vetter, S.W., Heizmann, C.W.: Crosstalk between calcium, amyloid beta and the receptor for advanced glycation endproducts in alzheimer's disease. Rev. Neurosci. **20**(2), 95–110 (2009)

26. Arancio, O., Zhang, H.P., Chen, X., et al.: RAGE potentiates abeta-induced perturbation of neuronal function in transgenic mice. EMBO J. **23**(20), 4096–4105 (2004). doi:10.1038/sj.emboj.7600415

# Machine Reading for Extraction of Bacteria and Habitat Taxonomies

Parisa Kordjamshidi[1,2]([✉]), Wouter Massa[2], Thomas Provoost[2],
and Marie-Francine Moens[2]

[1] Department of Computer Science, University of Illinois at Urbana-Champaign,
201 North Goodwin Avenue, Urbana, IL 61801-2302, USA
parisa.kordjamshidi@illinois.edu
[2] Department of Computer Science, KU Leuven,
Celestijnenlaan 200A, 3001 Heverlee, Belgium
wouter.massa@student.kuleuven.be,
{thomas.provoost,sien.moens}@cs.kuleuven.be

**Abstract.** There is a vast amount of scientific literature available from various resources such as the internet. Automating the extraction of knowledge from these resources is very helpful for biologists to easily access this information. This paper presents a system to extract the bacteria and their habitats, as well as the relations between them. We investigate to what extent current techniques are suited for this task and test a variety of models in this regard. We detect entities in a biological text and map the habitats into a given taxonomy. Our model uses a linear chain Conditional Random Field (CRF). For the prediction of relations between the entities, a model based on logistic regression is built. Designing a system upon these techniques, we explore several improvements for both the generation and selection of good candidates. One contribution to this lies in the extended flexibility of our ontology mapper that uses an advanced boundary detection and assigns the taxonomy elements to the detected habitats. Furthermore, we discover value in the combination of several distinct candidate generation rules. Using these techniques, we show results that are significantly improving upon the state of art for the BioNLP Bacteria Biotopes task.

## 1 Introduction

A vast amount of scientific literature is available about bacteria biotopes and their properties [1]. Processing this literature can be very time-consuming for biologists, as efficient mechanisms to automatically extract information from these texts are still limited. Biologists need information about ecosystems where certain bacteria live in. Hence, having methods that rapidly summarize texts and list properties and relations of bacteria in a formal way becomes a necessity. Automatic normalization of the bacteria and biotope mentions in the text against certain ontologies facilitates extending the information in ontologies and databases of bacteria. Biologists can then easily query for specific properties or relations, e.g. which bacteria live in the gut of a human or in which habitat *Bifidobacterium Longum* lives.

© Springer International Publishing Switzerland 2015
A. Fred et al. (Eds): BIOSTEC 2015, CCIS 574, pp. 239–255, 2015.
DOI: 10.1007/978-3-319-27707-3_15

The Bacteria Biotopes subtask (BB-Task) of the BioNLP Shared Task (ST) 2013 is the basis of this study. It is the third event in this series, following the same general outline and goals of the previous events [2]. BioNLP-ST 2013 featured six event extraction tasks all related to "Knowledge base construction". It attracted wide attention, as a total of 38 submissions from 22 teams were received.

The BB-Task consists of three subtasks. In the first subtask habitat entities need to be detected in a given biological text and the entities must be mapped onto a given ontology. The habitat entities vary from very specific concepts like '*formula fed infants*' to very general concepts like '*human*'. The second subtask is focused on the extraction of two relations: a *loc* and a *partof* relation. These relations need to be predicted between a given set of entities (bacteria, habitats and geographical locations). *loc* relations occur between a bacterium and a habitat or geographical location, *partof* relations only occur between habitats. The third subtask is an extended combination of the two other subtasks: entities need to be detected in a text and relations between these entities need to be extracted. In this paper we focus on the first two subtasks.

We first describe related work done in context of the BioNLP-ST (Sect. 2). We then discuss our methodology for the two subtasks (Sect. 3). Next, we discuss our experiments and compare our results with the official submissions to BioNLP-ST 2013 (Sect. 4). We end with a conclusion (Sect. 5).

## 2   Related Work

The BB-task along with the experimental dataset has been initiated for the first time in the BioNLP Shared Task 2011 [3]. Three systems were developed in 2011 and five systems for its extended version proposed in the 2013 shared task [1]. In 2011 the following systems participated in this task. TEES [4] was proposed by UTurku as a generic system which uses a multi-class Support Vector Machine classifier with linear kernel. It made use of Named Entity Recognition patterns and external resources for the BB model. The second system was JAIST [5], specifically designed for the BB-task. It uses CRFs for entity recognition and typing and classifiers for coreference resolution and event extraction. The third system was Bibliome [6], also specifically designed for this task. This system is rule-based, and exploits patterns and domain lexical resources.

The three systems used different resources for Bacteria name detection which are the List of Prokaryotic Names with Standing in Nomenclature (LPNSN), names in the genomic BLAST page of NCBI and the NCBI Taxonomy, respectively. The Bibliome system was the winner for detecting the Bacteria names as well as for the coreference resolution and event extraction. The important factor in their outperformance was exploiting the resources and ontologies. They found useful matching patterns for the detection of entities, types and events. Using their manually drawn patterns and rules performed better than other task participant systems, in which learning models apply more general features.

In the 2013 edition of this task, the event extraction is defined in a similar way but an extension to the 2011 edition considered biotope normalization using a large ontology of biotopes called OntoBiotope. The task was proposed in three subtasks to which we pointed in Sect. 1. Five teams participated in these subtasks. In the first subtask all entities have to be predicted, even if they are not involved in any relation. The participated systems performed reasonably well. However, the difficulty of this task has been boundary detection.

The participating systems obtained a very low recall for the relation extraction even when the entities and their boundaries are given (subtask 2 and 3). The difficulty of the relation extraction is partially due to the high diversity of bacteria and locations. The many mentions of different bacteria and localization in the same paragraph makes it difficult to select the right links between them. The second difficulty lies in the high frequency of anaphora. This makes the extraction of the relations beyond sentence level difficult. The strict results of the third task were very poor, due to struggling with the difficulties of both previous tasks i.e., boundary detection and link extraction.

For detecting entities (subtask 1), one submission [7] worked with generated syntactical rules. Three other submissions [8–10] used an approach similar to ours. They generated candidates in an initial phase from texts. These candidates were subsequently selected by trying to map them onto the ontology. Two submissions [8,10] generated candidates by extracting noun phrases. One submission [9] used a CRF model to generate candidates, as we do in this work. However, we test candidates more thoroughly and consider every continuous subspan of tokens in each candidate instead of just the candidate itself, which explains our improved results.

For the relation extraction with given entities (subtask 2), there were four submissions. One system from LIMSI [9] relied solely on the fact that the relation was seen in the training set which fails to yield a reasonable accuracy. A second system BOUN [10] extracted relations using only simple rules, e.g. in a specific paragraph they created relations between all locations and the first bacterium in that paragraph. A third system IRISA [8] used a nearest neighbor approach. Another system was TEES [11] (an improved version of the UTurku participation in 2011) which provided the best results. However, the results were still poor.

One reason for this lies in the limited scope of candidates that the submitted systems considered, e.g. TEES [11] and IRISA [8] only examined relations between a habitat and location that occur in the same sentence. One of our contributions lies in considering more possible relations, including relations across sentences. This is confirmed by a much better recall, as can be seen in Sect. 4.3.

## 3   Methodology

In this section we lay out our developed system. For each of subtasks 1 and 2, we first discuss the goal of the subtask, followed by an explanation of our used methodology. The performance of our model is discussed in the next section (Sect. 4).

242    P. Kordjamshidi et al.

## 3.1 Subtask 1: Entity Detection and Ontology Mapping

The goal of this subtask is to detect habitat entities in texts and map them onto concepts defined by the OntoBiotope-Habitat ontology. For each entity the name, the location in the text and the corresponding ontology entry need to be predicted. E.g. the expected output for a text consisting of the single sentence *"This organism is found in adult humans and formula fed infants as a normal component of gut flora."* is:

```
T1 Habitat 27 33 adult humans
T2 Habitat 44 63 formula fed infants
T3 Habitat 44 51 formula
T4 Habitat 89 92 gut
N1 OntoBiotope Annotation:T1 Ref:MBTO:00001522
N2 OntoBiotope Annotation:T2 Ref:MBTO:00000308
N3 OntoBiotope Annotation:T3 Ref:MBTO:00000798
N4 OntoBiotope Annotation:T4 Ref:MBTO:00001828
```

Four habitat entities are found in this sentence and they are mapped onto four different ontology entries.

Figure 1 gives an overview of the followed approach. We first search in the text for token spans (candidates) that might contain one or more entities (Sect. 3.1). These generated candidates are given to a Candidate Selection module, that searches substrings within the candidate for entities (Sect. 3.1). This Candidate Selection module uses an Ontology Mapper (Sect. 3.1), finding the ontology entry that matches closest to a given substring. Additionally it returns a dissimilarity value to give an indication of how close the match is. Based on this dissimilarity, we can decide to classify part of a candidate as the given entry or not.

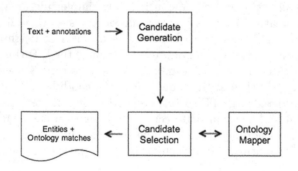

**Fig. 1.** Overview of the followed approach in subtask 1.

**Candidate Generation.** The Candidate Generation module generates token spans from a given input text. The goal of the Generation module is to quickly reduce a large text to a candidate set that can be analysed more efficiently. First the text is split into sentences and tokens, then every sentence is mapped onto a set of candidates. We use the given annotation files of the Stanford Parser [12] to split the texts and tokenize the sentences. Sentences are assumed to be independent in the model, i.e. we do not use information from one sentence in another sentence.

*Conditional Random Fields.* To generate candidates, we use Conditional Random Fields (CRF) [13]. In particular, we choose a linear chain CRF; previous research shows that these perform well for various natural language processing tasks, especially Named Entity Recognition [14]. In contrast to general purpose noun phrase extractors used by some other existing models for this task, a CRF can easily exploit the information of the given annotated files as features.

A CRF model is an undirected probabilistic graphical model $G = (V, E)$ with vertices $V$ and edges $E$. The vertices represent a set of random variables with the edges showing the dependencies between them. The set of observed random variables is denoted by $X$ and the unknown/output random variables are denoted by $Y$. This model represents a probability distribution over a large number of random variables by a product of local functions that each depend on a small subset of variables, called factors.

A CRF generally defines a probability distribution $p(\mathbf{y}|\mathbf{x})$, where $\mathbf{x}, \mathbf{y}$ are specific assignments of respective variables $X$ and $Y$ as follows:

$$p(\mathbf{y}|\mathbf{x}) = \frac{1}{Z(\mathbf{x})} \prod_{\Psi_F \in G} \Psi_F(x_F, y_F) \tag{1}$$

where $x_F$ are those observed variables that are part of factor $F$ and similarly for $y_F$ and $Y$. $\Psi_F : V^n \to \mathbb{R}$ is the potential function associated with factor $F$ and is defined in terms of the features $f_{Fk}(x_F, y_F)$ as:

$$\Psi_F(x_F, y_F) = \exp\left\{\sum_k \lambda_{Fk} f_{Fk}(x_F, y_F)\right\} \tag{2}$$

The parameters of the conditional distribution $\lambda_{Fk}$ are trained with labelled examples. Afterwards, using the trained model the most probable output variables can be calculated for a given set of observed variables. $Z(\mathbf{x})$ is a normalization constant and is computed as:

$$Z(\mathbf{x}) = \sum_{\mathbf{y}} \prod_{\Psi_F \in G} \Psi_F(x_F, y_F) \tag{3}$$

CRFs can represent any kind of dependencies, but the most commonly used model, particularly in the NLP tasks such as Named Entity Recognition is the Linear-chain model. In this work, we use the linear chain implementation in Factorie [15]. Linear chain CRFs consider the dependency between the labels of the adjacent words. In other words, each local function $f_k(y_t, y_{t-1}, x_t)$ represents the dependency of each output variable $y_t$ in location $t$ in the chain to its previous output variable $y_{t-1}$ and the observed variable $x_t$ at that location. The global conditional probability then is computed as the product of these local functions [13]. With the usual assumption that all local functions share parameters and feature functions, its log-linear form is now written as:

$$p(\mathbf{y}|\mathbf{x}) = \frac{1}{Z(\mathbf{x})} \exp\left\{\sum_t \sum_k \lambda_k f_k(y_t, y_{t-1}, x_t)\right\} \tag{4}$$

where the normalization constant is derived in an analogous manner to equation (3) for the case of sequential dependencies.

In our model, every token is an observed variable. The biological entity labels (e.g. 'Bacterium') of the tokens are the output variables (or labels).

We now discuss the features that we use, along with the label set into which the tokens are classified.

*CRF Features.* The following features are used for each token:

- Token string
- Stem
- Length token
- Is capitalized (binary)
- Token is present in the ontology (binary)
- Stem is present in the ontology (binary)
- Category of the token in the Cocoa annotations
- Part-of-speech tag
- Dependency relation to the head of the token

The stem is calculated using an online available Scala implementation[1] of Porter's stemming algorithm [16]. The part-of-speech tag and the dependency relation to the head are added using the available annotation files from the Stanford Parser [12]. Cocoa[2] is a dense annotator for biological text. The Cocoa annotations cover over 20 different semantic categories like 'Processes' and 'Organisms'.

*CRF Labels: Extended Boundary Detection Tags.* We use five different labels for the tokens. The used labels are:

- Start: The token is the first token of an entity.
- Center: The token is in the middle of an entity.
- End: The token is the last token of an entity.
- Whole: The token itself is an entity.
- None: The token does not belong to an entity.

The most immediate alternative to this is the traditional IOB labeling [17]. An even more simple possibility is a binary labeling that just indicates if a token belongs to an entity mention or not. The more elaborated proposed labeling generally performs better in our tests.

---

[1] https://github.com/aztek/porterstemmer
[2] http://npjoint.com

**Ontology Mapper.** The Ontology Mapper maps a string onto the ontology entry with the lowest dissimilarity. The dissimilarity between an ontology entry and string is calculated by comparing the string with the name, synonyms and plural of the name and synonyms of the entry with respect to a certain comparison function. The plurals are calculated simply by just adding '*s*' or '*es*' to the end of the singular form.

To compare two strings they are split into tokens. The tokens from the two strings are matched to minimize the sum of the relative edit distance between the matched tokens. If not all tokens can be matched i.e. the number of tokens in the two strings are different, 1.0 is added to the sum for each remaining token. As a measure for relative edit distance, we use the Levenshtein distance [18] divided by the sum of the lengths of the strings to get a number between 0.0 and 1.0.

**Candidate Selection.** The Candidate Selection module receives spans of tokens as input, it searches within these spans for ontology entries. For each span every continuous subspan of tokens is tested with the Ontology Mapper. This means that we select $\frac{n(n+1)}{2}$ subspans for every token span with $n$ tokens. If for a subspan a dissimilarity lower than a specific bound is reached, we classify this subspan as an entity. E.g. for the token span '*formula fed infants*', six subspans are selected: '*formula*', '*fed*', '*infants*', '*formula fed*', '*fed infants*' and '*formula fed infants*'. '*formula*' and '*formula fed infants*' are found in the ontology and we classify these as entities.

Based on cross-validation experiments on the training and development set, we decided to take as maximal dissimilarity 0.1, i.e. the subspan must be very close to an ontology entry. This very strict parameter allows us to be less strict in the Candidate Generation module: every entity that has a minimum probability of 0.1 to contain entities will be tested. The sensitivity of our results with respect to this measure is further discussed in Subsect. 4.3 and Table 2.

**Additional Improvements.**

*Dashed Words* Not all entities in the texts consist of one or more tokens, some entities are only a part of a token. E.g. in the token '*tick-born*', '*tick*' is an entity. To handle these cases we search for all the words that contain one or more dashes. These words are split and every part is matched against the ontology. These parts are easy to match because they are usually just nouns in singular form.

*Extending the Ontology.* Mappings from phrases onto ontology entries are given in the training and development set. These phrases are usually similar to the name or a synonym of the ontology entry. However in rare cases the phrase is not similar to the name or a synonym. Based on the assumption that the given mapping is correct we can extend the ontology. We do this by adding the phrase as a new synonym to the ontology entry. Some submissions to the BioNLP-ST 2013 task used this approach as well [9,10].

*Correcting Boundaries.* An important part of the task is to predict the correct boundaries of the entities. E.g. for the noun phrase '*blood-sucking tsetse fly*', it is not sufficient to predict '*fly*' or '*tsetse fly*'. The whole noun phrase is the correct entity in this case. This particular example is hard because '*blood-sucking tsetse fly*' does not occur in the ontology. To handle this case we add to each found entity the dependent words that precede the entity. These dependent words can be extracted by using the given parser annotations. E.g. from the phrase '*blood-sucking tsetse fly*' the entity '*tsetse fly*' is selected, '*blood-sucking*' is added to it because its headword is '*fly*'.

*Filter out Parents.* Many generated candidates refer to the same entity, it is required that we predict every entity only once. E.g. in the phrase '*person with untreated TB*' the entities '*person*' and '*person with untreated TB*' are detected. They refer both to the same habitat, '*person*' is just a more general term to describe '*person with untreated TB*'. That is why we filter '*person*' out. We can do this by using the parent/child relations given in the ontology. The ontology entry '*person*' is a parent (a more general term) of the entry '*person with untreated TB*', so we only predict the phrase '*person with untreated TB*' in this case. Because the ontology is a deep graph of entities, we test this parent/child relationship recursively.

### 3.2   Subtask 2: Relation Extraction

In this subtask relations need to be extracted from a text based on annotated entities in the text. There are three types of entities: habitats, geographical entities and bacteria. Two types of relations exists: *loc* and *partof*. *loc* relations are always between a bacterium and a habitat or geographical location. *partof* relations occur between two habitats. We handle these two relations independently. In the training and development set combined, *loc* and *partof* relations are responsible for respectively 81 % and 19 % of the relations.

We used a similar approach for both relation types. We will describe our approach for *loc* relations. Our model consists of two modules. A first module generates sets of relation candidates from the text using simple rules (Sect. 3.2). These sets are then forwarded to a second module that trains for each set a separate model (Sect. 3.2). Figure 2 shows a visualization of our approach.

**Candidate Generation.** The Candidate Generation module reduces the set of all possible relations, i.e. all combinations of bacteria and locations, to multiple smaller sets of candidate relations. Every set is created by using a generation rule. These smaller sets are then forwarded to the Candidate Selection module that will try to identify if a candidate relation is really a relation or not.

We use generation rules for two reasons. On one side we decrease the overall number of candidates by a significant amount. On the other side we group similar types of relations to build more specific models. For every set of candidate relations defined by a generation rule, we build a separate model to test

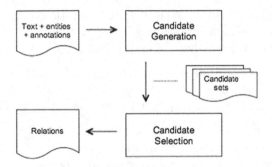

**Fig. 2.** Overview of the followed approach in subtask 2.

these relations. A good candidate generation method generates a relatively large number of correct relations while keeping the number of wrong relations to a minimum.

We tested 5 different candidate generation rules for *loc* relations:

- **All Possible:** All combinations of bacteria and locations are possible.
- **Same Sentence:** The bacterium and location occur in the same sentence. This assumption is used by two submissions: [8,11].
- **Previous Bacteria:** The bacterium is the first bacterium that occurs *before* the location in the text.
- **Next Bacteria:** The bacterium is the first bacterium that occurs *after* the location in the text.
- **Paragraph Subject:** The text is split into paragraphs. The bacterium is the first bacterium that occurs in the paragraph of the location. This is used by one submission: [10].

The results from Sect. 4.3 are achieved by combining the '*Same sentence*' and '*Previous bacteria*' generation rule, which yields the best performance.

**Candidate Selection.** The Candidate Generation module forwards different sets of candidate relations to the Candidate Selection module. This Candidate Selection module builds for every set a separate logistic regression model (using the Factorie toolkit [15]). We use these logistic regression models as binary classifiers (is a relation or not). In the training phase, the models are trained based on positive and negative relations extracted from example texts. In the testing phase, each set of candidate relations is tested by their separate model.

The model uses the following features based on the two involved entities in a relation:

- The type of the entity
- Surface form
- Is capitalized (binary)
- Stem of each entity token

- Category of each entity token in the Cocoa annotations
- Part-of-speech tag of each entity token
- Dependency relation to the head of each token

None of the above features combine information from both the bacterium and location. We tested some features that do this, but without a significant influence on F1, as we saw a slightly better precision with a small drop in recall. The tested features are:

- Token distance between bacterium and location
- Length of syntactic path between bacterium and location
- The depth of the tree that contains the syntactic path
- Whether the bacterium or location occurs first

*Alternative Models.* Besides this model we also tested a nearest neighbor model. In this, we compare a candidate with a seen example based on the sequence of part-of-speech tags that occur on the syntactic path between the bacterium and location. Between these two sequences of tags the edit distance is calculated. Finally, the candidate is classified as a relation if the closest seen example with respect to this distance encodes a real relation.

In another approach we used two language models based on the tokens between the bacterium and location, where a separate model for positive and negative relations was built. Here, a candidate is classified as a relation if the probability that the candidate is generated by the positive model is higher than the probability for the negative model.

Both alternative models failed to achieve reasonable performance.

### 3.3  Subtask 3: Relation Extraction Without Gold Entities

Subtask three is very challenging and state-of-the-art systems perform very poorly on this task. Some of the authors of this paper have committed a stand-alone investigation focusing on the relation extraction part. They have implemented and evaluated a model that jointly learns the entity classes and their relations. The model is based on a structured max-margin approach, i.e., a structured support vector machine that integrates structured constraints between the entity labels and their relations. A background to this model can be found in [19,20] and further specified with regard to the bacteria-biotopes task in [21]. For reasons of comparison, we report the results of using this model in the Experiments section.

## 4  Experiments

In the first subsection we describe the data set and the resources. The subsections thereafter then present the results and discussions.

## 4.1   Data Set

The data set consists of public available documents from web pages from bacteria sequencing projects and from the MicrobeWiki encyclopedia [1]. The data is divided into a training, a development and a test set. The solution files of the training and development set are provided. The solution files of the test set are not available, but it is possible to test a solution with an online evaluation service[3] with a minimal time of 15 min between two submissions. During the contest the minimal time between two submissions has been 24 h. We limited our use of the online evaluation service to keep our results comparable with the contest submissions.

The data consists of 5,183 annotated entities and 2,260 annotated relations. The data was manually annotated twice followed by a conflict resolution phase [1]. Table 1 gives an overview of the data distribution. The training and development set is the same for both subtasks, but the test set is different.

**Table 1.** Summary statistics of the data set.

|            | Training/Dev | Testset 1 | Testset 2 |
|------------|--------------|-----------|-----------|
| Documents  | 78           | 27        | 26        |
| Words      | 25,828       | 7,670     | 10,353    |
| Entities   | 3,060        | 877       | 1,246     |
| Relations  | 1,265        | 328       | 667       |

## 4.2   Used Ontology

In the first subtask the OntoBiotope-Habitat ontology[4] is used. This ontology contains 1,756 habitat concepts. For each concept an id, the name and exact and related synonyms are given. Additionally if a concept can be described by a more general concept, an *is_a* relation is given. The ontology entry *'dental caries'* is for example:

```
id: MBTO:00001830
name: dental caries
related_synonym:''tooth decay'' [TyDI:30379]
exact_synonym: ''dental cavity'' [TyDI:30380]
is_a: MBTO:00002063 ! caries
```

## 4.3   Results

The results are presented separately for the two subtasks of entity detection and relation extraction.

---

[3] http://genome.jouy.inra.fr/~rbossy/cgi-bin/bionlp-eval/BB.cgi
[4] http://bibliome.jouy.inra.fr/MEM-OntoBiotope/OntoBiotope_BioNLP-ST13.obo

**Entity Detection and Ontology Mapping.** The score is calculated by mapping the predicted entities onto the entities of the reference solution. Entities are paired in a way that the sum of the dissimilarities are minimized. The dissimilarity between a predicted entity and a reference entity is based on boundary accuracy and the semantic similarity between the ontology concepts. Based on this optimal mapping of entities the Slot Error Rate (SER) is calculated. A perfect solution has a SER score of 0, if no entities are predicted a score of 1 is obtained. The SER is calculated as follows:

$$SER = \frac{S + I + D}{N} \tag{5}$$

- $S$: number of substitutions, based on the dissimilarity between the matched entities.
- $I$: number of insertions, the number of predicted entities that could not be paired.
- $D$: number of deletions, the number of reference entities that could not be paired.
- $N$: number of entities in the reference solution.

*Improvement Effects.* We implemented four variations to improve our model (see Sect. 3.1). The highest improvement is achieved by correcting the boundaries and filtering out redundant parents. Although handling dashed words gives only a slight improvement, it is definitely worth to use it because it increases the number of found entities without creating much incorrect entities. Extending the ontology improves our solution only by a very small margin.

*Influence of the Maximal Dissimilarity.* As explained in Sect. 3.1, the Candidate Selection module receives spans of tokens as input and searches within these spans for ontology entries. For a specific subspan of tokens, the Ontology Mapper returns the ontology entry that best matches, together with a dissimilarity measure. Based on cross-validation experiments we picked 0.1 as maximal dissimilarity, i.e. we classify all subspans with a lower dissimilarity as 0.1 as a found entity.

Table 2 shows the SER score together with the number of Substitutions, Insertions and Deletions (using 10 fold cross validation on the training and development set) for several values of maximal dissimilarity. For a range of low thresholds, only a very small variation in the number of Substitutions and Deletions is observed. However, the number of Insertions increases steadily with an increasing maximal dissimilarity. This is because we allow subspans to be less and less similar to the ontology entries, causing an increasing number of wrongly extracted entities.

*Comparison with Contest Submissions.* Testing our model with the online evaluation service, we obtained a SER score of 0.36 which is significantly better than all submissions to BioNLP-ST 2013. The best result of the contest is a SER score of 0.46 (IRISA).

**Table 2.** Influence of the maximal dissimilarity on entity detection performance.

| Dissimilarity | Sub | Ins | Del | SER |
|---|---|---|---|---|
| 0.05 | 212 | 195 | 181 | 0.38 |
| 0.10 | 210 | 197 | 180 | 0.38 |
| 0.15 | 212 | 208 | 180 | 0.39 |
| 0.20 | 211 | 212 | 180 | 0.39 |
| 0.25 | 227 | 236 | 173 | 0.41 |
| 0.30 | 230 | 249 | 169 | 0.41 |
| 0.35 | 319 | 497 | 141 | 0.61 |

We also improved the precision and F1 compared to all submissions. Recall, precision and F1 were respectively 0.68, 0.73 and 0.70. The IRISA submission scored a higher recall but a lower precision than our model. Table 3 shows our scores together with the scores of the submissions to BioNLP-ST 2013.

**Table 3.** Subtask 1 results compared to contest submissions.

| Participant | SER | Recall | Precision | F1 |
|---|---|---|---|---|
| IRISA | 0.46 | **0.72** | 0.48 | 0.57 |
| Boun | 0.48 | 0.60 | 0.59 | 0.59 |
| LIPN | 0.49 | 0.61 | 0.61 | 0.61 |
| LIMSI | 0.66 | 0.35 | 0.62 | 0.44 |
| Ours | **0.36** | 0.68 | **0.73** | **0.70** |

Some reasons why we outperform the others are:

- With a CRF model it is easy to consider any information through the addition of features. However, many systems that use a CRF to generate candidates are based on a general purpose noun phrase extractor, and do not use the biological annotations that are supplied.
- We search within each candidate for matches, which makes it possible that a candidate contains multiple entities.
- We redefine the boundaries of an entity by using the head annotations from the given Stanford parser annotated data.

The main weakness of our model is that an entity needs to be very close to a name or synonym of an ontology entry to be detected. We picked a value of 0.1 as maximal dissimilarity. This means that entities that do not occur in the ontology or are described by an unknown synonym can not be found. We implemented an improvement by correcting the boundaries to lower the impact of this weakness. In this way, words that are not seen in the ontology can be part of an entity if its head word occurs in the ontology.

## Relation Extraction

*Baseline Model.* To better analyse the performance of our approach, we have first built a baseline model. This model predicts *loc* relations between all bacteria and locations that occur in the same sentence and no *partof* relations. The results are presented in Table 5. Considering the achieved scores in BioNLP-ST 2013, this model performs dramatically better. It outperforms all submissions based on F1 due to a much higher recall. But the precision of one submission (TEES) is clearly better (0.82).

This baseline model predicts 53 % of the *loc* relations. Based on the fact that this baseline model only predicts relations within the same sentence, we know that about half of the *loc* relations occur in the same sentence, for the other half multiple sentences need to be examined.

*Performance on Different Relation Types.* We use a similar approach for *partof* relations as for *loc* relations. Table 4 shows the performance of our model for the prediction of one relation type separately and the prediction of both types jointly. We see a very low precision if we only predict *partof* relations, this is due to the fact that we recall many relations wrongly and there are only few true *partof* relations in the texts. When we combine our *loc* and *partof* model the result is worse than the *loc* model on itself. The *partof* model decreases the overall precision of our model much more compared to the gain in recall.

**Table 4.** Relation extraction results for the different relation types.

| Model | Recall | Precision | F1 |
|---|---|---|---|
| Localization | 0.59 | **0.50** | **0.54** |
| PartOf | 0.09 | 0.15 | 0.12 |
| Combined | **0.68** | 0.35 | 0.46 |

*Comparison with Contest Submissions.* We tested our solution with the available online evaluation service and receive a F1 of 0.67 which is significantly better than all submissions to BioNLP-ST 2013. The best result of the contest achieved a F1 of 0.42 (TEES). Our recall and precision are respectively 0.71 and 0.63. This recall is much higher than all the contest submissions, one submission (TEES) scored a better precision (0.82). Table 5 shows our achieved results together with the scores of the official submissions to BioNLP-ST 2013.

Some reasons why we outperform the others are:

– We use a combination of generation rules, the contest submissions were mainly limited to one specific generation rule.
– We do not predict *partof* relations in our final model due to low accuracy and overall negative impact.

**Table 5.** Subtask 2 results compared to contest submissions.

| Participant | Recall | Precision | F1 |
|---|---|---|---|
| TEES-2.1 | 0.28 | **0.82** | 0.42 |
| IRISA | 0.36 | 0.46 | 0.40 |
| Boun | 0.21 | 0.38 | 0.27 |
| LIMSI | 0.04 | 0.19 | 0.06 |
| Baseline | 0.43 | 0.47 | 0.45 |
| Ours | **0.71** | 0.63 | **0.67** |

*Bacterium Model.* The logistic regression model achieves significantly better results than the baseline model and all contest submissions. However, many of the used features have only very little influence. We remark that almost comparable results can be achieved by a model that always predicts true unless the bacterium name starts with '*bacteri*'. This sort of model is of course not generic and largely overfits the data. It works well because it succeeds in excluding a significant amount of false relations. Labeled entities occur in surface forms '*bacterium*', '*bacterial infections*', . . . These forms occur relatively often in texts, but they rarely appear in *loc* relations. The reason for this is that when the word '*bacterium*' appears in a text, it usually does not refer to the general concept but to a specific bacterium discussed previously in the text. However, to avoid overfitting it is preferred to use such patterns in the data by including relevant features, rather than implementing strict decision rules based on them. In the case of the above characteristic, the name of the specific bacterium entity is added as a feature in our system.

**Relation Extraction without Gold Entities.** Although the focus of this paper is on the detection of the entities and their taxonomy classes we performed some experiments on task 3 which is the extraction of the localization relations without having the gold entities. As described above we used a structured learning model proposed in [Kordjamshidi and Moens, 2015]. The results given in Table 6 show that our approach improves the state-of-the-art results. Full evaluation details are found in [21].

**Table 6.** Joint model vs. Task-3 participants (TEES and LIMSI) evaluated on test set by the online system of the BioNLP-ST 2013 task; relations without gold entities; strict evaluation which is punished by missing PartOf relations, also see [21].

| System | P | R | F |
|---|---|---|---|
| TEES | 0.18 | 0.12 | 0.14 |
| LIMSI | 0.12 | 0.04 | 0.06 |
| JoinModel | 0.311 | 0.171 | 0.221 |

# 5 Conclusion

In this paper we discussed an approach for the first two subtasks of the Bacteria Biotopes task of BioNLP-ST 2013. For the first subtask (entity detection and ontology mapping) we implemented a model based on Conditional Random Fields. In this system, candidates are generated from the text and thoroughly inspected to find matches within the ontology. We also devised several improvements for the boundary detection of entities. Our model achieved significantly better results than all official submissions to BioNLP-ST 2013.

For the second subtask (relation extraction) we generated candidates with multiple generation rules (e.g. all bacteria and locations that occur in the same sentence). To select a candidate we used a logistic regression model. Because we used a combination of generation rules we achieved a much higher recall and therefore a much better score than all official submissions to BioNLP-ST 2013.

In spite of these pronounced gains, we think there is still room for improvement, especially for the second subtask. One potential improvement of our model will be to consider long distance dependencies between the bacterium and location, more contextual features and additional background knowledge from external resources. In this direction, using a joint learning framework has already improved the recognition of entities and their relations when extracting bacteria and their habitat. We will extend these lines of research in the future.

**Acknowledgements.** This research is supported by grant 1U54GM114838 awarded by NIGMS through funds provided by the trans-NIH Big Data to Knowledge (BD2K) initiative (www.bd2k.nih.gov) and by Research Foundation Flanders (FWO) (grant G.0356.12). The content is solely the responsibility of the authors and does not necessarily represent the official views of the National Institutes of Health. Also we would like to thank the reviewers for their insightful comments and remarks.

# References

1. Bossy, R., Golik, W., Ratkovic, Z., Bessières, P., Nédellec, C.: BioNLP shared task 2013 - an overview of the bacteria biotope task. In: Proceedings of the BioNLP Shared Task 2013 Workshop, Sofia, Bulgaria. ACL, pp. 161–169 (2013)
2. Nédellec, C., Bossy, R., Kim, J.D., Kim, J.J., Ohta, T., Pyysalo, S., Zweigenbaum, P.: Overview of BioNLP shared task 2013. In: Proceedings of the BioNLP Shared Task 2013 Workshop, Sofia, Bulgaria. ACL, pp. 1–7 (2013)
3. Bossy, R., Jourde, J., Bessieres, P., van de Guchte, M., Nedellec, C.: BioNLP shared task 2011 - bacteria biotope. In: Proceedings of BioNLP Shared Task 2011 Workshop. ACL, pp. 56–64 (2011)
4. Bjorne, J., Salakoski, T.: Generalizing biomedical event extraction. In: Proceedings of BioNLP Shared Task 2011 Workshop. ACL (2011)
5. Nguyen, N.T.H., Tsuruoka, Y.: Extracting bacteria biotopes with semi-supervised named entity recognition and coreference resolution. In: Proceedings of BioNLP Shared Task 2011 Workshop. ACL (2011)

6. Ratkovic, Z., Golik, W., Warnier, P., Veber, P., Nedellec, C.: Task bacteria biotope-the Alvis system. In: Proceedings of BioNLP Shared Task 2011 Workshop. ACL (2011)
7. Bannour, S., Audibert, L., Soldano, H.: Ontology-based semantic annotation: an automatic hybrid rule-based method. In: Proceedings of the BioNLP Shared Task 2013 Workshop, Sofia, Bulgaria. ACL, pp. 139–143 (2013)
8. Claveau, V.: IRISA participation to BioNLP-ST 2013: lazy-learning and information retrieval for information extraction tasks. In: Proceedings of the BioNLP Shared Task 2013 Workshop, Sofia, Bulgaria. ACL, pp. 188–196 (2013)
9. Grouin, C.: Building a contrasting taxa extractor for relation identification from assertions: biological taxonomy & ontology phrase extraction system. In: Proceedings of the BioNLP Shared Task 2013 Workshop, Sofia, Bulgaria. ACL, pp. 144–152 (2013)
10. Karadeniz, I., Özgür, A.: Bacteria biotope detection, ontology-based normalization, and relation extraction using syntactic rules. In: Proceedings of the BioNLP Shared Task 2013 Workshop, Sofia, Bulgaria. ACL, pp. 170–177 (2013)
11. Björne, J., Salakoski, T.: TEES 2.1: automated annotation scheme learning in the BioNLP 2013 shared task. In: Proceedings of the BioNLP Shared Task 2013 Workshop, Sofia, Bulgaria. ACL, pp. 16–25 (2013)
12. Klein, D., Manning, C.D.: Fast exact inference with a factored model for natural language parsing. In: Advances in Neural Information Processing Systems (NIPS), vol. 15, pp. 3–10. MIT Press (2003)
13. Sutton, C., McCallum, A.: An Introduction to conditional random fields for relational learning. In: Getoor, L., Taskar, B. (eds.) Introduction to Statistical Relational Learning. MIT Press (2007)
14. Lei, J., Tang, B., Lu, X., Gao, K., Jiang, M., Xu, H.: A comprehensive study of named entity recognition in chinese clinical text. J. Am. Med. Inform. Assoc. **21**, 808–814 (2014)
15. McCallum, A., Schultz, K., Singh, S.: FACTORIE: probabilistic programming via imperatively defined factor graphs. In: Neural Information Processing Systems (NIPS) (2009)
16. Porter, M.: An algorithm for suffix stripping. Program **14**, 130–137 (1980)
17. Ramshaw, L.A., Marcus, M.P.: Text chunking using transformation-based learning. In: Proceedings of the 3rd ACL Workshop on Very Large Corpora, Cambridge, MA, USA, pp. 82–94 (1995)
18. Levenshtein, V.: Binary codes capable of correcting deletions, insertions and reversals. Sov. Phys. Dokl. **10**, 707 (1966)
19. Kordjamshidi, P., Moens, M.F.: Global machine learning for spatial ontology population. J. Web Semant. **30**, 3–21 (2015)
20. Kordjamshidi, P., Moens, M.F.: Designing constructive machine learning models based on generalized linear learning techniques. In: NIPS Workshop on Constructive Machine Learning (2013)
21. Kordjamshidi, P., Roth, D., Moens, M.F.: Structured learning for spatial information extraction from biomedical text: bacteria biotopes. BMC Bioinform. **16**, 129 (2015)

# Validation Study of a Wave Equation Model of Soft Tissue for a New Virtual Reality Laparoscopy Training System

Sneha Patel and Jackrit Suthakorn[✉]

Center for Biomedical and Robotics Technology (BART LAB),
Department of Biomedical Engineering, Faculty of Engineering,
Mahidol University, Salaya, Thailand
sneha.pat@student.mahidol.ac.th,
jackrit.sut@mahidol.ac.th

**Abstract.** Despite the benefits of laparoscopic procedures for the patients, this technique comes with a number of environmental limitations for the surgeon, which therefore require distinctive psychomotor skills. VR training systems aim to improve these skills. For effective transference of skills from these training systems, it is important to mimic the surgical environment; including the soft tissue models. This study introduces a novel two dimensional wave equation model to mimic the interactions between soft tissue and laparoscopic tools. This model accounts for mechanical and material properties of the soft tissue. This study also proposes a new face validation technique, for an objective analysis of the developed model as a viable soft tissue model. The statistical analyses and computational cost support the use of wave equation as a replacement for present models. In the future, this model will be applied to a novel VR surgical training system for an enhanced training experience.

**Keywords:** Soft tissue model · Surgical training · Two dimensional wave equation · Finite element analysis (FEA) · Computer based models · Virtual reality (VR) training

## 1 Introduction

The medical industry is motivated by innovation in procedures, devices, and drugs. Laparoscopic surgeries, an innovation from the recent decades, are gaining popularity due to the benefits for patients. Some benefits of this procedure are: shorter recovery period, reduced blood loss, and less scarring, which are all results of the smaller incisions used in these surgeries [1, 2]. The smaller incisions result in various environmental constraints that hinder the performance of inexperienced surgeons; some examples of these are: 2-dimensional view of operating area, limited hand-eye coordination, increased tremor due to long, inflexible tools and restricted movement [3, 4].

© Springer International Publishing Switzerland 2015
A. Fred et al. (Eds): BIOSTEC 2015, CCIS 574, pp. 256–271, 2015.
DOI: 10.1007/978-3-319-27707-3_16

## 1.1    Need for Laparoscopic Surgery Training

The environmental constraints in laparoscopic surgeries can result in some serious complications; for example: bleeding, infection, visceral injury or death. These complications are directly related to the amount of practice or experience of the surgeon; for example there is a higher likelihood of complications in the surgeon's first ten procedures [5]. Moreover, the smaller incisions entail distinctive psychomotor skills that vary significantly from those used in open surgeries; therefore requiring intensive and repetitive training to acquire the required skills and reduce the likelihood of complications [6, 7].

## 1.2    Present Surgical Training Systems

The significance of training in laparoscopic surgeries has brought about extensive research towards the development of training systems for teaching hospitals [4]. The conventional training systems are either live patients or cadaveric humans. However, these systems come with various legal, ethical and cost issues, resulting in teaching hospitals' move towards the use of inanimate training systems [4, 8].

Inanimate training systems are popular due to their adaptability to the needs of the user, and the system's capability of repetitive training without risking the lives of patients [8–10]. There are two major inanimate training systems: synthetic material and virtual reality (VR) models. The VR training system is being extensively researched in this area due to its promise of realistic simulations and user interactions [4, 8, 10].

In the following sections, the authors include their observations of previous works in the field of VR training systems and the soft tissue models used. Subsequently, the authors introduce their novel soft tissue model, and a novel face validity test that aims to provide objective results. Lastly, statistical analyses are performed, followed by a discussion of the results.

## 2    Previous Works

In recent years, VR training systems have been extensively researched and developed for use in surgical training; inspired by the flight simulators used for training pilots [10, 11]. These systems aim to provide medical students or young surgeons essential laparoscopic skills while limiting the risks on patients. Another application of these systems is surgical planning with patient specific data [12, 13].

VR surgical training systems range in price from US$5,000 to US$200,000 [14]. These systems provide skills, which span from basic skills to complete laparoscopic procedures [14, 15]. Some commercially available training systems include: MIST VR, SIMENDO simulator, LAP Mentor, and LapSim. These training systems are designed to refine the user's dexterity within the constrained environment of a laparoscopic procedure, while removing the risks associated with on-patient training [3, 16]. VR training systems are popular due to their promise of intensive and repetitive training without risking the lives of patients, while providing objective performance assessment to the user [3, 4, 10].

## 2.1    Soft Tissue Models in Present VR Training Systems

There are various components of the VR training system that work together to create the ideal system for effective training in a surgical task or procedure. This paper highlights one such component; the visual representation of soft tissues within the VR environment. The models of soft tissue are important because they represent the environment of a laparoscopic procedure and how laparoscopic tools' manipulations would affect the soft tissue; therefore accurate representation of this environment would allow the users to transfer the skills acquired from the training system to the operation room (OR) [8, 17, 18]. Not only the realism of this model is important, but also the rate at which the manipulations take place; for this application, the soft tissue models should be transformable in real-time [18, 19]. Researchers, also, have to take into consideration the effects of the computational cost on the user's ability to interact with the environment [10].

VR training systems, today, use mass-spring model, finite element models, or mesh-free models to model soft tissue in the environment. These models demonstrate the obstacles associated with modeling soft tissue in VR as a result of the need for balance between accuracy and computational cost.

Mass-spring model is the most commonly used soft tissue model in VR surgical training systems [20]. This is because of its low computational cost, which makes it the ideal model for real-time manipulations. This model uses the Kelvin-Voigt model and therefore implements springs and dashpots to model the viscoelastic behavior of the soft tissue [19].

A more accurate representation of the soft tissue in VR is the finite element model but the constraint of this model is its computational cost, which makes it less than ideal for real-time manipulations [19, 20]. On the other hand, mesh-free models are specifically designed to meet the requirements for real time manipulation in a surgical training device. In this model, the nodes are not connected in a mesh, therefore permitting the cutting and reconnecting of the model [19].

The present VR training system at BART LAB uses a simplified spring-mass model [21]; therefore this study aims to go beyond and enhance the user interface through the development of novel soft tissue models.

## 2.2    VR Training System at BART LAB

Inspired by da Vinci Skill Simulator by Intuitive Surgical, at our lab, a researcher has developed a surgical training device. The Skill Simulator is a supplementary product available with the da Vinci robotic surgical system and provides training to users. The system has a realistic virtual environment but lacks a human interaction aspect. Therefore the BART LAB VR training system focuses on the human-robot interface [21, 22]. The BART LAB training system is displayed in Fig. 1 (a); it includes two Phantom OMNI devices and a tool holder to mimic the physical setup of a laparoscopic procedure. On the other hand, Fig. 1(b) demonstrates the VR environment of the training system.

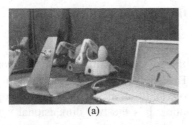
(a)

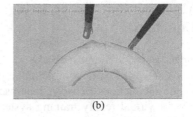
(b)

**Fig. 1.** The (a) physical and (b) graphical interface of our VR training system.

## 2.3 Limitations of Present Soft Tissue Models

VR training systems are gaining popularity among teaching hospitals due to their promise of greater patient safety, but they are still are not widely used due to their limitations. As discussed in the earlier subsection, soft tissue models play an important role in the success of a VR training system and its ability to teach the distinctive psychomotor skills that are required for laparoscopic procedures. A limitation of these models is their ability to simulate, realistically, the interactions between internal organs and laparoscopic tools [8].

The soft tissue models used in VR training systems are either simplified or are computationally expensive. These models propose a need for a novel soft tissue model that can mimic mechanical, material, and visual properties, while maintaining the need for real-time manipulations. This paper presents a novel soft tissue model, using a two dimensional wave equation to improve the realism of the training system and consequently increasing the transference of skills acquired on the training system.

## 3 Approach

The computer used for the development of the novel computer based soft tissue model and validation study has the following specifications: Intel Core 2 Duo 2.66 GHz processor, NVIDIA GeForce 9400 GT graphics card, 4 GB RAM and a 160 GB hard disk. An overview of the approach of this study can be seen in Fig. 2, which is discussed in further detail in the next two sections (Sects. 4 and 5).

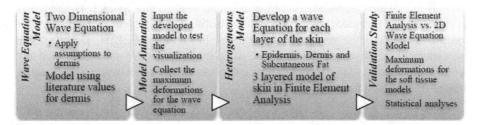

**Fig. 2.** Approach of this study: (1) to develop a novel model of a computer based soft tissue and (2) to perform a validation study on the novel model.

## 4  Modeling Soft Tissue Using Two-Dimensional Wave Equation

Engineers study vibrations in elastic, flexible threads and membranes using wave equations; a type of partial differential equation. This study aims to use this equation to model the interactions between laparoscopic tools and soft tissue in a laparoscopic surgery for virtual reality training systems. Equation 1 is the two dimensional wave equation, which is used to model vibrations in membranes; in this equation u(x,y,t) is the displacement function, T is the tension per unit length, and ρ is the density of the membrane. Equation 1 is based on Newton's Second Law.

$$\frac{\delta^2 u}{\delta t^2} = c^2 \left( \frac{\delta^2 u}{\delta x^2} + \frac{\delta^2 u}{\delta y^2} \right); c^2 = \frac{T}{\rho} \tag{1}$$

Using initial and border conditions, Kreyszig et al. present a solution for the two dimensional wave equation, as seen in Eq. 2. This solution is developed using double Fourier series and therefore $B_{mn}$ Eq. 3, is the Euler formula of the Fourier coefficients, f (x,y), which is also the function representing the initial displacement. The solution is developed to model a drum membrane's response to being hit by a drumstick [23]. In Eqs. 2-3 the variables are defined as follows: $B_{mn}$ is the Euler formula, $B^*_{mn}$ is the relationship that takes into consideration the initial velocity, a and b are the boundary conditions, and $\lambda_{mn}$ is the eigenvalue of this equation.

$$u_{mn}(x, y, t) = (B_{mn} \cos \lambda_{mn} t + B^*_{mn} \sin \lambda_{mn} t) \sin \frac{m\pi x}{a} \sin \frac{n\pi y}{b},$$
$$where \ m = 1, 2, 3 \ldots, n = 1, 2, 3 \ldots, and \ \lambda_{mn} = c\pi \sqrt{\frac{m^2}{a^2} + \frac{n^2}{b^2}} \tag{2}$$

$$B_{mn} = \frac{4}{ab} \int_0^b \int_0^a f(x, y) \sin \frac{m\pi x}{a} \sin \frac{n\pi y}{b} dx dy \tag{3}$$

Structural and mechanical engineers use wave equations to study the effects of vibrations on various structures; for example beams, rods, cables and plates. These studies idealize the structures as they are considered homogeneous and isotropic materials, which are composed of continuous chains of mass and spring [24]. Engineers can therefore use this model to determine the feasibility and strength of a structure; for example to observe the response of buildings in earthquakes [25]. Despite these applications of the wave equation in engineering, the authors have not encountered any previous applications of this model to mimic the behavior of soft tissue in virtual reality surgical training systems.

## 4.1   2D Wave Equation of Dermis

The skin is a viscoelastic, non-homogeneous and anisotropic material. The mechanical and material behaviors of the skin are substantially dependent on the collagen and elastic fibers of the dermis; therefore this is the layer of the skin discussed thoroughly, in this paper [26–28]. In this section, the development of the two dimensional wave equation of dermis is discussed because of the role this tissue plays in the mechanical behavior of skin [26]. Other layers of the skin that are observed in this study are the epidermis and the subcutaneous fat. The three layers are used to develop a heterogeneous material, much like the skin, to model the soft tissue.

**Assumptions of Wave Equation.** There are five key assumptions that are applied to the 2D wave equation to get the solution in Eq. (2). These assumptions are applied to the dermis, to demonstrate limited effects on the mechanical properties of the tissue.

1. Mass of dermis per unit area is constant
2. The dermis is flexible therefore bends without resistance.
3. The dermis is stretched and fixed throughout its boundary; as it is held in place by bones and connective tissues. This stretching results in a uniform tension, T, per unit length, which is constant during motion.
4. The deformation of the membrane is small compared to the size of the dermis, which is plausible since the area of deformation is smaller than the dermis that covers the entire body.
5. The membrane is thin; this is the rationale for modeling a single layer of skin, the dermis, using this equation. Multiple two dimensional wave equations are used to model all of the layers of the skin to show how they would interact to create a specific manipulation or deformation.

**Literature Values for the Wave Equation Model.** Here we look at the literature values of the mechanical and material properties of the dermis, listed in Table 1, which are required for the development of a wave equation model of the dermis. These properties will define the boundary and initial conditions of the soft tissue. As mentioned in the assumptions, multiple two dimensional wave equations are developed for the layers of the skin therefore a similar list is compiled for the epidermis and subcutaneous fat.

**Table 1.** Mechanical and material properties of dermis, a layer of the skin.

| Properties of Dermis | Values |
| --- | --- |
| Area of Dermis | $60mm \times 60mm = 3600mm^2$ |
| Thickness of Dermis | $1mm$ [29] |
| Volume of Dermis | $60mm \times 60mm \times 1mm = 3600mm^3$ $1.8 \times 10^{-8} \frac{g}{mm^2}$ [27] |

*(Continued)*

**Table 1.** (*Continued*)

| Properties of Dermis | Values |
|---|---|
| Weight of Dermis | $\therefore Weight = \left(1.8 \times 10^{-8} \frac{g}{mm^2}\right) \times 3600 mm^2 = 6.48 \times 10^{-5} g$ |
| Density of Dermis | $\rho = \frac{m}{v} = \frac{6.48 \times 10^{-5} g}{3600 mm^3} = 1.8 \times 10^{-8} \frac{g}{mm^3} = 18 \frac{g}{m^3}$ |
| Prestress (along the fibers) | 0.024 MPa [28] |
| Prestress (across the fibers) | 0.0093 MPa [28] |
| Prestress[a] | $F = \sqrt{0.024^2 + 0.0093^2} = 2.57 \times 10^{-2} MPa$ |
| Tension | $T = (2.57 \times 10^{-2})(60 \times 10^{-3}) = 1.54 \times 10^{-3} \frac{N}{m}$ |

[a]Prestress is the variable T from Eq. 1, and takes into account the effects of the natural forces, for example the result of connective tissue and bones holding the skin in place.

**Applying Properties of Dermis to the 2D Wave Equation.** This study proposes a simple 2D wave equation model to mimic a simple interaction between soft tissue and laparoscopic tools. Therefore the 2D wave equation is used to model the deformations the soft tissue experiences when a laparoscopic tool pushes down on it and after the tool is removed, the tissue's return to its original form.

The size of the soft tissue is $60mm \times 60mm \times thickness of dermis$ This size is based on the diameter of the tool, $5 - 8mm$ which would have a deformation in an area no larger than that suggested above. Using the 2D wave equation solution, from Kreyszig et al. [23], and the literature values for the dermis, the authors have developed a wave equation model of the soft tissue as seen in Eq. 5 whereas Eq. 4 is the Euler formula of the solution. The expression of $B_{mn}^*$ is zeroed out by the assumption that there is no initial velocity. The solution of the Euler formula of f(x,y) is found using a symbolic math toolbox and is plugged into the 2D wave equation solution (Eq. 5).

$$B_{1,1} = \frac{4}{0.06 \times 0.06} \int_0^{0.060} \int_0^{0.060} (5x(1 - x) \times y(1 - y)) \times \sin\frac{\pi x}{0.06} \sin\frac{\pi x}{0.06} dxdy = 6.11 \times 10^{-6} \quad (4)$$

$$u_{mn}(x, y, t) = \left((6.11 \times 10^{-6}) \cos(0.686t)\right) \times \sin\frac{\pi x}{0.06} \sin\frac{\pi y}{0.06} \quad (5)$$

**Animation of Wave Equation Model of Dermis.** The authors have developed an animated model of the dermis in a symbolic math toolbox, using the 2D wave equation solution in Eq. 5. This animation aims to verify the model as a representation of an interaction between the soft tissue and a laparoscopic tool, as discussed in the previous subsection. The resulting animation is demonstrated in Fig. 3; through multiple screenshots. As can be seen from the images, the model starts out by reacting to a push to create a deformation, and following the deformation, the model starts to return back to its original shape.

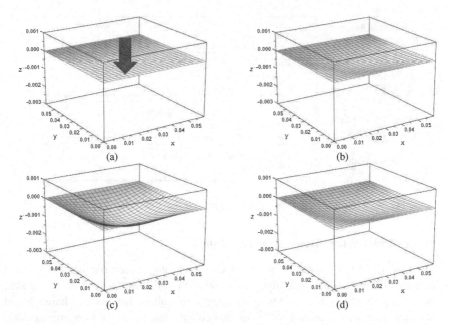

**Fig. 3.** Here are screenshots from the animation of the two dimensional wave equation of the dermis. The first three images (figures a-c) depict the model's response to the tissue being pushed down by a laparoscopic tool. In figure a, the arrow demonstrates the direction of the virtual force causing the deformation. On the other hand, the last image (figure d) represents the model's return to its original shape after the tool and applied force are removed. The unit along x, y, and z axes is meters, therefore resulting in the small values for the deformation over time.

## 5    Validation Study of Novel Soft Tissue Model

Presently, validation studies of novel soft tissue models in VR surgical training systems are assessed with face validation. Face validity, is performed on many VR surgical training systems and provides a subjective analysis of the realism of the soft tissue model in the system. These studies collect the opinions of users of the system through questionnaires and Lickert scores of 1-5 [30–32].

### 5.1    Validation Study Design

The flowchart in Fig. 4 shows an overview of the novel validation study that is performed in this study. As discussed earlier, the wave equation is developed for three layers of the skin; epidermis, dermis and subcutaneous fat. Therefore when developing a finite element analysis (FEA) model, the authors use a heterogeneous material to compare between the two models, at each layer. The authors have chosen FEA model because it is the most accurate computer based model for mimicking soft tissue behavior.

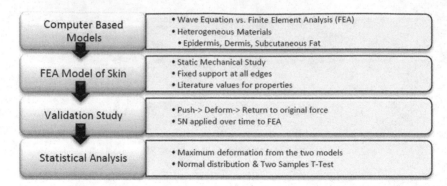

| Computer Based Models | • Wave Equation vs. Finite Element Analysis (FEA)<br>• Heterogeneous Materials<br>• Epidermis, Dermis, Subcutaneous Fat |
| FEA Model of Skin | • Static Mechanical Study<br>• Fixed support at all edges<br>• Literature values for properties |
| Validation Study | • Push-> Deform-> Return to original force<br>• 5N applied over time to FEA |
| Statistical Analysis | • Maximum deformation from the two models<br>• Normal distribution & Two Samples T-Test |

**Fig. 4.** Flowchart representing the process of the face validation study.

## 5.2    FEA Model Setup: Heterogeneous Material

The setup of the FEA model is essential to developing a comparable model and therefore an effective validation study. As can be seen in Fig. 4, the FEA model of skin has fixed support at all edges. Also, the literature values for the mechanical and material properties of the dermis are incorporated into this model, which are: density, damping factor, Young's modulus, Poisson's ratio, and tensile yield strength [26, 28, 33]. Figure 5 is an exploded view of the developed heterogeneous material, which shows the three layers of the skin, observed throughout this research. A static mechanical study is performed, in which a total force of 5 N is applied at increments of 0.05 N/s and this force is removed at the same increment, over a time period of 100 s. Maximum deformations of the skin and each of its layers is collected over time to see changes in the heterogeneous material and how they compare to those observed in the wave equation model.

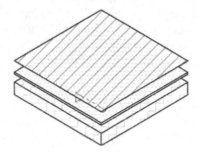

**Fig. 5.** Here, the hatched layer is the epidermis, the dotted layer is the dermis and the layer with the zigzag pattern represents the subcutaneous fat.

## 6   Results of Validation Study

Maximum deformations of the two models at hand, FEA and wave equation, allow for a comparison of mechanical and material properties of the two tissue models. In Fig. 6, there is a comparison of the two models of dermis; therefore it shows the resulting deformation and then the tissue returning to its original shape.

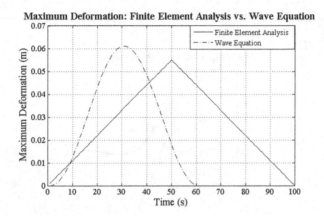

**Fig. 6.** Graphical representation to display a comparison of maximum deformations between the FEA and wave equation models.

## 7   Statistical Analysis

To evaluate the results from the two models, FEA and wave equation, the authors use two methods of statistical analysis: normal distribution and two sample t-test. For the two analyses, the maximum deformation data from the two models, for each layer of the skin, are used. These two studies help identify the similarities and differences between the models as they are used to model the same soft tissue; the three layers of the skin.

### 7.1   Normal Distribution Study

Figure 7 looks at the normal distribution studies based on the collected data. The normal distributions are performed for the two models, wave equation and FEA. Therefore this study observes the behavior of each layer as a result of interaction with laparoscopic tools. The graphs in Fig. 7 suggest high variability between the two models. The calculations for this statistical analysis are performed in Microsoft Excel.

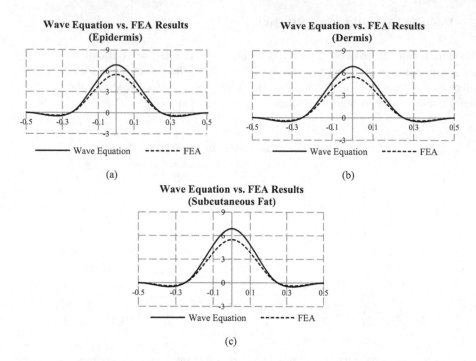

Fig. 7. Normal distribution to compare the wave equation and FEA models of the three layers of skin.

## 7.2 Two Sample T-Test

Table 2 shows a list of important results from the two sample t-test. In Microsoft Excel t-test is performed, while assuming equal variance, with a significance level of: $\alpha = 0.05$, therefore the null hypothesis is that the means are equal for the two models, whereas the alternative hypothesis is that the means are not equal. The expected result is the acceptance of the null hypothesis because the models represent the same soft tissue. This is determined using the two relationships, displayed in Table 2: $p - value > 0.05$ and $tstat < tcritical$. If they are true, the null hypothesis can not be rejected.

Table 2. Two sample t-test results for epidermis, dermis and subcutaneous fat

|  | T-test Results | $p - value > 0.05$ | $tstat < tcritical$ |
|---|---|---|---|
| **Epidermis** | | | |
| P (Two Tail) | 0.364 | | |
| t Stat | -0.912 | True | True |
| t Critical Value | 1.975 | | |

(Continued)

**Table 2.** (*Continued*)

|  | T-test Results | $p - value > 0.05$ | $tstat < tcritical$ |
|---|---|---|---|
| **Dermis** | | | |
| P (Two Tail) | 0.395 | | |
| t Stat | 0.911 | True | True |
| t Critical Value | 1.975 | | |
| **Fat** | | | |
| P (Two Tail) | 0.363 | | |
| t Stat | 0.853 | True | True |
| t Critical Value | 1.975 | | |

# 8 Discussion

In this validation study, two computer based models of soft tissue are compared to justify the use of a novel model in VR surgical training systems. Aforementioned, the authors compare the behavior of two computer based models of soft tissue, the wave equation models with respect to FEA models.

Two statistical analyses are used to verify wave equation as a valid model of soft tissue. The normal distribution study results demonstrate high variability for models at each of the layers of the skin, therefore suggesting high correlation, which is expected as the study looks at two different models of the same tissue that makes up skin. A two sample t-test is performed to verify the results from the normal distribution study. This is determined using the relationships between the p-value and the significance value, and the t-stat and t critical values. This study shows that the null hypothesis, the means of the samples are equal, can not be rejected. This analysis therefore further validates the use of wave equation as a model of soft tissue because it shows similarities between the means of the two models' maximum deformations. As can be seen from the sections above, the statistical analyses are performed for the three layers of the skin, therefore allowing an accurate observation of the behaviors of the different soft tissues of the skin and how they compare for the two models.

Another data used to compare the two models is the computational costs associated with the wave equation and FEA models. Both of the models are tested on the computer that is described in an earlier section. The wave equation model is solved in 6 s whereas the FEA model takes approximately 4 min.

The FEA is used to evaluate the novel soft tissue model as it is one of the most accurate computer based models available. Based on the statistical analyses and the computational cost of the two models, wave equation is considered a feasible alternative to presently used soft tissue models.

The validation study described in this study makes certain assumptions or simplifications, which, if addressed, can provide more accurate results. For example, the heterogeneous material is simplified as it does not take into consideration the effects of additional components, i.e. hair follicle, connective tissue and blood vessels, towards the mechanical behavior of the soft tissue. Along with that the thickness of these models are consistent throughout the model, but in biological tissue this property varies and is irregular. This simplification can be rectified using blocks of soft tissue with various thicknesses to create a single block of irregularly shaped tissue. Another improvement that could enhance the results from this study would be to solve the problem of shifting peaks between the two models' maximum deformation; as can be seen in Fig. 6. This change could show more correlation between the two models as a result of similar forces and rate of reaction.

Despite the use of wave equation to model skin, in this study, the authors aim to develop an easily modifiable model which can be altered to fit the mechanical and material properties of various soft tissues. This feature would allow the manipulability of the model to meet the needs of the surgeon, medical residents or the specific laparoscopic procedure.

## 8.1    Additional 2D Wave Equation Models

Due to the success of the wave equation models in comparison to the FEA model of soft tissue, the authors propose two variations of the wave equation models as shown in Fig. 8; Fig. 8(a) is the deformation resulting when the tool collides at an angle whereas Fig. 8(b) is when two tools interact with the tissue at the same time.

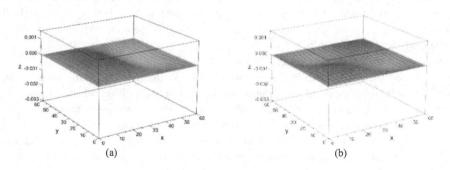

(a)    (b)

**Fig. 8.** Variations of the wave equation model of soft tissue and tools' interaction.

## 8.2    Future Application

The aim of the research team is to apply this model of soft tissue to a VR laparoscopic surgery training system. The user interface would also include: laparoscopic tools (i.e. needle holder and suture) and guidelines for correctly performing the task at hand. This system intends to not only provide realistic visual and mechanical representation of biological tissue in a VR environment but also to provide realistic haptic interaction during the training. The haptic interaction is offered through Phantom Omni. The training system will provide the user with intensive and repetitive training with objective assessment therefore limiting the risks of on-patient training.

To support the need for surgical training and provide the same level of healthcare nationwide, our team has designed a tele-surgical and surgical training system, as shown in Fig. 9, which is designed by our research group. This system aims to not only

**Fig. 9.** Proposed, preliminary design of the telesurgical and surgical training system at BART LAB.

allow surgeons to provide their skills remotely but also encourage training of young surgeons or medical students.

## 9  Conclusions

Laparoscopic surgeries' popularity can be attributed to the advantages of this procedure for the patient. However, there are a number of limitations for the surgeons, which arise due to the smaller incisions used in this procedure. This paper, therefore, suggests a novel soft tissue model with the aim to provide an improved training experience on a novel VR training system; a system that shows promise for complementing with on-patient training. The new face validation study in this paper supports the use of the 2D wave equation as a soft tissue model. Along with that, the computational cost is significantly lower for the novel model with respect to one of the most accurate computer based models, the FEA model. In future studies, the researchers intend to apply this novel method to a new VR training system, which will allow the user to obtain intensive and repetitive training while obtaining an objective performance feedback.

**Acknowledgements.** The authors would like to thank Thailand National Research University Grant through Mahidol University for their financial support. Secondly, the authors would like to thank Prof. Chumpon Wilasrusmee, M.D., R.N., Ramathibodi Hospital, Mahidol University, for his invaluable input towards the development of a virtual reality surgical training system and insights into the needs of surgeons. Lastly, we would like to thank BART LAB's Miss Nantida Nillahoot for her design of the telesurgical/training system in Fig. 9. The first author would also like to take this opportunity to thank Aditya Birla Group's Pratibha Scholarship, and Biomedical Engineering Scholarship (BMES) from the Department of Biomedical Engineering, Mahidol University, for financial aid towards her graduate education. The first author would, finally, like to thank her colleagues at BART LAB for their continuous support and assistance.

# References

1. Basdogan, C., Ho, C.-H., Srinivasan, M.A.: Virtual environments for medical training: graphical and haptic simulation of laparoscopic common bile duct exploration. IEEE/ASME Trans. Mechatron. **6**(3), 269–285 (2001)
2. Nguyen, N.T., Goldman, C., Rosenquist, C.J., Arango, A., Cole, C.J., Lee, S.J., Wolfe, B. M.: Laparoscopic versus open gastric bypass: a randomized study of outcomes, quality of life, and costs. Ann. Surg. **234**(3), 279–291 (2001)
3. Bashankaev, B., Baido, S., Wexner, S.D.: Review of available methods of simulation training to facilitate surgical education. Surg. Endosc. **25**(1), 28–35 (2011). doi:10.1007/s00464-010-1123-x
4. Roberts, K.E., Bell, R.L., Duffy, A.J.: Evolution of surgical skills training. World J. Gastroenterol.: WJG **12**(20), 3219–3224 (2006)
5. Gallagher, A.G., McClure, N., McGuigan, J., Ritchie, K., Sheehy, N.P.: An ergonomic analysis of the fulcrum effect in the acquisition of endoscopic skills. Endoscopy **30**(7), 617–620 (1998)
6. See, W.A., Cooper, C.S., Fisher, R.J.: Predictors of laparoscopic complications after formal training in laparoscopic surgery. JAMA **270**(2), 2689–2692 (1993)
7. Wherry, D.C., Rob, C.G., Marohn, M.R., Rich, N.M.: An external audit of laparoscopic cholecystectomy performed in medical treatment facilities of the department of Defense. Ann. Surg. **220**(5), 626–634 (1998)
8. Munz, Y., Kumar, B.D., Moorthy, K., Bann, S., Darzi, A.: Laparoscopic virtual reality and box trainers: is one superior to the other? Surg. Endosc. **18**(3), 485–494 (2004). doi:10.1007/s00464-003-9043-7
9. Madan, A.K., Frantzides, C.T., Park, W.C., Tebbit, C.L., Kumari, N.V., O'Leary, P.J.: Predicting baseline laparoscopic surgery skills. Surg. Endosc. **19**, 101–104 (2005)
10. Kneebone, R.: Simulation in surgical training: educational issues and practical implications. Med. Educ. **37**(3), 267–277 (2003)
11. Rosen, K.: The History of Simulation. In: Levine, A., DeMaria, S., Schwartz, A., Sim, A. (eds.) The Comprehensive Textbook of Healthcare Simulation. LNCS, pp. 5–49. Springer, New York (2013)
12. DiMaio, S.P., Salcudean, S.E.: Needle insertion modeling and simulation. IEEE Trans. Robot. Autom. **19**(5), 864–875 (2003). doi:10.1109/TRA.2003.817044
13. Auer, L.M., Radetzky, A., Wimmer, C., Kleinszig, G., Schroecker, F., Auer, D.P., Delingette, H., Davies, B., Pretschner, D.P.: Visualization for planning and simulation of minimally invasive neurosurgical procedures. In: Taylor, C., Colchester, A. (eds.) MICCAI 1999. LNCS, vol. 1679, pp. 1199–1209. Springer, Heidelberg (1999)
14. Sutherland, L.M., Middleton, P.F., Anthony, A., Hamdorf, J., Cregan, P., Scott, D., Maddern, G.J.: Surgical simulation: a systematic review. Ann. Surg. **243**(3), 291–300 (2006). doi:10.1097/01.sla.0000200839.93965.26
15. Ali, M.R., Mowery, Y., Kaplan, B., DeMaria, E.J.: Training the novice in laparoscopy. More challenge is better. Surg. Endosc. **16**(12), 1732–1736 (2002). doi:10.1007/s00464-002-8850-6
16. Carter, F.J., Schijven, M.P., Aggarwal, R., Grantcharov, T., Francis, N.K., Hanna, G.B., Jakimowicz, J.J.: Work Group for, E., Implementation of, S., Skills Training, P.: Consensus guidelines for validation of virtual reality surgical simulators. Surg. Endosc. **19**(12), 1523–1532 (2005). doi:10.1007/s00464-005-0384-2
17. Seymour, N.E.: VR to OR: a review of the evidence that virtual reality simulation improves operating room performance. World J. Surg. **32**, 7 (2008). doi:10.1007/s00268-007-9307-9

18. Niroomandi, S., Alfaro, I., Cueto, E., Chinesta, F.: Accounting for large deformations in real-time simulations of soft tissues based on reduced-order models. Comput. Methods Programs Biomed. **105**(1), 1–12 (2012). doi:10.1016/j.cmpb.2010.06.012

19. Basdogan, C., De, S., Kim, J., Muniyandi, M., Kim, H., Srinivasan, M.A.: Haptics in minimally invasive surgical simulation and training. IEEE Comput. Graphics Appl. **24**(2), 56–64 (2004)

20. Brown, J., Sorkin, S., Latombe, J.C., Montgomery, K., Stephanides, M.: Algorithmic tools for real-time microsurgery simulation. Med. Image Anal. **6**(3), 289–300 (2002)

21. Itsarachaiyot, Y.: Haptic Interaction of Laparoscopic Surgery in Virtual Environment. Mahidol University (2012)

22. Itsarachaiyot, Y., Pochanakorn, R., Nillahoot, N., Suthakorn, J.: Force acquisition on surgical instruments for virtual reality surgical training system. In: 2011 International Conference on Computer Control and Automation (ICCCA 2011), pp. 173–176. IEEE, Jeju Island, South Korea, 1-3 May 2011

23. Kreyszig, E., Kreyszig, H., Norminton, E.J.: Partial differential equations (PDEs). In: Corliss, S. (ed.) Advanced Engineering Mathematics, pp. 540–585. John Wiley & Sons Inc., New York (2011)

24. Beards, C.: The vibration of continuous structures. In: Structural Vibration: Analysis and Damping. vol. 4, pp. 129–156. Butterworth-Heinemann, Burlington, MA, (1996)

25. Sánchez-Sesma, F.J., Palencia, V.J., Luzón, F.: Estimation of local site effects during earthquakes: an overview. ISET J. Earthq. Technol. **39**(3), 167–193 (2002)

26. Silver, F.H., Freeman, J.W., DeVore, D.: Viscoelastic properties of human skin and processed dermis. Skin Res. Technol. Off. J. Int. Soc. Bioeng. Skin **7**(1), 18–23 (2001)

27. MacLaughlin, J., Holick, M.F.: Aging decreases the capacity of human skin to produce vitamin D3. J. Clin. Investig. **76**(4), 1536–1538 (1985). doi:10.1172/JCI112134

28. Hendriks, F.M.: Mechanical behaviour of human skin in vivo: a literature review. Koninklijke Philips Electronics N.V., Nat. Lab. Unclassified report, pp. 1–46 (2001)

29. Silver, F.H., Seehra, G.P., Freeman, J.W., DeVore, D.: Viscoelastic properties of young and old human dermis: a proposed molecular mechanism for elastic energy storage in collagen and elastin. J. Appl. Polym. Sci. **86**(8), 1978–1985 (2002). doi:10.1002/app.11119

30. McDougall, E.M.: Validation of surgical simulators. J. Endourol./Endourological Soc. **21**(3), 244–247 (2007). doi:10.1089/end.2007.9985

31. Kenney, P.A., Wszolek, M.F., Gould, J.J., Libertino, J.A., Moinzadeh, A.: Face, Content, and Construct Validity of dV-Trainer, a Novel Virtual Reality Simulator for Robotic Surgery. Urology **73**(6), 1288–1292 (2009). doi:10.1016/j.urology.2008.12.044

32. Gavazzi, A., Bahsoun, A.N., Van Haute, W., Ahmed, K., Elhage, O., Jaye, P., Khan, M.S., Dasgupta, P.: Face, content and construct validity of a virtual reality simulator for robotic surgery (SEP Robot). Ann. R. Coll. Surg. Engl. **93**(2), 152–156 (2011). doi:10.1308/003588411X12851639108358

33. Gibney, M.A., Arce, C.H., Byron, K.J., Hirsch, L.J.: Skin and subcutaneous adipose layer thickness in adults with diabetes at sites used for insulin injections: implications for needle length recommendations. Curr. Med. Res. Opin. **26**(6), 1519–1530 (2010). doi:10.1185/03007995.2010.481203

# Bio-inspired Systems and Signal Processing

# Exploring the Relationship Between Characteristics of Ventilation Performance and Response of Newborns During Resuscitation

Huyen Vu[1]([✉]), Trygve Eftestøl[1], Kjersti Engan[1], Joar Eilevstjønn[2], Ladislaus Blacy Yarrot[3], Jørgen E. Linde[4,6], and Hege Ersdal[5,6]

[1] Department of Electrical Engineering and Computer Science, University of Stavanger, Stavanger, Norway
vu.huyen@uis.no
[2] Strategic Research, Laerdal Medical AS, Stavanger, Norway
[3] Research Institute, Haydom Lutheran Hospital, Haydom, Manyara, Tanzania
[4] Department of Pediatrics, Stavanger University Hospital, Stavanger, Norway
[5] Department of Anesthesiology and Intensive Care, Stavanger University Hospital, Stavanger, Norway
[6] Department of Health Sciences, University of Stavanger, Stavanger, Norway

**Abstract.** Birth asphyxia is one of the leading causes of newborn deaths in low resource settings. In non-breathing newborns, ventilation should commence within the first minute after birth. Ventilation signals were studied and parameterized to reflect the characteristics of the provided ventilation. The effectiveness of ventilation was characterized by changes in Apgar score and heart rate. A framework for exploring the association between ventilation parameters and the effectiveness of ventilation is proposed. A statistical hypothesis test method was used to calculate p-values for different patient groups for some ventilation parameters. The results show some low p-values indicating the possible correlation between the corresponding ventilation parameters and the outcome of the treatment.

## 1 Introduction

According to the International Liaison Committee on Resuscitation guidelines for neonatal resuscitation [1], initation of positive pressure ventilation in the first minute after birth is critical for a non-breathing newborn. However, ventilation of newborns is challenging because it is time critical and involves complicated interactions between newborn pathophysiology and the clinical treatment. Determining beneficial characteristics of different ventilation parameters is necessary for clinicians in order to improve neonatal treatment and survival.

The International Liaison Committee on Resuscitation and the World Health Organization provide healthcare workers with guidelines for neonatal resuscitation [1,2]. However, the optimal values of pressure, volume and frequency during positive pressure ventilation of newborns are still unclear. Software systems to extract

A. Fred et al. (Eds): BIOSTEC 2015, CCIS 574, pp. 275–290, 2015.
DOI: 10.1007/978-3-319-27707-3_17

ventilation parameters and to provide decision support during mechanical ventilation have been studied [3–5]. The interaction between mechanical ventilation parameters and the physiology of the lungs has been investigated, concluding that it was difficult to select an effective respirator setting because the results were highly dependent on the characteristics of the newborns population [6]. Although mechanical ventilation has been widely studied, very little research has been done to understand the significantly different initial situations of manual positive pressure ventilation during resuscitation at birth [7].

"Apgar score" is an universal scoring system to evaluate the clinical status of a newborn baby usually performed at 1, 5, 10 and 20 min at birth [8]. The evaluation is based on five subjective factors: appearance (skin color), pulse (heart rate), grimace response (or reflex irritability), activity and muscle tone, and respiration (breathing rate and effort). Each factor is scored on a scale from 0 to 2 (the best score). Thus, the total score is from 0 to 10. The 1-minute score reflects the immediate condition of the baby after birth, whereas the 5-minute score shows how the cardio-respiratory transition from intrauterine to extrauterine life progresses. Heart rate is another indicator of status of newborns. An increase in heart rate after ventilation is believed to be the result of effective ventilation. Heart rate can be monitored during resuscitation, thus giving a more immediate indicator of response to therapy compared to the Apgar score.

In this work, five ventilation parameters were derived from bag mask ventilation pressure and flow signals from recordings of resuscitation of newborns. The hypothesis is that certain characteristics of these ventilation parameters could relate to specific neonatal conditions and improvement in outcome and thus be useful in guiding further treatment. In our previous work, we have studied the infants status reflected by the change in Apgar score as well as the change in heart rate. The change in Apgar score [10] after ventilation compared to before ventilation and also the change in heart rate [9] from before to after ventilation were used to group patients into two groups with high responses (increase in Apgar and/or heart rate) and low responses. An exploratory framework was defined and used to study the ventilation parameters [9,10]. In this paper, we compare results from an exploratory framework similar to what was adopted in our previous work studying both the changing in Apgar score and heart rate applied on more patients than in the previous papers. We also combine these two ways of infant's status evaluation to study the correlation between ventilation parameters and different outcome evaluation methods. The ultimate objective is to identify the determinant factors for beneficial neonatal outcome of ventilation after birth.

## 2    Materials and Methods

### 2.1    Dataset

This exploratory analysis is based on data from the "Safer Births" project at Haydom Lutheran Hospital in Northern Tanzania [11]. Safer Births is a research and development collaboration between Stavanger University Hospital, University of Stavanger, Laerdal Global Health, Haydom Lutheran Hospital, Muhimbili

National Hospital, and some other international research institutions. Haydom is a resource limited rural hospital with a great shortage in health care staff. During the study period, basic newborn resuscitations (i.e. stimulation, suction, and bag mask ventilation) and Apgar scoring were predominantly conducted by midwives, always observed by trained research assistants recording the findings on a data collection form. The implementation of the research project was approved by the National Institute for Medical Research (NIMR) in Tanzania and the Regional Committee for Medical and Health Research Ethics (REK) in Norway.

The Laerdal Newborn Resuscitation Monitor (LNRM) is a resuscitation monitor designed for research use in low resource settings where newborn resuscitations usually are performed by a single care provider. The whole set up is presented in Fig. 1.

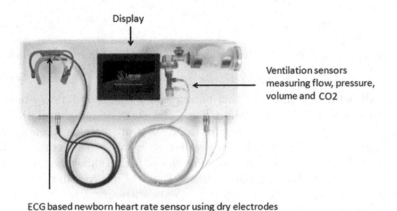

**Fig. 1.** Laerdal newborn resuscitation monitor.

LRNMs were employed in the labour ward of Haydom to measure various physiological data such as ECG signals through dry-electrode ECG measured on the thorax/abdomen, exhaled $CO_2$ concentration, airway pressure and flow signals. The ECG signal is sampled at 500 Hz, the $CO_2$ signal is sampled at 20 Hz, the pressure and flow signals are sampled at 100 Hz. A flow-sensor (Acutronic Medical Systems AG) is arranged between the face mask and the resuscitator bag. The airway adapter also connects two plastic tubes with the LNRM: one tube draws a small sample of exhaled air (50 /min) for standard $CO_2$ measurement (Masimo Sweden AB), and one tube is used for standard pressure measurement.

The dataset in the previous works [9,10] contains recordings of 218 infants collected between July 2013 to June 2014. In this paper, we experiment with a dataset of 354 infants collected between July 2013 to February 2015. Quality control and management of all research data were performed on a daily basis by local research staff.

## 2.2 Processing and Parameterization of Ventilation Signals

To characterize the ventilation given by healthcare workers, we detected bag-mask ventilation events by using two signals: the airway pressure and the flow signals from the ventilation sensors. We define five ventilation parameters: average ventilation frequency, average peak inspiratory pressure (PIP), average expired volume, initial peak inspiratory pressure, and ventilation time percentage (the percentage of time of ventilation sequences in the total time of ventilation including pauses).

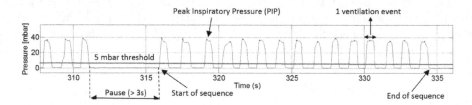

**Fig. 2.** Ventilation event, PIP, pause and ventilation sequence detection.

**Airway Pressure Signal.** In this paper, the term "ventilation event" corresponds to pressing the ventilation bag. The start of one ventilation event is detected when the value of pressure increases from baseline then exceeds a threshold of 5 mbar. The PIP of each ventilation event is the maximum value of the pressure signal. A "pause" is defined as the period of time when the pressure signal value is lower than 5 mbar for more than 3 s. A "ventilation sequence" represents several continuous ventilation events without pauses. The detection of ventilation event, PIP, pause and ventilation sequence is illustrated in Fig. 2. The "total resuscitation time" is the time from the first to the last ventilation event including pauses as shown in Fig. 3. Among the five ventilation parameters, four of them are derived from the pressure signal and can be described as follows:

– The average ventilation frequency $(f_{v_{av}})$ is the ratio of total number of ventilation events $(n_{v_i})$ over the sum of duration of each ventilation sequence $(t_i)$. This parameter represents how fast healthcare workers press the ventilation bag.

$$f_{v_{av}} = \frac{\sum_i n_{v_i}}{\sum_i t_i} \tag{1}$$

– The average PIP $(PIP_{av})$ is the "weighted average" of mean values of PIPs of ventilation sequences $(\overline{PIP})$ where the weight is the duration of each ventilation sequence $(t_i)$, which means that the long sequences dominate to the $PIP_{av}$ more than the short sequences. This parameter shows the average peak inspiratory pressure applied for ventilation.

$$PIP_{av} = \frac{\sum_i \overline{PIP}_i \cdot t_i}{\sum_i t_i} \tag{2}$$

- Initial peak inspiratory pressure ($PIP_{init}$) is the first average PIP value of the first ventilation sequence that has the duration longer than 0.5 s. This parameter represents the initial peak inspiratory pressure to open the lung.
- Ventilation time percentage ($VT_{PRC}$) is the proportion of the sum of the duration of all ventilation sequences in the total resuscitation time ($T_r$). This parameter shows the percentage of the time the rescuer is spending on ventilation during the whole ventilation procedure. For example, $VT_{PRC} = 60\%$ means that 60 % of the total resuscitation time is spent for ventilation effort and 40 % of the time on other resuscitative actions.

$$VT_{PRC} = \frac{\sum_i t_i}{T_r} \cdot 100\% \tag{3}$$

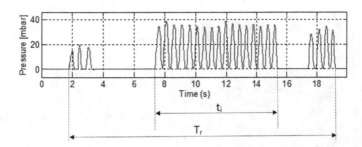

**Fig. 3.** Total resuscitation time is the duration from the first ventilation event to the last one.

**Volume Waveform.** The volume waveform is integrated from the flow signal which is measured by the hot-wire flow sensor (Acutronic Medical Systems AG) as in Eq. 4.

$$V = \int F(t)dt \tag{4}$$

The expired volume is the amount of air going back through the flow sensor after one inflation. The expired volume is the volume drop from the maximum value to zero or to a non zero value when there is mask leakage. The average expired volume ($ExV_{av}$) is the mean of all expired volume values. Figure 4 shows one inflation cycle.

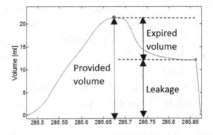

**Fig. 4.** Volume waveform of one ventilation cycle.

## 2.3  Proposed Framework for Data Exploration

We want to investigate if certain patterns of the above mentioned parameters show association with specific neonatal outcome measures. A resuscitation attempt on a newborn with a very bad initial condition should possibly be performed differently than on a newborn with a much better initial condition. Thus, the initial condition of the newborn is used to divide the dataset into more homogeneous subgroups. Therefore a wide range of subgroups of babies with varying initial conditions are explored. Furthermore, the exploration aims to identify parameter characteristics corresponding to a positive response to the therapy. Thus, the subsets are split into two groups according to high and low responses to the therapy. The parameter values for these two groups are then compared.

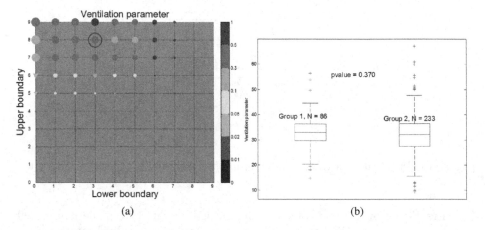

(a)                                      (b)

**Fig. 5.** (a) An example of LU-plot. (b) Box plot of the corresponding LU coordinate.

**Defining Patient Subsets According to Initial Conditions.** To investigate how the ventilation parameters affect various subsets of patients, the data set was analyzed with various ranges of initial conditions. A specific range is specified by setting a lower (L) and upper boundary (U) of an initial condition. The babies with initial values in this range are included and categorized into the two response groups based on outcome measures. The ventilation parameter values for these two groups are then compared by applying the statistical test. The corresponding p-value provides the indication of the significance of the ventilation parameter. The results for a given parameter is visualized in a LU-coordinate plot where the L and U shows initial values indicated by the horizontal and vertical axes respectively. Each subset is represented by a (L,U) coordinate point in such a visualization. A similar approach was introduced in a different data analysis [13]. Figure 5 shows an example of LU-plot where a circle or triangle represents the (L,U) coordinate of the corresponding analyzed subset of patients. For example, the red-circled coordinate where L = 3, U = 8 includes a subset of babies having the initial condition in the range from 3 (lower boundary) to 8 (upper boundary).

**Defining High and Low Response Groups by Outcome Measures.** The first outcome measure is based on delta Apgar score ($\Delta AP$) which is the change from the initial Apgar score at 1 min (AP1) to the final Apgar score at 5 min (AP5) defined in Eq. 5. Group 1 has $\Delta AP \leq$ threshold (less improvement from AP1 to AP5) and group 2 has $\Delta AP >$ threshold (better improvement from AP1 to AP5). Therefore, group 2 is considered to be the best group (or the group with most positive outcome). For example, if the Apgar score at 1 min and Apgar score at 5 min of a baby are 3 and 7 respectively, then the $\Delta AP$ is 4. And if the chosen threshold $\Delta AP = 2$, then the baby is categorized into group 2.

$$\Delta AP = AP5 - AP1 \tag{5}$$

The second outcome measure is based on delta heart rate $\Delta HR$ as presented in Eq. 6. The initial heart rate ($HR_{ini}$) is the heart rate at the beginning of the first ventilation event. The final heart rate ($HR_{fin}$) is measured after the last ventilation event. A change in heart rate from the initial to the final one after ventilation is likely to be associated with the quality of the ventilation. Group 1 has $\Delta HR \leq$ threshold and group 2 has $\Delta HR >$ threshold. Similarly, group 2 is considered to be the high response group because these patients will have a higher increase in heart rate during ventilation compared to group 1 which is the low response group.

$$\Delta HR = HR_{fin} - HR_{ini} \tag{6}$$

The ventilation parameter values of the subset indicated by the red-circled coordinate in Fig. 5a are split into response groups 1 and 2 as shown in the box plot in Fig. 5b. There are 86 ventilation parameter values of 86 patients in group 1, the low response group while there are 233 ventilation parameter values of 233 patients in group 2, the high response group.

### 2.4 Comparing Ventilation Parameter Values for High and Low Response Groups

For the patients in the high and low response groups, the ventilation parameter values are compared using hypothesis testing. P-values from the significance tests were used to represent the discriminative capability of different ventilation parameters. A low p-value implies the difference in medians of the two groups and that the corresponding ventilation parameter might have an important effect on the result of the resuscitation attempt.

The low p-values ($< 0.05$) are illustrated as triangles otherwise as circles. The upward triangles represent a higher median value of group 2 compared to group 1 and vice verse. The size of each point on the LU-plot is proportional to the number of patients in the smallest group, thus illustrating the sample size of the data used for statistical testing. For example, group 1 has 34 patients and group 2 has 45 patients, then the size of the circle or the triangle is proportional to the size of group 1 that has the smallest number of patients.

## 3   Experiments

Due to other resuscitation activities, e.g. stimulation, suction and chest compression often performed during or in between ventilation sequences, the ECG signals can be disturbed and consequently be unreliable. We only included heart rate information when the ECG signal quality is good. There are some cases where we have information about Apgar score but do not have information about heart rate before and/or after ventilation and vice verse. The number of patients for each experiment is illustrated in Table 1.

**Table 1.** Number of patients for each experiment.

|   | Experiments | Number of patients |
|---|---|---|
| 1 | Apgar score 1 min and $\Delta AP$ | 354 |
| 2 | $HR_{ini}$ and $\Delta HR$ | 216 |
| 3 | Apgar score 1 min and $\Delta HR$ | 165 |
| 4 | $HR_{ini}$ and $\Delta AP$ | 186 |

According to Eq. 5, a threshold value is used to categorize patients into high and low response groups. Thus, this threshold value is critical in determining a baby belonging to group 1 or group 2. To find a suitable value of $\Delta AP$ threshold, a histogram representing number of patients with different Apgar score at 1 min and Apgar score at 5 min is shown in Fig. 6a. The blue diagonal line on the histogram in Fig. 6a represents the $\Delta AP$ equal to 0 where the Apgar score at 5 min is equal to Apgar score at 1 min. The line also separates the histogram into two parts: the left one with $\Delta AP > 0$ (Apgar score at 5 min > Apgar score

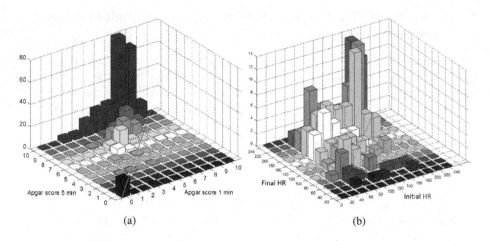

(a)                                    (b)

**Fig. 6.** (a) Apgar score histogram. (b) Heart rate histogram.

at 1 min) and the right one with $\Delta AP < 0$ (Apgar score at 5 min $<$ Apgar score at 1 min). The lines in parallel with the main diagonal represent different $\Delta AP$ thresholds and this is a visualization of how threshold values separate our data into groups with high and low responses. Threshold $\Delta AP = 2$ was chosen to balance the number of patients in each group.

Similarly, to find an appropriate value of $\Delta HR$ threshold for the experiment, a histogram is used to visualize the distribution of heart rate as illustrated in Fig. 6b. Points on the blue diagonal in Fig. 6b have $\Delta HR = 0$ corresponding to $HR_{fin} = HR_{ini}$. The blue diagonal also separates the histogram into two parts: the left corresponding to $\Delta HR > 0$ and the right part corresponding to $\Delta HR < 0$. From the histogram, we chose threshold of $\Delta HR = 20$ to balance the number of patients in the high and low response groups in our experiment.

We used Wilcoxon rank-sum function in Matlab for hypothesis test since the data is non-normally distributed [12]. Four experiments were performed: (1) Apgar score at 1 min as initial condition and $\Delta AP = 2$ as outcome, (2) $HR_{ini}$ as initial condition and $\Delta HR = 20$ as outcome, (3) Apgar score at 1 min as initial condition and $\Delta HR = 20$ as outcome, and (4) $HR_{ini}$ rate as initial condition and $\Delta AP = 2$ as outcome. In all experiments, we used the condition that the number of babies in each group had to be equal or larger than 20 to exclude hypothesis test with too few data.

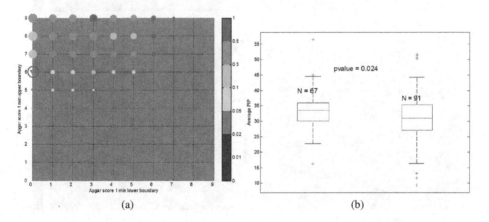

(a)                                        (b)

**Fig. 7.** Results of Experiments with Apgar Score. (a) LU-plot for average PIP ($PIP_{av}$), $\Delta AP$ threshold = 2. The low p-values ($< 0.05$) are illustrated as triangles otherwise as circles. The upward-pointing triangles indicate the higher medians and the downward-pointing triangles indicate the lower medians of group 2 (the improved group with higher $\Delta AP$) in comparison with group 1. The size of each point in the LU-plot is proportional to the minimum number of patients between two groups. (b) Box plot for the red-circled point corresponding to a subset of patients having Apgar score 1 min in the range indicated by the coordinate of that point in the LU-plot (Apgar score 1 min ranging from 0 to 6). P-value obtained from statistical significance test on $PIP_{av}$ for this subset is 0.024. The median value of group 2 (including infants in that subset having $\Delta AP > 2$) is 31.01, which is lower than the median value 33.53 of group 1 (infants with $\Delta AP <= 2$), thus the downward-pointing triangle is used.

## 4    Results

### 4.1    Experiments with Apgar Score

Fig. 7a depicts the LU-plot of $PIP_{av}$ using the threshold $\Delta AP = 2$ and Fig. 7b shows the corresponding box plot of the red-circled point in the LU-plot (the color bar next to the LU-plot represents ranges of p-values). The number on each box plot represents the number of babies in each group. The box plot shows the difference in medians of two outcome groups, specifically, group 2 has the lower median value of $PIP_{av}$. The results of other ventilation parameters are shown

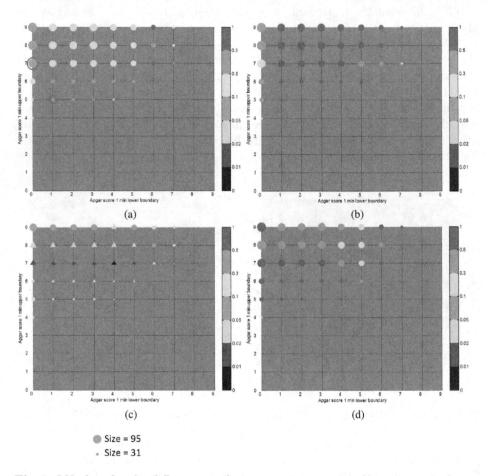

Size = 95

Size = 31

**Fig. 8.** LU-plots for the different ventilation parameters using Apgar score 1 min as initial condition and $\Delta AP$ as outcome. (a) Initial PIP ($PIP_{init}$). (b) Ventilation frequency ($f_{v_{av}}$). (c) Average expired volume ($ExV_{av}$). (d) Ventilation time percentage ($VT_{PRC}$). The size of two circled points in (a) is illustrated at the bottom. For bigger circle, group 1 has 95 newborns, group 2 has 170 newborns, the size of the circle is proportional to the size group 1. For smaller circle, group 1 has 31 newborns, group 2 has 62 newborns, the size of the circle is proportional to the size group 1.

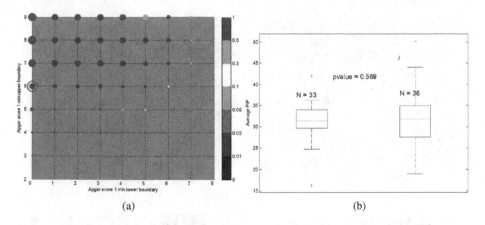

(a)                                    (b)

**Fig. 9.** Results of experiments using Apgar Score with only new data. (a) LU-plot for average PIP ($PIP_{av}$). (b) Box plot for the red-circled point.

in Fig. 8. Low p-values were found for $PIP_{av}$, $PIP_{init}$ and $ExV_{av}$. From the experiment with Apgar score, the p-value points on the LU-plot of $PIP_{av}$ are similar to the result in the previous paper [10]. Specifically, we found low p-values for $PIP_{av}$, $PIP_{init}$ and $ExV_{av}$ parameters. However, the blue color in the LU-plot of $PIP_{av}$ is in the higher range of p-value in comparison with the same coordinate x = 0 and y = 6 in the previous paper [10]. Thus, we experimented with only new data (136 patients) to see the effect of new data in the whole dataset. Figure 9 shows the changing of results with new data which indicates the need of more data in using our proposed data analysis framework in order to obtain valid or trustworthy results.

## 4.2    Experiments with Heart Rate

Figure 10 shows LU-plot for each ventilation parameters where $HR_{ini}$ is used as initial condition and $\Delta HR$ as outcome. Low p-values were found for $PIP_{av}$, $VT_{PRC}$ and $ExV_{av}$.

## 4.3    Experiments Combining Apgar Score at 1 min and $\Delta HR$

Figure 11 shows LU-plots for each ventilation parameters. The initial condition is defined by Apgar score at 1 min represented by the x-y axis. The outcome is defined as $\Delta HR > 20$. The only parameter that shows some low p-values is $ExV_{av}$.

## 4.4    Experiments Combining $HR_{ini}$ and $\Delta AP$

The LU-plots for all the ventilation parameters are depicted in Fig. 12. The $HR_{ini}$ was used for defining the subgroups. The outcome was defined by $\Delta AP$. Low p-values were found for $ExV_{av}$, $PIP_{av}$, and $PIP_{init}$.

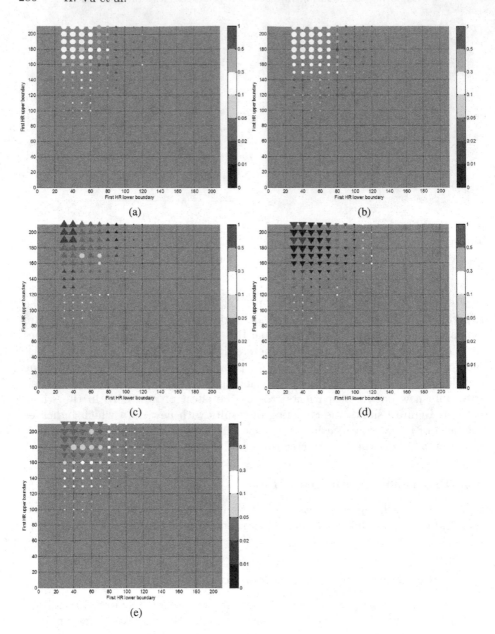

**Fig. 10.** Results of experiments using $HR_{ini}$ as initial condition and $\Delta HR$ as outcome. (a) Initial PIP ($PIP_{init}$). (b) Ventilation frequency ($f_{v_{av}}$). (c) Average expired volume ($ExV_{av}$). (d) Ventilation time percentage ($VT_{PRC}$). (e) Average PIP ($PIP_{av}$).

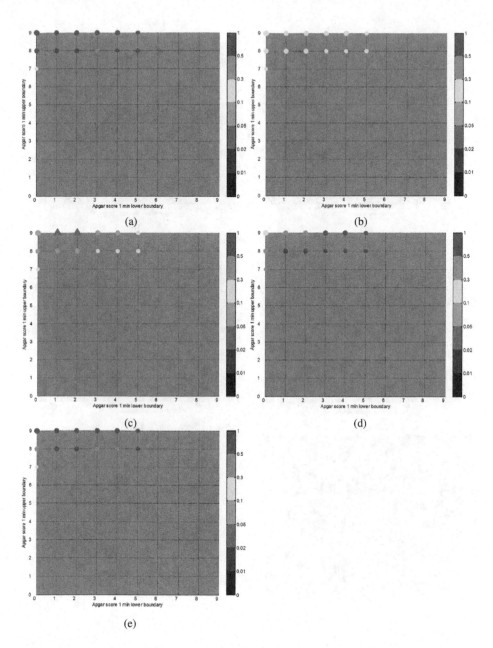

**Fig. 11.** Results of experiments with Apgar Score at 1 min as initial condition and $\Delta HR$ as outcome. (a) Initial PIP ($PIP_{init}$). (b) Ventilation frequency ($f_{v_{av}}$). (c) Average expired volume ($ExV_{av}$). (d) Ventilation time percentage ($VT_{PRC}$). (e) Average PIP ($PIP_{av}$).

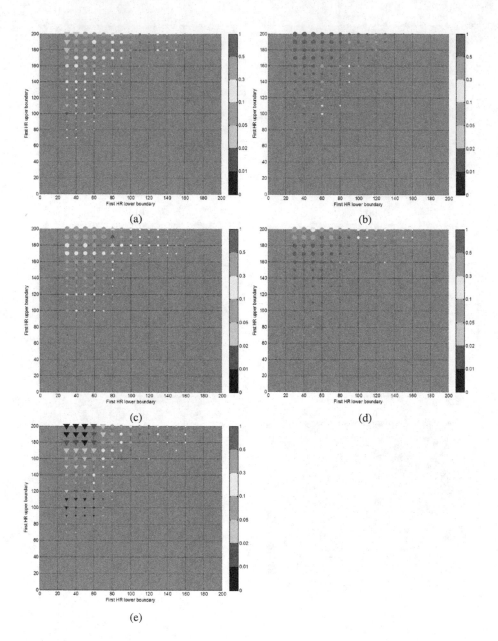

**Fig. 12.** Results of experiments with $HR_{ini}$ and $\Delta AP$. (a) Initial PIP ($PIP_{init}$). (b) Ventilation frequency ($f_{v_{av}}$). (c) Average expired volume ($ExV_{av}$). (d) Ventilation time percentage ($VT_{PRC}$). (e) Average PIP ($PIP_{av}$).

# 5  Discussion

According to our knowledge, the dataset of our work is unique. It is difficult to include a comparison with other works since we are not aware of any similar work. From the heart rate experiment, we found low p-values for $PIP_{av}$, $ExV_{av}$ and $VT_{PRC}$. For $VT_{PRC}$, the direction of the triangles is downward, meaning the median value of the high group is lower than the low response group. This could possibly be explained that the importance of stimulation, i.e. rubbing back of the newborn, which was not included in the analysis. In future work we want to explore the significance of stimulation in relation to ventilation for different groups of troubled newborns.

In different experiments, we found low p-values for different ventilation parameters. Some experiments have low p-value for the same parameters, while some are slightly different. However, low p-value were found in all experiments for $ExV_{av}$ which could indicate this parameter correlate more with changes in Apgar score and heart rate than the other parameters that were tested.

# 6  Conclusions

In this paper, we extend our previous works by using new data and compare the result from Apgar score and heart rate. Some ventilation parameters such as average peak inspiratory pressure (PIPav), average expired volume ($ExV_{av}$), $PIP_{av}$, $PIP_{init}$ and ventilation time percentage ($VT_{PRC}$) have shown low p-value in the analysis. These findings might indicate that these parameters could be determinant factors for beneficial positive pressure ventilation in resuscitation attempts of non-breathing newborns. As this is a study using exploratory data analysis, some of the low p-values might have appeared by chance. Thus, the findings needs to be further investigated by using new data.

For future work, we might include the stimulation time in addition to the ventilation time when we derive the ventilation time percentage ($VT_{PRC}$) parameter. There is also a possibility to combine some ventilation parameters to investigate their joint effect on the resuscitation outcome.

# References

1. Neonatal Resuscitation: 2010 International Consensus on Cardiopulmonary Resuscitation and Emergency Cardiovascular Care Science With Treatment Recommendations, Resuscitation (2010)
2. Guidelines on basic newborn resuscitation, WHO Library Cataloguing-in-Publication Data, ISBN: 978-92-4-150369-3 (2011)
3. Ciurea, B.M., Palade, D., Kostrakievici, S.: Lung ventilators parameters determination. U.P.B. Sci. Bull. **73**(2), 107–120 (2011)
4. Tehrani, F.: Efficient Decision Support Systems - Practice and Challenges in Biomedical Related Domain, Computerized Decision Support Systems for Mechanical Ventilation, ISBN: 978-953-307-258-6 (2011)

5. Schulze, K., Stefanski, M., Soulos, T., Masterson, J., Kim, Y.I., Rey, H.R.: Computer Analysis of Ventilatory Parameters for Neonates on Assisted Ventilation, Engineering in Medicine and Biology Magazine. IEEE, ISSN:0739–5175 (1984)
6. Ramsden, C.A., Reynolds, E.O.: Ventilator settings for newborn infants. Arch. Dis. Child. **62**(9), 529–538 (1987)
7. Perlman, J., Kattwinkel, J., Wyllie, J.: Neonatal resuscitation: in pursuit of evidence gaps in knowledge. Resuscitation **83**(5), 545–550 (2012)
8. Apgar score. http://en.wikipedia.org/wiki/Apgar_score
9. Vu, H., Eftestøl, T., Engan, K., Eilevstjønn, J., Linde, J., Ersdal, H.: Analysis of heart rate changes in newborns to investigate the effectiveness of bag-mask ventilation. In: Proceedings of Computing in Cardiology Conference (2014)
10. Vu, H., Eftestøl, T., Engan, K., Eilevstjønn, J., Linde, J., Ersdal, H.: Exploratory analysis of ventilation signals from resuscitation data of newborns. In: Proceedings of Biosignals (2015)
11. Safer Birth Research and Developemtn to Save Newborn lives. http://www. saferbirths.com/
12. Matlab Wilcoxon rank sum test. http://se.mathworks.com/help/stats/ranksum. html
13. Eftestøl, T., Maloy, F., Engan, K., Kotu, L.P., Woie, L., Orn, S.: A texture-based probability mapping for localisation of clinically important cardiac segments in the myocardium in cardiac magnetic resonance images from myocardial infarction patients. In: 2014 IEEE International Conference on Image Processing (2014)

# A Novel Application of Universal Background Models for Periocular Recognition

João C. Monteiro$^{(\boxtimes)}$ and Jaime S. Cardoso

INESC TEC and Faculdade de Engenharia,
Universidade do Porto Campus da FEUP, Rua Dr. Roberto Frias, 378,
4200-465 Porto, Portugal
dee12007@fe.up.pt, jaime.cardoso@inescporto.pt

**Abstract.** In recent years the focus of research in the fields of iris and face recognition has turned towards alternative traits to aid in the recognition process under less constrained acquisition scenarios. The present work assesses the potential of the periocular region as an alternative to both iris and face in such conditions. An automatic modeling of SIFT descriptors, using a GMM-based Universal Background Model method, is proposed. This framework is based on the Universal Background Model strategy, first proposed for speaker verification, extrapolated into an image-based application. Such approach allows a tight coupling between individual models and a robust likelihood-ratio decision step. The algorithm was tested on the UBIRIS.v2 and the MobBIO databases and presented state-of-the-art performance for a variety of experimental setups.

**Keywords:** Biometrics · Iris segmentation · Unconstrained environment · Gradient flow · Shortest closed path

## 1 Introduction

Over the past few years face and iris have been on the spotlight of many research works in biometrics. The *face* is a easily acquirable trait with a high degree of uniqueness, while the *iris*, the coloured part of the eye, presents unique textural patterns resulting from its random morphogenesis during embryonic development [1]. These marked advantages, however, fall short when low-quality images are presented to the system. It has been noted that the performance of iris and face recognition algorithms is severely compromised when dealing with non-ideal scenarios such as non-uniform illumination, pose variations, occlusions, expression changes and radical appearance changes [1]. Several recent works have tried to explore alternative hypothesis to overcome this problem, either by developing more robust algorithms or by exploring new traits to allow or aid in the recognition process [28].

The *periocular* region is one of such unique traits. Even though a true definition of the periocular region is not standardized, it is common to describe it as the region in the immediate vicinity of the eye [10]. Periocular recognition

© Springer International Publishing Switzerland 2015
A. Fred et al. (Eds): BIOSTEC 2015, CCIS 574, pp. 291–307, 2015.
DOI: 10.1007/978-3-319-27707-3_18

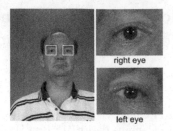

**Fig. 1.** Example of periocular regions from both eyes, extracted from a face image [28].

can be motivated as a representation in between face and iris recognition. It has been shown to present increased performance when only degraded facial data [7] or low quality iris images [26] are made available, as well as promising results as a soft biometric trait to help improve both face and iris recognition systems in less constrained acquisition environments [2] (Fig. 1).

Periocular biometrics is a recent area of research, proposed by the first time in a feasibility study by Park et al. [12]. In this pioneer work, the authors suggested the periocular region as a potential alternative to circumvent the significant challenges posed to iris recognition systems working under unconstrained scenarios. In recent years, a number of relevant works have helped further the potential of periocular recognition in the field of biometrics. Padole and Proença [10] explored the effect of scale, pigmentation and occlusion, as well as gender, and propose an initial region-of-interest detection step to improve recognition accuracy. Ross et al. [21], following a recent trend in biometric research, explored information fusion based on several feature extraction techniques, to handle the significant variability of input periocular images. Tan et al. [26] studied the benefits of periocular recognition when highly degraded regions result from the traditional iris segmentation step. A more in depth analysis of the state-of-the-art in periocular recognition can be found in our previous work [8] and in the thorough review work of Santos and Proença [23].

The present work serves as an extension of [8], where we first proposed an innovative approach to periocular recognition under less ideal acquisition conditions. Such proposal is based on the idea of maximum *a posteriori* adaptation of Universal Background Model, as proposed by Reynolds for speaker verification [19]. We evaluate the proposed algorithm on two datasets of color periocular images acquired under visible wavelength (VW) illumination. Multiple noise factors such as varying gazes/poses and heterogeneous lighting conditions are characteristic to such images, thus representing the main challenge of the present work. In addition to the aforementioned objectives, we also aimed to assess the performance of the proposed algorithm with variable distances between the captures individuals and the acquisition apparatus. As no objective definition exists concerning which anatomical structures compose the periocular region, this analysis might provide significant insight for further research in the field of periocular recognition.

# 2    Proposed Methodology

## 2.1    Algorithm Overview

The proposed algorithm is schematically represented in Fig. 2. The two main blocks - *enrollment* and *recognition* - refer to the typical architecture of a biometric system. During enrollment a new individual's biometric data is inserted into a previously ex-istent database of individuals. Such database is probed during the recognition process to assess either the validity of an identity claim - *verification* - or the $k$ most probable identities - *identification* - given an unknown sample of biometric data.

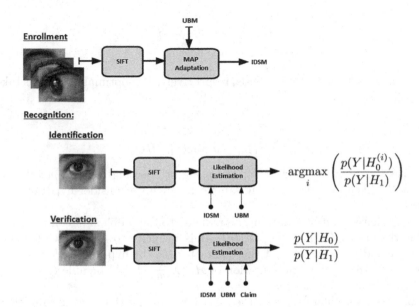

**Fig. 2.** Graphical representation of the main steps in both the enrollment and recognition (verification and identification) phases of the proposed periocular recognition algorithm.

During the enrollment, a set of $N$ models describing the unique statistical distribution of biometric features for each individual $n \in \{1, \ldots, N\}$ is trained by maximum *a posteriori* (MAP) adaptation of an Universal Background Model (UBM). The UBM is a representation of the variability that the chosen biometric trait presents in the universe of all individuals. MAP adaptation works as a specialization of the UBM based on each individual's biometric data. The idea of MAP adaptation of the UBM was first proposed by Reynolds [19], for speaker verification, and will be further motivated in the following sections.

The recognition phase is carried out through the projection of the features extracted from an unknown sample onto both the UBM and the individual specific models (IDSM) of interest. A likelihood-ratio between both projections outputs the final recognition score. Depending on the functioning mode of the system - verification or identification - decision is carried out by thresholding or maximum likelihood-ratio, respectively.

## 2.2 Universal Background Model

Universal background modeling is a common strategy in the field of voice biometrics [13]. It can be easily understood if the problem of biometric verification is interpreted as a basic hypothesis test. Given a biometric sample $Y$ and a claimed ID, $S$, we define:

$$H_0 : Y \text{ belongs to } S$$
$$H_1 : Y \text{ does } \underline{\text{not}} \text{ belong to } S$$

as the null and alternative hypothesis, respectively. The optimal decision is taken by a *likelihood-ratio test*:

$$\frac{p(Y|H_0)}{p(Y|H_1)} \begin{cases} \geq \theta & \text{accept } H_0 \\ \leq \theta & \text{accept } H_1 \end{cases} \tag{1}$$

where $\theta$ is the decision threshold for accepting or rejecting $H_0$, and $p(Y|H_i), i \in \{0, 1\}$ is the likelihood of observing sample $Y$ when we consider hypothesis $i$ to be true.

Biometric recognition can, thus, be reduced to the problem of computing the likelihood values $p(Y|H_0)$ and $p(Y|H_1)$. It is intuitive to note that $H_0$ should be represented by a model $\lambda_{hyp}$ that characterizes the hypothesized individual, while, alternatively, the representation of $H_1$, $\lambda_{\overline{hyp}}$, should be able to model *all the alternatives to the hypothesized individual*.

From such formulation arises the need for a model that successfully covers the space of alternatives to the hypothesized identity. The most common designation in literature for such a model is *universal background model* or *UBM* [20]. Such model must be trained on a large set of data, so as to faithfully cover a representative user space and a significant amount of sources of variability. The following section details the chosen strategy to model $\lambda_{\overline{hyp}}$ and how individual models, $\lambda_{hyp}$, can be adapted from the UBM in a fast and robust way.

## 2.3 Hypothesis Modeling

On the present work we chose Gaussian Mixture Models (GMM) to model both the UBM, i.e. $\lambda_{\overline{hyp}}$, and the individual specific models (IDSM), i.e. $\lambda_{hyp}$. Such models are capable of capturing the empirical probability density function (PDF) of a given set of feature vectors, so as to faithfully model their intrinsic statistical properties [19]. The choice of GMM to model feature distributions in biometric data is extensively motivated in many works of related areas. From the most common interpretations, GMMs are seen as capable of representing broad "hidden" classes, reflective of the unique structural arrangements observed in the analysed biometric traits [19]. Besides this assumption, Gaussian mixtures display both the robustness of parametric unimodal Gaussian density estimates, as well as the ability of non-parametric models to fit non-Gaussian data [18]. This duality, alongside the fact that GMM have the noteworthy strength of generating smooth parametric densities, confers such models a strong advantage as generative model of choice. For computational efficiency, GMM models are often trained using diagonal covariance matrices. This approximation is often found in biometrics literature, with no significant accuracy loss associated [29].

All models are trained on sets of Scale Invariant Feature Transform (SIFT) keypoint descriptors [6]. This choice for periocular image description is thoroughly motivated in literature [11,21], mainly due to the observation that local descriptors work better than their global counterparts when the available data presents non-uniform conditions. Furthermore, the invariance of SIFT features to a set of common undesirable factors (image scaling, translation, rotation and also partially to illumination and affine or $3D$ projection), confer them a strong appeal in the area of unconstrained biometrics.

### 2.4   $H_1$: UBM Parameter Estimation

To train the Universal Background Model a large amount of "impostor" data, i.e. a set composed of data from all the enrolled individuals, is used, so as to cover a wide range of possibilities in the individual search space [25]. The training process of the UBM is simply performed by fitting a $k$-mixture GMM to the set of PCA-reduced feature vectors extracted from all the "impostors".

If we interpret the UBM as an "impostor" model, its "genuine" counterpart can be obtained by adaptation of the UBM's parameters, $\lambda_{\overline{hyp}}$, using individual specific data. For each enrolled individual, $ID$, an *individual specific model* (IDSM), defined by parameters $\lambda_{hyp}$, is therefore obtained. The adaptation process will be outlined in the following section.

### 2.5   $H_0$: MAP Adaptation of the UBM

IDSMs are generated by the *tuning of the UBM parameters* in a maximum *a posteriori* (MAP) sense, using individual specific biometric data. This approach provides a tight coupling between the IDSM and the UBM, resulting in better performance and faster scoring than uncoupled methods [29], as well as a robust and precise parameter estimation, even when only a small amount of data is available [25]. This is indeed one of the main advantages of using UBMs. The determination of appropriate initial values (i.e. seeding) of the parameters of a GMM remains a challenging issue. A poor initialization may result in a weak model, especially when the data volume is small. Since the IDSM are learnt only from each individual data, they are more prone to a poor convergence that the GMM for the UBM, learnt from a big pool of individuals. In essence, UBM constitutes a good initialization for the IDSM.

The adaptation process, as proposed by Reynolds [19], resembles the Expectation-Maximization (EM) algorithm, with two main estimation steps. The first is similar to the expectation step of the EM algorithm, where, for each mixture of the UBM, a set of sufficient statistics are computed from a set of $M$ individual specific feature vectors, $X = \{\mathbf{x}_1 \ldots \mathbf{x}_M\}$:

$$n_i = \sum_{m=1}^{M} p(i|\mathbf{x}_m) \tag{2}$$

$$E_i(\mathbf{x}) = \frac{1}{n_i} \sum_{m=1}^{M} p(i|\mathbf{x}_m)\mathbf{x}_m \tag{3}$$

$$E_i(\mathbf{x}\mathbf{x}^t) = \frac{1}{n_i} \sum_{m=1}^{M} p(i|\mathbf{x}_m)\mathbf{x}_m\mathbf{x}_m^t \tag{4}$$

where $p(i|\mathbf{x}_m)$ represents the probabilistic alignment of $\mathbf{x}_m$ into each UBM mixture. Each UBM mixture is then adapted using the newly computed sufficient statistics, and considering diagonal covariance matrices. The update process can be formally expressed as:

$$\hat{w}_i = [\alpha_i n_i / M + (1 - \alpha_i)w_i]\,\xi \tag{5}$$

$$\hat{\boldsymbol{\mu}}_i = \alpha_i E_i(\mathbf{x}) + (1 - \alpha_i)\boldsymbol{\mu}_i \tag{6}$$

$$\hat{\Sigma}_i = \alpha_i E_i(\mathbf{x}\mathbf{x}^t) + (1 - \alpha_i)(\boldsymbol{\sigma}_i\boldsymbol{\sigma}_i{}^t + \boldsymbol{\mu}_i\boldsymbol{\mu}_i{}^t) - \hat{\boldsymbol{\mu}}_i\hat{\boldsymbol{\mu}}_i{}^t \tag{7}$$

$$\boldsymbol{\sigma}_i = \mathrm{diag}(\Sigma_i) \tag{8}$$

where $\{w_i, \boldsymbol{\mu}_i, \boldsymbol{\sigma}_i\}$ are the original UBM parameters and $\{\hat{w}_i, \hat{\boldsymbol{\mu}}_i, \hat{\boldsymbol{\sigma}}_i\}$ represent their adaptation to a specific speaker. To assure that $\sum_i w_i = 1$ a weighting parameter $\xi$ is introduced. The $\alpha$ parameter is a data-dependent adaptation coefficient. Formally it can be defined as:

$$\alpha_i = \frac{n_i}{r + n_i} \tag{9}$$

where $r$ is generally known as the *relevance factor*. The individual dependent adaptation parameter serves the purpose of weighting the relative importance of the original values and the new sufficient statistics in the adaptation process. For the UBM adaptation we set $r = 16$, as this is the most commonly observed value in literature [19]. Most works propose the sole adaptation of the mean values, i.e. $\alpha_i = 0$ when computing $\hat{w}_i$ and $\hat{\sigma}_i$. This simplification seems to bring no nefarious effects over the performance of the recognition process, while allowing faster training of the individual specific models [3].

## 2.6    Recognition and Decision

After the training step of both the UBM and each IDSM, the recognition phase with new data from an unknown source is somewhat trivial. As referred in previous sections, the identity check is performed through the projection of the new test data, $X_{test} = \{\mathbf{x}_{t,1}, \ldots, \mathbf{x}_{t,N}\}$, where $\mathbf{x}_{t,i}$ is the $i$-th PCA-reduced SIFT vector extracted from the periocular region of test subject $t$, onto both the UBM and either the claimed IDSM (in verification mode) or all such models (in identification mode). The recognition score is obtained as the average likelihood-ratio of all keypoint descriptors $\mathbf{x}_{t,i}, \forall i \in \{1..N\}$. The decision is then carried out by checking the condition presented in Eq. (1), in the case of verification, or by detecting the maximum likelihood-ratio value for all enrolled IDs, in the case of identification.

This is a second big advantage of using UBM. The ratio between the IDSM and the UBM probabilities of the observed data is a more robust decision criterion than relying solely on the IDSM probability. This results from the fact that some subjects are more prone to generate high likelihood values than others, i.e. some people have a more "generic" look than others. The use of a likelihood ratio with an universal reference works as a normalization step, mapping the likelihood values in accord to their global projection. Without such step, finding a global optimal value for the decision threshold, $\theta$, presented in Eq. 1 would be a far more complex process.

## 3    Experimental Results

In this section we start by presenting the datasets and the experimental setups under which performance was assessed. Further sections present a detailed analysis regarding the effect of model complexity and fusion of color channels in the global performance of the proposed algorithm.

### 3.1    Tested Datasets

The proposed algorithm was tested on two noisy color iris image databases: UBIRIS.v2 and MobBIO. Even though both databases were designed in an attempt to promote the development of robust iris recognition algorithms for images acquired under VW illumination, their intrinsic properties make them attractive to study the feasibility of periocular recognition under similar conditions. The following sections detail their main features as well as the reasoning behind their choice.

**UBIRIS.v2 Database.** Images in UBIRIS.v2 [17] database were captured under non-constrained conditions (at-a-distance, on-the-move and on the visible wavelength), with corresponding realistic noise factors. Figure 3 depicts some examples of these noise factors (reflections, occlusions, pigmentation, etc.). Two acquisition sessions were performed with 261 individuals involved and a total of 11100 300 × 400 color images acquired. Each individual's images were acquired at variable distances with 15 images per eye and per season.

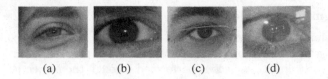

Fig. 3. Examples of noisy image from the UBIRIS.v2 database.

**MobBIO Database.** The MobBIO multimodal database [24] was created in the scope of the 1st Biometric Recognition with Portable Devices Competition 2013, integrated in the ICIAR 2013 conference. The main goal of the competition was to compare various methodologies for biometric recognition using data acquired with portable devices. We tested our algorithm on the iris modality present on this database. Regarding this modality the images were captured under two alternative lighting conditions, with variable eye orientations and occlusion levels, so as to comprise a larger variability of unconstrained scenarios. Distance to the camera was, however, kept constant for each individual. For each of the 105 volunteers 16 images (8 of each eye) were acquired. These images were obtained by cropping a single image comprising both eyes. Each cropped image was set to a $300 \times 200$ resolution. Figure 4 depicts some examples of such images.

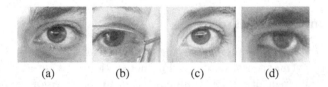

Fig. 4. Examples of iris images in the MobBIO database.

The MobBIO database presents a face modality which has also been explored for comparative purposes in the present work. Images were acquired in similar conditions to those described above for iris images, with 16 images per subject. Examples of such images can be observed in Fig. 5.

Fig. 5. Examples of face images in the MobBIO database.

## 3.2    Evaluation Metrics

Performance was evaluated for both verification and identification modes. Regarding the former we analyzed the *equal error rate* (EER) and the *decidability index* (DI). The EER is observed at the decision threshold, $\theta$, where the errors of falsely accepting and falsely rejecting $H_0$ occur with equal frequency. The global behavior of both types of errors is often analyzed through receiver operating characteristic (ROC) curves. On the other hand, the DI quantifies the separation of the "genuine" and "impostor" likelihood score distributions, as follows:

$$DI = \frac{|\mu_g - \mu_i|}{\sqrt{0.5(\sigma_g^2 + \sigma_i^2)}} \qquad (10)$$

where $(\mu_g, \sigma_g)$ and $(\mu_i, \sigma_i)$ are the mean and standard deviation of the genuine and impostor score distributions, respectively.

For identification we analyze cumulative match curves (CMC). These curves represent the rate of correctly identified individuals, by checking if the true identity is present in the $N$ highest ranked identities. The $N$ parameter is generally referred to as rank. That allows us to define the *rank-1 recognition rate* as the value of the CMC at $N = 1$.

## 3.3    Experimental Setups

Our experiments were conducted in three distinct experimental setups, two of them regarding the UBIRIS.v2 database and the remaining one the MobBIO database:

1. In the first setup, for the UBIRIS.v2 images, six samples from 80 different subjects were used, captured from different distances (4 to 8 meters), with varying gazes/poses and notable changes in lighting conditions. One image per individual was randomly chosen as probe, whereas the remaining five samples were used for the UBM training and MAP adaptation. The results were cross-validated by changing the probe image, per subject, for each of the six chosen images.
2. Many works on periocular biometrics evaluate their results using a well-known subset of the UBIRIS.v2 database, used in the context of the NICE II competition [14]. This dataset is divided in train and test subsets, with a total of 1000 images from 171 individuals. In the present work we choose to use test subset, composed by 904 images from 152 individuals. Only individuals with more than 4 available images were considered, as 4 images were randomly chosen for training and the rest for testing. Results were cross-validated 10-fold. The train dataset composed by the remaining 96 images from 19 individuals was employed in the parameter optimization step described in further sections.

3. Concerning the MobBIO database, 8 images were randomly chosen from each of the 105 individuals for the training of the models, whereas the remaining 8 were chosen for testing. The process was cross-validated 10-fold. For comparative purposes a similar experiment was carried out on face images from the same 105 individuals, using the same $8 + 8$ image distribution.

As both databases are composed by color images, each of the RGB channels was considered individually for the entire enrollment and identification process. For the parameter optimization described in the next section images were previously converted to grayscale.

## 3.4    Recognition Results

The results obtained for both databases and experimental setups are represented through ROC and CMC curves on Figs. 6(a) to (f). A comparison with some state-of-the-art algorithms in the UBIRIS.v2 database is also presented in Table 1. In this table results are grouped according to the experimental setup of each reported work and also the studied trait: $P$ - Periocular, $I$ - Iris or $P + I$ - Fusion of both traits.

**Table 1.** Comparison between the average obtained results with both experimental setups for the UBIRIS.v2 database and some state-of-the-art algorithms.

| Work | Setup | Traits | $R_1$ | EER | $D_i$ |
|---|---|---|---|---|---|
| Proposed | 1 | $P$ | **97.73 %** | **0.0452** | 4.9795 |
| Proposed | 2 | $P$ | **88.93 %** | **0.0716** | **3.6141** |
| Moreno et al. [9] | 1 | $P$ | 97.63 % | 0.1417 | – |
| Tan et al. [26] | 2 | $P + I$ | 39.4 % | – | – |
| Tan et al. [27] | 2 | $P + I$ | – | – | 2.5748 |
| Kumar et al. [5] | 2 | $I$ | 48.01 % | – | – |
| Proença et al. [16] | 2 | $I$ | – | $\approx 0.11$ | 2.848 |

Besides testing each of the RGB channels individually, a simple sum-rule score-level fusion strategy [4] was also considered. It is easily discernible, from the observation of Fig. 6, that the fusion of information from multiple color channels brings about a significant improvement in performance for all the tested datasets. When comparing the results obtained with this approach with some state-of-the-art algorithms a few points deserve further discussion. First, the proposed algorithm is capable of achieving and even surpassing state-of-the-art performance in multiple experimental setups. Concerning the most common of such setups (2), it is interesting to note that a few works attempted to explore the UBIRIS.v2 dataset for iris recognition. The obtained performance has been considered "discouraging" in the work by Kumar et al. [5]. Comparing the rank-1

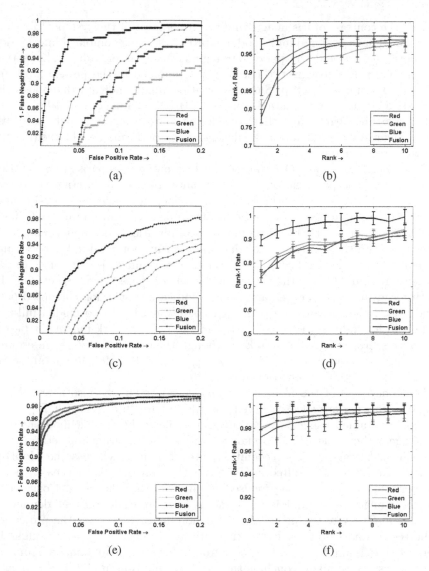

**Fig. 6.** ROC and CMC curves for: (a-b) UBIRIS.v2 database, setup (1); (c-d) UBIRIS.v2 database, setup (2) and (e-f) MobBIO database, setup (3);. ROC curves present the average results of cross-validation, whereas CMCs present the average value and error-bars for the first 10 ranked IDs in each setup.

recognition rate obtained with our algorithm (88.93 %) with the 48.1 % reported in the former work, we conclude that the periocular region may represent a viable alternative to iris in images acquired under visible wavelength (VW) illumination. Such acquisition conditions are known to increase light reflections from the cornea, resulting in a sub-optimal signal-to-noise ratio (SNR) in the sensor, lowering the contrast of iris images and the robustness of the system [15]. More recent works have explored multimodal approaches, using combined information from both the iris and the periocular region. Analysis of Table 1 shows that none of such works reaches the performance reported in the present work for the same experimental setup. Such observation might indicate that most discriminative biometric information from the UBIRIS.v2 images might be present in the periocular region, and that considering data from the heavily noisy iris regions might only result in a degradation of the performance obtained by the periocular region alone.

Concerning the MobBIO database, an alternative comparison was carried out to analyze the potential of the periocular region as an alternative to face recognition. The observed performance for periocular images was considerably close to that using full-face information, with rank-1 recognition rates of 98.98 % and 99.77 % respectively. These results are an indication that, under more ideal acquisition conditions, there is enough discriminative potential in the periocular region alone to rival with the full face in terms of recognition performance. In scenarios where some parts of the face are purposely disguised (scarves covering the mouth for example) this observation might indicate that a non-corrupted periocular region can, indeed, overperform recognition with the occluded full-face images. Such conditions were not tested in the present work but might be the basis for an interesting follow-up. Even though the observed results are promising, it must be noted that the noise factors present in the MobBIO database are still far from a highly unconstrained scenario.

The robustness of the likelihood-ratio decision step was also assessed. We compared the performance observed for the scores obtained with Eq. 1 and the scores obtained using only its numerator, i.e. only the likelihood of each test image without the UBM normalization. For the experimental setup (2) we obtained an average rank-1 recognition rate of 43.6 %, whereas the MobBIO experimental setup (3) resulted in 90.5 % for the same metric. It is easily noted that performance is less compromised in the MobBIO database. Considering only the numerator of Eq. 1 is the same as considering a constant denominator value for every tested image. As the denominator represents the projection of the tested images on the UBM, this alternative decision strategy might be interpreted as assuming a constant background for every tested image. From the observed results we might conclude that such assumption fits better the images from the MobBIO database. We also note that for more challenging scenarios, where the constant background assumption fails, the use of background normalization produces a significant improvement in performance.

A few last considerations regarding the discriminative potential of the proposed algorithm may be taken from the observation of Fig. 7. On each row we analyze the 4 highest ranked models for the images presented in the first column. The first two rows depict correct identifications. It is interesting to note how each of the 4 highest ranked identities in the second row correspond to individuals wearing glasses. Such observation seems to indicate that the proposed modeling process is capable of describing high-level global features, such as glasses. Furthermore, the fact that the correct ID was guessed also demonstrates its capacity of distinguishing between finer details separating individual models. The third and fourth rows present some test images whose ID was not correctly assessed by the algorithm. In the third row we present a case where even though the correct ID and the most likely model were not correctly paired, the correct guess still appears in the top ranked models. We note that even a human user analyzing the four highest ranked models would find it very difficult to detect significant differences. The fourth row presents the extreme case where none of the top ranked models correspond to the true ID. It is worth noting how the test images presented in the third and fourth rows are very similar to a large number of images present in other individual's models. This observation leads to the hypothesis that *some users are easier to identify than others inside a given population*, an effect known as the *Doddington zoo effect* [22]. It also shows that the proposed algorithm is capable of narrowing the range of possible identities to those subjects who "look more alike".

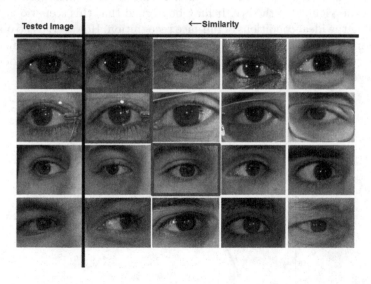

**Fig. 7.** Identification results for rank-4 in the UBIRIS.v2 database. The first column depicts the tested images while the remaining 4 images exemplify representative images from the 4 most probable models, after the recognition is performed. The blue squares mark the true identity (Color figure online).

## 3.5   Effect of Acquisition Distance on Performance

An additional test was carried out to assess how the distance between the acquisition apparatus and the captured individual affects the performance of the proposed algorithm. With this in mind an alternative evaluation setup was designed. We chose the UBIRIS.v2 database for this analysis, as it is comprised by images acquired at variable distances (4 − 8 meters). Images were divided into training and testing as defined in Sect. 3.3 for thes training setup 2. During the test process, the test images were divided into 5 different groups, according to the acquisition distance at which they were captured. Figure 8 depicts the evolution of the decidability index values with respect to the acquisition distance parameter, $d_a$. It is interesting to note that images acquired at farther distances present higher recognition rates, with exception of the images acquired at the closest distance. It was expected that farther images presented consistently better performance, as more anatomical landmarks (eyebrows, for example) become accessible to the recognition block, thus increasing the amount of discriminative information present in each tested image. The fact that the closest images present an increased performance might be connected either to a peculiarity of the tested data or to the fact that, in closer images, the iris contributes with a non-negligible role to the recognition process. The possibility that some of the closer images present enough detail in the iris region to improve the performance in such conditions seems legitimate. It is, however, also discernible that even in this possible semi-iris recognition scenario, performance is still significantly lower than the one observed when the periocular region is set to incorporate a larger portion of the vicinity of the eye. It may be argued that this scenario is rapidly approaching the ideal conditions of full-face recognition. However, as no objective description of the periocular region is standardized in the research community, these observations might help in the process of establishing an uniform definition of such region.

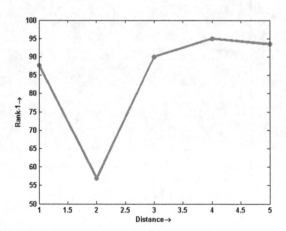

**Fig. 8.** Evolution of the rank-1 recognition rate values observed for increasing distance between the captured individual and the acquisition apparatus in the UBIRIS.v2 database.

## 4    Conclusions and Future Work

In the present work we propose an automatic modeling of SIFT descriptors, using a GMM-based UBM method, to achieve a canonical representation of individual's biometric data, regardless of the number of detected SIFT keypoints. We tested the proposed algorithm on periocular images from two databases and achieved state-of-the-art performance for all experimental setups. Periocular recognition has been the focus of many recent works that explore it as a viable alternative to both iris and face recognition under less ideal acquisition scenarios.

Even though we propose the algorithm for periocular recognition, the framework can be easily extrapolated for other image-based traits. To the extent of our knowledge, GMM-based UBM methodologies were solely explored for speaker recognition so far. The proposed work may, thus, represent the first of a series of experiments that explore its main advantages in the scope of multiple trending biometric topics. For example, the fact that any number of keypoints triggers a recognition score may be relevant when only partial or occluded data is available for recognition. Scenarios, like the one described in the last section, where faces are purposely occluded may be an interesting area to explore.

Besides from the conceptual advantages of the proposed algorithm, a few technical details may be improved in further works. Exploring further color channels besides the RGB space could bring benefits to the proposed algorithm. Regarding fusion, exploring individual specific parameters instead of a global parametrization, would enable the algorithm to be trained to counter the Doddington zoo effect. As not all people are as easy to identify, fitting the properties of the designed classification block to adapt to different classes of individuals seems like an interesting idea.

Finally, and regarding the training setup, some questions might be worthy of a more thorough research. In the case of voice recognition it is common to train two separate UBMs for male and female speakers. Extrapolating this idea to image-based traits, multiple UBMs trained on homogeneous sets of equally or similarly zoomed images might improve the results when more realistic and dynamic conditions are presented to the acquisition system. In a related topic it is also not consensual whether the left and right eyes, due to the intrinsic symmetry of the face, should be considered in a single model or as separate entities. All the aforementioned questions demonstrate how much the present results can be improved, leaving some promising prospects for future works.

**Acknowledgements.** The first author would like to thank Fundação para a Ciência e Tecnologia (FCT) - Portugal the financial support for the PhD grant SFRH/BD/87392/2012.

# References

1. Bakshi, S., Kumari, S., Raman, R., Sa, P.K.: Evaluation of periocular over face biometric: A case study. Procedia Eng. **38**, 1628–1633 (2012)
2. Joshi, A., Gangwar, A.K., Saquib, Z.: Person recognition based on fusion of iris and periocular biometrics. In: 2012 12th International Conference on Hybrid Intelligent Systems (HIS), pp. 57–62. IEEE (2012)

3. Kinnunen, T., Saastamoinen, J., Hautamaki, V., Vinni, M., Franti, P.: Comparing maximum a posteriori vector quantization and gaussian mixture models in speaker verification. In: IEEE International Conference on Acoustics, Speech and Signal Processing, pp. 4229–4232. IEEE (2009)
4. Kittler, J., Hatef, M., Duin, R.P., Matas, J.: On combining classifiers. IEEE Trans. Pattern Anal. Mach. Intell. **20**(3), 226–239 (1998)
5. Kumar, A., Chan, T.S.: Iris recognition using quaternionic sparse orientation code (qsoc). In: 2012 IEEE Computer Society Conference on Computer Vision and Pattern Recognition Workshops, pp. 59–64 (2012)
6. Lowe, D.G.: Distinctive image features from scale-invariant keypoints. Int. J. Comput. Vis. **60**(2), 91–110 (2004)
7. Miller, P.E., Rawls, A.W., Pundlik, S.J., Woodard, D.L.: Personal identification using periocular skin texture. In: 2010 ACM Symposium on Applied Computing, pp. 1496–1500. ACM (2010)
8. Monteiro, J.C., Cardoso, J.S.: Periocular recognition under unconstrained settings with universal background models. In: Proceedings of the International Conference on Bio-inspired Systems and Signal Processing (BIOSIGNALS) (2015). http://www.inescporto.pt/~jsc/publications/conferences/2015JC MonteiroBIOSIGNALS.pdf
9. Moreno, J.C., Prasath, V., Proença, H.: Robust periocular recognition by fusing local to holistic sparse representations. In: Proceedings of the 6th International Conference on Security of Information and Networks, pp. 160–164. ACM (2013)
10. Padole, C.N., Proenca, H.: Periocular recognition: Analysis of performance degradation factors. In: 5th IAPR International Conference on Biometrics, pp. 439–445 (2012)
11. Park, U., Jillela, R.R., Ross, A., Jain, A.K.: Periocular biometrics in the visible spectrum. IEEE Trans. Inform. Forensics Secur. **6**(1), 96–106 (2011)
12. Park, U., Ross, A., Jain, A.K.: Periocular biometrics in the visible spectrum: A feasibility study. In: IEEE 3rd International Conference on Biometrics: Theory, Applications, and Systems, pp. 1–6 (2009)
13. Povey, D., Chu, S.M., Varadarajan, B.: Universal background model based speech recognition. In: IEEE International Conference on Acoustics, Speech and Signal Processing, pp. 4561–4564 (2008)
14. Proença, H.: NICE:II: Noisy iris challenge evaluation - part ii (2009). http://nice2. di.ubi.pt/
15. Proença, H.: Non-cooperative iris recognition: Issues and trends. In: 19th European Signal Processing Conference, pp. 1–5 (2011)
16. Proença, H., Santos, G.: Fusing color and shape descriptors in the recognition of degraded iris images acquired at visible wavelengths. Comput. Vis. Image Underst. **116**(2), 167–178 (2012)
17. Proença, H., Filipe, S., Santos, R., Oliveira, J., Alexandre, L.A.: The ubiris.v2: A database of visible wavelength iris images captured on-the-move and at-a-distance. IEEE Trans. Pattern Anal. Mach. Intell. **32**(8), 1529–1535 (2010)
18. Reynolds, D.: Gaussian mixture models. In: Encyclopedia of Biometric Recognition, pp. 12–17 (2008)
19. Reynolds, D., Quatieri, T., Dunn, R.: Speaker verification using adapted gaussian mixture models. Digital Sign. Proces. **10**(1), 19–41 (2000)
20. Reynolds, D.A.: An overview of automatic speaker recognition technology. In: 2002 IEEE International Conference on Acoustics, Speech, and Signal Processing (ICASSP), vol. 4, pp. IV-4072. IEEE (2002)

21. Ross, A., Jillela, R., Smereka, J.M., Boddeti, V.N., Kumar, B.V., Barnard, R., Hu, X., Pauca, P., Plemmons, R.: Matching highly non-ideal ocular images: An information fusion approach. In: 2012 5th IAPR International Conference on Biometrics (ICB), pp. 446–453. IEEE (2012)
22. Ross, A., Rattani, A., Tistarelli, M.: Exploiting the doddington zoo effect in biometric fusion. In: IEEE 3rd International Conference on Biometrics: Theory, Applications, and Systems, 2009. BTAS 2009, pp. 1–7. IEEE (2009)
23. Santos, G., Proença, H.: Periocular biometrics: An emerging technology for unconstrained scenarios. In: 2013 IEEE Workshop on Computational Intelligence in Biometrics and Identity Management (CIBIM), pp. 14–21. IEEE (2013)
24. Sequeira, A.F., Monteiro, J.C., Rebelo, A., Oliveira, H.P.: MobBIO: a multimodal database captured with a portable handheld device. In: Proceedings of International Conference on Computer Vision Theory and Applications (VISAPP) (2014)
25. Shinoda, K., Inoue, N.: Reusing speech techniques for video semantic indexing [applications corner]. IEEE Sign. Proces. Mag. $30(2)$, 118–122 (2013)
26. Tan, C.W., Kumar, A.: Towards online iris and periocular recognition under relaxed imaging constraints. IEEE Trans. Image Proces. $22(10)$, 3751–3765 (2013)
27. Tan, T., Zhang, X., Sun, Z., Zhang, H.: Noisy iris image matching by using multiple cues. Pattern Recogn. Lett. $33$, 970–977 (2011)
28. Woodard, D.L., Pundlik, S.J., Lyle, J.R., Miller, P.E.: Periocular region appearance cues for biometric identification. In: IEEE Computer Society Conference on Computer Vision and Pattern Recognition Workshops, pp. 162–169 (2010)
29. Xiong, Z., Zheng, T., Song, Z., Soong, F., Wu, W.: A tree-based kernel selection approach to efficient gaussian mixture model-universal background model based speaker identification. Speech Commun. $48(10)$, 1273–1282 (2006)

# Fusion and Comparison of IMU and EMG Signals for Wearable Gesture Recognition

Marcus Georgi$^{(\boxtimes)}$, Christoph Amma, and Tanja Schultz

Cognitive Systems Lab, Institute for Anthropomatics and Robotics,
Karlsruhe Institute of Technology, Karlsruhe, Germany
{marcus.georgi,christoph.amma,tanja.schultz}@kit.edu

**Abstract.** We evaluate the performance of a wearable gesture recognition system for arm, hand, and finger motions, using the signals of an Inertial Measurement Unit (IMU) worn at the wrist, and the Electromyogram (EMG) of muscles in the forearm. A set of 12 gestures was defined, similar to manipulatory movements and to gestures known from the interaction with mobile devices. We recorded performances of our gesture set by five subjects in multiple sessions. The resulting data corpus is made publicly available to build a common ground for future evaluations and benchmarks. Hidden Markov Models (HMMs) are used as classifiers to discriminate between the defined gesture classes. We achieve a recognition rate of 97.8 % in session-independent, and of 74.3 % in person-independent recognition. We give a detailed analysis of error characteristics and of the influence of each modality to the results to underline the benefits of using both modalities together.

**Keywords:** Wearable computing · Gesture recognition · Inertial Measurement Unit · Electromyography

## 1 Introduction

Mobile and wearable computing have become an integral part of our everyday lives. Smartwatches and mixed-reality-glasses are getting popular and widely available, promoting the idea of an immersive usage with micro-interactions. The interaction with such devices differs from the interaction with mobile phones and tablet computers, devices that already gained functionality allowing for their usage as a replacement for conventional computers in a wide range of usage scenarios. For glasses and watches, the usage of on-screen keyboards becomes cumbersome, if not impossible. Therefore, alternative interaction paradigms have to be used, allowing an intuitive handling of these devices.

Gestures performed with the hand and fingers can resemble actual physical manipulations connected to spatial tasks, like navigation on a map or manipulation of a picture. Hence, they are a beneficial complementary modality to speech recognition, as these are tasks not easily solved using only spoken language [12]. For mobile usage a system using hardware that can be worn like clothing or an accessory would be favourable.

© Springer International Publishing Switzerland 2015
A. Fred et al. (Eds): BIOSTEC 2015, CCIS 574, pp. 308–323, 2015.
DOI: 10.1007/978-3-319-27707-3_19

Different approaches were proposed to sense both hand and finger movements in a mobile environment without placing sensors directly at the hand of a user. They were based on the usage of cameras, body-worn [11] or wrist-worn [8], on the measurement of tendon movement [14] or on the usage of Electromyography (EMG) [9,15–17] and Inertial Measurement Units (IMUs) [2,3,5,6].

This paper presents a recognition framework for gesture interfaces, based on Electromyography and an Inertial Measurement Unit, both being wearable sensor systems.

We will systematically evaluate the performance of this system in differentiating between gestures, using the IMU and EMG individually, as well as the multimodal recognition performance. Additionally, the contributions of both modalities to the overall results will be presented, with focus on the benefits for specific types of movement. This will clarify the advantages and disadvantages of IMUs and EMG, and will validate their combined usage for gesture recognition.

The performance of this system will be evaluated both for session-independent and person-independent recognition, which is of importance for practical and usable gesture based interfaces. Session-independent recognition surveys how well the system can be attuned to a user. A system with high session-independent accuracy can be used by one person without training it each time it is used. This makes the system ready to use by just mounting the sensors and starting the recognition. Person-independent recognition is relevant in regard to an envisioned system that can be used without prior training by some new user. Instead, it would use exemplary training data from a selected, representative group, making explicit training sessions superfluous. Such a system could nonetheless still benefit from further adaption to an individual user.

The evaluations of this paper will be based on gestures that were designed to resemble actual physical manipulations, as well as gestures known from the interaction with mobile devices. They were not designed to be optimally distinguishable, but rather to represent useful gestures for real-world interfaces.

The recorded dataset is publicly available to be used by others as a common ground for the development and evaluation of gesture recognition systems based on IMUs and EMG. It can be found under http://csl.anthropomatik.kit.edu/ 2907.php.

## 1.1 Related Work

The simultaneous usage of an IMU and EMG for gesture based interfaces was also evaluated in a small set of other studies, that are listed and compared in Table 1. The comparison of the various different approaches to the task of gesture recognition based only on reported recognition accuracies is hard, as different gesture sets were designed and a variety of recording setups and hardware was used.

In [10] IMUs and EMG are used for the automatic recognition of sign language subwords, specifically of Chinese sign language (CSL) subwords. On a vocabulary of 121 subwords a high accuracy of 95.78 % is achieved, but only session-independent accuracy was evaluated. Additionally, it is hard to estimate

**Table 1.** Comparison of investigated research questions. This table shows what kind of gesture recognition tasks were evaluated in the work related to this paper, and which modalities were used.

| | Session-dependent | | | Session-independent | | | Person-independent | | |
|---|---|---|---|---|---|---|---|---|---|
| | EMG | IMU | Combined | EMG | IMU | Combined | EMG | IMU | Combined |
| [10] | - | - | - | - | - | x | - | - | - |
| [17] | - | - | x | - | - | - | - | - | - |
| [18] | x | x | x | - | - | - | - | - | x |
| [4] | x | x | x | - | - | - | - | - | - |
| [16] | - | - | - | x | - | - | x | - | - |
| [9] | x | - | - | - | - | - | x | - | - |
| [15] | - | - | - | - | - | - | x | - | - |
| This paper | - | - | - | x | x | x | x | x | x |

the transferability of these results to other tasks, as a very task specific recognizer design is used, in contrast to a more general design, as it is used in this paper and most other studies.

In [17] two different approaches for gesture recognition are presented. One uses a SVM to distinguish between 17 static gestures with 96.6 % accuracy. It discriminates these 17 gesture classes based on EMG and uses an additional orientation estimation based on IMU signals to distinguish whether the gestures were performed with a hanging, raised or neutral arm, increasing the number of classes that can be interpreted by a factor of three. To decode dynamic gestures, they combine the features of both modalities to a single feature vector. Again, they achieve a high accuracy of 99 % for a set of nine gestures. However, both session-independent and person-independent performance are not evaluated for both techniques, and the contribution of the individual modalities is not discussed.

Zhang et al. [18] show, that it is possible to construct a robust, person-independent, gesture based interface using an IMU and EMG to manipulate a Rubik's Cube. They report a person-independent accuracy of 90.2 % for 18 different gesture classes. One has to note that this gesture set was designed to include only three different static hand postures that were hard to detect with only an IMU, and that, similar to [17], the direction of movement with the whole arm then increased the number of gestures by a factor of six. In comparison, we use a set of twelve different gestures, with dynamic postures and movement. Additionally, they discuss the contribution of the individual modalities IMU and EMG in a CSL recognition task similar to [10], but do this only for session-dependent recognition.

Chen et al. [4] also compare the session-dependent gesture recognition performance for various gesture sets when using 2D-accelerometers and two EMG channels individually as well as in combination. When using both modalities combined, they report accuracy improvements of $5 - 10\%$ on various gesture sets with up to 12 classes. Again, this evaluation is not done for the session- and person-independent case.

As we evaluate single-modality performance for person-independent recognition, we also evaluate the performance when using only EMG. Only very few studies have reported on this so far [9,15,16], often with lacking results, or only on small gesture sets.

Samadani and Kulić [15] use eight dry EMG sensors in a prototypic commercial armband from Thalmic Labs[1] to capture EMG readings from 25 subjects. In person-independent recognition, they achieve $49\%$ accuracy for a gesture set with 25 gestures. Additionally, they report $79\%$, $85\%$, and $91\%$ accuracy for select gesture sets with 10, 6, and 4 gestures, respectively.

## 2  Dataset

### 2.1  Gesture Set

To get a reliable performance estimate for practical interfaces, we define a set of gestures that is a reasonable choice for gesture based interfaces. Different studies, e.g. [1,7], show that spatial gestures can convey unique information not included in linguistic expressions when used to express spatial informations. Therefore we use spatial gestures, as an interface based on such gestures could be beneficial for mobile and wearable usage when used complementary to speech recognition.

Hauptmann et al. [7] evaluate what kind of gestures were performed by subjects trying to solve spatial tasks. In a Wizard of Oz experiment they collected statistical values of the amount of fingers and hands that were used to solve tasks related to the graphical manipulation of objects.

We compared their statistical values with gestures commonly occurring during the actual manipulation of real objects or whilst using touchscreens or touchpads. On the one hand, we hope that users can intuitively use these gestures due to both their level of familiarity, as well as their similitude to physical interactions. On the other hand, we assume that interfaces like virtual or augmented reality glasses might benefit from such gestures by allowing a user to manipulate displayed virtual objects similar to real physical objects.

As a result we defined the list of gestures in Table 2 to be the list of gestures to be recognized. These gestures involve both movements of the fingers, as well as of the whole hand.

### 2.2  Experimental Procedure

**Inertial Measurement Unit.** A detailed description of the sensor we use to capture acceleration and angular velocity of the forearm during the experiments can be found in [2]. It consists of a 3D accelerometer, as well as a 3D gyroscope. We transmit the sensor values via Bluetooth and sample at 81.92 Hz.

---

[1] Thalmic Labs Inc., www.thalmic.com.

**Table 2.** All gestures defined to be recognized.

| # | Gesture name | Description | Interaction equivalence |
|---|---|---|---|
| 1 | Flick Left (FL) | Flick to the left with index finger extended | Flicking left on a touchscreen |
| 2 | Flick Right (FR) | Flick to the right with index finger extended | Flicking right on a touchscreen |
| 3 | Flick Up (FU) | Flicking upwards with index and middle finger extended | Scrolling upwards on a touchscreen |
| 4 | Flick Down (FD) | Flicking downwards with index and middle finger extended | Scrolling downwards on a touchscreen |
| 5 | Rotate Left (RL) | Grabbing motion followed by turning the hand counterclockwise | Turning a knob counterclockwise |
| 6 | Rotate Right (RR) | Grabbing motion followed by turning the hand clockwise | Turning a knob clockwise |
| 7 | Flat Hand Push (PSH) | Straightening of the hand followed by a translation away from the body | Pushing something away or compressing something |
| 8 | Flat Hand Pull (PLL) | Straightening of the hand followed by a translation towards the body | Following the movement of something towards the body |
| 9 | Palm Pull (PLM) | Turning the hand whilst cupping the fingers followed by a translation towards the body | Pulling something towards oneself |
| 10 | Single Click (SC) | Making a single tapping motion with the index finger | Single click on a touchscreen |
| 11 | Double Click (DC) | Making two consecutive tapping motions with the index finger | Double click on a touchscreen |
| 12 | Fist (F) | Making a fist | Making a fist or grabbing something |

**EMG Sensors.** To record the electrical activity of the forearm muscles during the movements of the hand, two *biosignalsplux* devices from Plux[2] are used. These devices allow the simultaneous recording of up to eight channels simultaneously per device. Both devices are synchronized on hardware level via an additional digital port. One additional port is reserved to connect a reference or ground electrode. Each bipolar channel measures an electrical potential using two self-adhesive and disposable surface electrodes. The devices sample at 1000 Hz and recorded values have a resolution of 12 bit.

**Subject Preparation.** Each subject that was recorded received an explanation of what was about to happen during the experiments. Afterwards 32 electrodes were placed in a regular pattern on both the upper and lower side of their forearm. This pattern can be seen in Fig. 1. The position of the electrodes near the elbow was chosen, as the flexors and extensors for all fingers apart from the thumb are mainly located in this area. (*M. extensor digitorum communis, M. extensor digiti minimi, M. extensor pollicis brevis, M. flexor digitorum profundus* and *M. flexor digitorum superficialis*). These muscles are oriented largely parallel to the axis between elbow and wrist. Therefore we applied the electrodes as eight parallel rows around the arm, each row consisting of four electrodes. From each row, the two upper, as well as the two lower electrodes form a single, bipolar channel. With eight rows and two channels per row we get 16 EMG channels in total. The reference electrodes for each device were placed on the elbow.

We decided to place the electrodes in a regular pattern around the forearm instead of placing them directly on the muscles to be monitored. This resembles the sensor placement when using a wearable armband or sleeve like in [17] or [16], that can easily be worn by a user. This kind of sensor placement does lead to redundant signals in some channels, as some muscles are recorded by multiple electrodes. However, it should prove useful for future work on how to further improve the session-and person-independent usage, as it allows for the compensation of placement variations using virtual replacement strategies or some blind source separation strategy. Placement variations can hardly be avoided with sensor sleeves that are to be applied by a user.

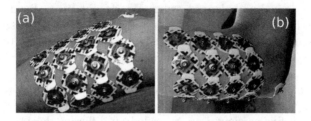

**Fig. 1.** Electrode placements on the (a) upper and (b) lower side of the forearm.

[2] PLUX wireless biosignals S.A., www.plux.info.

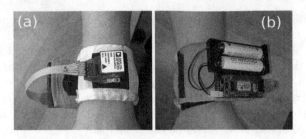

**Fig. 2.** Inertial sensor device mounted at the wrist on top of the forearm (a). Power supply and controller board underneath the arm (b).

The inertial sensor device was fixed to the forearm of the subject using an elastic wristband with Velcro. The sensor itself was mounted on top of the forearm, which can be seen in Fig. 2. No hardware is placed on the hand and fingers themselves, all sensing equipment is located at the forearm.

The whole recording setup is not yet intended to be practical and wearable, but can be miniaturized in the future. IMUs can already be embedded in unobtrusive wristbands and [17] show, that the manufacturing of armbands with integrated EMG sensor channels is possible. Also the release of the *Myo* from Thalmic Labs showed the recent advances in the development of wearable sensing equipment.

**Stimulus Presentation and Segmentation.** To initiate the acquisition of each gesture, a stimulus is presented to a subject. Figurative representations for all the gestures defined in Table 2 were created. Additionally a short text was added, describing the movements of the gesture.

The start and end of each gesture acquisition had to be manually triggered by the subjects by keypress to generate the segmentation groundtruth.

### 2.3 Data Corpus

In one recording session, 15 repetitions of the twelve defined gestures were recorded in random, but balanced order. We do not use a fixed order of gestures to force the subjects to comprehend and perform each gesture individually, rather than to perform a repetitious pattern. Additionally, this makes the movements of a gesture less dependent on the context of the gestures prior to and after it.

Five different subjects were asked to participate in such recording sessions. Their age was between 23 and 34; four of them were male, one was female. For each subject, five sessions were recorded on different days. Therefore, each gesture was recorded 75 times by each subject, which sums up to 900 gesture performances per subject. In total, 25 sessions with 4500 gesture repetitions are present in the data corpus. The corpus also contains the data samples we recorded during the relaxation phases, as well as the data samples that were transmitted between the actual gesture acquisitions. It might therefore also be used in future work to investigate other tasks and topics, like automatic segmentation, gesture sequence recognition, or the effects of muscle fatigue and familiarization. The data corpus is publicly available and can be found under http://csl.anthropomatik.kit.edu/2907.php.

## 3  Gesture Recognition

### 3.1  Preprocessing and Feature Extraction

For preprocessing, we normalize the signals of both sensors using Z-Normalization. For the IMU, this normalization decreases the impact of movement speed on the recordings. Furthermore, it reduces the influence of gravitation on the accelerometer signals, that we assume to be a largely constant offset, as the normalization removes baselineshifts. It is characteristic for EMG signals to fluctuate around a zero baseline. Therefore the mean of the signal should already be almost zero and mean normalization only removes a shift of the baseline. But the variance normalization has a rather large impact, as it makes the signal amplitude invariant to the signal dampening properties of the tissue and of skin conductance, up to a certain degree.

After preprocessing, we extract features on sliding windows for both modalities. As the different sensor systems do not have the same sampling rate, we choose the number of samples per window in accordance to the respective sampling rate, so that windows for both modalities represent the same period of time. This allows for the early fusion of the feature vectors for each window to one large feature vector.

Similar to [2], we use the average value in each window of each IMU channel as a feature. As the average computation for each window is effectively a smoothing operation and could therefore lead to information loss, we added standard deviation in each window as a second feature for each channel of the IMU.

We also compute standard deviation as a feature on each window of each EMG channel. [17] state that, like Root Mean Square (RMS), standard deviation is correlated to signal energy, but not as influenced by additive offsets and baselineshifts.

Averaging operations on large windows often reduce the influence of outliers by smoothing the signal, whereas feature extraction on small windows increases the temporal resolution of the feature vectors. As it is hard to predict the influence of different window sizes on the recognition results, we evaluated different window sizes in the range of 50 ms to 400 ms. We did not evaluate longer windows, as some of the recorded gestures were performed in under 600 ms. The mean duration of the gestures is about 1.1 s. We got the best results for windows with a length of 200 ms and chose an overlap of half a window size, namely 100 ms.

In conclusion we compute for each window a 28 dimensional feature vector. The first twelve dimensions are mean and standard deviation for each of the six channels of the IMU. The remaining 16 dimensions are the standard deviation

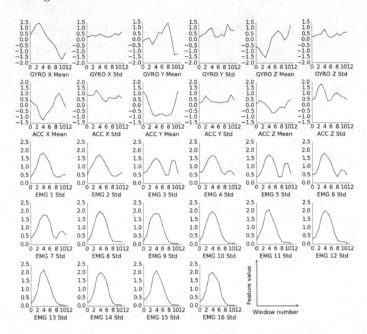

**Fig. 3.** Computed features for the gesture *fist*. All subplots represent one of the 28 feature vector dimensions. The raw signals were z-normalized and windowed into windows of 200 ms length with an overlap of 100 ms between consecutive windows. On each window one value for each dimension was computed. The first twelve subplots show features for the IMU channels, namely mean and standard deviation for the accelerometers and gyroscopes. The remaining 16 subplots show standard deviation for each of the EMG channels.

of each EMG channel. For single modality evaluations, we only use the features of one modality whilst omitting the features of the other one.

Figure 3 shows the set of features computed for the signals recorded when making a fist. Each subplot shows the progression of values of one dimension of the feature vector over all extracted windows.

## 3.2    Gesture Modeling

In our system, continuous density Hidden Markov Models (HMMs) are used to model each of the gesture classes. For an introduction to HMMs refer to [13]. We chose linear left-to-right topologies to represent the sequential nature of the recorded signals and Gaussian Mixture Models (GMMs) to model the observation probability distribution. Empirically we found topologies with five states and GMMs with five Gaussian components to deliver the best performance. The Gaussian components have diagonal covariance matrices to avoid overfitting to the data, as the number of free model parameters is then only linearly dependent on the number of features, in contrast to the quadratic dependence for full matrices. Provided a sufficiently large dataset, in future work the usage of unrestricted covariance matrices might further improve the gesture modeling by representing the correlation between the different channels. To fit the models to the training data, we initialize them using kMeans and use Viterbi Training afterwards.

## 4    Results

We evaluate the performance of our gesture recognition system for session-independent and person-independent recognition by determining its accuracy in discriminating the twelve different gesture classes of Table 2, with chance being 8.33 %. The gesture labels used in the graphics of this section follow the short names introduced in Table 2.

### 4.1    Session Independent

For the session-independent evaluation, testing is done using cross-validation individually for each subject. The training set for each validation fold consists of all sessions but one of a subject. The remaining session is used as the test set. We achieve 97.8 %($\pm$1.79 %) recognition accuracy as an average for all evaluated subjects.

The first plot in Fig. 4 displays the individual recognition performance for each subject. With all of the five subjects achieving more than 94 % accuracy, the session-independent recognition yields very satisfying results. Additionally, Fig. 4 shows the recognition accuracy when modalities are used individually. This will be discussed later on.

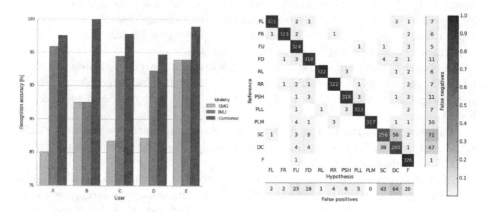

**Fig. 4.** Left: Session-independent recognition accuracy for all recorded subjects. For each subject, the performance when using IMU or EMG individually, as well as in combination, are shown. Right: Confusion matrix for session-independent gesture recognition. The matrix shows what hypotheses were given for different gestures that were fed to the recognizer. The labels follow Table 2.

The second plot in Fig. 4 shows the confusion matrix for the different gesture classes. It has a strong diagonal character, which is fitting the high overall accuracy. A block formation at the crossing between the rows and columns corresponding to *single click* and *double click* illustrates that these gestures are occasionally confused with each other. Only one small movement of the index finger differentiates both gestures. This could easily be confused with involuntary finger movements at the beginning and the end of the gesture recording.

**Modality Comparison.** Figure 4 visualizes the recognition performance when we use only the feature subset of one of the sensors for the recognition. One can see that the accuracy is lower when the sensing modalities are used individually. Nonetheless rather high results were achieved using both modalities separately, on average 92.8 %(±2.88 %) with the IMU and 85.1 %(±5.09 %) with the EMG sensors. Therefore both of the systems are suitable choices for hand gesture recognition.

To further analyze the advantages and disadvantages of the IMU and EMG, Fig. 5 shows the confusion matrices for session-independent recognition using the individual modalities. We expect the IMU and EMG to be complementary and to differ in their performance for certain gesture classes, due to them monitoring different aspects of the performed movements.

The first matrix in Fig. 5 shows the confusion matrix for single modality recognition using the IMU. It is overall rather similar to the one for multi modality recognition in Fig. 4. The most prominent difference is, that more false positives were reported for *single* and *double click*, and for *fist*. Whilst the *clicks* are often confused with one another, *fist* is more often given as a hypothesis for the *flick* and *rotate* classes. As making a fist only involves minor arm movements, we assume that the gesture has largely unpredictable IMU features. Thus the HMM for *fist* is less descriptive. This is consistent with the expectation, that gesture recognition based solely on IMU signals might prove problematic for small-scale hand gestures that do not involve arm movements.

In contrast, one would expect movements involving large rotational or translational movements of the whole arm to be easily discriminated. Accordingly, *rotate right* and *rotate left*, as well as *flat hand push* and *flat hand pull* have a low false positive count.

The second matrix in Fig. 5 shows the confusion matrix when using only EMG features. As for the multimodal case, we see most of the confusions for the *double* and *single click* gestures.

Also some other less frequent misclassifications occur when using EMG individually. As we anticipate that the direction of a movement is largely encoded in the IMU signal, we expect that gestures are confused, that differ mostly in the direction of their movement. This is the case with *flick left*, *flick right*, *flick up* and *flick down*, all being performed mainly with extended index and middle finger. Also *flat hand push* and *pull*, as well as *palm pull* are sometimes confused. As the accuracy is still rather high, one can assume that the direction of movement does in fact imprint a pattern on the activity signals.

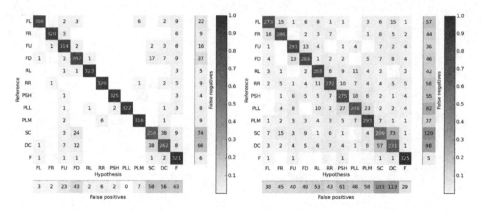

**Fig. 5.** Confusion matrices for session-independent gesture recognition. Left: The recognizer used only the features extracted from the IMU signals. Right: The recognizer used only the features extracted from the EMG signals. The labels follow Table 2.

Making a fist produces very distinctive features in all feature space dimensions for EMG, as all the monitored muscles contract. Accordingly, in contrast to IMU based recognition, only few false positives and negatives are reported for the gesture *fist*.

## 4.2 Person Independent

For person-independent evaluation, the recordings of all subjects but one are used as training data in the cross-validation, leaving the recordings of the remaining subject as test data. The same model and preprocessing parameters as for session-independent testing are used.

We achieve a mean accuracy of 74.3 % ($\pm$4.97 %) for person-independent recognition. The results for each test run, and therefore for each subject, are shown in Fig. 6, together with the results when using the modalities individually.

The second plot in Fig. 6 shows the confusion matrix for person-independent recognition. As expected, the matrix does not show a diagonal character as pronounced as in the confusion matrices for session-independent recognition. The gestures *flat hand push* and *pull*, *palm pull*, *fist* and *rotate left* and *right* are the classes with the lowest false negative count. The classes, together with the *flick left* and *right* classes, also have the lowest false positive count. Person-independent recognition seems therefore already reliable for a specific gesture subset. But especially the *flick* gestures, as well as the *click* gestures are often confused with others. Presumably, the inter-person variance is the highest for these classes.

**Modality Comparison.** Also for person-independent recognition the performance of the individual modalities is evaluated.

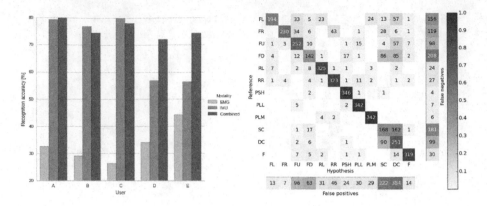

**Fig. 6.** Left: Person-independent recognition accuracy. Each bar depicts the achieved accuracy with the recordings of the respective subject used as test samples. For each subject, the performance when using IMU or EMG individually, as well as in combination, are shown. Right: Confusion matrix for person-independent gesture recognition. The labels follow Table 2.

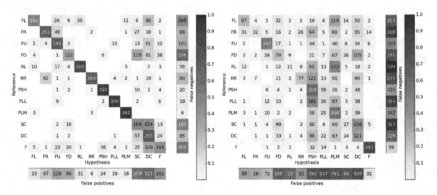

**Fig. 7.** Confusion matrices for person-independent gesture recognition. Left: The recognizer used only the features extracted from the IMU signals. Right: The recognizer used only the features extracted from the EMG signals. The labels follow Table 2.

Figure 6 compares the best results for each individual subject as well as the average result for both modalities. Clearly the IMU with an average accuracy of 70.2 %(±11.21 %) outperforms the EMG modality with only 33.2 %(±6.06 %) accuracy on average. But the multimodal recognition can still benefit from the combination of both sensor modalities.

The first matrix in Fig. 7 shows the misclassifications when using only features extracted from the IMU signals. The confusion matrix is largely similar to the one resulting from a combined usage of both modalities. This underlines that for this system, EMG has rather small influence on person-independent recognition results. But some differences have to be addressed. The gesture *double click* shows a higher false positive count and is often given as the hypothesis for other gestures. As was mentioned before, we assume that especially the EMG signals are descriptive for *fist*. It therefore consistent to this expectation, that the most prominent degradation of performance is visible for the *fist* gesture, which was performing rather well when using IMU and EMG together. Both its false positive and negative count are now rather high.

The diagonal character of the second confusion matrix in Fig. 7, representing recognition using only EMG features, is rather weak. Interestingly, no overly predominant attractor classes are present. Also no distinct block formations are visible, meaning that there is no pronounced clustering of different gesture groups in feature space. Instead, inter-person variance leads to many confusions scattered across the confusion matrix. The gestures *flick upwards* and *fist* both perform exceedingly well using only EMG, thereby explaining why they also perform well using both modalities combined.

We expected person-independent gesture recognition with EMG to be a hard task, as EMG signals vary from person to person and produce patterns over the different channels that are hard to compare. This explains, why only a few studies exist to person-independent, solely EMG based gesture recognition. Still, with 33.2 % accuracy we perform better than chance with 8.33 %.

## 5    Conclusion

In this paper we present a system for recognizing hand and finger gestures with IMU based motion and EMG based muscle activity sensing. We define a set of twelve gestures and record performances of these gestures by five subjects in 25 sessions total. The resulting data corpus was made publicly available. We built a baseline recognition system using HMMs to evaluate the usability of such a system. We achieve a high accuracy of 97.8 % for session-independent and an accuracy of 74.3 % for person-independent recognition, which still has to be considered rather high for person-independent gesture recognition on twelve classes. With that, we show both the feasibility of using IMUs and EMG for gesture recognition, and the benefits from using them in combination. We also evaluate the contribution of the individual modalities, to discuss their individual strengths and weaknesses for our task.

**Acknowledgements.** This research was partly funded by Google through a Google Faculty Research Award.

## References

1. Alibali, M.W.: Gesture in spatial cognition: Expressing, communicating, and thinking about spatial information. Spat. Cogn. Comput. **5**(4), 307–331 (2005)
2. Amma, C., Georgi, M., Schultz, T.: Airwriting: a wearable handwriting recognition system. Pers. Ubiquit. Comput. **18**(1), 191–203 (2014). http://link.springer.com/10.1007/s00779-013-0637-3

3. Benbasat, A.Y., Paradiso, J.A.: An inertial measurement framework for gesture recognition and applications. In: Wachsmuth, I., Sowa, T. (eds.) GW 2001. LNCS (LNAI), vol. 2298, pp. 9–20. Springer, Heidelberg (2002)
4. Chen, X., Zhang, X., Zhao, Z., Yang, J., Lantz, V., Wang, K.: Hand gesture recognition research based on surface emg sensors and 2d-accelerometers. In: 2007 11th IEEE International Symposium on Wearable Computers, pp. 11–14 (2007). http://dblp.org/db/conf/iswc/iswc2007.html#ChenZZYLW07
5. Cho, S.J., Oh, J.K., Bang, W.C., Chang, W., Choi, E., Jing, Y., Cho, J., Kim, D.Y.: Magic wand: a hand-drawn gesture input device in 3-d space with inertial sensors. In: Ninth International Workshop on Frontiers in Handwriting Recognition, 2004 IWFHR-9 2004, pp. 106–111 October 2004
6. Hartmann, B., Link, N.: Gesture recognition with inertial sensors and optimized dtw prototypes. In: Proceedings of the IEEE International Conference on Systems, Man and Cybernetics, Istanbul, Turkey, 10–13 October 2010. pp. 2102–2109. Proceedings of the IEEE International Conference on Systems, Man and Cybernetics, Istanbul, Turkey, 10–13 October 2010, IEEE (Oct 2010). http://dblp.org/db/conf/smc/smc2010.html#HartmannL10
7. Hauptmann, A.G., McAvinney, P.: Gestures with speech for graphic manipulation. Int. J. Man Mach. Stud. **38**(2), 231–249 (1993). http://linkinghub.elsevier.com/retrieve/doi/10.1006/imms.1993.1011
8. Kim, D., Hilliges, O., Izadi, S., Butler, A.D., Chen, J., Oikonomidis, I., Olivier, P.: Digits: Freehand 3d interactions anywhere using a wrist-worn gloveless sensor. In: Proceedings of the 25th Annual ACM Symposium on User Interface Software and Technology, pp. 167–176. UIST 2012, ACM, New York, NY, USA (2012). http://doi.acm.org/10.1145/2380116.2380139
9. Kim, J., Mastnik, S., André, E.: Emg-based hand gesture recognition for realtime biosignal interfacing. In: Proceedings of the 13th International Conference on Intelligent User Interfaces. IUI 2008, vol. 39, pp. 30–39. ACM Press, New York, NY, USA (2008). http://portal.acm.org/citation.cfm?doid=1378773.1378778
10. Li, Y., Chen, X., Tian, J., Zhang, X., Wang, K., Yang, J.: Automatic recognition of sign language subwords based on portable accelerometer and emg sensors. In: International Conference on Multimodal Interfaces and the Workshop on Machine Learning for Multimodal Interaction, pp. 17–1. ICMI-MLMI 2010, ACM, New York, NY, USA (2010). http://portal.acm.org/citation.cfm?id=1891926
11. Mistry, P., Maes, P., Chang, L.: WUW-wear Ur world: a wearable gestural interface. In: Proceedings of CHI 2009, pp. 4111–4116 (2009). http://dl.acm.org/citation.cfm?id=1520626/npapers://c80d98e4-9a96-4487-8d06-8e1acc780d86/Paper/p10196
12. Oviatt, S.: Ten myths of multimodal interaction. Commun. ACM **42**(11), 74–81 (1999)
13. Rabiner, L.: A tutorial on hidden markov models and selected applications in speech recognition. Proc. IEEE **77**(2), 257–286 (1989)
14. Rekimoto, J.: GestureWrist and GesturePad: unobtrusive wearable interaction devices. In: Proceedings Fifth International Symposium on Wearable Computers (2001)
15. Samadani, A.A., Kulic, D.: Hand gesture recognition based on surface electromyography. In: 2014 36th Annual International Conference of the IEEE Engineering in Medicine and Biology Society (EMBC) (2014)

16. Saponas, T.S., Tan, D.S., Morris, D., Balakrishnan, R.: Demonstrating the feasibility of using forearm electromyography for muscle-computer interfaces. In: Proceedings of the SIGCHI Conference on Human Factors in Computing Systems. pp. 515–524. CHI 2008, ACM Press, New York, New York, USA (2008). http://portal.acm.org/citation.cfm?doid=1357054.1357138

17. Wolf, M.T., Assad, C., Stoica, A., You, K., Jethani, H., Vernacchia, M.T., Fromm, J., Iwashita, Y.: Decoding static and dynamic arm and hand gestures from the jpl biosleeve. In: 2013 IEEE Aerospace Conference, pp. 1–9 (2013). http://ieeexplore.ieee.org/stamp/stamp.jsp?tp=&arnumber=6497171&isnumber=6496810

18. Zhang, X., Chen, X., Li, Y., Lantz, V., Wang, K., Yang, J.: A framework for hand gesture recognition based on accelerometer and emg sensors. IEEE Trans. Syst., Man Cybern., Part A: Syst. Humans 41(6), 1064–1076 (2011)

# Integrating User-Centred Design in the Development of a Silent Speech Interface Based on Permanent Magnetic Articulography

Lam A. Cheah[1(✉)], James M. Gilbert[1], Jose A. Gonzalez[2], Jie Bai[1], Stephen R. Ell[3], Michael J. Fagan[1], Roger K. Moore[2], Phil D. Green[2], and Sergey I. Rychenko[1]

[1] School of Engineering, The University of Hull, Kingston upon Hull, UK
{l.cheah,j.m.gilbert,j.bai,m.j.fagan,s.i.rynchenko}@hull.ac.uk
[2] Department of Computer Science, The University of Sheffield, Sheffield, UK
{j.gonzalez,r.k.moore,p.green}@sheffield.ac.uk
[3] Hull and East Yorkshire Hospitals Trust, Castle Hill Hospital, Cottingham, UK
srell@doctors.org.uk

**Abstract.** A new wearable silent speech interface (SSI) based on Permanent Magnetic Articulography (PMA) was developed with the involvement of end users in the design process. Hence, desirable features such as appearance, portability, ease of use and light weight were integrated into the prototype. The aim of this paper is to address the challenges faced and the design considerations addressed during the development. Evaluation on both hardware and speech recognition performances are presented here. The new prototype shows a comparable performance with its predecessor in terms of speech recognition accuracy (i.e. ~ 95 % of word accuracy and ~ 75 % of sequence accuracy), but significantly improved appearance, portability and hardware features in terms of miniaturization and cost.

**Keywords:** Assistive speech technology · User-centred design · Silent speech interface · Permanent magnetic articulography · Magnetic sensors

## 1 Introduction

Speech is an important part of human communication and plays a vital role in our social and work life. There are many situations in which people wish to communicate through speech but where it is either impossible (i.e. medical condition) or not desirable (i.e. communicating in private or in noisy environment). Patients whose voice box has to be removed because of throat cancer, trauma, destructive throat infections or neurological problems will inevitably lose their ability to speak. Therefore, they may experience a severe impact on their lives which can lead to social isolation and depression [1]. Conventional speech restoration methods after laryngectomy (e.g. oesophageal speech, the electrolarynx and speech valves) have limitations in terms of quality of speech and usability [1, 2]. Moreover, in the case of implanted speech valves, frequent valve replacement is required within a time span of 3-4 months, because of the growth of

© Springer International Publishing Switzerland 2015
E. Tambouris et al. (Eds.): EGOV 2015, LNCS 9248, pp. 324–337, 2015.
DOI: 10.1007/978-3-319-27707-3_20

biofilm coating over time [3–5]. To address these shortcomings, a novel approach has been introduced: silent speech interfaces (SSIs). SSIs are devices that enable speech communication in the absence of audible acoustic signals. To do that, SSIs exploit other non-acoustic information generated during the speech production process. These alternative sources can range from brain activity to articulator movements. To extract these forms of information several types of SSIs using different modalities have been proposed so far [6]. Permanent Magnetic Articulography (PMA) is a type of SSI and it is based on sensing the magnetic field variations from a set of permanent magnets attached to the articulators (i.e. lips and tongue) during speech [1, 2]. Contrary to other similar SSIs such as Electromagnetic Articulography (EMA), PMA does not provide explicit information regarding the position of the attached magnets. Instead, the measured PMA data is the summation of the magnetic field patterns associated to a particular articulatory gesture. As will be shown later, this is not a limitation of PMA as long as captured articulator data is used for pattern recognition (e.g. speech recognition). In this case, pattern recognition techniques can be employed to recognize the PMA patterns associated to the particular speech sounds.

Although there are obvious advantageous in using SSIs, there are still challenges in the form of the processing software (e.g. efficiency, robustness and reliable speech generation) and hardware (e.g. portability, light weight, unobtrusiveness and wearability). Preliminary investigation on the influential factors of the SSIs' implementation had been presented in [6], based upon criteria such as ability to operate in silence and noisy environments, usability by laryngectomees, issues of invasiveness, market readiness and cost. The focus of this paper is on the hardware challenges facing the PMA-based SSI system. A number of significant steps have been taken in order to develop a wearable system that is appropriate for everyday use. A novel embodiment comprising miniaturized sensing modules and a wireless headset that is compact and comfortable is proposed in this work.

The rest of this paper is organized as follow. The next section overviews the PMA technique and its development to date. Section 3 describes the design of the $2^{nd}$ generation system and the associated challenges, followed by the performance evaluation in Sect. 4. The final section concludes and provides an outlook for future research.

## 2   Overview on PMA-based SSI

A PMA-based device, the Magnetic Voice Output Communication Aid (MVOCA), is developed within the DiSArM (Digital Speech Recovery from Articulator Movement, www.hull.ac.uk/speech/disarm) project, aiming to restore speech communication ability for patients who have undergone surgical removal of the larynx.

In a nutshell, the current MVOCA device consists of multiple magnetic sensors mounted onto a lightweight headset for detection, a set of permanent magnets, four on the lips (ø1 mm × 5 mm), one at tongue tip (ø2 mm × 4 mm) and one at tongue blade (ø5 mm × 1 mm) as illustrated in Fig. 1. Information on magnet placement was described in [2]. These magnets are temporarily attached using Histoacryl surgical tissue adhesive (Braun, Melsungen, Germany). Eventually, these magnets will be surgically implanted for long term usage. The acquired measurements are pre-conditioned by the control unit prior to further signal processing.

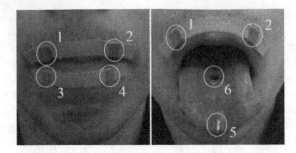

**Fig. 1.** Placement of six magnets with diameter and length of 1 mm × 5 mm for lips (pellets 1-4), 2 mm × 4 mm for tongue tip (pellet 5) and 5 mm × 1 mm for tongue blade (pellet 6).

To date, all the experiments were carried out using the 1$^{st}$ generation MVOCA, which consisted of five tri-axial Honeywell HMC2003 magnetic sensors, mounted on a pair of safety glasses, as shown in Fig. 2. The fluctuation in the magnetic field is captured on the 15 PMA channels and recorded onto a PC via ADLink DAQ-2206 analogue-to-digital converter (ADC), a PCI-based card with 16-bit linear encoding. Before describing the 2$^{nd}$ generation MVOCA device in next section, related background works on PMA are briefly outlined as follow:

- In earlier work [2, 7], the viability of isolated-word and connected digits recognition tasks using the PMA technology were presented.
- Investigation into the performance across multiple speakers conducted in [8].
- A feasibility study of direct speech synthesis bypassing the intermediate recognition step was reported in [9].
- More recently, extensive investigation into effectiveness of PMA data in terms discriminating the voicing, place and manner of articulation of English phones was presented in [10].

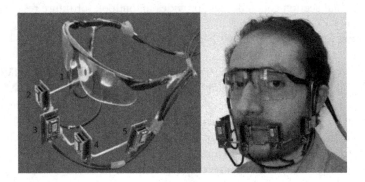

**Fig. 2.** MVOCA headset (1$^{st}$ generation) - five magnetic sensors mounted on a frame that attached onto a pair of safety glasses. Appearance of the device worn by user.

# 3   System Description

## 3.1   Design Challenges and Considerations

In order, to make the MVOCA device more usable and desirable, the 1st generation prototype has undergone several design cycles over the last 12 months. This is because the earlier MVOCAs [2, 7, 8] were not satisfactory, particularly in their appearance, comfort and ergonomic factors for the user, despite encouraging performance. The focus of the development task was to consider underexplore user-inclusive requirements by using a qualitative methodology, including informal opinion survey, focus group and user observation. These approaches were commonly used in other user-centred design studies [11–13].

Through discussion with the panels (i.e. laryngectomees) of the focus group and data from the survey questionnaires of 50 potentials users and their families/friends, the appearance of the device was seen as the major factor affecting acceptability. In fact, other researches also indicated that unobtrusive appearance is considered a highly desirable feature for any assistive device [12, 14]. Six possible configurations were presented in the survey. Those resembling a Bluetooth earpiece or a pair of spectacles were preferred by the majority of the potential users, while the device resembling a headset microphone was marginally acceptable by approximately 25 % of the respondents. On the other hand, devices that might obstruct the view of the mouth in anyway (in full or partially, such as the 1st generation MVOCA as illustrated in Fig. 2) were deemed unacceptable. Moreover, through focus group meetings and observation studies (participants had given their consent and the studies were approved by The University of Hull ethics committee), valuable feedback was gained and has greatly influenced the creation of a user-centred design prototype. Critical design questions were raised during prototype development, in term of headset appearance, portability, weight, ease of use and cost.

In addition, the survey questionnaires also identified other desirable features, such as software features (see Table 1) and speech quality (see Table 2), by their preferred ranking. As indicated in Table 1, the quality of reconstructed speech is highly rated, whereas the issue of delay between reconstructed sound and lips movement is least prioritized. In term of speech quality, this was further subdivided into the characteristics listed in Table 2. Both intelligibility and naturalness of speech are considered equally important, but the ability to convey emotion into the reconstructed speech is least preferred. It should be noted that respondents to the survey may have had some difficulty interpreting the meaning of some of these terms since, for instance, they may not be aware of the extent of emotion present in normal speech. The non-hardware related features will not be discussed in this paper but will be addressed separately in our future work.

## 3.2   New MVOCA Device

Based on the information gathered from the potential users (as presented in Sect. 3.1), a new prototype has been developed. Key components of the 2nd generation MVOCA

**Table 1.** Desirable software features.

| Ranking | Software feature | Description |
|---|---|---|
| 1st | Speech quality | Measuring the quality of reconstructed speech (see Table 2) |
| 2nd | Speech mode | Ability to communicate in fluent speech (ranging from isolated words to fluent speech) |
| 3rd | Vocabulary | Size and range of words available in the database (ranging from a small context specific vocabulary to unrestricted vocabulary) |
| 4th | Speaking delay | Synchronization between lips movement and synthesized voice (ranging from speaking a complete phrase before any speech output to no delay) |

**Table 2.** Desirable speech qualities.

| Ranking | Speech quality | Description |
|---|---|---|
| 1st | Intelligibility | Ability to communicate intelligibly (i.e. ranging from barely intelligible to a BBC newsreader) |
| 2nd | Naturalness | Ranging from a monotonic electronic voice to natural speech |
| 3rd | Personification | The choice of using own or preferred voice (ranging from another appropriate voice to the user's own voice) |
| 4th | Ability to convey emotion | Ability to include emotions (ranging from no emotional content to full emotion content) |

prototype consists a set of four tri-axial Anisotropic Magnetoresistive (AMR) magnetic sensors (Honeywell HMC5883L), a control unit and a power source (rechargeable 7.4 V Lithium Ion battery). These components are mounted on a customized headset, as illustrated in Fig. 3. Two headsets design were developed: (1) attached onto a headband (see Fig. 4a), and (2) onto a pair of spectacles (see Fig. 4b). The headsets (excluding the pair of spectacles or headband) were fabricated using rapid prototyping technology and their building materials were VeroWhitePlus RGD835 and VeroBlue RGD840. A set of six Neodymium Iron Boron (NdFeB) permanent magnets are attached onto the lips and tongue as illustrated in Fig. 1. Each magnetic sensor has three orthogonal sensing elements to measure the three spatial components of the magnetic field. Sensor1-3 (a total of 9 channels) are used to capture magnetic field variations caused by articulatory movements and digitize it with 12-bit resolution. Sensor4 on the other hand is used for background cancellation that is for compensating the effect of earth's magnetic field on the articulography signals captured by other sensors, in order to enhance the signal-to-noise (SNR) of the articulatory signals.

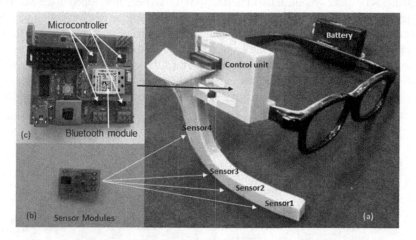

**Fig. 3.** Overview of the 2$^{nd}$ generation MVOCA system, (a) MVOCA headset with (b) sensor modules, (c) control unit and battery.

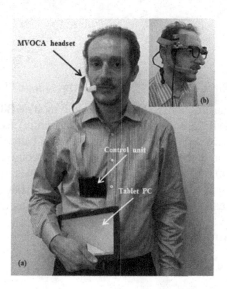

**Fig. 4.** Two MVOCA headset designs, (a) mounted on a headband and (b) attached on a pair of glasses. Appearance of the devices when worn by a user.

An operational block diagram of the 2$^{nd}$ generation MVOCA is shown in Fig. 5. Each magnetic sensor communicates to a low-power ATmega328P microcontroller (housed inside the control unit) through an I$^2$C interface, and a handful of control signals (i.e. SE, S0, and S1) are used in managing the acquisition process where samples are acquired at

100 Hz for each channel. These samples (total of 12 PMA channels) are then transmitted to a computer/tablet PC wirelessly via Bluetooth or via a USB connection for further processing. A bespoke graphical user interface (GUI) has been developed in the MATLAB environment and used mainly for on-line recognition testing or demonstration purposes. All necessary speech processing and recognition algorithms were embedded into the GUI and running in the background. If the acquired PMA signal correctly matched an articulation gesture from the pre-stored training dataset, thus the corresponded utterance will be identified. A text-to-speech synthesizer is used to generate a playback audio as an output for the identified utterance, via an audio device (e.g. integrated speaker of a computer).

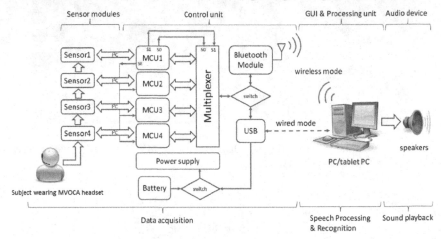

**Fig. 5.** Simplified MVOCA (2$^{nd}$ generation) operational block diagram.

For wireless data transmission, a class 2 Bluetooth module BTM411 (housed inside the control unit) and USB transceiver (attached on computer) are used. In the wireless case, the MVOCA device will acquire its power from a battery rather than from the computer via USB (wired mode). The average power consumption of the current MVOCA prototype from a 5 V (regulated from 7.4 V) supply is ~ 104 mA, which means that it can run continuously for ~ 10 h on a full charge (total 1080mAh). The battery can be removed from the headset for charging using a freestanding charger.

## 4   Experiments and Results

### 4.1   Test Speaker

As stated previously [7], the current MVOCA is a speaker-dependent device, i.e. all associated headset measurements and training parameters were calibrated towards particular individual. In this work, the data used for evaluating the latest MVOCA prototype were collected from a male native English speaker who is proficient in the usage of the MVOCA interface. Although inter-speaker performance has proven possible [8] the headset design and measurements would require individually tailored for optimal performance. In particular, the headset was specifically designed according to the speaker's anatomy.

## 4.2  Data Recording

A continuous speech recognition task consisting of the identification of sequences of English digits is chosen in this work to evaluate the performance of the latest MVOCA prototype. This was chosen because the limited size of vocabulary enables whole-word model training from relatively sparse data and also because of the simplicity of the language model involved. The algorithm used to generate the random digits sequences was the one underlying the TIDigits database of connected digits [15]. The longest digit sequence consists of seven individual digits. During the training, both zero and oh (the two representations of 0) were denoted as separate items.

The data used for training the speech recognizer were collected in six independent sessions, i.e. two sessions using each of the different $2^{nd}$ generation MVOCA headsets (Fig. 4(a) and (b)) and the remained two sessions using the $1^{st}$ generation headset (see Fig. 3). Each training session consisted of a total of 385 utterances containing 1265 individual digits. Furthermore, within each session, five different datasets were recorded: four of them (three spoken datasets and one mouthed dataset) were used for training and the remained mouthed dataset was used for testing purpose. The reason behind this configuration is to try to mimic a realistic scenario where the voice of the patient is recorded before the operation happens for personalizing the speech synthesizer, while after the operation only articulography PMA (mouthed) data can be obtained.

## 4.3  Experimental Setup

To achieve optimal recording performance, all experiments in this paper were conducted inside a sound-proof room, where the audio signal was recorded with a shock-mounted AKG C1000S condenser microphone and a dedicated USB sound card (Lexicon Lambda). A Matlab-based GUI was created to provide visual prompt of the digit sequences to the speaker at regular interval of 5 s during the recording session. The GUI also used provides simultaneous recording of both audio signal (sampled at 48 kHz) and PMA data (sampled at 100 Hz) as illustrated in Fig. 6.

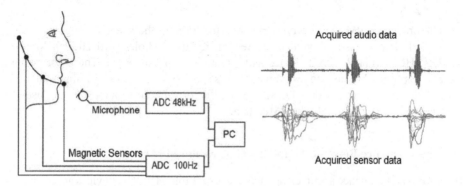

**Fig. 6.** Acoustic and PMA data streams were recorded in parallel into PC via a bespoke GUI. Both data streams were then synchronized prior to pre-processing.

Since both data streams were measured from separate modality, synchronization between the two data streams was necessary to compensate for any small deviation from the ideal sampling frequencies of the analog-to-digital converters (ADC). To do that an automatic timing alignment mechanism was used to realign both data streams by generating start-stop markers in addition to both audio and PMA data streams. The measured PMA data were transferred to a PC via USB connection. Since the speaker's head was not restrained, large movements could potentially distort the recorded data and thus degrade the recognition performance. Hence, background cancellation was applied to compensate for any movement induced interference against the desired PMA signals.

### 4.4    HMM Training and Recognition

The acquired PMA data used for speech recognition was first low-pass filtered (i.e. removal of 50 Hz noise) and normalized as described in [7]. Two different conditions (i.e. *Sensor* and *SensorD*) were computed in connected digits recognition experiments, and they relate to a specific configuration of the data used for model training and testing:

- *Sensor*: training and testing directly on the 9 channels of PMA data.
- *SensorD*: as above, plus the first time derivatives (related to articulator velocity, $D$ stands for "delta").

An overview of the two conditions is presented in Table 3. The second-order derivatives (i.e. delta-delta parameters) were not included as part of the feature vector since, as shown in our previous works [7, 8], they did not produced significant improvement in performance.

**Table 3.**  Vector sizes for different experimental conditions.

| Condition | Original | 1st delta | Vector size |
|-----------|----------|-----------|-------------|
| *Sensor* | √ | | 9 |
| *SensorD* | √ | √ | 18 |

The processed PMA data were then used for training the speech recognizer using HTK [16]. The acoustic model in the recognizer uses whole-word Hidden Markov Models (HMMs) [17] with 25 states and 5 Gaussians per state [7]. These parameters were not optimal, but the suggested parameters settings were known for their performances from our previous works [7, 8]. For clarification, audio signals were not used to train the recognizer, but only the PMA data.

### 4.5    Performance Between 1st and 2nd Generation MVOCAs

Both word and sequence accuracy results across multiple MVOCA devices are presented in Fig. 7. The results reflect the averaged value of the data (i.e. *Sensor* and *SensorD*) collected on two independent training sessions on each of the 1st and 2nd generation MVOCA devices. The data were analyzed independently session-by-session, and the recognition rates averaged across the sessions. Merging all the data from different sessions for recognition would seem a more attractive approach, but this might lead to inconsistent outcomes as very precise repetitive magnets placement are required on each training session. Nonetheless this could be overcome, as the magnets will be surgically implanted in the final MVOCA for long term usage. Investigations into session-independent approach on other SSIs technique were presented in [18, 19].

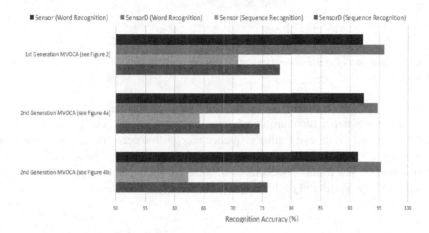

**Fig. 7.** Comparison of word and sequence accuracies of connected digits between 1st and 2nd generation MVOCAs.

As seen in Fig. 7, it is obvious that *SensorD* performs significantly better than using *Sensor* data alone. Similar trends were also reported in [7]. Moreover, the results showed a comparable performance between the 1st and 2nd generation MVOCAs. Hence, this suggests that the newer MVOCA can have better hardware features (i.e. appearance, light weight and portability) but without compromising its recognition performance by using miniaturized components (i.e. sensors and data acquisition unit).

Figure 8 illustrates that the inclusion of mouthed data in the training dataset improves the recognition accuracy, particularly in terms of sequence recognition. A comparison between using mixed data (spoken and mouthed data) and non-mixed data (spoken only data) as part of training dataset was investigated. The darker bars relate to mixed training data (spoken and mouthed data) and the light ones to non-mixed training data (spoken only data). The results presented in Fig. 8 were trained and tested using only *SensorD* data from the 2nd generation MVOCA devices, as they provided better performance as illustrated in Fig. 7. Although further investigation is needed, we recognized the importance of mixing both spoken and mouthed data in any training session.

So far, the results in Figs. 7 and 8 suggest that a *SensorD* data trained using a mixture of spoken and mouthed data generally performed better. A follow up test was conducted to explore the relationship between quantity of training data and the recognition performances (i.e. word and sequence recognition). Figure 9 illustrates that an increased number of training sessions yields an improvement in the performance in both word and sequence recognitions through the reduction of in word error rate (WER). Fewer training

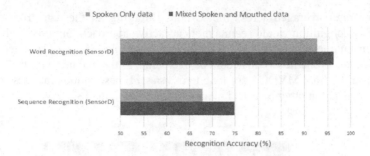

**Fig. 8.** Comparison of training dataset (mixed or non-mixed data) used in the recognition of connected digits.

sessions (i.e. ≥ 2 sets) is needed to achieve reasonable performance in word recognition as compared to sequence recognition, whereas it appears significantly more training sets would be required to achieve similar sequence recognition performance. It also appears that even for word recognition, the inclusion of further training data sets could reduce the WER further. The training sessions were not extended because of the speaker fatigue and increased the likelihood of the magnets becoming detached.

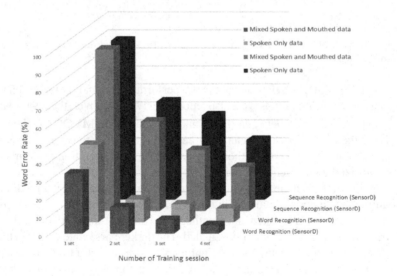

**Fig. 9.** Decrease in word error rate (WER) with the increase in training sessions.

## 4.6  Hardware Comparison Between 1st and 2nd Generation MVOCAs

So far, the challenge is to satisfy the design objective to improve the MVOCA's appearance, without compromising the device's performance. A summary of the key features of the latest MVOCA system is presented in Table 4. Two versions of MVOCA headsets were designed (see Fig. 4), both headsets aim to provide the desirable features such as light weight, comfort and unobtrusive appearance as suggested by the survey question-naires. The current designs significantly reduces the unattractive appearance of the previous headset (see Fig. 2), thus this would improve the acceptability to the end user and ultimately improves its usage.

Table 4. Hardware specifications of the $2^{nd}$ generation MVOCA.

| | Specification | Parameter |
|---|---|---|
| **Sensor Modules** | Type | Anisotropic Magnetoresistive |
| | Dimension | 12 x 12 x 3 mm$^3$ |
| | Sensitivity | 440 LSb/gauss |
| | Sampling rate | 100 Hz/sensor |
| | No. channels | 12 (3 per sensor) |
| **Control Unit** | Microcontroller | Low power ATmega328P |
| | Dimension | 50 x 60 x 15 mm$^3$ |
| | Operating voltage | 5 V |
| | Power source | Lithium Ion battery |
| | Transceiver | Bluetooth/USB |
| **Headset** | Material | VeroBlue/VeroWhitePlus resin |
| | Total weight | 160g (including battery & control unit) |

In addition, significant improvements were made in term of the hardware miniaturizations and portability, as previous generation relied on a PCI-based data acquisition card, thus restricted it to a desktop PC/workstation which is highly immobile and bulky. Although the magnetic sensors HMC2003 are high precision sensors, they are significant larger in size ($24 \times 45 \times 10$ mm$^3$) and required higher operation voltage (i.e. 12 V), thus making them non-power efficient. In the current prototype, magnetic sensors HMC5883L were chosen because of their compactness, low operation voltage, low cost and wide sensitivity range. As for signal conditioning, low-powered microcontrollers were used. By utilizing a Bluetooth modules and a tablet PC (i.e. mobile processing unit), the current MVOCA will be highly portable and practical for everyday use. In addition, the cost of the prototyping is relatively low, as the MVOCA only utilized commercial off-the-shelf (COTS) components. Moreover, by shrinking the size of electronics, this inevitably reduces the overall weight of the headset, and making it more appealing as a wearable assistive speech technology.

On the other hand, this would mean the omission of higher precision components (i.e. magnetic sensors) used in the previous prototype, a reduction in the numbers of sensors and the use of a lower sampling rate. However, from the results presented in Fig. 7, these concerns would appear to be irrelevant as the performances are comparable between $1^{st}$ and $2^{nd}$ generation MVOCAs. This could be that the articulator movements during speech are slow and therefore a lower sampling rate (i.e. 100 Hz) might be sufficient. In addition, reduction in the number of sensors was possible because there were excess of information available from previous MVOCA, thus some sensors can be made redundant.

# 5 Conclusions

The preliminary evaluation of the new MVOCA prototype shows comparable recognition performances to the previous system, but providing much more desirable hardware features such as portability, hardware miniaturization, improved appearance and lower cost. Nonetheless, there are still many challenges ahead before MVOCA can be practically operated outside laboratory environments on a day-to-day basis. Encouraged by the results obtained so far, extensive work is needed to create a viable wearable assistive communication aid. Potential future works may include enhancing overall MVOCA appearance, reducing power consumption and implementing real-time features (i.e. reducing latency in processing and decision making). On the other hand, to address the desirable features on speech quality as discussed in Sect. 3, investigation work on speech synthesis (similar to the work in [9]) from PMA data has started and preliminary results obtained are very encouraging.

**Acknowledgements.** The work is an independent research funded by the National Institute for Health Research (NIHR)'s Invention for Innovation Programme. The views stated are those of the authors and not necessary reflecting the thoughts of the sponsor.

# References

1. Fagan, M.J., Ell, S.R., Gilbert, J.M., Sarrazin, E., Chapman, P.M.: Development of a (silent) speech recognition system for patients following laryngectomy. Med. Eng. Phys. **30**(4), 419–425 (2008)
2. Gilbert, J.M., Rybchenko, S.I., Hofe, R., Ell, S.R., Fagan, M.J., Moore, R.K., Green, P.D.: Isolated word recognition of silent speech using magnetic implants and sensors. Med. Eng. Phys. **32**(10), 1189–1197 (2010)
3. Ell, S.R., Mitchell, A.J., Parker, A.J.: Microbial colonisation of the Groningen speaking valve and its relationship to valve failure. Clin. Otolaryngol. **20**, 555–556 (1995)
4. Ell, S.R., Mitchell, A.J., Clegg, R.T., Parker, A.J.: Candida: the cancer of silastic. J. Laryngol. Otol. **110**(3), 240–242 (1996)
5. Heaton, J.M., Parker, A.J.: Indwelling trachea-oesophageal voice prostheses post-laryngectomy in Sheffield, UK: a 6-year review. Acta Otolaryngol. **114**, 675–678 (1994)
6. Denby, B., Schultz, T., Honda, K., Hueber, T., Gilbert, J.M., Brumberg, J.S.: Silent speech interfaces. Speech Commun. **52**(4), 270–287 (2010)
7. Hofe, R., Ell, S.R., Fagan, M.J., Gilbert, J.M., Green, P.D., Moore, R.K., Rybchenko, S.I.: Small-vocabulary speech recognition using silent speech interface based on magnetic sensing. Speech Commun. **55**(1), 22–32 (2013)
8. Hofe, R., Bai, J., Cheah, L.A., Ell, S.R., Gilbert, J.M., Moore, R.K., Green, P.D.: Performance of the MVOCA silent speech interface across multiple speakers. In: Proceedings of the 14th INTERSPEECH, pp. 1140-1143, Lyon, France (2013)
9. Hofe, R., Ell, S.R., Fagan, M.J., Gilbert, J.M., Green, P.D., Moore, R.K., Rybchenko, S.I.: Speech synthesis parameters generation for the assistive silent speech interface MVOCA. In: Proceedings of the 12th INTERSPEECH, pp. 3009–3012, Florence, Italy (2011)

10. Gonzalez, J.A., Cheah, L.A., Bai, J., Ell, S.R., Gilbert, J.M., Moore, R.K., Green, P.D.: Analysis of phonetic similarity in a silent speech interface based on permanent magnetic articulography. In: Proceedings of the 15th INTERSPEECH, pp. 1018–1022, Singapore (2014)

11. Bright, A.K., Conventry, L.: Assistive technology for older adults: psychological and socio-emotional design requirements, In: Proceedings of the 6th International Conference on Pervasive Technologies Related to Assistive Environments, pp. 1–4, Rhodes, Greece (2013)

12. Hirsch, T., Forlizzi, J., Goetz, J., Stoback, J., Kurtx, C.: The ELDer project: Social and emotional factors in the design of eldercare technologies. In: Proceedings on the 2000 Conference of Universal Usability, pp. 72–79, Arlington, USA (2000)

13. Martin, J.L., Murphy, E., Crowe, J.A., Norris, B.J.: Capturing user requirements in medical devices development: the role of ergonomics. Physiol. Meas. 27(8), R49–R62 (2006)

14. Cook, A.M., Polgar, J.M.: Assistive Technologies: Principles and Practice, 3rd edn. Mosby, London (2008)

15. Leonard, R.G.: A database for speaker-independent digit recognition. In: International Conference on Acoustic, Speech and Signal Processing, pp. 328–331, San Diego, USA (1984)

16. Young, S., Everman, G., Gales, M., Hain, T., Kershaw, D., Liu, X., Moore, G., Odell, J., Ollason, D., Povery, D., Valtchev, V., Woodland, P.: The HTK Book (for HTK version 3.4.1), Cambridge University Press, Cambridge (2009)

17. Rabiner, L.R.: A tutorial on Hidden Markov models and selected applications in speech recognition. Proc. IEEE 77, 257–286 (1989)

18. Maier-Hein, L., Metze, F., Schultz, T., Waibel, A.: Session independent non-audible speech recognition using surface electromyography. In: Automatic Speech Recognition and Understanding Workshop, pp. 331–336, Cancun, Mexico (2005)

19. Wand, M., Schultz, T.: Session-independent EMG-based speech recognition. In: Proceedings of the 4th International Conference on Bio-inspired Systems and Signal Processing, pp. 295–300, Rome, Italy (2011)

# Health Informatics

# Who Is 1011011111...1110110010? Automated Cryptanalysis of Bloom Filter Encryptions of Databases with Several Personal Identifiers

Martin Kroll$^{(\boxtimes)}$ and Simone Steinmetzer

Research Methodology Group, University of Duisburg-Essen,
Lotharstraße 65, 47057 Duisburg, Germany
martin.kroll@uni-due.de

**Abstract.** We provide the first efficient cryptanalysis of Bloom filter encryptions of a database containing more than one personal identifier. The cryptanalysis is fully automated and shows several drawbacks of existing encryption methods based on Bloom filters. In particular, the special representation of the hash functions as linear combinations of two hash functions $f$ and $g$ is exploited in order to detect Bloom filter encryptions of single bigrams (so-called *atoms*). The assignment of atoms to bigrams is obtained via a modification of an algorithm which was originally proposed for the automated cryptanalysis of simple substitution ciphers. Using our approach, we were able to reconstruct 77.7 % of the identifier values correctly. We point to further improvements of the basic Bloom filter approach that are worth being investigated with respect to their privacy guarantees in future work.

**Keywords:** Bloom filter · Privacy-preserving record linkage · Anonymity · Hash function · Cryptographic attack

## 1 Introduction

Record linkage between databases containing information about individuals is popular in a large number of medical applications, for example the identification of patient deaths [1], the evaluation of disease treatment [2] and the linkage of cancer registries in follow-up studies in epidemiology [3]. In many applications data sets are merged using personal identifiers such as forenames, surnames, place and date of birth. Due to privacy concerns this has to be done via privacy-preserving record linkage (PPRL). However, since personal identifiers often contain typing or spelling errors, encrypting the identifier values and linking only those that match exactly does not provide satisfactory results. Therefore, to allow for errors in encrypted personal identifiers, in many European countries encrypted phonetic codes, such as Soundex codes, are commonly used, especially by cancer registries. As the performance of these codes is still non satisfactory, several novel privacy-preserving record linkage methods have been suggested during the last years. For example Schnell et al. [4] developed a method based on

© Springer International Publishing Switzerland 2015
A. Fred et al. (Eds): BIOSTEC 2015, CCIS 574, pp. 341–356, 2015.
DOI: 10.1007/978-3-319-27707-3_21

Bloom filters. Bloom-filter-based record linkage has already been used in medical applications in a number of different countries [5–8].

Another frequently applied privacy-preserving record linkage method uses anonymous linking codes [9]. The basic principle of an anonymous linking code is to standardize all particular identifiers of a record (removal of certain characters and diacritics, use of upper case letters), to concatenate them to a single string and finally to put this single string into a cryptographic hash function. By combining this principle with Bloom filters, Schnell et al. [10] first developed a novel error-tolerant anonymous linking code, called Cryptographic Longterm Key (CLK). Instead of encrypting every single identifier from a record of several identifiers through a Bloom filter, multiple identifiers are stored in one single Bloom filter, called CLK. Tests on several databases showed that CLKs yield good linkage properties, superior to well-known anonymous linking codes [10].

Only recently Randall et al. [7] presented a study on 26 million records of hospital admissions data and showed that privacy-preserving record linkage with Bloom filters built from multiple identifiers is applicable to large real-world databases without loss in linkage quality.

However, only little research on the security of Bloom filters built from more than one identifier has yet been published (see Subsect. 2.2). In several countries, this lack of research prevents the widespread use of Bloom filter encryptions for real-world medical databases (such as cancer registries) where the anonymity of the individuals has to be guaranteed. For example, in its *Beyond 2011 Programme* the British Office for National Statistics investigated several methods for linking sensitive data sets [11]. The investigators came to the conclusion that none of the '(...) recent innovations, such as bloom filter encryption (...)' can be recommended because they '(...) have not been fully explored from an accreditation perspective'. Thus, research showing drawbacks of the recent Bloom filter techniques is important because it guides the direction for future research and might motivate further development of the recent procedures.

In this paper, we intend to investigate this issue in detail by giving the first convincing cryptanalysis of Bloom filter encryptions built from more than one identifier.

## 2    Background

In 1970, Bloom [12] introduced a novel approach that permits the efficient testing of set membership through a probabilistic space-efficient data structure. A *Bloom filter* is a bit array of length $L$, which is initialized with zeros only. Let $S \subseteq \mathcal{U}$ be a subset of a universe $\mathcal{U}$. Then $S$ can be stored in a Bloom filter $\mathcal{B} = \mathcal{B}(S) = (b_0, \ldots, b_{L-1})$ in the following way: Each element $s \in S$ is mapped via $k$ different hash functions $h_0, \ldots, h_{k-1} : S \longrightarrow \{0, \ldots, L-1\}$ and all the corresponding bit positions $b_{h_0(s)}, \ldots, b_{h_{k-1}(s)}$ are set to one. Once a bit position is set to one by an element $s \in S$, this value no longer changes.

In order to test whether an item $u \in \mathcal{U}$ from the universe is contained in $S$, $u$ is hashed through the $k$ hash functions $h_0, \ldots, h_{k-1}$ as well. Consequently, if all bit positions $b_{h_0(u)}, \ldots, b_{h_{k-1}(u)}$ in the Bloom filter are set to one, then $u \in S$ holds with high probability. However, in this case false positive classifications can occur when all the ones on the positions $h_0(u), \ldots, h_{k-1}(u)$ are caused by elements from $S$ distinct from $u$. Then the test indicates $u \in S$ although this is not the case. Otherwise, if at least one bit position in the two Bloom filters varies, $u$ clearly is no member of $S$. This latter case is illustrated in Fig. 1.

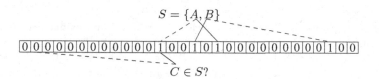

**Fig. 1.** A Bloom filter storing the elements $A$ and $B$ from a set $S$. Two hash functions (illustrated by solid and dashed respectively) are used. Set-membership of another element $C$ can be checked by hashing $C$ through the same two hash functions. In this example, it is guaranteed that $C$ is not a member of $S$ since one of the positions in the array to which $C$ is hashed equals zero.

## 2.1    PPRL with Bloom Filters Built from Multiple Identifiers

In [4] Bloom filters were used in privacy-preserving record linkage for the first time. This approach was expanded to Cryptographic Longterm Keys in [10].

In common PPRL protocols two data owners A and B agree on a set of identifiers that occur in both of their databases. Next, these identifiers are standardized, then padded with blanks at the beginning and the end, and finally split into substrings of two characters. Each substring of the first identifier corresponding to a record is mapped to the first Bloom filter via several hash functions. Afterwards, each substring of the second identifier, corresponding to the same record, is mapped through another set of hash functions to the first Bloom filter as well. This procedure is repeated until all identifiers of the first record are stored in the first Bloom filter. Next, all identifiers corresponding to the second record of the database are mapped through the utilized hash functions to a second Bloom filter and so on. Performing this procedure for all entries of the database results in a set of Bloom filters where each Bloom filter is built from multiple identifiers. The similarity (e.g., Jaccard similarity) of the resulting Bloom filters is a very good approximation of the (Jaccard) similarity of the unencoded bigram sets and thus the unencoded identifier values. This latter fact is illustrated through Fig. 2 in the case of one identifier.

Because of the specific structure of Bloom filters, record linkage based on Bloom filters built from multiple identifiers allows for errors in the encrypted data. Therefore, they can be applied to linking large data sets such as national medical databases [7].

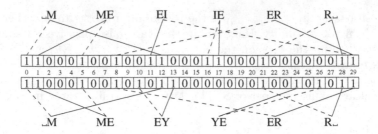

**Fig. 2.** The similarity of Bloom filters yields a very good approximation to the similarity of unencoded bigram sets. The bigram sets $A = \{\text{\_M, ME, EI, IE, ER, R\_}\}$ and $B = \{\text{\_M, ME, EY, YE, ER, R\_}\}$ of the strings \_MEIER\_ and \_MEYER\_ have a Jaccard similarity of $\frac{|A \cap B|}{|A \cup B|} = \frac{4}{8} = 0.5$. The Jaccard similarity of binary vectors is defined as $\frac{n_{11}}{n_{01} + n_{10} + n_{11}}$ where $n_{ij}$ is the number of bit positions that are equal to $i$ for the first and equal to $j$ for the second vector. In our example, we get a value of $\frac{8}{15} \approx 0.53$.

## 2.2  Extant Research: Attacks on Bloom Filters of One or More Identifiers

To the best of our knowledge, only two ways of attacking Bloom filters of one identifier and one way of attacking Bloom filters of multiple identifiers are known so far.

The first cryptanalysis of Bloom filters was published in 2011. Kuzu et al. [13] sampled 20,000 records from a voter registration list and encrypted the substrings of two characters from the forenames through 15 hash functions and Bloom filters of length 500 bits. Their attack consisted in solving a constraint satisfaction problem (CSP). Through a frequency analysis of the forenames and the Bloom filters and by applying their CSP solver to the problem, Kuzu et al. were able to decipher approximately 11 % of the data.

In contrast, Niedermeyer et al. [14] proposed an attack on 10,000 Bloom filters built from encrypted German surnames that were considered to be a random sample of a known population. For the generation of the Bloom filters 15 hash functions and Bloom filter length 1,000 were used. Then they conducted a manual attack based on the frequencies of the substrings of length two, which they derived from the German population. Thus, Niedermeyer et al. deciphered the 934 most frequent surnames of 7,580 different ones, which corresponds to approximately 12 % of the data set. However, their attack is not limited to the most frequent names and could be extended to the decipherment of nearly all names.

In 2012 Kuzu et al. [15] proposed an attack on Bloom filters built from multiple identifiers. They applied their constraint solver to forename and surname, as well as forename, surname, city and ZIP code, of 50,000 randomly selected records from the North Carolina voter registration list. However, they were not able to mount a successful attack. Thus, Kuzu et al. supposed that combining multiple personal identifiers into a single Bloom filter would offer a protection mechanism against frequency attacks. Although they suspected that their attack did not uncover all vulnerabilities of the Bloom filter encodings, they showed that the CSP for multiple identifiers is intractable to solve by their constraint solver.

## 2.3  Our Contribution

In this paper we present a fully automated attack on a database containing forenames, surnames and the relevant place of birth as well. All records are considered to be a random sample of a known population. We suppose that the attacker only knows some publicly available lists of the most common forenames, surnames and locations. The attack is based on analyzing the frequencies and the combined occurence of substrings of length two from the identifiers of these lists. Furthermore, we are interested in re-covering as many identifiers as possible. Our cryptanalysis was implemented using the programming languages Python and C++.

# 3  Encryption

In this section some basic notation is introduced and the encrypting procedure is described.

In record linkage scenarios, strings are usually standardized through transformations such as capitalization of characters or removal of diacritics [16].

After this *preprocessing step* all strings contain only tokens from some predefined alphabet $\Sigma$. Throughout this article, we use the canonical alphabet $\Sigma := \{A, B, \ldots, Z, \_\}$, where $\_$ denotes the padding blank. Thus, for example the popular German surname Müller is transformed to $\_$MUELLER$\_$ in the preprocessing step. As usual, we denote substrings of two characters with *bigrams* and the set containing all the bigrams with $\Sigma^2$, i.e.

$$\Sigma^2 = \{\_\_, \_A, \ldots, \_Z, A\_, \ldots, Z\_, AA, \ldots, ZZ\}.$$

The Bloom filter encryption of a record from a database is created by storing the bigram set associated with this record into a Bloom filter. The bigram set associated with a record is defined as the set containing the bigrams from all the identifiers. Here, a distinction between the bigrams occuring in different identifiers has to be made. Thus, if the set of identifiers is denoted with $\mathcal{I}$, the bigram set of a record is a subset of $\mathcal{I} \times \Sigma^2$.

For example, if we have $\mathcal{I} = \{\texttt{surname}, \texttt{forename}\}$ and the database contains a record, Peter Müller, the bigram set associated with this record would contain the bigrams $\_P_f$, $PE_f$, $ET_f$, $TE_f$, $ER_f$, $R\_f$, $\_M_s$, $MU_s$, $UE_s$, $EL_s$, $LL_s$, $LE_s$, $ER_s$ and $R\__s$ (the subscript $f$ indicates the bigrams occuring in the forename identifier, the subscript $s$ the ones occuring in the surname identifier).

Next, this bigram set is stored into a Bloom filter $(b_0, \ldots, b_{L-1})$ of length $L$ by means of $k$ independent hash functions

$$h_i : \mathcal{I} \times \Sigma^2 \rightarrow \{0, \ldots, L-1\}$$

for $i = 0, \ldots, k - 1$. In practice, one could alternatively use different hash functions $h_i : \Sigma^2 \to \{0, \ldots, L-1\}$ for the distinct identifiers in order to guarantee that the hash values for distinct identifiers are not the same.

Further, as in [14] we introduce the term *atom* for the specific Bloom filters which occur as the fundamental building blocks of the encryption method.

**Definition 1 (Atom).** *Let $L, k \in \mathbb{N}$ and some hash functions $h_0, \ldots, h_{k-1}$ be defined as above. Then, a Bloom filter*

$$\mathcal{B} := (b_0, \ldots, b_{L-1}) \in \{0, 1\}^L$$

*is termed an* atom *if there exists a bigram $\beta \in \mathcal{I} \times \Sigma^2$ such that $b_j = 1 \Leftrightarrow h_i(\beta) = j$ for some $i = 0, \ldots, k - 1$. Such a Bloom filter is called the* atom realized by the bigram $\beta$ and denoted with $\mathcal{B}(\beta)$.

Thus, atoms are special Bloom filters. Since each bigram is hashed via each $h_i$ for $i = 0, \ldots, k - 1$, at most $k$ positions in an atom can be set to one.

By combining the atoms of the underlying bigram set of a record with the bitwise OR operation, the Bloom filter of a record is composed as

$$\mathcal{B}(\texttt{record}) = \bigvee_{\beta \in \mathcal{S}_{\text{record}}} \mathcal{B}(\beta),$$

where $\bigvee$ denotes the bitwise OR operator.

Note that the same bigram from $\Sigma^2$ is hashed differently if it occurs in distinct identifiers. This is illustrated in Fig. 3 for the example of the bigram ER which occurs in the record `Peter Müller` both in the surname and the forename identifier.

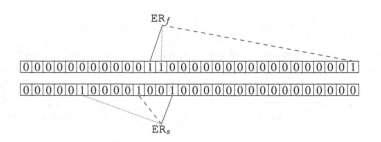

**Fig. 3.** Two different atoms of the bigram ER. These atoms are realized when instances of ER occur in distinct identifiers (the forename and surname identifier in this example).

Mapping each bigram of the forename `Peter` with $k$ hash functions results in six atoms; for the surname `Müller`, we get eight atoms. Thus, the separate Bloom filters for these identifiers might be composed as illustrated in Fig. 4.

|   |                              |                      |
|---|------------------------------|----------------------|
|   | 0000100000 ... 0000000010    | $\mathcal{B}(\_P_f)$ |
| V | 0001000001 ... 0100000100    | $\mathcal{B}(PE_f)$  |
| V | 0101010101 ... 0001010101    | $\mathcal{B}(ET_f)$  |
| V | 0001000010 ... 0001000010    | $\mathcal{B}(TE_f)$  |
| V | 0100010001 ... 0000000100    | $\mathcal{B}(ER_f)$  |
| V | 0101010101 ... 0000000001    | $\mathcal{B}(R\_f)$  |
|   | 0101110111 ... 0101010111    | $\mathcal{B}(\texttt{Peter})$ |

|   |                              |                      |
|---|------------------------------|----------------------|
|   | 0000000100 ... 0000000001    | $\mathcal{B}(\_M_s)$ |
| V | 0010000000 ... 0100000000    | $\mathcal{B}(MU_s)$  |
| V | 0000100000 ... 0010000010    | $\mathcal{B}(UE_s)$  |
| V | 1000000010 ... 0010000000    | $\mathcal{B}(EL_s)$  |
| V | 0100001000 ... 0100001000    | $\mathcal{B}(LL_s)$  |
| V | 1000000100 ... 0001000000    | $\mathcal{B}(LE_s)$  |
| V | 1001001001 ... 0000100100    | $\mathcal{B}(ER_s)$  |
| V | 0010001000 ... 0000000010    | $\mathcal{B}(R\_s)$  |
|   | 1111101111 ... 0111101111    | $\mathcal{B}(\texttt{Müller})$ |

**Fig. 4.** Bloom filters of the forename `Peter` and the surname `Müller`, composed of the atoms belonging to the underlying bigrams.

|   |                              |                      |
|---|------------------------------|----------------------|
|   | 0101110111 ... 0101010111    | $\mathcal{B}(\texttt{Peter})$ |
| V | 1111101111 ... 0111101111    | $\mathcal{B}(\texttt{Müller})$ |
|   | 1111111111 ... 0111111111    | $\mathcal{B}(\texttt{entire record})$ |

**Fig. 5.** The Bloom filter of the record `Peter Müller` is obtained by applying the bitwise OR operation to the Bloom filter encryptions of the separate identifiers.

The final Bloom filter for the record `Peter Müller` is composed by appling the bitwise OR operation to the separate Bloom filter encryptions of the distinct identifiers. This is demonstrated in Fig. 5.

In practice, the Bloom filter encryption of a record might contain a mixture of string valued identifiers (such as forename, surname or place of birth) and also numerical identifiers, such as date of birth. However, in this paper we restrict ourselves to the case of string valued attributes only, albeit our cryptanalysis proposed below is not limited to such attributes.

## Assumptions

In many record linkage scenarios, it is supposed that a semi-trusted third party conducts the record linkage between two encrypted databases. In this paper we assume a data set containing Bloom filters built from multiple identifiers that is sent to such a semi-trusted third party. This third party acts as the adversary and tries to infer as much information as possible from the record encryptions. We further suppose that the attacker has knowledge of the encryption process.

For our experiment we generated 100,000 Bloom filters built from standardized German forenames, surnames and cities according to the distribution in the population. The identifiers were truncated after the tenth letter, padded with blanks, respectively, and were broken into bigrams. Then the bigrams were hashed through $k = 20$ hash functions into Bloom filters of length $L = 1,000$. As proposed in [4,10], we used the so-called *double hashing scheme* for the generation of $k$ hash functions from two hash functions $f$ and $g$. This double hashing scheme is defined via the equation

$$h_i = (f + i \cdot g) \mod L \quad \text{for } i = 0, \ldots, k - 1 \tag{1}$$

and was originally proposed in [17] as a simple hashing method for Bloom filters yielding satisfactory performance results.

In our cryptanalysis we assume that the adversary knows that the hash values are generated in accordance with Eq. (1). It is self-evident that she must not have direct access to the hash functions $f$ and $g$ since this would permit the adversary to check whether a specific bigram is contained in a given Bloom filter.

Note that the double hashing scheme has also been used for the generation of Bloom filters by Kuzu et al. [15]. However, in that paper the knowledge of the *double hashing scheme* was not exploited in their cryptanalysis.

## 4   Cryptanalysis

This section provides a detailed description of the deciphering process. At first we try to detect the atoms that are contained in the given Bloom filters. Then, we assign bigrams to these atoms by means of an optimization algorithm. Finally, the original attributes are reconstructed from the atoms.

Our approach for the development of a fully automated attack is based on previous results on the automated cryptanalysis of simple substitution ciphers presented by Jakobsen [18]. We give a short account of Jakobsen's results in order to motivate our procedure.

### 4.1   Automated Cryptanalysis of Simple Substitution Ciphers

The encryption of a plaintext message through a *simple substitution cipher* is defined by a permutation of the underlying alphabet $\Sigma$. For instance, the message HELLO_LISBON with tokens from the alphabet $\Sigma = \{\_, A, B, \ldots, Z\}$ could be encrypted as RVUUYJUOWAYL. It is well known that this kind of encryption can be broken easily by means of a frequency analysis. However, just replacing the i-th frequent character in the ciphertext with the i-th frequent character in the underlying language will usually not lead to the correct decipherment (even for longer messages). This fact is commonly compensated for by taking bigram frequencies into consideration as well.

The expected bigram frequencies can be obtained from a training data set composed of the underlying language and stored in a quadratic matrix $E$ (in the above example a $27 \times 27$ matrix), where the entry $e_{ij}$ is equal to the relative proportion of the bigram $c_i c_j$ in the training text corpus and $c_i$ denotes the i-th character of the alphabet. Analogously, the bigram frequencies of the ciphertext can be stored in a matrix $D$.

The algorithm proposed by Jakobsen [18] was intended to find a permutation $\sigma_{opt}$ of the alphabet such that the objective function $f$ defined via

$$f(\sigma) := \sum_{i,j} |d_{\sigma(i)\sigma(j)} - e_{ij}| \tag{2}$$

is minimized. The algorithm starts with the initial permutation that reflects the best assignment between single characters in the plaintext and the ciphertext with respect to their relative frequency. In each step of the algorithm two elements of the currently best permutation $\sigma_{opt}$ are swapped, leading to a new candidate permutation $\sigma$. If $f(\sigma) < f(\sigma_{opt})$ holds, the current permutation is updated to $\sigma$, otherwise $\sigma$ is discarded and a new candidate $\sigma$ is generated by swapping two other elements of $\sigma_{opt}$. This is repeated until no swap leads to a further improvement of the objective function $f$. Throughout this paper we use the same strategy as Jakobsen in [18] in order to determine the elements of the current permutation to be swapped. For a more detailed description of Jakobsen's method in the case of simple substitution ciphers we refer the reader to his original paper [18]. Figure 2 in [18] shows that a ciphertext of length 600 built by a simple substitution cipher can be entirely broken by this method.

It is clear that some modification of Jakobsen's original algorithm is necessary in order to make it applicable in our setting as well. In particular, the definitions of the matrices $D$ and $E$ must be changed. Their adopted definitions are introduced in Subsect. 4.3.

## 4.2   Atom Detection

As in [14], the basic principle of our approach consists in the detection of atoms, which represent the encryption of one single bigram only. Since the Bloom filter of a string is created by the superposition of at least a few atoms, the reconstruction of the atoms given only a set of Bloom filters turns out to be difficult. Note that this task cannot be solved in a satisfactory manner if Bloom filters are considered isolatedly or in small groups because in this case too many binary vectors will be wrongly classified as atoms.

Let us give a short motivation for our novel method aiming at atom detection. If the bitwise AND operation is applied to a set of Bloom filters that have one bigram $\beta$ in common, at least all positions set to one by $\beta$ are equal to one in the result. However, for prevalent bigrams it should be expected that all the other positions are set to zero if a sufficient number of Bloom filters are considered, i.e., the result would be exactly the atom induced by the bigram $\beta$.

Of course, if an adversary has access to a set of Bloom filters, she does not a priori know which Bloom filters have a bigram in common. This obstacle can be avoided as follows: Under the assumption that the double hashing scheme is being used, the adversary is able to determine for each combination of bit positions from Eq. (1) the set of Bloom filters for which all these positions are set to 1. Then, the bitwise AND operation is applied to the set of these Bloom filters. If the result coincides with the atom, it is considered to be the realization of a bigram by the adversary.

The resulting set of atoms was further reduced by discarding atoms of Hamming weight $\sum_{i=0}^{999} b_i$ equal to 1, 2, 4 or 5 and keeping only atoms of Hamming weight equal to 8, 10 or 20. Otherwise, too many binary vectors would have been classified incorrectly as atoms. The probability that an atom has Hamming weight less than 8 in our setting is equal to 0.008. This value can be derived in analogy to Lemma A.1 and the subsequent example in [14].

We denote the number of atoms found by $n$. For our specific data set we got $n = 1,776$. This result seems reasonable, because the total number of possible atoms is bounded from above by 2,187 and obviously not all of these atoms, in particular atoms realized by rare bigrams, occur in our simulated data. As we checked later on, 1,337 of the 1,776 extracted conjectured atoms were indeed true atoms, that is to say atoms generated by one of the 2,187 bigrams. The subsequent analysis demonstrates that this percentage of correct atom detection is sufficient for a successful cryptanalysis. For each atom $\alpha$ we determined the set of Bloom filters containing this atom, i.e. Bloom filters for which all bit positions of the atom are set to 1. We denote the atoms with $\alpha_1, \ldots, \alpha_{1776}$ according to decreasing frequency. In order to give an illustrative example, we assert that in the Bloom filter No. 850 (which looks like $1011011111 \ldots 1110110010$) the atoms $\alpha_5$, $\alpha_8$, $\alpha_{14}$, $\alpha_{15}$, $\alpha_{29}$, $\alpha_{33}$, $\alpha_{36}$, $\alpha_{46}$, $\alpha_{55}$, $\alpha_{106}$, $\alpha_{110}$, $\alpha_{123}$, $\alpha_{138}$, $\alpha_{169}$, $\alpha_{194}$, $\alpha_{197}$, $\alpha_{218}$, $\alpha_{254}$, $\alpha_{309}$, $\alpha_{313}$, $\alpha_{317}$, $\alpha_{334}$, $\alpha_{335}$, $\alpha_{396}$, $\alpha_{398}$, $\alpha_{453}$, $\alpha_{607}$, $\alpha_{668}$, $\alpha_{705}$, $\alpha_{782}$, $\alpha_{821}$, $\alpha_{960}$ and $\alpha_{1131}$ were detected.

In the subsequent section we explain how correlations between the occurences of atoms in the Bloom filters and bigrams in a training data set can be used to give adequate definitions of the matrices $D$ and $E$ that serve as the input of Jakobsen's algorithm.

### 4.3   Correlation of Atoms and Bigrams

A naive assignment of bigrams to atoms is possible only for few frequent bigrams at most. For example, if German surnames, given names and birth locations are considered together, usually the most frequent bigram is $A_{\llcorner}f$ (the bigram $A_{\llcorner}$ in the forename identifier) such that the most frequent atom is likely to be the encryption of this bigram. The absolute frequencies of the 10 most frequent bigrams in the considered training data are illustrated in Fig. 6.

Except for the first few bigrams, the bigram frequencies are too close together such that naive matching is not promising for automatic decipherment.

In the example of Bloom filter No. 850 already introduced above, this naive assignment would lead to the conjecture that the corresponding record contains the following bigrams: $N_{\llcorner l}$, $R_{\llcorner s}$, $CH_s$, $N_{\llcorner f}$, $HE_l$, $_{\llcorner\llcorner l}$, $SC_s$, $S_{\llcorner f}$, $E_{\llcorner l}$, $_{\llcorner}L_f$, $BE_s$, $NI_f$, $AR_s$, $_{\llcorner}W_f$, $_{\llcorner}P_f$, $NG_s$, $IR_f$, $ET_s$, $MI_s$, $NI_s$, $VE_l$, $OS_l$, $NS_s$, $UN_s$, $AT_s$, $_{\llcorner}V_s$, $LH_l$, $OW_l$,

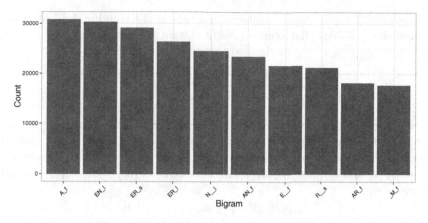

**Fig. 6.** Absolute frequencies of the 10 most frequent bigrams in our training data set.

$AA_s$, $ZB_l$, $RR_l$, $DY_f$ and $MR_s$. However, from this list of bigrams it is obviously impossible to reconstruct any meaningful information.

For this reason, we also took correlations between bigrams into account. For example, for records sampled from the population of Germany the appearance of the bigram $CH_s$ in a record makes the appearance of the bigram $SC_s$ in the same record more likely because the trigram $SCH$ frequently appears in German surnames.

We model this kind of information on the correlation of atoms and bigrams by means of two matrices $D$ and $E$. Assume that the attribution values of the records built from tokens of the alphabet $\Sigma = \{ \_, A, B, \ldots, Z \}$ are to be encrypted. Thus, for each (string valued) identifier we have 729 possible bigrams. Since the same bigram is encrypted differently for each identifier we have to distinguish between different instances of the same bigram. In our setting we denote the bigram $\beta$ for the surname, forename and location identifier with $\beta_s$, $\beta_f$ and $\beta_l$, respectively. Altogether, the set $\Sigma^2$ containing all possible bigrams consists of $3 \cdot 729 = 2{,}187$ elements.

Let us now introduce the matrix $E$ containing information about the expected bigram correlations obtained from the training data set. Note that the training data should be as similar to the encrypted data as possible, e.g. a random sample from the same underlying population as the encrypted data. If the prevailing Bloom filters are known to contain encryptions of records from the German population, an attacker would try to get access to a comparable database containing the same identifiers. Thus, the choice of the reference dataset should depend on the context. For example, we would expect our cryptanalysis to be less successful when the Bloom filters mainly encrypt German names whereas the training data consists of a random sample from the French population. The attribute values of this training data set are preprocessed analogously to the preprocessing routine before the encryption process. Then, the bigram sets for all the attribute values are created. We denote the bigrams with $\beta_1, \ldots, \beta_{2187}$ according to decreasing frequency. Let $T$ be the total number of records in the training data set and $t_{ij}$ the number of records that contain both bigram $\beta_i$ and bigram $\beta_j$. Then the matrix $E = (e_{ij})_{i,j=1,\ldots,2187}$ is defined via

$$e_{ij} = \begin{cases} t_{ij}/T & \text{if } i \neq j, \\ 0 & \text{if } i = j. \end{cases}$$

The matrix $D$ is formed in a similar way on the basis of joint appearances of atoms in the Bloom filters. Let $N$ be the number of Bloom filters for which atoms have been extracted. We denote the number of Bloom filters that contain both atom $\alpha_i$ and atom $\alpha_j$ by $b_{ij}$. The matrix $D = (d_{ij})_{i,j=1,\ldots,2187}$ is defined through

$$d_{ij} = \begin{cases} b_{ij}/N & \text{if } i \neq j \text{ and } i,j \leq 1776, \\ 0 & \text{if } i = j \text{ or } \max(i,j) > 1776. \end{cases}$$

The procedure suggested by Jakobsen which was described above can now directly be applied to the matrices $D$ and $E$. The pseudocode for this can be found in Algorithm 1.

The progress of the optimization algorithm is illustrated in Fig. 7.

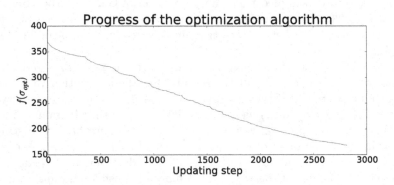

**Fig. 7.** Progress of the optimization algorithm for our data set. The initial value of the objective function is 370.99 and 2,812 updating steps were performed. The final value of the objective function $f(\sigma_{\text{opt}})$ was equal to 168.5.

The result of the algorithm will be the final assignment between atoms and bigrams defined by a permutation $\sigma_{\text{opt}} \in S_{2187}$ and the assignment rule $\alpha_{\sigma_{\text{opt}}(i)} \rightarrow \beta_i$. This assignment is used to reconstruct the original bigram sets encrypted in the Bloom filters.

For example, the bigrams $ER_s$, $R\_s$, $CH_s$, $N\_f$, $HE_l$, $\_\_l$, $SC_s$, $\_S_f$, $E\_l$, $HE_s$, $\_K_l$, $RL_l$, $AR_l$, $Z\_s$, $ON_f$, $SI_f$, $\_F_s$, $IS_s$, $LS_l$, $HW_l$, $SO_f$, $RU_l$, $UR_s$, $IM_f$, $KA_l$, $MO_f$, $AV_f$, $FI_s$, $UH_l$, $HH_l$, $SR_l$, $UZ_l$ and $MR_s$ were assigned to the Bloom filter No. 850.

In the following section we describe how attribute values were reassembled from the reconstructed bigram sets.

---

**Algorithm 1.** Optimization algorithm.

---

**Input:** $D, E$ as defined in Sect. 4.3

**Output:** $\sigma_{\text{opt}} \in S_{2187}$ minimizing

$$f(\sigma) = \sum_{i,j} |d_{\sigma(i)\sigma(j)} - e_i e_j|$$

1: $\sigma_{\text{opt}}(i) = i \ \forall i$                                              ▷ Initialization
2: $\min \leftarrow f(\sigma_{\text{opt}})$
3: $a, b \leftarrow 1$
4: **repeat**
5:      $\sigma \leftarrow \sigma_{\text{opt}}$
6:      $a \leftarrow a + 1$
7:      **if** $a + b \leq 2187$ **then**
8:          $\sigma(a) \leftarrow \sigma_{\text{opt}}(b), \ \sigma(b) \leftarrow \sigma_{\text{opt}}(a)$
9:      **else**
10:          $a \leftarrow 1, \ b \leftarrow b + 1$
11:      **if** $f(\sigma) < f(\sigma_{\text{opt}})$ **then**                         ▷ Update
12:          $\min \leftarrow f(\sigma)$
13:          $\sigma_{\text{opt}} \leftarrow \sigma$
14:          $a, b \leftarrow 1$
15: **until** $b = 2187$

---

## 4.4  Reconstruction of Attribute Values

In order to reconstruct the original attribute values of the records, we separated the bigrams belonging to different identifiers for each Bloom filter.

In the example of Bloom filter No. 850, we obtained the bigrams $N\_f$, $\_S_f$, $ON_f$, $SI_f$, $SO_f$, $IM_f$, $MO_f$, $AV_f$ for the forename identifier, the bigrams $ER_s$, $R\_s$, $CH_s$, $SC_s$, $HE_s$, $Z\_s$, $\_F_s$, $IS_s$, $UR_s$, $FI_s$, $MR_s$ for the surname identifier and finally the bigrams $HE_l$, $\_l$, $E\_l$, $\_K_l$, $RL_l$, $AR_l$, $LS_l$, $HW_l$, $RU_l$, $KA_l$, $UH_l$, $HH_l$, $SR_l$, $UZ_l$ for the location identifier. From this list it is already possible to guess the original identifier values at first glance.

Our fully automated approach to reconstructing the original identifier values was to compare the obtained bigram sets with a list of bigram sets generated from reference lists of surnames, names and locations. For Bloom filter No. 850, for example, an adversary would correctly obtain that this Bloom filter encrypts a record belonging to the person `Simon Fischer` from the German city `Karlsruhe`.

## 4.5  Results

By using the approach described above, we were able to reconstruct 59.6 % of the forenames, 73.9 % of the surnames and 99.7 % of the locations correctly. For 44 % of the 100, 000 records all the identifier values were recuperated successfully.

## 5    Conclusion

In this paper we demonstrated a successful fully automated attack on Bloom filters built from multiple identifiers. We were able to recover approximately 77.7 % of the original identifier values. In contrast to the assumptions in [14,15], that storing all identifiers in a single Bloom filter makes it more difficult to attack, we needed only moderate computational effort and publicly available lists of forenames, surnames, and locations to reconstruct the identifiers. Note that there is no huge impact of the size of the database containing the Bloom filters. For our cryptanalysis it is sufficient to perform the attack on a subset of the given Bloom filters (100,000 as in our example should be adequate in most cases). Then for the remaining Bloom filters it would be sufficient to check for the atoms contained in those and to reconstruct the attribute values, since most assignments of atoms to bigrams are already known. Thus, the time needed for cryptanalysis is linear in the number of input Bloom filters. The time needed for the detection of atoms is $O(L^2)$ since there are $L$ possible values for the hash functions $f$ and $g$ in Eq. (1). Furthermore, the detection of atoms could easily be parallelized to make the computation faster. In addition, values of $L$ significantly larger than $L = 1,000$ as considered in this paper would have negative effects on the time needed for performing the linkage between two databases (note that in the large scale study reported in [7] a Bloom filter length of only 100 was considered). Thus, the most time consuming step in our cryptanalysis should be the optimization algorithm presented in Subsect. 4.3. Indeed, in the chosen parameter setup this procedure took about 402 min on a notebook with 2.80 GHz Intel® Core running Ubuntu 14.04 LTS.

To sum up, we do not recommend the usage of Bloom filters built from one or more identifiers, generated with the double hashing scheme, in applications where high security standards are required. However, we applied our attack in a very special scenario, because the generated databases were encrypted using the double hashing scheme. In case of arbitrary hash functions, i.e. without the restriction on their generation from two hash functions in accordance with Eq. (1), the detection of atoms becomes much harder since an iterative approach is no longer feasible in this case. To be more precise, the number of atoms increases significantly from less than $L^2 = 10^6$ to $\sum_{j=1}^{20} \binom{1000}{j} \geq 3 \cdot 10^{41}$. However, we think that using independent hash functions alone will not be sufficient to ensure security, since in this case other approaches (maybe related to or at least inspired through work from the area of Frequent Itemset Mining [19]) are promising to detect at least the most frequent atoms automatically. The development of such a more general method for atom detection will be part of future work.

Niedermeyer et al. [14] proposed several methods such as fake injections, salting or randomly selected hash values to harden the Bloom filters. Hence, we are confident that methods like those proposed by Niedermeyer et al. show promise in the prevention of attacks like the one presented in this paper and might be suitable for PPRL of sensitive personal data. Further investigations in this direction will also be part of future work.

**Acknowledgements.** Research of both authors was supported by the research grant SCHN 586/19-1 of the German Research Foundation (DFG) awarded to the head of the Research Methodology Group, Rainer Schnell. We thank him and the three anonymous reviewers for their helpful comments.

# References

1. Jones, M., McEwan, P., Morgan, C.L., Peters, J.L., Goodfellow, J., Currie, C.J.: Evaluation of the pattern of treatment, level of anticoagulation control, and outcome of treatment with warfarin in patients with non-valvar atrial fibrillation: a record linkage study in a large British population. Heart **91**(4), 472–477 (2005)
2. Newman, T.B., Brown, A.N.: Use of commercial record linkage software and vital statistics to identify patient deaths. J. Am. Med. Assoc. **4**(3), 233–237 (1997)
3. Van den Brandt, P.A., Schouten, L.J., Goldbohm, R.A., Dorant, E., Hunen, P.M.H.: Development of a record linkage protocol for use in the Dutch cancer registry for epidemiological research. Int. J. Epidemiol. **19**(3), 553–558 (1990)
4. Schnell, R., Bachteler, T., Reiher, J.: Privacy-preserving record linkage using Bloom filters. BMC Med. Inform. Decis. **9**(41), 1–11 (2009)
5. Kuehni, C.E., Rueegg, C.S., Michel, G., Rebholz, C.E., Strippoli, M.-P.F., Niggli, F.K., Egger, M., von der Weid, N.X.: Cohort profile: the swiss childhood cancer survivor study. Int. J. Epidemiol. **41**(6), 1553–1564 (2012)
6. Rocha, M. C. N.: Vigilância dos óbitos registrados com causa básica hanseníase: caracterização no Brasil (2004–2009) e investigação em Fortaleza, Ceará (2006–2011). Master thesis, Universidade de Brasília (2013)
7. Randall, S.M., Ferrante, A.M., Boyd, J.H., Bauer, J.K., Semmens, J.B.: Privacy-preserving record linkage on large real world datasets. J. Biomed. Inform. **50**, 205–212 (2014)
8. Schnell, R., Richter, A., Borgs, C.: Performance of different methods for privacy preserving record linkage with large scale medical data sets. In: Presentation at the International Health Data Linkage Conference, Vancouver (2014)
9. Herzog, T.N., Scheuren, F.J., Winkler, W.E.: Data Quality and Record Linkage Techniques. Springer, New York (2007)
10. Schnell, R., Bachteler, T., Reiher, J.: A novel error-tolerant anonymous linking code. Working Paper NO. WP-GRLC-2011-02, German Record Linkage Center, Nürnberg (2011)
11. Office for National Statistics: Beyond: Matching anonymous data (M9). Methods and Policies, Office for National Statistics, London (2011)
12. Bloom, B.H.: Space/time trade-offs in hash coding with allowable errors. Commun. ACM **13**(7), 422–426 (1970)
13. Kuzu, M., Kantarcioglu, M., Durham, E., Malin, B.: A constraint satisfaction cryptanalysis of bloom filters in private record linkage. In: Fischer-Hübner, S., Hopper, N. (eds.) PETS 2011. LNCS, vol. 6794, pp. 226–245. Springer, Heidelberg (2011)
14. Niedermeyer, F., Steinmetzer, S., Kroll, M., Schnell, R.: Cryptanalysis of basic bloom filters used for privacy preserving record linkage. J. Priv. Confidentiality **6**(2), 59–79 (2014)
15. Kuzu, M., Kantarcioglu, M., Durham, E., Toth, C., Malin, B.: A practical approach to achieve private medical record linkage in light of public resources. J. Am. Med. Assoc. **20**(2), 285–292 (2012)

16. Randall, S.M., Ferrante, A.M., Boyd, J.H., Semmens, J.B.: The effect of data cleaning on record linkage quality. BMC Med. Inform. Decis. **13**(64), 1–10 (2013)
17. Kirsch, A., Mitzenmacher, M.: Less hashing, same performance: building a better bloom filter. Random Struct. Algor. **33**(2), 187–218 (2008)
18. Jakobsen, T.: A fast method for the cryptanalysis of substitution ciphers. Cryptol. **19**(3), 265–274 (1995)
19. Borgelt, C.: Frequent item set mining. WIREs Data Min. Knowl. Discov. **2**, 437–456 (2012)

# Patient Feedback Design for Stroke Rehabilitation Technology

Daniel Tetteroo[✉], Lilha Willems, and Panos Markopoulos

Eindhoven University of Technology, Den Dolech 2, 5612 AZ
Eindhoven, The Netherlands
{d.tetteroo,p.markopoulos}@tue.nl, l.l.willlems@hotmail.com

**Abstract.** The use of technology in stroke rehabilitation is increasingly common. An important aspect in stroke rehabilitation is feedback towards the patient, but research on how such feedback should be designed in stroke rehabilitation technology is scarce. Therefore, in this paper we describe an exploratory process on the design, implementation and evaluation of a patient feedback module for TagTrainer: an interactive stroke rehabilitation technology. From this process, and from previous literature, we derive five guidelines for patient feedback design in stroke rehabilitation technology. Finally, we illustrate how these guidelines can be used to evaluate existing patient feedback solutions.

**Keywords:** Design guidelines · Assistive technology · Arm-hand rehabilitation · Stroke · Data visualization

## 1 Introduction

Stroke prevalence is on the rise, due ageing of the population [4, 7, 12]. As a result, the health system is under severe pressure due to an increasing ratio of stroke patients to therapists. It has been suggested that patients could receive better healthcare and have a higher quality of life by using interactive technologies for rehabilitation. According to [13] the use of technology in the rehabilitation has four main benefits:

- It can create opportunities for patients to train more often
- It can provide a variety of exercises
- It can enable the patient to practice in absence of the therapist
- It reduces workload of the paramedical staff

Additionaly, interactive technology is well suited for motivating, involving, and immersing stroke patients in their rehabilitation [1]. Interactive technology has been an area of research for getting stroke patients more involved in their rehabilitation, which they may otherwise find tedious and not stimulating due to its intense and repetitive nature [9].

One way interactive technologies can get stroke patients more involved and motivated to perform exercises is by incorporating the use of feedback [1, 16]. It has been shown that in game design feedback plays a crucial role in achieving more effective

© Springer International Publishing Switzerland 2015
A. Fred et al. (Eds): BIOSTEC 2015, CCIS 574, pp. 357–371, 2015.
DOI: 10.1007/978-3-319-27707-3_22

engagement [1]. Feedback makes users aware of their progress towards goals and how their actions impact their progress. It provides users with a means to accomplish their goals and when this information is provided effectively, it enables users to independently learn and improve their performance. In addition, [14] has shown that feedback enhances learning and self-efficacy when it is positive and encouraging.

The use of effective feedback is therefore an important means for improving stroke rehabilitation and enabling patients to practice independently. Research suggests that the (extrinsic) feedback provided by technology carries special importance for stroke patients due to their compromised intrinsic feedback system as a result of the stroke [13, 14]. Extrinsic feedback, when provided properly, can improve stroke patients' learning and increase their active involvement, motivation, confidence and self-efficacy [13, 16].

Although the positive effects of feedback on the stroke recovery process are somewhat understood, the design of feedback systems targetted specifically at stroke patients is an underexplored, but emergent and important area of research. Therefore, we present an investigation into how to provide effective feedback to stroke patients using different options for feedback content and modality. We addressed this question by performing an exploratory case study with seven stroke patients using the TagTrainer system [11]. In this paper we present the design process of a feedback module for TagTrainer. From there, we present guidelines for providing feedback with stroke rehabilitation technology that have been derived from the design and evaluation of this module. Finally, we show how these guidelines can be applied in the evaluation of other patient feedback interfaces for stroke rehabilitation technology.

## 2   Related Work

### 2.1   Recovery After Stroke

There is ample research on the effects of feedback on motor learning in non-disabled persons. However, to which extent these findings apply to stroke patients is largely unknown. For one, the intrinsic feedback system of stroke patients is impaired [13–15], while this system plays an important role in motor learning of non-disabled persons. In addition, it has been suggested by [6] that stroke may cause learning deficits in the patients. This suggestion implies that stroke patients require a different approach for learning motor skills; a task that is simple for a non-disabled person may be complex for a stroke patient.

Movement recovery after stroke is usually attributed to two mechanisms [2, 6]:

- True recovery: this occurs when the same muscles for a certain activity are once again used as before the stroke.
- Compensation: this occurs when alternative muscles are used as a strategy, different than before the stroke, to perform an activity.

According to [6] learning is required for both mechanisms to occur and in order to achieve this, rehabilitation should emphasize on learning different techniques to reach a certain goal and not just repetition of the same movements.

## 2.2 Feedback in Stroke Rehabilitation

In general there has been little research on the effects of feedback in motor learning following stroke. Depending on the impairments caused by the stroke, different feedback is needed to accommodate the patients' capabilities and to facilitate motor learning. The feedback should be adjusted to the patient's stage of learning [13].

It is commonly accepted that three factors play an important role in transmitting feedback to stroke rehabilitation patients: focus of attention, feedback content and feedback scheduling [14].

[2] conducted a study with the objective to determine if the manipulation of the attentional focus may lead to arm motor recovery during a repetitive pointing training intervention. In their experiment participants were either provided with Knowledge of Results (KR) that directs attention to performance outcomes (external focus of attention) after every 5th trial, or with Knowledge of Performance (KP) that directs attention to arm movement patterns (internal focus of attention), concurrently and on a fading schedule. The results showed that the motor improvements in stroke patients whom received KP reflect true recovery, in contrast with those who received the KR feedback. This suggests that if the goal of rehabilitation is true recovery, stroke patients may benefit more from KP feedback. However, in their review study, [14] found research that suggests that feedback inducing external attentional focus may be more effective to improve performance of task execution after a stroke. In the same study, [14] found that additional verbal KR is redundant when KR information is inherent to the task. When this is not the case, they found that summary or average feedback benefits motor learning of stroke patients.

For stroke rehabilitation, [1] believe that feedback concerning failure should be more conservative, and successful engagement should be rewarded and encouraged. Furthermore [13] argues that it is important to give feedback concerning motor control as this enhances learning, positively influences motivation, self-efficacy, and compliance. Correct performance feedback increases motivation while incorrect performance feedback facilitates learning.

As for feedback scheduling, [14] found that providing reduced feedback to stroke patients may enhance learning. Apart from this finding, little is known on how feedback scheduling influences learning in stroke patients.

## 2.3 Designing for Impaired Persons

Older people form the majority of stroke patients. In the U.S. nearly 75 % of stroke patients are over the age of 65. In fact after the age of 55 the chances of stroke doubles with every decade [5]. Given these statistics it is important to consider how older people interact with technology, as they generally experience a decline in sensory, cognitive, and motor functions that can interfere when interacting with technology [5, 8]. In addition, even though stroke is prominent in older people, it can occur at any age and its consequences can also induce cognitive- and motor impairments amongst younger stroke patients.

The study of [8] yielded several guidelines for designing websites for older people. When applying those guidelines for use in arm-hand rehabilitative technology the following guidelines need to be considered:

- Language should be simple, clear, and to the point. Important information should be highlighted. Irrelevant information causes too much distraction for users with cognitive impairments.
- Text design should be static, and presented in a readable format with high contrast. With age the color- and contrast sensitivity declines [5].
- Graphics should relevant and easy to understand.
- Navigation cues should clear and provide current location of the page.

These guidelines contribute to dealing with limitations in vision and cognition, by which stroke patients are often affected. Especially for cognition it is important that the interface is simple and intuitive, and to contain the proper affordances to reduce the workload of information processing [5].

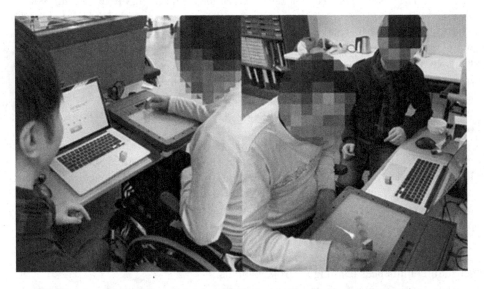

**Fig. 1.** A patient and therapist using TagTrainer in arm-hand rehabilitation therapy.

In their study, [5] also recommend providing the same, redundant information using different modalities in order to compensate for visual and auditory limitations. In addition to these limitations, during stroke rehabilitation redundancy is also necessary for cognitive limitations. It is not uncommon for therapists to repeat instructions to their patients and/or actively guide patients' attention towards important information.

To determine how to provide effective feedback to stroke patients, a feedback module system for stroke rehabilitation technology was designed and evaluated with stroke patients in a case study involving the TagTrainer stroke rehabilitation platform.

# 3    Case Study: Patient Feedback for TagTrainer

## 3.1    TagTrainer

TagTrainer [11] is a technology for arm-hand training in stroke rehabilitation. It allows patients to manipulate physical objects (e.g. lift, place, rotate) on one or more interactive tabletop surfaces– called 'TagTiles' – that are connected to a computer. TagTrainer allows therapists to use objects of daily life for rehabilitation training, since the Tag-Tiles are able to detect the presence, position and orientation of these objects as long as these are tagged appropriately with RFID tags. TagTrainer guides patients through an exercise by lighting up areas on the TagTile boards where objects have to be placed, moved or picked up from. Furthermore, the system provides both written and spoken instructions. Finally, TagTrainer collects quantitative performance data (e.g. speed of execution, number of repetitions) through the RFID sensors in the TagTile boards, as well as qualitative performance data (torso compensation, shoulder compensation, accuracy of object placement on the TagTile boards) through the RFID sensors in the board and accelerometers attached to the patient's torso and shoulders.

An example TagTrainer exercise would ask a patient to repeat a few times the following steps:

- Pick up a cup with her left hand and place it on aTagTile.
- Pick up the cup from a TagTile and raise it to the level of another TagTile, positioned at a 90-degree angle to the first TagTile.
- Touch the upper TagTile with the cup and put it back on the lower TagTile.

While rich performance data is collected by TagTrainer, none of this information is currently presented to patients. Therapists still have to manually guide patients through exercises and provide feedback about their performance (see Fig. 1). Given the importance of feedback for the recovery process of stroke patients, we set out to develop a patient feedback module for TagTrainer based on existing literature and user research.

## 3.2    User Research

In order to get a better idea of how rehabilitation sessions are set up, how patients are instructed and supported by therapists, how therapists determine appropriate feedback, and how feedback is currently provided to patients, we conducted two unstructured interviews with stroke rehabilitation therapists and observed an arm-hand training session at a stroke rehabilitation clinic (Adelante Centre of Expertise in Rehabilitation and Audiology, Hoensbroek, NL).

**Interviews with Therapists.**  Two unstructured interviews were performed with stroke rehabilitation therapists at the before mentioned rehabilitation clinic. The goal of the interviews was to get insight into how therapists set up training sessions for their patients, and how they guide and work with them during these sessions.

At Adelante, the rehabilitation process is strictly patient-centered. The therapist sits down with the patient and asks the patient about the problems (s)he encounters. Together with the patient, the therapist will determine the goals the patient wants to achieve. According to one therapist establishing goals helps to keep the patient focused. When the patient's goal cannot be achieved straight away, the therapist will divide it into smaller sub-goals that are easier to achieve.

During training sessions, therapists usually provide encouraging verbal feedback. Additionally, they may make use of mirrors, physical guidance or other materials if the situation requires it. Therapists do not apply a systematic approach in giving feedback because of the differences between individual patients. Feedback that works for one patient may not be suitable for another. Therefore, feedback is tailored to the patient by employing a trial-and-error approach. Finally the feedback given differs amongst the therapists and is based on their previous experiences (i.e. observed best practices).

One therapist indicated that during rehabilitation it should be clear to patients why they must invest in certain tasks, and that feedback should primarily concern the quality (speed, fluency, and trajectory) of movement. The other therapist stressed the importance of keeping patients motivated with feedback. Both therapists agreed that it is important that patients experience success and are able to achieve their goal. Therefore the therapists would sometimes relax on giving 'negative' feedback and give more encouraging feedback instead.

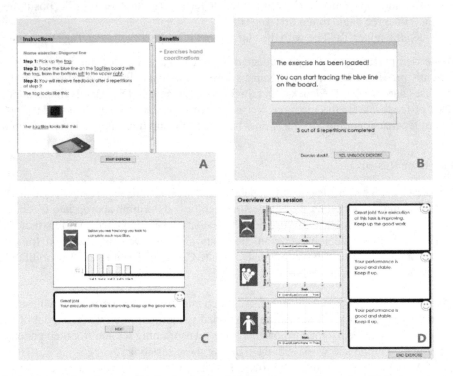

**Fig. 2.** Screenshots from the feedback module: A) before the start of a session, B) during exercise execution, C) after each block of 5 trials, D) at the end of the exercise.

**Observation of Arm-hand Training Session.** The first author embedded herself in the Adelante stroke rehabilitation clinic to observe a one-hour arm-hand training session. During the training session, one therapist attended to five patients. Three of them were practicing daily tasks independently and did not receive feedback from the therapist.

The other two patients practiced different tasks that were repetitive in nature. These patients received more attention from the therapist, who helped them in placing a harness around their affected hand. Once the patients started performing their exercises, the therapist predominantly gave positive verbal feedback on their performance. Other times, the therapist gave tips on how the patients could execute the task more easily.

At the end of the training session, the therapist requested the patients to rate their performance on a scale from 1 to 10. In addition to that, patients needed to make and write down a plan of which exercises they planned to practice with their affected arm over the coming weekend. The plan needed to be very specific on time, location and duration of the exercises. The overall goal was that they would use their affected arm at least one hour a day during the weekend.

**Design Implications.** From the user research, we distilled the following implications for the feedback module design:

- For patient involvement and motivation the goal and purpose of the exercise performed should be clear to the patient.
- Provided feedback should be primarily positive and tailored to the patient's needs.
- The design should allow for trial-and-error approach for giving feedback.

### 3.3 Designing Patient Feedback

Although TagTrainer supports stroke patients in their training by providing them with instructions for the execution of exercises, the current system does not offer them feedback on their performance. Therefore, we designed a patient feedback module for TagTrainer that would provide stroke patients with relevant feedback on their performance. Here we present the final design, and then discuss relevant experiences gathered during the design process and from evaluations.

**Method.** The feedback module was designed in a user-centered iterative process consisting of three consecutive design-implement-evaluate cycles. Initial design choices were based on suggestions from previous research involving non-disabled persons (e.g. [10, 14]) and older people in general (e.g. [5]), and the user research reported on earlier in this paper.

During the evaluation sessions, patients performed an exercise on the TagTrainer board. The exercise required the participants to trace a diagonal line 5 times with the affected arm using a small ($2 \times 2x2$ cm) wooden cube. While the exercise was performed, the feedback module was presented on a separate display in front of the user. The same exercise was used throughout the design process.

After executing the exercise, the participants were asked in an open interview questions concerning their understanding of the displayed information, which information felt to be missing or redundant, and the perceived value of the information presented.

**Participants.** The feedback software was evaluated with seven stroke patients undergoing general arm-hand rehabilitation at the Adelante Centre of Expertise in Rehabilitation and Audiology, Hoensbroek, NL. The age of the participants ranged between 50 and 83 years. The time since their stroke and the evaluation ranged between weeks to a couple of months. All participants were affected in their motor skills, mostly affecting their upper extremities and gate, and some participants were affected in their balance, memory and visual capabilities.

**Design.** The feedback module designed aims to provide relevant feedback throughout the patients' training session. A session consists of a movement activity that a patient needs to practice repeatedly for either a fixed number of times, or for a given duration. Sessions consist of individual trials: single units of meaningful movement that the patient needs to repeat during the session (e.g., the exercise described above).

The feedback module was designed to guide patients during the entire exercise session and thus consists of four main screens. These screens are shown (1) before the start of an exercise session (2) during the session (3) after every block of 5 trials, and 4) at the end of the session respectively (see Fig. 2).

The screen shown at the start of the session (Fig. 2-A) contains the exercise instructions and its benefits. The instructions are presented as a series of short sentences that are formulated in simple language and displayed in a readable font. Benefits are shown here because the therapists interviewed during the initial design phase suggested that it will motivate users to engage in the exercise.

While the patient is executing a block of 5 trials, the feedback module shows the patient instructions for the exercise and the patient's progress in completing the block (Fig. 2-B). The reduced feedback scheduling of 5 trials was chosen for providing feedback during the session, because [2] reported it to work well for stroke patients.

After each block of 5 trials the feedback module shows bar graphs with performance information about those 5 trials (Fig. 2-C). The bar graphs denote the duration of every trial, and shoulder- and torso-compensations performed during the trials. The time measure was included because therapists indicated that it will challenge and motivate patients who are doing well in their rehabilitation. However, for those who are not, the therapists fear it will demotivate patients to perform the exercise. Shoulder- and torso compensation measures were included as they are major factors in arm-hand rehabilitation. However, as is the case for the time measure, these measures do not always apply to all patients. The graph type denoting the performance measures was decided upon together with the therapists. The actual numbers of the performance measures are not shown, as it is the pattern of the results that gives the most important information, according to the therapists interviewed in the initial design phase. Textual KP-feedback is given for each performance measure that either tells the user to keep doing what he is doing, or how he can improve his performance. In telling the user how to improve his performance it only gives feedback concerning the desired outcome and not how the user should perform the movement to get to the desired outcome. This induces an external locus of control and should lead to enhanced learning [10, 14, 17].

Finally, after a patient completes the entire session (s)he is presented with an overview of their performance on execution time, shoulder- and torso-compensation throughout the session, as well as with appropriate KP-feedback (Fig. 2-D).

**Evaluation Results.** Three factors play an important role in transmitting feedback to stroke rehabilitation patients: feedback content, feedback scheduling and focus of attention [14]. Given that the focus of attention and scheduling of the feedback were not varied in our study, we will not further report on these. However, in addition to 'traditional' feedback that is provided by therapists directly, we have used several modalities for transmitting our feedback to the participants and hence also report on results concerning this factor.

*Feedback Content.* In the initial design of the screen shown before the start of the exercise, performance results of earlier sessions were included to show progress and hence increase patient motivation. However this information was found to be irrelevant and distracting by the users as they were only concerned with the task at hand: performing the exercise. Therefore, the information in this screen was limited to the exercise instructions.

Instructions are shown again on the second screen, in case the patient does not remember exactly the nature of the task. The instructions on this screen are presented in a condensed form, in an attempt to reduce the amount of information the patient needs to process while performing the exercise. However, condensing the information comes at the expense of less clarity of the instructions and as a result one patient did not know anymore how to perform the exercise.

After each block of five trials, the patients are presented with graphs visualizing their performance for those 5 trials. Initially the graphs were all shown in one screen. However, during the evaluations one participant indicated that he did not understand the feedback information on this screen, even after explanation. When the participant was prompted to comment on the individual components on the screen, including the bar chart, it turned out that the participant did actually understand the information, despite his initial claim of not being able to do so. The problem was one of information density, and it was decided to spread the information by giving each graph its own screen that the users can leaf through. This adjustment was included in the final prototype and five participants indicated to perceive less problems understanding the information on the screen, compared to previous versions.

*Feedback Modality.* During the evaluations we observed that the system at times failed in directing the users' attention properly. Two participants were reading the instructions on the start screen and tried to execute the exercise on the TagTrainer board before they pressed the button to start the exercise. More generally, most participants were confused about when to look at the screen, and when to look at the TagTile board. One participant mentioned that he would prefer having the feedback on the TagTile board instead of a separate computer screen. He explained that during training he was focused on the exercise task on the board and did not feel inclined to constantly look back at the computer screen for performance feedback.

For six participants it was not immediately clear what was represented with the feedback shown as bar graphs (see Fig. 2-C). Participants found the icons depicting the different types of feedback were not self-explanatory enough. In addition, participants reported that they felt the information shown should be related to their personal context. It should explain, for example, why certain feedback information is important for them to know and what it says about their performance. During the evaluation five participants explicitly mentioned a desire for information about their performance that is relevant to their current situation.

Despite the participants not fully understanding the graphs, the accompanying summary text (i.e. KP-feedback) was clear. The participants stressed the importance of the text containing information about what is good or not good about their performance. However, written information posed a problem for four participants, as they were unable to comprehend the written information due to poor eyesight and cognitive limitations. However, once the written information was vocalized, these participants were able to grasp its meaning.

## 4    Towards Guidelines for Patient Feedback

Based on experiences from the design process and the results of the evaluations, we identified the following set of design guidelines for feedback in interactive stroke rehabilitation technology:

- **Provide Multimodal Information:** Account for sensory impairments that stroke patients might have. E.g., vocalizing textual information for the visually impaired [5]. In addition, providing multimodal information enhances understanding and learning. E.g., during the evaluations vocalizing textual information made it easier for stroke patients to understand the information.
- **Provide Stepwise Guidance:** Providing a stepwise guidance through exercise instructions alleviates the workload on the memory as it provides information in more digestible bits. Furthermore, providing a stepwise guidance through the feedback will also improve a patient's understanding of the system.
- **Provide Context Related Information:** The feedback information presented should be relevant to the patient's situation and performance. If the information is not relevant, the patient will lose motivation.
- **Prevent Information Overflow:** It is easier for stroke patients to process information when it is provided to them in smaller bits. A means of doing this is by only showing information that is relevant for the task at hand and by using short sentences and simple language.
- **Allow for Customization:** There is a great variety in the disabilities that stroke patients may have, and thus their individual needs for feedback show an equal variation. Therefore, it is crucial that feedback systems allow for customization of feedback modality, scheduling and content. So even though a stepwise guidance, short sentences, and simple language are recommended in general, there are patients who want and can understand more complex and related additional information about their performance. By tailoring feedback to the needs of individual patients, their motivation and involvement in the exercise can be increased [12].

**Fig. 3.**  A patient training with the REHAP system. Photo courtesy of victor donker.

# 5  Applying Our Guidelines to the REHAP System

In order to test the practical applicability of the guidelines that we propose, we evaluated another patient feedback system for stroke technology along our guidelines. The system that we used for this evaluation is the REHAP system [3], a modular solution for balance rehabilitation after stroke (see Fig. 3).

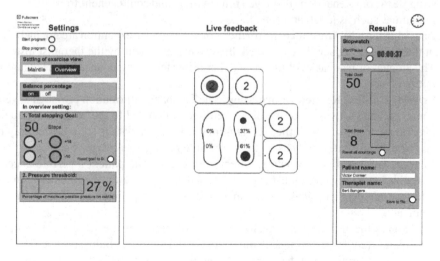

**Fig. 4.**  The REHAP patient feedback interface. Image courtesy of victor donker.

## 5.1 The REHAP System

The REHAP system consists of a set of pressure sensitive tiles, connected to a PC or laptop running a graphical user interface. Several smaller tiles can be connected to a main tile in various configurations, allowing for flexibility in the design of exercises. REHAP allows patients to train their lower extremities, and has mainly been designed for step and balance training (Fig. 4).

REHAP's patient interface is entirely visual (GUI) and is divided in three sections:

1. A section for setting training properties and feedback properties.
2. A central section that displays performance information.
3. A section displaying progress information.

The basic idea behind the patient feedback provided by REHAP is that patients get a visualization of the 'raw' input they provide through putting pressure on the system's tiles. The visualization consist of a virtual version of the tiles that are connected to the system, and each tile displays one or more 'touch points', representing the touch sensitive sensors within the tiles. The touch points are visualized as discs that grow in size with increased pressure.

## 5.2 Applying Our Guidelines to REHAP

- In order to test the practical applicability of our guidelines, we used them to evaluate the REHAP patient feedback interface.
- **Provide Multimodal Feedback.** The patient feedback interface of REHAP is entirely visually oriented. Both information on current performance, as well as longer term information on progress are displayed visually. Visually impaired patients might struggle to comprehend the displayed information, and could benefit from alternative modalities such as touch and sound.
- **Provide Stepwise Guidance.** Instructions on the exercise execution are not included in the patient interface, but provided directly to the patient by the therapist instead. The amount of feedback is limited, and therefore stepwise guidance in this regard is not required.
- **Provide Context Related Information.** The feedback information provided is very context related – the size of the discs are directly coupled to the pressure a patient puts on the force-sensitive tiles. Long term performance data is stored externally in an Excel file for later analysis, but is not visible during exercise execution.
- **Prevent Information Overflow.** Direct feedback on the patient's performance is visualized in a very minimalist and direct manner. The one-to-one link between the interface and the physical system makes it easy for patients to understand the feedback.

    Though the interface is organized to prominently feature the most important information (actual performance), other parts of the interface could have been hidden. For example, options for feedback settings do not necessarily need to be visible during exercise execution. Instead, these options could be hidden once the patient commences exercise execution.

- **Allow for Customization.** The REHAP interface provides multiple options for customization. Depending on the goal of the exercise, the interface can be modified to display just the main REHAB tile, or all connected tiles. Further, additional performance information (a percentage indicating the balance of a patient) can be switched on or off, depending on the cognitive capabilities and training goals of a particular patient. Finally, a slider can be used to set a threshold value that needs to be passed by the pressure sensors before the system registers an action.

## 5.3 Conclusion

This small case study has shown how our guidelines can be used to evaluate existing patient feedback systems for rehabilitation technology. Our findings largely correspond with the findings t1hat are reported from the studies that the creators of REHAP have performed themselves (see [3], and confirmed by their first author in person): In general, the REHAP patient feedback interface seems to be well-designed, but providing multimodal information could add to its usability for a wider group of patients.

# 6 Discussion

In this paper we have presented the design and evaluation of a patient feedback module for the TagTrainer rehabilitation technology. Although our evaluations have been performed with a limited number of participants, we believe that the experiences from our design process and the findings from our evaluations provide useful pointers to developers of interactive technologies for stroke rehabilitation. The guidelines that we have derived from our experiences, and from previous literature, seem to work well for evaluating the design of patient feedback interfaces. Still, future work could investigate the use of the guidelines as input for the *design* of new feedback systems for stroke rehabilitation technology.

Although we believe that our guidelines contain useful pointers for designers of interactive stroke rehabilitation technology, we realize that the list of guidelines is probably not complete. That is, these designers should also take into account guidelines that have been specified for other, related target groups such as older people (e.g., [5]). On the other hand, it is also likely that our guidelines are applicable beyond the domain of stroke rehabilitation. Many of the guidelines that we propose hold for physical rehabilitation in general (e.g., SCI, MS), and for feedback to users with cognitive limitations in general. Beyond their application as a tool for the design and evaluation of patient feedback systems for stroke rehabilitation technology, our guidelines provide a common vocabulary for researchers to discuss the design of such systems and allows them to draw meaningful comparisons.

The system that we have presented can be used to further explore and research the effects of feedback on stroke patients. The setup can also be used in practice, as it allows therapists to employ their trial-and-error approach in finding out what works for their patient.

# 7    Conclusion

Providing feedback to stroke patients about their performance in therapy is crucial to their recovery process. However, due to cognitive damage sustained by stroke, stroke patients' information processing is often impaired and their retention limited. The contributions of this paper are the design of a feedback module for TagTrainer, an interactive stroke rehabilitation technology, and a set of design guidelines for interactive stroke rehabilitation technology that are based on experiences we had, and evaluations we performed during our design process. Finally, we have shown how these guidelines can be applied in the evaluation of patient feedback systems for stroke rehabilitation technology.

The design guidelines that we propose, address the specific needs and account for the cognitive limitations that stroke patients might have. We invite the community to extend and validate these guidelines through their application in the design and evaluation of stroke rehabilitation technology, in order to improve the quality of technology supported stroke rehabilitation, and eventually the quality of the lives of people who are affected by stroke.

**Acknowledgements.**    We acknowledge the support of the Innovation-Oriented Research Programme 'Integral Product Creation and Realization (IOP IPCR)' of the Netherlands Ministry of Economic Affairs. In addition, we would like to thank Adelante Centre of Expertise in Rehabilitation and Audiology, and in particular the therapists and patients who participated in the design and evaluation of the feedback module.

# References

1. Burke, J.W., et al.: Optimising engagement for stroke rehabilitation using serious games. Vis. Comput. **25**(12), 1085–1099 (2009)
2. Cirstea, M.C., Levin, M.F.: Improvement of arm movement patterns and endpoint control depends on type of feedback during practice in stroke survivors. Neurorehabil. Neural Repair. **21**(5), 398–411 (2007)
3. Donker, V. et al.: REHAP Balance Tiles: a modular system supporting balance rehabilitation. In: 2015 Proceedings of Pervasive Health. IEEE (2015)
4. Hochstenbach-Waelen, A., Seelen, H.A.: Embracing change: practical and theoretical considerations for successful implementation of technology assisting upper limb training in stroke. J Neuroeng Rehabil. **9**, 52 (2012)
5. Ijsselsteijn, W. et al.: Digital game design for elderly users. In: Proceedings of the 2007 Conference on Future Play, pp. 17–22. ACM, New York, USA (2007)
6. Krakauer, J.W.: Motor learning: its relevance to stroke recovery and neurorehabilitation. Curr. Opin. Neurol. **19**(1), 84–90 (2006)
7. Krebs, H.I., et al.: Robot-aided neuro-rehabilitation. IEEE Trans. Rehabil. Eng. Publ. IEEE Eng. Med. Biol. Soc. **6**(1), 75–87 (1998)
8. Kurniawan, S., Zaphiris, P.: Research-derived web design guidelines for older people. In: Proceedings of the 7th International ACM SIGACCESS Conference on Computers and Accessibility, pp. 129–135. ACM, New York, USA (2005)

9. Lövquist, E., Dreifaldt, U.: The design of a haptic exercise for post-stroke arm rehabilitation (2006)
10. Shea, C.H., Wulf, G.: Enhancing motor learning through external-focus instructions and feedback. Hum. Mov. Sci. **18**(4), 553–571 (1999)
11. Tetteroo, D.: TagTrainer: A Meta-design Approach to Interactive Rehabilitation Technology. In: Dittrich, Y., Burnett, M., Mørch, A., Redmiles, D. (eds.) IS-EUD 2013. LNCS, vol. 7897, pp. 289–292. Springer, Heidelberg (2013)
12. Timmermans, A., et al.: Technology-assisted training of arm-hand skills in stroke: concepts on reacquisition of motor control and therapist guidelines for rehabilitation technology design. J. Neuroeng. Rehabil. **6**(1), 1 (2009)
13. Timmermans, A.A.A.: Technology-Supported Training of Arm-Hand Skills in Stroke. Eindhoven University of Technology (2010)
14. Van Vliet, P.M., Wulf, G.: Extrinsic feedback for motor learning after stroke: what is the evidence? Disabil. Rehabil. **28**(13–14), 831–840 (2006)
15. Winstein, C.J.: Knowledge of results and motor learning–implications for physical therapy. Phys. Ther. **71**(2), 140–149 (1991)
16. Wulf, G., et al.: Motor skill learning and performance: a review of influential factors. Med. Educ. **44**(1), 75–84 (2010)
17. Wulf, G., Lewthwaite, R.: Effortless Motor Learning? An External Focus of Attention Enhances Movement Effectiveness and Efficiency. Effortless Attention: A New Perspective in Attention and Action. MIT Press, Cambridge (2010)

# A Method and Tool for Strategic Hospital Planning

Dominique Brodbeck[1]([✉]), Markus Degen[1], Andreas Walter[2], Serge Reichlin[3], and Christoph Napierala[4]

[1] University of Applied Sciences and Arts Northwestern Switzerland,
Muttenz, Switzerland
`dominique.brodbeck@fhnw.ch`
[2] Inselspital, Bern University Hospital, Bern, Switzerland
[3] University Hospital Basel, Basel, Switzerland
[4] University of Lucerne and Siemens Schweiz AG, Lucerne, Switzerland

**Abstract.** We developed a visualization tool and a methodology to support strategic planning of hospital service portfolios. Hospitals in Switzerland are reimbursed with a fixed fee per case. The fixed-fee model makes medical services comparable from a financial point of view. In order to take advantage of this model, the data that characterizes the medical services must be operationalized. The method that we developed, centers around a visual metaphor that provides the basis for strategic thinking. It is complemented by a visualization tool that allows visualization, analysis, and modification of service portfolios. Special features enable the tool to be used during live planning sessions. We describe the method, the tool, and its application in strategy workshops for infrastructure planning, reorganization, and resource optimization decisions.

**Keywords:** Strategic hospital planning · DRG · Visual analytics

## 1 Introduction

The Swiss healthcare system is continuously undergoing change since the implementation of the new Swiss health insurance law in 1996. Most significantly, in 2012 inpatient financing was changed from a system based on cost per case to a system based on a fixed fee per case. Such a significant change has a far-reaching impact on the overall system, especially on the hospital sector. Hospital management, but also national and regional policy makers, are forced to plan and manage in different dimensions than before.

The idea behind a fixed-fee model is that hospital cases can be classified into groups of similar cases, and that these groups can then be treated like products that are comparable, and that are reimbursed with the same fixed amount of money, because it is assumed that they have the same cost structure. The classification rules are based on diagnoses, medical procedures, demographic patient information, and other case-specific data. The combination of all these

© Springer International Publishing Switzerland 2015
A. Fred et al. (Eds): BIOSTEC 2015, CCIS 574, pp. 372–389, 2015.
DOI: 10.1007/978-3-319-27707-3_23

groups, or products, makes up the medical service portfolio of a hospital. This comparability of medical intervention from a financial point of view is new and opens a variety of new possibilities. In particular, it allows policy makers and hospital management to make decisions based on factual information from the analysis of current service portfolios, as well as the simulation of medical service portfolios into the future. Such flexible analysis possibilities are crucial to make sound infrastructure decisions that will meet patient needs in the next years.

Characterizing and comparing service portfolios in such a way quickly produces large amounts of high-dimensional data. The portfolios need to be visualized, compared, interpreted, and modified by analysts, as well as by groups of managers in live workshop settings. Common spreadsheet programs are too general to cover these tasks well. In order to take full advantage of the possibilities offered by the new reimbursement system, there is a need for tools that allow the various stakeholders to analyze and communicate the data in a flexible and efficient way.

This paper describes a method and a corresponding visualization tool that supports management levels in discussing strategic decisions and future developments based on current restrictions and conditions.

## 2   Background

Diagnosis-related groups (DRGs) is a patient classification system that links similar types of cases that a hospital treats, to the resources that the hospital uses for the treatment. DRG systems were pioneered in the USA, but are now in use in many other countries [1].

The SwissDRG system is a variation of German DRG but refers to Swiss reference values [2]. Like in all DRG systems, single cases are classified into a specific class according to their diagnoses (using ICD10), treatment types (e.g. hip surgery), and other case attributes. Each group has a unique code, which enables a precise attribution of a case's revenues to its costs. This results in a shift from a more cost-oriented daily allowance to a product-oriented view, where optimal management along the patient pathway gains in relevance (e.g. in [3,4]). This shift opens the potential for providing a basis for evidence-based decisions in an operative and above all in a strategic perspective.

The dimensions of interest for strategic analysis are: average length of stay, cost weights, and the overall base rate. The average length of stay (ALOS) is used in a DRG system as an output measurement for complexity of the case [5]. The other dimension is the cost weight, a relative indicator for the severity of the case, which serves as a measure for resource intensity. Cost weight is recalculated every year based on cost data provided by hospitals to the Case Mix Office (i.e. SwissDRG in Switzerland). Each DRG is assigned specific values for these two indicators, and all the values for all the DRGs are stored in a reference catalog.

The base rate (which is basically determined by negotiation between the hospital and insurance companies) is multiplied with the cost weight attribute to determine the monetary value of a case that is classified into a particular

DRG. Besides revenues from private insurance and specialized pharmaceuticals or technologies for special treatments (e.g. specialized oncological treatments) these revenues represent the main income that a hospital can generate from their patients.

As theoretically the cases behind any DRG code should on average represent similar cases in each hospital, these elements make hospitals comparable and allow a benchmark-oriented approach. Comparisons can be achieved by using one hospital's cases, and then comparing them either with the reference values of the DRG catalog, or to the portfolio of a peer hospital, both at the service and at the cost level. Most importantly, this allows comparing the efficiency of hospitals, but it will also enable quality or other assessments.

The services that a hospital offers are influenced by many factors. In most administrative districts (i.e. in our Swiss case corresponding to cantons), service portfolios or service requirements are defined by policy makers. However not all districts decide to provide a full scope of medical services, but might delegate the remaining activities to other providers that can be situated beyond their control or area of direct responsibility. This is complicated by the fact that with the new reimbursement system, patients also gained the freedom to choose their hospital of treatment independently of their canton of residency.

On top of that, there are a number of national policies that impose further restrictions, e.g., the highly specialized medicine act that restricts the number of centers for very complicated treatments, or the changing outpatient health care provision that still heavily depends on resident physicians, who are however overaged and struggle to find replacing practice holders. All these factors push hospitals towards stratifying patient portfolios and focusing on selected medical areas, in order to improve their economic situation actively.

Modern hospitals must engage in strategic discussions about visions, cooperations, specialization, centralization and further infrastructure or organizationally relevant questions. These decisions affect their market position and help them to cope with growing and dynamic competitors both at administrative district or regional levels. For example, hospitals that today provide maximal service levels, will have to decide whether they will focus on more specific therapies that generate high cost weights [6], or whether they will continue offering a broad range of health care services, focusing on a me-too strategy. The analysis of pancreas and esophagus carcinoma surgery in Germany as another example has shown that it can be highly beneficial for hospitals to treat a high number of benign cases instead of focusing on the malign cases with high cost weights but bad average length of stay management. This is the case even if the minimal numbers are reached that would be required by the official rules and regulations.

Ultimately, Swiss hospitals will have to adapt their current management models. In order to take the right decisions in this complex environment, hospital managers need to rely on data, and models that are based on that data. Collecting the data is usually not a problem anymore, as modern hospital information systems are well equipped for that job and are commonly in use today. Projecting the data into the strategic models and operationalizing it however, requires new approaches.

# 3    Methods

## 3.1    Strategic Model

As outlined above, the key dimensions of a hospital case are its cost weight - as determined by the DRG into which the case was classified - and the length of stay. Since the analysis of individual cases is too low-level for the kind of strategic questions that need to be supported, the cases are aggregated into groups according to their DRG code. These groups can be considered as the services that a hospital performs.

Working with inpatient service portfolios from various hospitals of various sizes, types, and different geographic areas has shown that plotting the normalized average of the cases' cost weights (also called case mix index CMI) of such a service, versus the deviation of the average of the lengths of stay (also called ALOS) from the DRG catalog reference (CHALOS), produces a graph that is easy to interpret in the context of strategic questions (Fig. 1). Each service is plotted as a bubble with the size of the bubble proportional to the number of cases, and the color mapped to any of the other available service attributes (e.g. Major Diagnostic Category MDC, profit, cost, department) depending on the focus of the analysis.

The plot can be separated into four quadrants that each have a distinct strategic meaning, similar to the BCG growth-share matrix used in a strategy or marketing context [7]. This allows to analyze the strategic positioning of a hospital's inpatient portfolio using the four quadrants (A-D). The quadrants have the following interpretation:

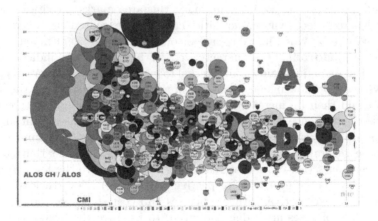

**Fig. 1.** The visual metaphor of the strategic model: a bubbleplot that plots the case mix index (CMI) of a medical service vs. how well the length of stay compares to the country average (ALSO CH / ALOS). This creates four quadrants A, B, C, and D, with different strategic interpretations. The size of the bubble is proportional to the case count for this service. The color shows Major Diagnostic Category MDC.

**Quadrant A** represents the area where the grouped cases generate on average a high cost weight, and result in an average length of stay that is better (i.e. shorter) than the DRG catalog value. We therefore expect the bubbles in this quadrant to be profitable and to represent the services where the hospitals portfolio performs better than the benchmark. Strategically, these activities support the hospital in creating a brand, where the hospital can distinguish itself for quality and performance, and thus should aim at increasing case numbers and building relationships to referring medical doctors. Further options could be to create a good infrastructure that would attract above-standard surgeons, and centralize the activities of surrounding hospitals in this area. In general, these kind of services should be expanded and require focus.

In **quadrant B** the case mix indices (CMIs) are lower than the average of the hospital, but the average length of stay in the hospital is better than the catalog value. The cases in this part of the grid require detailed analysis of costs. The right infrastructure is of special importance because economic margins are tight. Strategic decisions like outsourcing or PPP (private public partnerships) models should be considered. In general, this area is expected to be at least cost covering and the portfolio overall should be kept constant.

At the bottom left, **quadrant C** represents cases where ALOS ratio is below 100 %, i.e. worse than the benchmark and where a low CMI is generated. Because the CMI is low, these activities are potentially loss generating. The hospital has to thoroughly investigate its patient pathways and focus on workflow management, trying to cooperate closely with rehabilitation or care management. Furthermore, direct measures to lower costs need to be taken through implementing for example intermediate care units (IMC) or similar. Additionally, cooperations with resident medical doctors and other hospitals that could take over those cases, need to be investigated. In summary, this area marks activities where the hospital has to reflect, why its activities cannot be executed cost-effectively, or if they could be provided in a day-care or outpatient management setting.

The last **quadrant D** is characterized by high case mix indices and low ALOS ratios. As in quadrant C, an internal focus needs to concentrate on patient pathways and workflow management. From a long term perspective, the number of cases could then be increased. This can be realized through the creation of specialized competence centers, with the aim of attracting cases from surrounding hospitals and referrals from resident physicians. Interesting improvements can also be generated from applying a sound case management that addresses high-outlier issues (i.e. cases that remain in hospital above the high trim point length of stay according to the respective DRG code). Nevertheless this area has a dualistic perspective. Either processes and cost structures are optimized, or the number of cases are reduced in order to improve the overall economical situation.

While the above description is necessarily rough and exemplary, it clearly indicates the added value of such a graphical representation. It serves as a map of reference for discussions without the need to refer to quantitative tables with many dimensions.

## 3.2    Applied Strategy

In this section, we expand the argument towards how the strategic model described above fits into a strategy development framework from a practical perspective.

Strategy has become an important dimension also in public administrations [8], but public hospitals are oftentimes still very much driven by public management behaviors. Classic strategic analysis has two main dimensions: an internal and an external one. Both perspectives apply different methods on which we will not expand here as there is abundant literature available on the subject. For the external dimension, Michael E. Porters book Competitive Strategy: Techniques for Analysing Industries and Competitors [9] from 1980 is still the reference, and for the internal dimension, SWOT analyses are often used.

One strength of our chosen approach is that it combines internal and external analysis through its interpretative graphic space. In other methods this is typically done in two steps. Even though we mainly describe a hospital's internal portfolio, there are also external consequences. For instance, a strong portfolio in quadrant A could mean that from a business segment point of view, the hospital has a strong external position on the market, that it wants to further strengthen. This rather crude example hints at the double benefit of our analysis in that it provides a holistic approach right from the beginning. This unique feature provides a substantiated discussion base, and clarifies strategic challenges or advantages at an early stage of the process.

In the following, we describe the strategic process with the concrete example of a regional hospital group in Switzerland.

The visualization in Fig. 2 clearly highlights differences between the same services within one hospital group. In this specific case we compare orthopedic services (e.g. knee replacements) in a baseline and a reference hospital. One can easily see that process quality largely differs between the two sites. Strategically

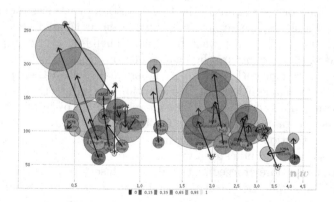

**Fig. 2.** A comparison of orthopedic services (e.g. knee replacements) of a baseline and a reference hospital (transparent bubbles). The color shows the proportion of private patients (darker color equals more private patients) (Color figure online).

this observation allows for an improved evaluation of strategic measures by comparing it to the competitor positioning. The toolset allows for a straightforward analysis between hospital units or between competitors, indicated by the arrows between the same service groups.

Nevertheless one has to distinguish both perspectives as the latter one is a consecutive or clearly distinctive step in the day to day controlling process of a healthcare institution. But both an internal or external benchmark oriented approach largely enhance strategic analysis and add purpose and weight to the discussions. It has also the advantage that it provides the possibility to reposition a portfolio by means of the toolset, within a scenario that replicates competitor or internal peer performance. This in itself allows for a progress in the current organization, as results are not solely discussed looking at them internally. This can hint at process changes or improvements without needing further analysis.

In addition, coloring allows to assess a fourth dimension. In this case we represent the proportion of private patients in our current facility. We can therefore also derive the revenue strength of each group and derive its future potential. Herein we can also get a glimpse of the importance of combining internal and external analysis in one overview as this can also hint at a the exposure towards health policy change or changes within the health financing system. Again one has to distinguish the operational level from the strategy development we are aiming at.

The example above describes a Swiss regional hospital group that wants to focus and create clinical competence areas (i.e. light towers) and thus improve their strategic setting. The aim of the analytic work is to act economically more sensitive and improve delivered quality within all service centers (i.e. single hospitals or departments). One of the secondary goals of the exercise was to optimize patient groups per site, i.e. having the right patients in the appropriate infrastructure. Which for example means that some of the patients that stay in an acute treatment unit should rather be lying in rehabilitation centers where the opportune treatments would be applied.

One of the major restrictions in the free definition of services though remain in the regional policies, where hospitals have been attributed certain activity areas that they can or do not want to refuse. Therefore every hospital keeps a certain base service structure, but can focus on areas of activity that they either deliver more efficiently or with a higher quality than their peers within the hospital group. Nonetheless a strategic option can still be to focus on certain clinical segments. This could for instance mean to get rid of activities where one can clearly see substantial differences in the benchmark visual analysis. This can be where the institution differs in comparable services, or where the current low number of cases already indicates a weakness.

Naturally, and before proceeding to reduce such services, one has to integrate demographic and epidemiological developments. These can be built into the portfolios either prior to the analysis by creating simulated portfolios, or by adapting the services directly during the analysis by using a sensitivity analysis.

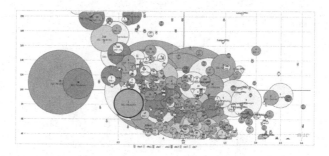

**Fig. 3.** Service portfolio of a hospital group (colored by site). Services are shown aggregated into the major diagnostic categories (MDC). Each bubble shows the number of cases (size of bubble) of an MDC (Color figure online).

Figure 3 shows the service portfolio of the hospital group that is composed of several sites. The services are shown aggregated into the major diagnostic categories (MDC) to avoid too much clutter by showing each DRG individually. Each bubble shows the number of cases (represented by the bubbles size) of an MDC, and the color represents the different sites.

The portfolio shows the typical characteristics of a general hospital in Switzerland that provides care for about 40'000 cases per year. The biggest bubbles are not surprisingly birth and the respective obstetrics services. Of more interest is the area in which the groups focus lays, in the Northwest sector (quadrant B). This indicates that the hospital requires a narrow cost management and a clear focus on processes. Above that it also indicates a high potential for outpatient care, as the case weights are rather low and the ALOS ratio (i.e. the relation of the effective length-of-stay and the reference value, also be called LOS compliance ratio) favorable.

A medically diverse set of cases lies below the ALOS ratio 100 % threshold (e.g. quadrant D - i.e. Southeast). Most probably, adequate case management from a process perspective would be an important driver to improve the situation. But also putting cases in the right diagnosis related group (DRG) or category would influence performance undoubtedly. An example is trying to dismiss elderly patients from more intensive care structures towards a rehabilitative hospital or department.

One can see that the short assessment of the current situation above, reveals an obvious need to further dissect the portfolio in order to get conclusive facts for decision making. Therefore the overall hospital portfolio has been made available according to their sites and their medical focus (MDC grouping). This allows for an improved in depth analysis for the decision makers, as the overall DRG view might be blurred by the fact that in current structures some DRGs appear in more than one hospital department and thus render a direct comparison difficult. Instead of MDCs one could also define subgroups of Episodes of Care (EoC), which will make the overview of a hospitals portfolio clinically oriented. However the basis for such regroupement is not necessarily the DRG reference but rather

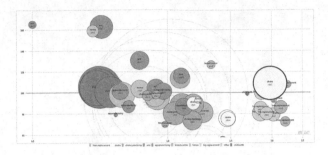

**Fig. 4.** The regrouped portfolio showing clinical specialities in the gastroenterology area (distinguished by color) by hospital site (bubbles of same color). Large differences between sites suggest e.g. concentration of activities into the best performing site, or cooperative approaches (Color figure online).

purely diagnostic or treatment driven. One can see that this view enhances the overview of clinical capabilities, and in our example, supports the concentration of certain activities in one site, thus shifting capabilities and infrastructure.

In the next step of the analysis, some simplifying steps were applied to the available DRG and cost data. First, DRG were clustered into groups according to service groups or episodes of care. And in a second step, the data was then made available per hospital site (per location) and per specialty (by EoC and partition, the latter indicating surgery vs. medical treatments).

Figure 4 highlights one part of the regrouped portfolio. The example only focuses on gastroenterology as an exemplary medical area. This permits to distinguish well between interventional and conservative measures that are decisive for any infrastructure, as well as patient pathways and process quality. Figure 4 shows one bubble per EoC, e.g. acute myocardial infarction (AMI) or hernia cases.

The data is cut by site and the visual analysis makes clear that the differences by hospital site (different hospitals of the group) are considerable. Looking at the Northwest (quadrant B) we can see that the performance is different for the AMI cases, especially regarding the ALOS ratio. This is an indication that processes are managed differently for the same type of cases by hospital site. AMI cases largely depend on competent case management and gold standard operating procedures. Seemingly this is not the case in all the sites.

The effect of such differences is not solely an economic one, but one can also distinguish clinical quality differences. From a purely economic perspective those cases lead to even less revenue, because the generated case weight for the comparable cases differs per site. Clinically those cases risk to rebounce to the hospital, not generating further revenue.

Summarizing the findings will lead the management to consider actions that would unify portfolios, for example into the best performing site or at least applying the same ways of working at all sites, with the expectancy that the ALOS ratio would align itself around the best performing site. On the other side, a cooperative approach could help distribute the case weight more appropriately.

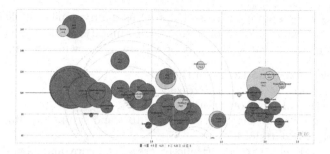

**Fig. 5.** The economic perspective of the portfolio in Fig. 4. Episode of care groups that generate benefits are colored in green, groups that generate losses in red (e.g. knee replacements or strokes) (Color figure online).

In Fig. 5 the corresponding economic perspective is shown to support the above statements. Looking at the same group of cases in quadrant D confirms the assumptions made above. We can clearly see episode of care groups that generate benefits (colored in green) above the threshold, and others below that dont (e.g. knee replacements or strokes).

It can be seen that the bubble which has the worst ALOS ratio also generates the lowest economic result (e.g. stroke site 4 or knee replacement site 6). Therefore it would be commendable to search for co-operations or try to consolidate these cases in one hospital. This would allow to reach critical mass and enable an improved overall treatment path by applying best practice and improving care models. It should at least be verified that referrals by resident physicians are made into the right site, and that on the long run there will be a focus within the hospital group.

An interesting part of the workshop was performed in the next step, in which several scenarios were investigated. The DRGee Tool supports this through a special screen (further described in the later section about the tool support) in which adaptations of the length of stay and of the number of cases can be simulated interactively, and are protocoled step-by-step. The results can then be shown and compared at any point of the simulation. Such a possibility enhances the workshops greatly and makes the scenario building straightforward. Mechanisms of strategic importance become instantly visible and understood. It also permits to proceed to a sensitivity analysis. This is done by indicating for example scenarios that display best or worst cases.

In the concrete case described in this section, we were able to confirm that the chosen scenario was a realistic one, and the hospital decided to adapt it accordingly.

## 3.3   Tool Support

In order to make it possible to apply this strategic model in practice, we developed an interactive visual tool called DRGee (as in "(DR)Gee! Look at this!")

that supports the planning process. The tool uses a dynamic bubble chart, similar to [10], but with a focus on dynamically modifying the underlying data, instead of temporal trend analysis. It provides the following core functionality:

- Load a collection of portfolios (a theme)
- Visualize a portfolio in a standardized way
- Allow selection of parts of a portfolio
- Calculate and display characteristic indicators and summary values of a portfolio or selected parts of it
- Allow the modification of a portfolio by creating a copy and editing individual services to play through what-if scenarios
- Provide the possibility to compare the differences between two or more portfolios

Figure 6 shows the main interface of the tool. The plot that was introduced with the strategic model is featured prominently in the center of the interface. The case-mix index is plotted against the ratio of catalog length-of-stay and actual length-of-stay. The bubbles represent medical services that a hospital performs, and the size of the bubbles is proportional to the number of cases for the service.

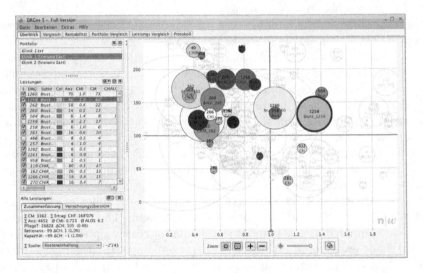

**Fig. 6.** The DRGee tool. The 4-quadrant plot is at the center stage. Services (bubbles) can be directly resized and repositioned with the mouse to simulate what-if scenarios. Controls along the left are for selecting portfolios, services within portfolios, and for displaying summary information. Tabs along the top are plugins that extend the basic functionality, mainly for comparing the various scenarios and documenting the analysis process.

Along the left side we find from top to bottom:

- The list of portfolios that is loaded (simple selection of a portfolio in the list displays it and makes it active, portfolios can be cloned here in order to modify them)

– A table of all the services defined in the selected portfolios, showing any
  number of attributes that characterize each service (services can be sorted,
  selected, and colored here to make them active and visible, deselected services
  are shown as ghosts in the background to preserve overall context)
– A display of characteristic indicators and summary values of a portfolio (e.g.,
  total number of cases, revenue, average case-mix index, average length of stay,
  etc.). There is a choice between the whole portfolio or just the selected subset.

All the necessary controls are contained in these three interface elements. The
menus in the menu bar are only used for high-level configuration of the tool
(e.g., base rate, currency, etc.) and infrequently used functions (e.g., data export,
printing, etc.). This was a deliberate design decision in order to ensure the dis-
coverability of the main functionality without having to resort to user manuals.

The software was implemented in Java for platform independence, and uses
only a handful of third party libraries, mainly for the look&feel, logging, and
PDF generation.

The architecture provides a dynamic extension mechanism, where modules
(plugins) can be compiled into self-contained (Java) jar files that are loaded
automatically and independently at starttime. Every plugin loaded this way is
placed into a separate tab in the main window of the application, as visible
along the top of the plot in Fig. 6. Encapsulating different features into plugins
allows to incrementally increase the functionality of DRGee without having to
touch the overall architecture and core functionality of the system. In addition,
it supports the easy packaging and delivery of different versions of DRGee with
tailored functionality.

### 3.4   Working with Scenarios

One of the key features of the tool is the possibility to modify existing portfolios
in order to play through what-if scenarios, and simulate how the characteristic values of a portfolio change. Each service (bubble) has two dimensions that
can be changed: length-of-stay compliance (y-axis) and case count (size of bubble). The case mix indices (x-axis) can not be modified, as they are fixed and
predefined by the DRG reference catalog.

To modify a bubble, it can simply be manipulated with the mouse pointer
(drag to new position, drag radius to new size). The values in the table are
adjusted accordingly. Direct manipulation techniques have the advantage that
they are intuitive and efficient if high precision is not required [11].

Comparing the newly created scenario with the original portfolio is challenging. There are three different approaches for visual comparison [12]: juxtaposition (showing objects side-by-side), superposition (showing objects overlaid in
the same space), and difference (showing the difference between the objects).

The DRGee tool supports all three approaches, each implemented as a separate plugin. Figure 7 shows two portfolios juxtaposed. This view works mainly
well for detecting a few strong outliers. Superposition was implemented by using
a small multiple [13] representation of the plot per bubble that changed. Each

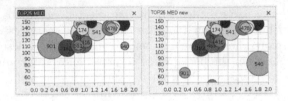

**Fig. 7.** Comparing portfolios by juxtaposition. Significant differences and overall trends are perceived quickly (e.g., 901 moved and shrank, 540 moved and grew).

small representation then only shows the changes to that particular bubble superimposed in the same plot (Fig. 8). The third approach represents the differences between two plots explicitly as arrows between the original and the new position (Fig. 9). This technique allows not only to identify changes in the portfolio but also shows trends (e.g., "DRGs in quadrant C tend to be smaller", "DRGs mostly move to quadrant A", etc.)

### 3.5    Workshop Use

A special requirement for the tool was that it had to support use in a live workshop setting. As a consequence the following additional aspects had to be taken into consideration:

**Animated Transitions.** One of the main differences between operating a tool oneself and watching someone else perform the operation is that not all the intentions and actions of the operator are visible, but very often only the results of the actions. Even with close attention, changes can be missed by the audience,

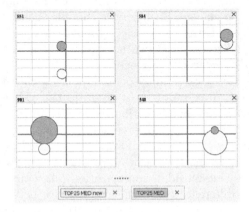

**Fig. 8.** Comparing portfolios by superposition. Individual differences of only the modified objects are shown as small multiples. This provides a complete and precise representation of all the changes, at the disadvantage of losing the overall context.

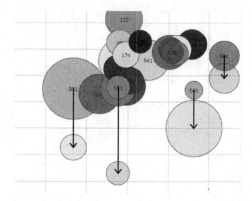

**Fig. 9.** Showing the differences between two portfolios explicitly. Changes in position are marked by arrows. The size of the new bubbles is shown transparently so as to not hide the original configuration in the background.

a phenomenon known as change blindness [14], leading to mental disconnect and discomfort. In order to address this problem, all the transitions in the tool (zooms, pans, size and position changes of bubbles, change of portfolios, etc.) are smoothly animated or designed as drag and drop interactions, leading the viewers to the next state of the visualization in a continuous way.

**Simple Interface.** Operating a tool live in front of an audience puts a certain pressure on the operator. It is therefore essential that the user interface is robust and the functionality reduced to a set of minimal yet powerful enough features. All the controls need to be visible so that the audience can perceive the series of actions performed by the operator.

The controls for navigating the bubble plot for instance are not implemented as continuous sliders, but are constrained to just a few discrete buttons (see Fig. 2 bottom, below the plot). The first button zooms to a fixed standardized view, and as such serves as the "home" button. The second button zooms to the full range that shows the complete portfolio, which provides the overview. From these two well-defined positions, in-between views can be obtained by using the "+" and "-" buttons that zoom the view by 20 % in the respective direction.

**Documentation of the Decision Process.** The strategic development of a hospital portfolio is an iterative process, and it is often necessary to go a step back and try another path. Traditional tools only provide linear undo/redo stacks, and only the final result is saved. In an analytical process however, especially if it is a collaborative effort shared between participants, the intermediate steps are important as well. Making this path visible and navigable is important to support the analytical reasoning process [15].

In the DRGee tool, we implemented a history in the form of a tree where every modification to the portfolio results in a new node (Fig. 10). Users can

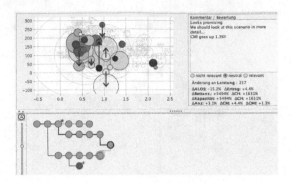

**Fig. 10.** The history of the analysis process is documented as a tree (bottom). Each node represents a different editing step of the portfolio. Nodes marked with a star have a comment attached. Nodes can be rated (green - relevant, red - not relevant). Any two nodes can be compared and their differences visualized (top left) and numerically characterized (top right) (Color figure online).

navigate to any node in the tree, see the portfolio's state at that point, and branch out from there. A comment can be added to a node to describe the reasons for the modification. Any two nodes in the tree can be compared, and the differences in the two states are shown both visually and numerically as the relative difference of the characteristic values.

The complete history is saved, and the history plugin therefore also serves as a documentation feature. Having access to an automatically generated, navigable and interactive documentation is very useful in the consolidation phase after the workshop is conducted. A viewer-only version of the DRGee tool can be distributed, loaded with the data and the history, to a wider audience for review.

## 4    Results and Discussion

The methodology, together with the supporting tool described in the previous section, was used in over forty different hospitals all around Switzerland over the course of four years. The type of hospitals ranged from smaller regional hospitals to hospital groups within a larger geographic region.

Over the course of this time, many workshops with hospital management were performed. Both the method and the tool were refined continuously, but typically the workflow looked like this:

1. Collect the basic data from the hospital and massage it into a set of portfolios relevant for the strategic theme that is of interest. Since the nature and quality of the data varies considerably between hospitals, we used a range of tools for data cleaning and consolidation. Typically these included a primary clean-up in delivery files (i.e. most of the time incomplete rows), load and enrichment of data in a database where consistency checks were performed,

and a final validation using known performance indicators for the respective hospital in order to ensure the consistency of the full data load. The final step consisted in creating the DRGee Viewer load files.

2. Analysis of the portfolios by healthcare experts. The tool's analysis capabilities proved very useful to first gain the general overview of the data. Following this, the portfolio comparison functionality allowed to check for outliers, or for yearly comparisons of two specific portfolios (e.g. yearly comparison of the same two clinics). A deeper analysis was then performed by using the superposition comparison. It was possible to directly display single DRG groups for example over time, and compare them in one overview allowing for a direct comparison. The latter oftentimes provided the basis for the simulation within the existing or a new portfolio.

3. Perform a strategy workshop with hospital management in which the tool is used to present the findings of the analysis, explore different scenarios, and discuss future strategies. These workshops typically included senior hospital management, controllers, or infrastructure planners, and are lead by a healthcare consultant that is familiar with the data. Typical questions revolved around what-if scenarios. For example, what happens if I alter the ALOS Ratio of a specific group by adapting my processes considerably? Or what are the effects of increasing the number of cases in a specific DRG or group? Participants found it very useful that they could directly see the effect of changing parameters (i.e. either the ALOS ratio or the number of DRGs per group) and validate and further discuss the results either in the various portfolio comparison views, or see the profitability effects in the summary values.

4. Provide the data used in the workshop and any resulting modifications to the portfolios to hospital management, together with a limited-functionality version of the tool for further discussion and communication. The recipients were able to profit by continuing the work beyond the guided workshops. The wider distribution of the tool, also to non-expert users, did not turn out to be a problem. The tool proved to be intuitive enough so that the required support was not significant.

In the workshops, the tool is used in front of a group of people. This prompts the question, if there is an opportunity to extend the tool to provide more explicit support for collaboration. There are various scenarios in which collaborative visualization can occur. In the space-time matrix that is used by [16] to categorize the design space, our implementation of DRGee currently corresponds to a co-located synchronous scenario, where a group of experts in the same room interact socially to create a common understanding of the data. Large interactive walls or multi-touch tabletop displays are technologies that are of interest in this context. However these devices are not yet widely available at the places where hospital decision makers work, which is why the practical potential of such approaches is still limited.

A web-based version of DRGee would facilitate a distributed asynchronous scenario, with access to the tool and the analysis results from arbitrary locations at any time. The focus would not be on social aspects of a large (mostly lay) audience such as investigated by systems like Many Eyes [17], but more on the asynchronous aspect of supporting collaborative analysis across time.

Examples of the strategic insights and decisions gained are:

- A group of hospitals reorganized their main operative theaters, because services were delivered far more efficiently in one hospital than the other. Consecutively the costs were reduced substantially, leading to an overall benefit within this sector of activity.
- The tool helped a smaller regional hospital to recognize its strategic fallacies, and thus proceed to a strategic partnership by giving up a certain part of their activities and re-focusing the hospitals overall market approach.

## 5   Conclusion

We have developed a tool to visualize and edit hospital service portfolios in live workshop settings. The tool is embedded in a strategy methodology that is used by hospital management and healthcare consultants. The method is actively used and has been successfully applied in more than forty hospitals around the country so far.

Strategic planning is difficult since the set of variables to consider is multidimensional and complex. Having a tool at hand that supports modelling, visualization, evaluation, and comparison of various approaches while self-documenting the individual steps, proved to be extremely helpful and often acted as a catalyzer in the planning workshops.

The supporting tool goes beyond graphics towards an interactive toolset that allows to understand complex situations intuitively and discuss strategic challenges. In this way the DRGee Viewer distinguishes itself from purely operative tools that are available in the market, as it is embedded in a logical framework. This makes it robust enough (by using the same portfolio structures at all levels) to provide a stable, comparable and evidence-based foundation for decisions.

The approach has the potential to be used at a more complex level, beyond a single hospital groups perspective, towards a more public health oriented dimension. This is where the value of such a tool could be useful for health service planning and provision.

Other directions in which the system will be extended are twofold:

- On the technical side, a migration to a web-based system is planned, to allow concurrent and collaborative work.
- It is planned to incorporate more data sources, such as quality data, demographic data and geographic conditions, and a comprehensive simulation within this new data landscape will be developed.

In summary, the method supported by the DRGee Viewer reflects claims in literature [18] that DRGs are not solely a tool for financing hospital services, but are also well suited for increasing transparency, inducing efficiency and supporting the hospital management in strategic decisions.

# References

1. Schreyögg, J., Stargardt, T., Tiemann, O., Busse, R.: Methods to determine reimbursement rates for diagnosis related groups (DRG): A comparison of nine european countries. Health Care Manage. Sci. **9**, 215–223 (2006)
2. Swiss DRG: Swiss DRG Homepage (Web). http://www.swissdrg.org/. Accessed 25 August 2014
3. Gocke, P., Debatin, J.F., Drselen, L.F.J.: Prozessmanagement und Controlling in der Diagnostischen Radiologie im Krankenhaus. Der Radiologe **42**, 332–343 (2002). 00012
4. Rohner, P.: Achieving impact with clinical process management in hospitals: an inspiring case. Bus. Process Manage. J. **18**, 600–624 (2012). 00003
5. Luke, R.D.: Dimensions in hospital case mix measurement. Inquiry **16**, 38–49 (1979)
6. Lüngen, M., Lauterbach, K.W.: Führen DRG zur Spezialisierung von Krankenhäusern? Gesundh ökon Qual manag **7**, 93–95 (2002)
7. Boston Consulting Group: The Product Portfolio (Web). http://tinyurl.com/prodportfolio. Accessed 26 August 2014
8. Andrews, R., Boyne, G.A., Walker, R.M.: Strategy content and organizational performance: an empirical analysis. Public Adm. Rev. **66**, 52–63 (2006)
9. Porter, M.E.: Competitive Strategy: Techniques for Analyzing Industries and Competitors, 1st edn. Free Press, New York (1980)
10. Gapminder: Gapminder Homepage (Web). http://www.gapminder.org/. Accessed 16 November 2014
11. Shneiderman, B.: Direct manipulation: a step beyond programming languages. Computer **16**(8), 57–69 (1983)
12. Gleicher, M., Albers, D., Walker, R., Jusufi, I., Hansen, C.D., Roberts, J.C.: Visual comparison for information visualization. Inf. Visual. **10**, 289–309 (2011)
13. Tufte, E.R., Graves-Morris, P.: The Visual Display of Quantitative Information, vol. 2. Graphics Press Cheshire, CT (1983)
14. Rensink, R.A., O'Regan, J.K., Clark, J.J.: To see or not to see: The need for attention to perceive changes in scenes. Psychol. Sci. **8**, 368–373 (1997)
15. Shrinivasan, Y.B., van Wijk, J.J.: Supporting the analytical reasoning process in information visualization. In: Proceedings of the SIGCHI Conference on Human Factors in Computing Systems CHI 2008, pp. 1237–1246. ACM, New York (2008)
16. Isenberg, P., Elmqvist, N., Scholtz, J., Cernea, D., Ma, K.L., Hagen, H.: Collaborative visualization: Definition, challenges, and research agenda. Inf. Visual. **10**, 310–326 (2011)
17. Viegas, F., Wattenberg, M., van Ham, F., Kriss, J., McKeon, M.: Manyeyes: a site for visualization at internet scale. IEEE Trans. Visual. Comput. Graphics **13**, 1121–1128 (2007)
18. Geissler, A., Quentin, W., Scheller-Kreinsen, D., Busse, R.: Introduction to DRGs in Europe: common objectives across different hospital systems. In: Diagnosis Related Groups in Europe: Moving Towards Transparency, Efficiency and Quality in Hospitals, pp. 9–21 (2011)

# Automatic Analysis of Lung Function Based on Smartphone Recordings

João F. Teixeira[1]([✉]), Luís F. Teixeira[2], João Fonseca[3], and Tiago Jacinto[3]

[1] Department of Electrical and Computer Engineering, University of Porto,
Porto, Portugal
jpfteixeira.eng@gmail.com
[2] Department of Informatics Engineering, University of Porto, Porto, Portugal
luisft@fe.up.pt
[3] Department of Health Information and Decision Sciences, University of Porto,
Porto, Portugal
jfonseca@med.up.pt, tajacinto@gmail.com

**Abstract.** Over 250 million people, worldwide, are affected by chronic lung conditions such as Asthma and COPD. These can cause breathlessness, a harsh decrease in quality of life and, if left undetected or not properly managed, even death. In this paper, we approached part of the lines of development suggested upon earlier work. This concerned the development of a system design for a smartphone lung function classification app, which would only use recordings from the built-in microphone. A more systematic method to evaluate the relevant combinations of methods was devised and an additional set of 44 recordings was used for testing purposes. The previous 101 were kept for training the models. The results enabled to further reduce the signal processing pipeline leading to the use of 6 envelopes, per recording, half of the previous amount. An analysis of the classification performances is provided for both previous tasks: differentiation into Normal from Abnormal lung function, and between multiple lung function patterns. The results from this project encourage further development of the system.

**Keywords:** Asthma · Breath · COPD · Machine learning · Signal processing · Smartphone · Spirometry

## 1 Introduction

Chronic respiratory diseases such as Asthma and Chronic Obstructive Pulmonary Disease (COPD) are incurable, yet treatable and their early detection is crucial to provide a better quality of life. Major risk factors include air pollution, tobacco smoking and occupational environments containing dust and chemicals. The World Health Organization (WHO) estimates that over 250 million people suffer from asthma and COPD [15] and more than 3 million people died of COPD in 2005 [16].

Spirometry is the measurement of breath, i.e., is the most popular noninvasive set of timed tests that enables to measure the mechanical properties of the lungs,

© Springer International Publishing Switzerland 2015
A. Fred et al. (Eds): BIOSTEC 2015, CCIS 574, pp. 390–402, 2015.
DOI: 10.1007/978-3-319-27707-3_24

also named pulmonary function [8]. The keystone test is the Forced Expiratory Maneuver (FEM) where the patient fully inspires and then forcefully exhales all the air available, as fast as possible.

The increasing use of smartphones has enabled the emergence of several health related systems. Their computational power is ever increasing and, equipped with multiple sensors, it is possible to develop disease prevention, diagnosis and monitoring applications.

This paper provides an extension of previous work [12], where the performance of several groups of methods, along with clinical parameters, were compared. The aim is to find the most relevant, most efficient and fastest combination to produce a smartphone app for measuring and classifying lung function. The system's input is restricted to the smartphone's built-in microphone, in order to avoid external components.

## 2   Background and Related Work

From the dawn of non-invasive lung function evaluation, which produced the standard spirometer, several improvements have been made throughout the years. The next logical step concerning portability and affordability involves lung function estimation with smartphones' microphones. Some studies have already been conducted in order to accurately measure the clinical parameters [5,17], and also considering robustness to ambient noise [11].

In previous work [12], a lung function estimation system was proposed containing signal processing and machine learning algorithms. It was based on 101 recordings, collected from 61 patients performing the forced expiratory maneuver (FEM). Due to the reduced and unbalanced number of instances (label-wise), the evaluation of all models was done in a 5-fold cross-validation scheme.

## 3   Dataset

The dataset for this work comprises the previous 101 recordings and a new set of 44 recordings, collected in the same conditions of the previous ones. Some recordings were gathered on a controlled environment with low background noise, however more than 80 % of the recordings experienced background noise such as physicians giving verbal incentive, talking voices and small machine noises at a short distance. The recordings were made using a Samsung GT-I9000. The first 101 instances constituted the training set while the new set was used for testing.

Due to data collection constraints, the new recordings only included the 4 most popular parameters. However, this is not very problematic since, as previously shown, the remaining clinical indicators are very unreliable and extremely patient cooperation dependent. In fact, this enables the optimization process for the signal processing pipeline to focus on the most relevant values.

Each recording is accompanied by the patient's anthropometric parameters (age, height, weight and gender), clinical parameters, and classification of the patients lung function provided by the recording physician. The clinical parameters were obtained by performing the FEM to a spirometers available. The classification types are normal and abnormal (obstruction, restriction or mixed).

The included patients were part of the clinical study Control and Burden of Asthma and Rhinitis (ICAR), patients attending the allergology clinic from CUF Porto Institute (ICP) or from CUF Porto Hospital (HCP). Data collection occurred between April 3rd 2014 and February 1st 2015.

# 4    Algorithms and System Architecture

## 4.1    Signal Processing

The system's input consists of microphone recordings which are AC coupled, uncalibrated signals that represent air pressure. The signal processing pipeline can be divided in four portions: automatic signal segmentation, signal pre-processing, envelope generation and envelope processing. Figure 1 shows the initial architecture for the signal processing part without the signal segmentation.

**Automatic Signal Segmentation.** The audio input was initially segmented in order to remove non expiration sounds, such as the inspiration portion of the maneuver and ambient noise. The definition and cropping the beginning of the sound was accomplished using a modified version of the Back-Extrapolation algorithm [7].

First, an LPC envelope of the signal is obtained, as it will be described further on the paper. Then the minimum value between inspiration and expiration peaks is found and the initial part is removed. Afterwards, the zero-time back-extrapolation is performed by finding the instant corresponding to the envelope's peak (PEF time), calculating the Time-Volume curve, drawing the tangent at the PEF time and finding where the tangent crosses the abscissas, which is the initial instant.

The ending at noise level was detected using a sliding window algorithm (5 % of signal's length, 25 % overlap) based on the magnitude ratio threshold of the maximum value (2 %).

**Signal Pre-Processing.** The recordings are limited in excursion and patients need to perform the expiratory maneuver at an arms length to avoid microphone saturation. Therefore, it seems relevant to compensate the pressure lost between the lips ($p_{lips}$) and smartphone ($p$), using an Inverse Radiation Model. Furthermore, this model also atones the reverberation effect from sound reflections around a person's body. Afterwards, $p_{lips}$ was converted to airflow at the lips ($u_{lips}$), using a Pressure to Flow Conversion Model. Both models were developed in similar fashion to [5].

**Envelope Generation.** The third stage employed several methods to calculate the signal envelopes, approaching different sound characteristics to obtain a comprehensively robust feature extraction. The algorithms' input consisted of both the segmented audio and the two resulting signals from the pre-processing stage, as all of them can be considering roughly proportional to air flow.

*Generic Envelope Extraction.* To obtain an envelope based on a time domain approach two methods were used: the Hilbert Transform and Shannon curves [6]. The first approach consists of calculating the signal's harmonic conjugate, with the Hilbert Transform, and to add it back to the signal, resulting on an envelope. The second approach involves calculating the Shannon Entropy and Energy envelopes of the signal. They act as non-linear transformations focusing either on the higher (Energy) and lower (Entropy) intensities of the signal. Both approaches output highly noisy curves that need subsequent smoothing.

*Linear Predictive Coding.* The audio input is segmented in windows of 31.25ms, with 50 % overlap. The white noise variance, or power, is obtained from the LPC model outputs. While the LPC filters can approximate the vocal tract [14], the succession of power values should be proportional to the exhalation power at the respective time and constitute a sampled envelope of the signal. The implementation included models of degrees 2, 4, 8, 16 and 32, which represents increasing vocal complexity.

*Mean of Resonances.* Similarly to LPC, the signal was buffered into 31.25 ms frames, with 50 % overlap. Each frame underwent a 256-point FFT operation using a hamming window, producing a spectrogram. All spectrogram values lower then 20 % the respective frames' maximum were considered noise and were consequently discarded. Resonances over 250ms, within the respective frequencies' 2 bin neighborhood were kept, preserving only relatively large and long frequencies, and taking into account the natural occurring frequency shift. The envelope was obtained by averaging the frames' saved resonances.

**Envelope Post-processing.** The several envelopes obtained were processed using different settings in order to find the best combination for the application. The envelopes were smoothed by a regular low pass filter (LPF) and, in parallel, were also approximated by a 4th order polynomial. To obtain the same sampling rate as the buffered methods, the Hilbert Transform and Shannon envelopes' results were downsampled accordingly. The non-approximated envelopes were also further processed using a Savitzky-Golay filter (SG) with order 3 and size 11 [10], as depicted on Fig. 1.

## 4.2   Parameter Extraction

For each recording, the spirometry parameters were calculated from each of the final envelopes. The measurements extracted were PEF, FVC, $FEV_1$ and $FEV_1/FVC$. The envelopes are viewed as Flow-Time curves, typical of spirometer reports.

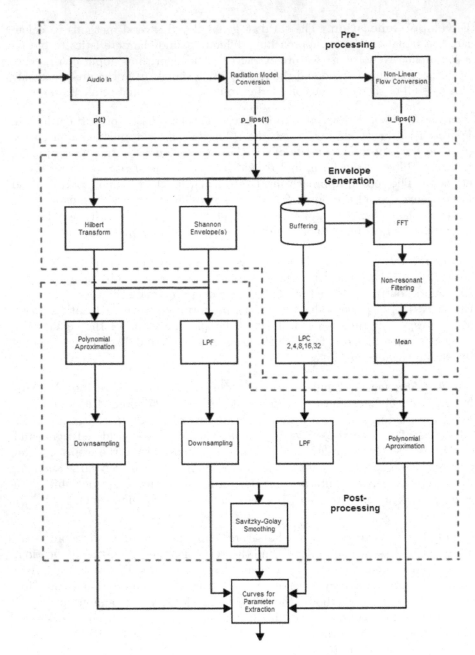

**Fig. 1.** Initial signal processing pipeline (P0).

PEF is defined as the Peak Expiratory Flow or the global maximum of the audio envelope. By integrating the envelope with respect to time the Volume-Time curve can be obtained. FVC is defined as the total volume expired of a FEM. $FEV_1$ is the total volume expired during the first second.

### 4.3    Machine Learning

The system's machine learning pipeline can be divided into two stages: the parameter regression and the classification. The first uses the parameters extracted from the curves to obtain an estimation of the respective clinical values, as given by spirometers. The second, devises models that can discern between the possible illness states, initially addressing the distinction of normal from abnormal lung function and then, normal from 2 types of pathologies.

**Regression Stage.** Every recording produces several envelopes and each one is used to extract clinical measurements. This information is used to produce a relatively robust estimation of the respective spirometer measurement. For instance, each set of PEF measurements computed from any recording is used as a batch input for the trained regression model to obtain an estimated PEF value. The process is then repeated for the other types of clinical measurements. The corresponding spirometer measurements acted as ground truth or regression targets. For this task, Regression Tree Bagging [1] and Random Forests (RF) [2] were used. The number of trees in the ensembles was reduced to 10 without significant loss in accuracy. Also, RF employed a selection size for the random feature subset of $n/3$ out of the total $n$ feature set.

**Classification Stage.** On this stage, the regressed parameters were the input of the learning models. Several different classification models were tested, namely: Decision Trees [3], either as one tree, Tree Bagging, Random Forest ($\sqrt{n}$ subset) and AdaBoost [4], Support Vector Machines (SVM) [13] and Naïve Bayes [9]. Although the tree ensemble methods used 70 trees, only 10 trees were grown for AdaBoost to avoid overfitting.

## 5    Experimental Approach

### 5.1    Regression Experiments

Unlike in the previous work, we aimed to devise a more systematic approach concerning the removal of the signal processing blocks. Hence, the experiments are based on random forward selection and model improvement tracking. Basically, a random path (sequence of methods) of the system's pipeline is used to predict the parameters' values and, iteratively, more are added. After each prediction the mean error and standard deviation is saved as well as the improvement of the respective path's addition (Fig. 2). The termination condition is adding 12 paths (maximum number of paths from previous work and to reduce cumulative path bias).

The *Avg(Avg)* and *Avg(Std)* refer to the averages of the mean value and standard deviation of the error, respectively. The *Score* (blue line) is calculated as an average of the parameter-wise weighted average between the mean ($w = 3$) and standard deviation ($w = 2$) of the error. The vertical lines indicate the minimum.

This process is repeated 100 times, ultimately providing a histogram of the relative influence of each path (Fig. 3). To evaluate which system blocks were more useful to include, the importance of a method was considered as the average

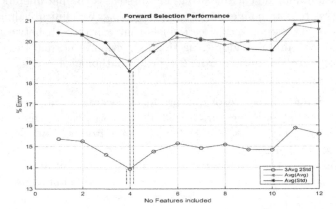

**Fig. 2.** Score progression with inclusion of system paths. Tree Bagging, initial dataset.

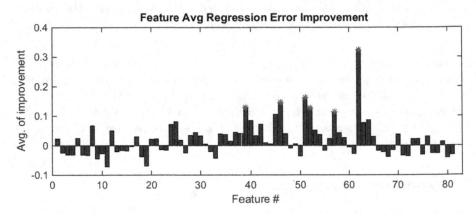

**Fig. 3.** Relative system path importance. Tree Bagging, initial dataset, 100 iterations. The 6 maximum values are represented as red dots (Color figure online).

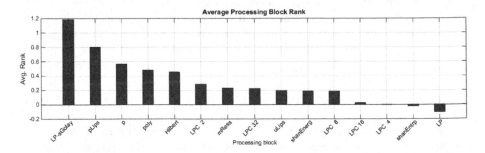

**Fig. 4.** Relative system block importance. Tree Bagging, initial dataset, 100 iterations.

of the importance of each path that uses that method (Fig. 4). Each model generated used a 5-fold cross validation dataset and was later tested with the new dataset.

## 5.2   Classification Experiments

The classification process was first devised as a Normal against Abnormal classification problem, referred to as two label experiments (TLE). Then, multiple label experiments (MLE) were conducted, where the models tried to distinguish between Normal lung function and Obstruction and Mixed pathologies. For both problems, the experiments varied on the feature space used. The experiments used:

1. Set A[1],
2. Set A and height,
3. Set A, height and age.

# 6   Results and Discussion

## 6.1   Regression Experiments

We proceeded to execute the analysis described in Sect. 5.1, several times. The ranking of relevance of the SP pipeline blocks, such as in Fig. 4, presented varying results for each 100 iterations run. The exception to these changes were the filtering options *LP-sGolay* and *Polynomial Fitting* that maintained a very high importance value, contrasting with using just the *LP* filter which tended

**Table 1.** Comparison of regression error average and standard deviation for the clinical parameters. P1 refers to results with the previous final SP pipeline and P2 to the new pipeline.

| Task | SP | Model | Average Error (%) | | | | Std. Dev. Error (%) | | | |
|------|----|----|------|------|------|------|------|------|------|------|
| | | | PEF | FVC | $FEV_1$ | Tiff | PEF | FVC | $FEV_1$ | Tiff |
| Train | P1 | Bag | 23.46 | 30.03 | 22.77 | 10.32 | 22.93 | 35.77 | 23.36 | 8.66 |
| | | RF | 22.26 | 28.25 | 24.05 | 9.59 | 22.98 | 29.14 | 26.40 | 8.20 |
| | P2 | Bag | 22.66 | 26.03 | 23.73 | 10.22 | 23.96 | 25.74 | 24.19 | 8.27 |
| | | RF | 22.43 | 28.62 | 24.29 | 10.21 | 24.55 | 27.98 | 24.98 | 8.10 |
| Test | P1 | Bag | 27.83 | 32.18 | 26.67 | 12.99 | 27.11 | 22.82 | 17.60 | 13.10 |
| | | RF | 27.21 | 32.53 | 25.53 | 13.74 | 27.34 | 23.19 | 17.56 | 13.56 |
| | P2 | Bag | 26.76 | 32.26 | 26.95 | 13.23 | 25.83 | 22.66 | 17.19 | 12.86 |
| | | RF | 27.37 | 32.57 | 25.88 | 13.61 | 26.18 | 23.30 | 17.51 | 13.29 |

---

[1] Set A: PEF, $FEV_1$, FVC, $FEV_1$/FVC.

to be of very low importance. The remaining methods, pre-processing and envelope generating functions, changed often relative importance value and order, which suggests that neither is particularly relevant for the generation of good attributes/paths.

Based on these results, and in order to further reduce the complexity of the system to be implemented, we opted for the following system:

– Maintain all the filtering models (discarding the direct use of $LP$ output)
– Maintain all the pre-processing outputs ($p$, $p_{lips}$, $u_{lips}$)
– Just use *Shannon Energy* (least complex envelope generator: $x^2$, $\sqrt{x}$ and $\log_{10}(x)$)

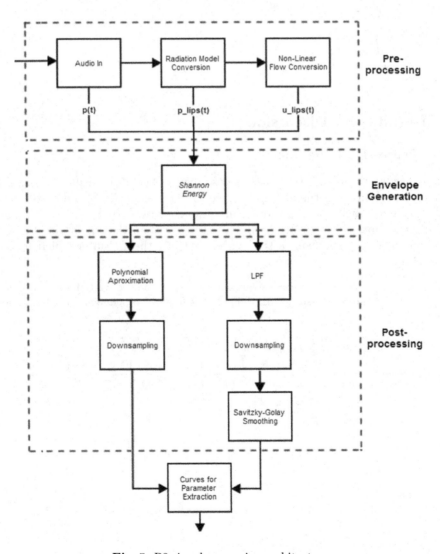

**Fig. 5.** P2 signal processing architecture.

This new pipeline (P2, Fig. 5), produces a total of 6 attributes per clinical parameter, half than the previous system (P1). With the pipeline selected, the models were retrained and tested. The results are shown in Table 1.

From these results a few remarks can be made. First, all values are very similar, either across models, signal processing pipeline and task. This is particularly interesting since, by halving the number of values to combine in regression, and consequently simplifying the pipeline, approximately the same results were obtained. Furthermore, the training version of Bagging with P2 pipeline for FVC managed to present significantly better results than the correspondent version for P1.

Another interesting fact is that the testing results quite resemble the training ones, with a maximum of 5 % increase in the average error rates and even significant improvement in standard deviation for FVC and $FEV_1$.

Like in previous work, Bagging and Random Forest present very similar results and an option cannot be chosen for deployment with sufficient confidence. As before, the Random Forest was chosen to be used for the subsequent experiments, so to reduce overfitting and shorten regression time, along with the P2 pipeline.

### 6.2   Classification Experiments

The classification experiments using P2 presented interesting results, shown on Table 2. As before, through the inclusion of height and age attributes the models accuracies managed to improve. However, unlike in the experiments of previous work, including just height introduced significant bias towards the *Normal* class, despite improving the accuracy (Ac.). This apparent contradiction can be explained by the instance label unbalance. In fact, both testing results have shown a clear bias for Example 1 and 2, leaving the precision (Prec.) and recall (Rec.) for the *Abnormal* class mostly at zeros (not on table). Either way, similarly to previous work, using Set A in conjunction with height and age provides the best results and thus, mainly the Example 3 results are considered for model evaluation.

At this stage, the model that provides best results is Tree Bagging, which has the best accuracy (88.1 % / 65.9 %) and has one of the best precision-recall tradeoffs, in the sense that a higher precision is desirable, implying lower bias towards the *Normal* class (prec. 90.4 % / 78.2 %).

Overall, the training results resembled those from previous work, reaching a difference in error rate of 4 % for the best method, which is hardly significant for the dataset's size. Surprisingly, however, the models did not manage to produce testing results that matched training predictions. Models with P1 pipeline were also tested but did not show significant change (accuracy improvement below 2 %, not shown). This suggests that the testing set is somewhat different from the training set.

Table 3 presents the results of the multiple label classification problem, using P2. Like before, the training results seemed to improve with the succession of experiments reinforcing the importance of using the anthropometric values with

**Table 2.** Two label classification problem (TLP) results (%) for the best performing models and single tree classifier.

| Method | Training | | | | | | Testing | | | | | |
|---|---|---|---|---|---|---|---|---|---|---|---|---|
| | Exp. 1 | | | Exp. 3 | | | Exp. 1 | | | Exp. 3 | | |
| | Ac. | Prec. | Rec. | Ac. | Prec. | Rec. | Ac. | Prec. | Rec. | Ac. | Prec. | Rec. |
| Single Tree | 53.4 | 68.7 | 61.9 | 83.1 | 87.5 | 88.7 | 56.8 | 60.9 | 89.2 | 65.9 | 80.9 | 60.7 |
| Tree Bagging | 62.3 | 69.4 | 83.1 | 88.1 | 90.4 | 92.9 | 59.0 | 61.9 | 92.8 | 65.9 | 78.2 | 64.2 |
| Random Forest | 68.3 | 70.9 | 92.9 | 86.1 | 86.0 | 95.7 | 61.3 | 62.7 | 96.4 | 63.6 | 66.6 | 85.7 |
| Adaboost | 68.3 | 71.4 | 91.5 | 80.2 | 82.2 | 91.5 | 63.6 | 63.6 | 100 | 65.9 | 66.6 | 92.8 |
| SVM | 70.3 | 70.3 | 100 | 87.1 | 90.2 | 91.5 | 63.6 | 63.6 | 100 | 63.6 | 63.6 | 100 |

the clinical ones. On the other hand, the accuracy of the test results were overall slightly reduced. However this is caused by a reduction of bias towards the *Normal* class. In fact, on the Tree Bagging experiment evolution, the accuracy was maintained but the precision for the *Normal* class rose from 66.6 % to 84.6 % while precision and recall values for the other classes were changed from null values.

**Table 3.** Multiple label classification problem (MLP) results (%) for the best performing models and single tree classifier.

| Method | Training | | Testing | |
|---|---|---|---|---|
| | Exp. 1 | Exp. 3 | Exp. 1 | Exp. 3 |
| Single Tree | 58.5 | 77.7 | 61.9 | 54.7 |
| Tree Bagging | 65.6 | 86.8 | 66.6 | 66.6 |
| Random Forest | 67.6 | 81.8 | 66.6 | 61.9 |
| Adaboost | 65.6 | 76.7 | 66.6 | 64.2 |
| SVM | 71.7 | 86.8 | 66.6 | 9.52 |

The testing results has also shown relatively low accuracy comparing to the training results, despite being more evident on Example 3 values.

Similarly to the TLP results, these also suggest that Tree Bagging is the best option among the tested. It provides the best accuracies in both training and testing and even slightly improves on the P1 results (accuracy 82.2 %) [12], even though not significantly.

## 7    Limitations and Future Work

### 7.1    Limitations

This project presented some issues concerning data collection that, once overcome, should enhance the learning models' performance and, consequently, the results.

A great portion of the recordings was gathered on a relatively fast paced clinical study where patients had to perform several respiration maneuvers before recording to the smartphone. This could have reduced the patient's cooperation level due to fatigue. On the other hand, the forced expiration maneuver itself is difficult to perform, specially when concerning these recordings where no mouthpiece was used. These factors also contributed to the reduced yield of properly executed recordings. The small dataset with little intra-patient samples is most likely the cause of the regression errors of over 20 %.

Additionally, since the spirometer and smartphone maneuvers were made separately there is no completely reliable ground truth.

## 7.2   Future Work

Further study of this technology is needed and some key features are proposed. In addition to collect further recordings, it is relevant to devise an algorithm to automatically detect poorly executed FEMs in order to immediately request a repetition during data collection. Also, an application based on the proposed architecture should be implemented.

## 8   Conclusion

In this paper, we approached part of the lines of development suggested upon earlier work. A more systematic method to evaluate the relevant combinations of methods was devised and an additional set of recordings was used for testing purposes.

The new evaluation method lead to a new pipeline (P2) which is simpler to implement and should be faster to execute, due to the decrease in envelope generating functions. This, in turn, further reduced the number of the envelopes processed to 6, half the amount of the previous system. The regression results from the P2 system are very similar to the previous ones (P1). Furthermore, models from either pipeline show test values with a maximum 5 % increase in error rate and even better standard deviation for some clinical parameters.

The classification experiments confirmed the usefulness of combining the clinical parameters with the patient's anthropometric data. The TLP and MLP results were comparable to previous work, even slightly improving on the MLP. The testing results were not expected to be so low and, considering the test results using the P1 pipeline, it suggests an issue with the testing data. Despite this, the remaining results encourage further development of the system.

**Acknowledgements.** This work was conducted with the support of the *Control and Burden of Asthma and Rhinitis* project (ICAR), with the grant PTDC/SAU-SAP/119192/ 2010. The authors would like to thank Bernardo Pinho for the development of an enhanced recording app and Ivânia Gonçalves, Rita Silva and Daniela Santos for the added effort of recording the patients in addition to their tasks on the patient screenings.

# References

1. Breiman, L.: Bagging predictors. Mach. Learn. **24**(2), 123–140 (1996). http://dx. doi.org/10.1007/BF00058655
2. Breiman, L.: Random forests. Mach. Learn. **45**(1), 5–32 (2001). http://link. springer.com/article/10.1023/A:1010933404324
3. Buntine, W.: Learning classification trees. Stat. Comput. **2**(2), 63–73 (1992). http://dx.doi.org/10.1007/BF01889584
4. Freund, Y., Schapire, R.E.: Experiments with a new boosting algorithm. In: Saitta, L. (ed.) ICML, pp. 148–156. Morgan Kaufmann, Bari, Italy (1996). http://web. eecs.utk.edu/~parker/Courses/CS425-528-fall12/Handouts/AdaBoost.M1.pdf
5. Larson, E.C., Goel, M., Boriello, G., Heltshe, S., Rosenfeld, M., Patel, S.N.: SpiroSmart: using a microphone to measure lung function on a mobile phone. In: 14th ACM International Conference on Ubiquitous Computing, p. 10. Pittsburgh, Pennsylvania, USA (2012)
6. Liang, H., Lukkarinen, S., Hartimo, I.: Heart sound segmentation algorithm based on heart sound envelogram. In: Computers in Cardiology 1997. vol. 24, pp. 105–108. IEEE (1997).http://ieeexplore.ieee.org/lpdocs/epic03/wrapper.htm? arnumber=647841
7. Miller, M.R., Hankinson, J., Brusasco, V., Burgos, F., Casaburi, R., Coates, A., Crapo, R., Enright, P., van der Grinten, C.P.M., Gustafsson, P., Jensen, R., Johnson, D.C., MacIntyre, N., McKay, R., Navajas, D., Pedersen, O.F., Pellegrino, R., Viegi, G., Wanger, J.: Standardisation of spirometry. Eur. Respir. J. **26**(2), 319–338 (2005). http://www.ncbi.nlm.nih.gov/pubmed/16055882
8. Pierce, R.: Spirometry: an essential clinical measurement. Aust. Fam. Physician **34**(7), 535–539 (2005). http://www.ncbi.nlm.nih.gov/pubmed/15999163
9. Russel, S., Norvig, P.: Artificial Intelligence: A Modern Approach, 2nd edn. Prentice Hall, Englewood Cliffs (2002)
10. Savitzky, A., Golay, M.J.E.: Smoothing and differentiation of data by simplified least squares procedures. Anal. Chem. **36**(8), 1627–1639 (1964). http://pubs.acs.org/doi/abs/10.1021/ac60214a047
11. van Stein, B.: A Mobile Smart Care platform Home spirometry by using the smartphone microphone. Master's thesis, Leiden University, Leiden, The Netherlands (2013)
12. Teixeira, J.F., Teixeira, L., Fonseca, J., Jacinto, T.: Lung function classification of smartphone recordings - comparison of signal processing and machine learning combination sets. In: Proceedings of the International Conference on Health Informatics, pp. 123–130. Lisbon, Portugal (2015). http://www.scitepress.org/ DigitalLibrary/Link.aspx?doi=10.5220/0005222001230130
13. Vapnik, V.N.: An overview of statistical learning theory. IEEE Trans. Neural Netw. Publ. IEEE Neural Netw. Counc. **10**(5), 988–999 (1999). http://www.ncbi.nlm.nih.gov/pubmed/18252602
14. Wakita, H.: Direct estimation of the vocal tract shape by inverse filtering of acoustic speech waveforms. Audio Electroacoust. IEEE Trans. **21**(5), 417–427 (1973)
15. Asthma: Fact sheet N307, November 2013. http://www.who.int/mediacentre/ factsheets/fs307/en/. Accessed: 26-06-2014
16. Chronic obstructive pulmonary disease (COPD): Fact sheet N315 (October 2013). http://www.who.int/mediacentre/factsheets/fs315/en/. Accessed: 26-06-2014
17. Xu, W., Huang, M.C., Liu, J.J., Ren, F., Shen, X., Liu, X., Sarrafzadeh, M.: mCOPD. In: Proceedings of the 6th International Conference on Pervasive Technologies Related to Assistive Environments - PETRA 2013, pp. 1–8. PETRA 2013, ACM Press, New York, New York, USA (2013). http://doi.acm.org/10.1145/ 2504335.2504383, http://dl.acm.org/citation.cfm?doid=2504335.2504383

# Towards a Safer and More Optimal Treatment of the Supracondylar Humerus Fracture

Mohamed Oussama Ben Salem[1,2]([✉]), Olfa Mosbahi[2], Mohamed Khalgui[2], and Georg Frey[3]

[1] Polytechnic School of Tunisia, University of Carthage, Tunis, Tunisia
bensalem.oussama@hotmail.com
[2] LISI Laboratory, INSAT, University of Carthage, Tunis, Tunisia
olfamosbahi@gmail.com, khalgui.mohamed@gmail.com
[3] Chair of Automation and Energy Systems, Saarland University,
Saarbrücken, Germany
georg.frey@aut.uni-saarland.de

**Abstract.** Treating the supracondylar humerus fracture, a very common elbow's injury, can be very challenging for pediatric orthopedic surgeons. Actually, using the pinning technique to treat it leads sometimes to many neurological and vascular complications. Furthermore, the medical staff faces a serious danger when performing such surgeries because of the recurrent exposure to harmful radiations emitted by the fluoroscopic C-arm. Considering these issues, a national project was launched to create a new robotic platform, baptized BROS, to automate the supracondylar humerus fracture's treatment and remedy the said issues. This chapter introduces this new robotic platform and uses a real case to prove the relevance and safety of BROS.

**Keywords:** Supracondylar humerus fracture · Robot-assisted surgery · e-Heath · Reconfiguration

## 1 Introduction

The field of robotics is expanding day after day. The ability of robots to replace, supplement or transcend human performance has had a profound influence on many fields of our society, spanning fields such as agriculture, military and especially medicine. Patients demand greater precision, less and minimally invasive procedures, and faster recovery times. The increasing life expectancy associated with a need for reducing costs and increasing efficiency have opened the door for new and innovative solutions in the medical robotic industry. The field of computer-assisted surgery is relatively new since the first clinical application of a robot was performed to a neurosurgery in 1985 [14]. Since then, many research centers around the world have developed a multitude of robotic surgical products to tackle new areas such as ophthalmology, radiology, urology, cardiothoracic and orthopedics [4].

© Springer International Publishing Switzerland 2015
A. Fred et al. (Eds): BIOSTEC 2015, CCIS 574, pp. 403–423, 2015.
DOI: 10.1007/978-3-319-27707-3_25

One of the most common injuries faced by pediatric orthopedic surgery is the supracondylar fracture of the humerus (or SCH). It accounts for 18 % of all pediatric fractures and 75 % of all elbow fractures [17]. It mainly occurs during the first decade of life and are more common among boys [16]. The current treatment of SCH may lead to many complications. The neurological ones consists in damages caused to the median nerve during the reduction of the fracture or during the open procedure. The study in [10] also reports some vascular complications, mostly consisting in the disruption of the brachial artery. All those complications are principally caused by the "blind" pinning the surgeons perform [9]. Even though they are usually using an image intensifier, the medical staff can't guess in advance the trajectory the pin will follow. Images are actually taken once the pin is inserted, which may cause the previously mentioned complications. Other inconvenient of the current treatment technique is the recurrent medical staff exposure to radiations when using the fluoroscopic C-arm [5]. These X-ray Radiations are harmful, and fluoroscopic examinations usually involve higher radiation doses than simple radiography. For example, a work in [20] showed that, for spine surgeons, radiation exposures may approach or exceed guidelines for cumulative exposure. Another research in [12] showed that the fluoroscopically assisted placement of pedicle screws in adolescent idiopathic scoliosis may expose the spine surgeon to radiation levels that exceed established lifetime dose equivalent limits.

Considering these constraints and issues, a new national project, baptized BROS (Browser-based Reconfigurable Orthopedic Surgery), has been launched to remedy these problems. BROS a new reconfigurable robotized platform dedicated to the treatment of supracondylar humeral fractures. It is capable of running under several operating modes to meet the surgeon's requirements and well-defined constraints. Thus, it can whether automatically perform the whole surgery or bequeath some tasks to the surgeon. BROS architecture is composed of a control unit, a browsing system with a middleware to perform image processing, two robotic arms to reduce the fracture and another one to insert pins in the fractured elbow.

This chapter is organized as follows: the next section describes useful preliminaries for the reader. Section 3 introduces a real case study of a surgery undergone by a patient suffering from SCH to show the limit of the current fracture treatment. We expose, in Sect. 4, our robotic platform and its functioning. Section 5 presents the developed middleware, while Sect. 6 introduces the control unit of BROS. We finish the paper in Sect. 7 by a conclusion and an exposition of our future works.

## 2   Background

We start, in this section, by presenting the robotic arm that we will use to implement BROS and the used software to configure it. We expose, thereafter, an overview about the different classifications of the supracondylar humerus fracture.

## 2.1   Platform and Environment

As the smallest robot from ABB, the IRB 120 offers all the functionality and expertise of the ABB range in a much smaller package. Like all ABB robots, the IRB 120 is a particularly agile 6-axis robot which, thanks to its compact turning radius, can be mounted closer to other equipment. Besides, it is ideal for a wide range of industries including the electronic, food and beverage, machinery, solar, pharmaceutical, medical and research sectors. With its lightweight but strong aluminum structure and small powerful engines, the IRB 120 weighs only 25 kg, which explains its rapid and precise acceleration. In fact, this featherweight has all the traditional features of ABB robots, including leading performance in terms of trajectory tracking and motion control. Thus, the IRB 120 won many manufacturers' spurs [3, 8].

IRB 120 can be programmed offline with RobotStudio ABB's software that allows to simulate an industrial manufacturing cell to find the optimal position of the robot and avoid costly downtime and production delays [6]. RobotStudio from ABB Robotics is a powerful off-line robot programming and simulation tool. What makes it unique is the fact that, when the code is fully developed off-line, it downloads to the actual controller with no translation stage, reducing time-to-market. RobotStudio is able to create the robot movements using graphical programming, edit and debug the robot system, and simulate and optimize existing robot programs. It is widely used in universities to educate engineering students in the capabilities and applications of robots, as well as in the automation industry by mechanical designers and robot programmers. RobotStudio is also used in remote maintenance and troubleshooting. It actually connects to the live system to take an instant virtual copy, and then goes off-line to enable the situation to be studied in depth. RobotStudio also features a RAPID Editor which enables the user to write a robot program. The user can watch a single robot execute the RAPID program in the graphical environment [7].

## 2.2   Classification of Supracondylar Humeral Fracture

Many classifications of the supracondylar humeral fractures were established. They are based on both the direction and the degree of displacement of the distal fragment [2]. The Lagrange classification system and the Gartland's are the most widely used. The first is the most widely used in the French literature. It divides these fractures into four types on the basis of antero-posterior and lateral radiographs [15]. In the English literature, the second is the most commonly used: the Gartland's classification is based on the lateral radiograph and fractures are classified, as illustrated in Fig. 1, according to a simple three-type system (Table 1) [18]. We adopt this classification in this paper.

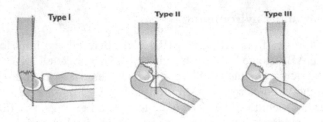

**Fig. 1.** Gartlands classification of supracondylar fractures of the humerus.

**Table 1.** Gartlands classification of supracondylar fractures of the humerus.

| Type | Radiologic characteristics |
|------|----------------------------|
| I | Undisplaced fractures |
| II | Displaced fracture with intact posterior hinge |
| III | Completely displaced fractures with no contact between the fragments |

## 3  Case Study

We expose, in this section, a true case of a patient suffering from a supracondylar humeral fracture who came to the Children Hospital of Bchir Hamza (Tunis). The patient who is a ten-year-old girl fell on her outstretched right hand on November 12th 2013. Once supported, the patient's elbow was placed in a brace in a 20-to-40 flexion to promote vascularization of the organ. She underwent a surgery on the same day. We were invited by Prof.Dr.med. Mahmoud Smida, the head of Child and Adolescent Orthopedics Service and our medical collaborator, to attend the intervention.

Treatment with single traction is not considered any more in modern centers due to a long required hospitalization and excellent current surgical results. The closed reduction with pinning is now the most used technique. It is performed under general anesthesia and fluoroscopic control. First, a radiography of the injured elbow is taken to determine the type of fracture. The latter was found a type III fracture according to Gartland's classification as show in Fig. 2. The patient, anesthetized, is then placed under the fluoroscopic image intensifier (Fig. 3).

The fracture is reduced in the frontal plane in extension and the elbow is bent while pushing forward the olecranon. The surgeon repeatedly rotated the image intensifier rather than the limb and took a total of 9 images to verify the reduction profile. The limb is immobilized once a satisfying reduction is obtained (Fig. 4). Two percutaneous pinning are finally performed in the distal fragment as illustrated in Fig. 5 to fix the bone and avoid any risk of cubitus varus (a common deformity in which the extended forearm is deviated towards midline of the body). To avoid any vascular or nerve injury during the insertion of the two pins, 15 fluoroscopic images were taken.

**Fig. 2.** The fracture's radiography.

**Fig. 3.** The patient installed under the fluoroscopic image intensifier.

**Fig. 4.** Limb immobilization.

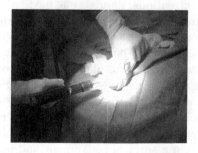

**Fig. 5.** Percutaneous pinning.

During this surgery, a total of 24 fluoroscopic images was taken, which involves high doses of radiation to the medical staff, especially since such interventions are performed 5 times per day on average. The second pin had to be removed and reinserted since it didn't straightaway follow the right trajectory, which can lead to some complications. To remedy these problems, we launch a new project, baptized BROS, which consists in a robotized platform to automatically perform such surgeries or assist the surgeon by limiting his exposition to radiations and bypassing the blind pinning issue.

## 4   BROS

BROS is a new and original robotic platform. This project was launched to remedy the two most important difficulties the medical staff is facing: the blind pining and the recurrent exposure to radiations.

We present in this section the BROS's architecture and its operating modes.

### 4.1   Architecture

BROS is a robotic platform dedicated to humeral supracondylar fracture treatment. It is able to reduce fractures, block the arm and fix the elbow bone's fragments by pinning. It also offers a navigation function to follow the pins' progression into the fractured elbow.

BROS is, as shown in the class diagram hereafter, composed of a browser (BW), a control unit (UC), a middleware (MW), a pining robotic arm (P-BROS) and 2 blocking and reducing arms (B-BROS1 and B-BROS2). The said components are detailed hereafter.

**Browser.** The browser, which is a Medtronics's product and called FluoroNav, is a combination of specialized surgical hardware and image guidance software designed for use with a StealthStation Treatment Guidance System. Together, these products enable a surgeon to track the position of a surgical instrument in the operating room and continuously update this position within one or more still-frame fluoroscopic images acquired from a C-Arm. The advantages of this virtual navigation over conventional fluoroscopic navigation include: (i) the ability to navigate using multiple fluoroscopic views simultaneously, (ii) the ability to remove the C-Arm from the operative field during navigation, (iii) significant reduction in radiation exposure to the patient and staff.

In addition, the FluoroNav System allows the surgeon to: (i) simulate and measure instrument progression or regression along a surgical trajectory, (ii) save instrument trajectories, and display the angle between two saved trajectories or between a saved trajectory and the current instrument trajectory, (iii) measure the distance between any two points in the cameras field of view, (iv) measure the angle and distance between a surgical instrument and a plane passing through the surgical field (such as the patient midplane).

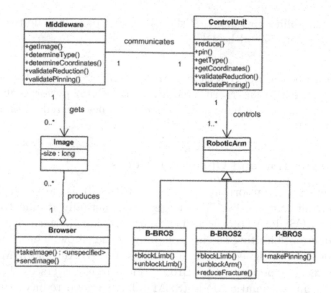

**Fig. 6.** BROS's class diagram.

Primary hardware components in the FluoroNav System include the FluoroNav Software, a C-Arm Calibration Target, a reference frame, connection cables, and specialized surgical instruments.

**Control Unit.** The CU ensures the smooth running of the surgery and its functional safety. It asks the supracondylar fracture's type to the middleware, and then computes, according to it, the different coordinates necessary to specify the robotic arms' behaviors concerning the fracture's reduction, blocking the arm and performing pinning. The surgeon monitors the intervention progress thanks to a dashboard installed on the CU.

**Middleware.** The middleware is a software installed on the browser and which acts as a mediator between the CU and the BW. It is an intelligent component that provides several features of real-time monitoring and decision making. The middleware contains several modules which are fully explained in Sect. 5: (i) an image processing module, (ii) a controller, (iii) a communication module with the CU.

**Pining Robotic Arm.** The pining robotic arm, P-BROS, inserts two parallel Kirschner wires according to Judet technique [13] to fix the fractured elbow's fragments. To insure an optimal postoperative stability, BROS respects the formula:

$$S = \frac{B}{D} > 0.22 \tag{1}$$

where $S$ is the stability threshold, $B$ the distance separating the two wires and $D$ the humeral palette's width [21].

**Blocking and Reducing Robotic Arms.** B-BROS1 blocks the arm at the humerus to prepare it to the fracture reduction. B-BROS2 performs then a closed reduction to the fractured elbow before blocking it once the reduction is properly completed.

### 4.2  Reconfiguration and Operating Modes

Reconfiguration is an important feature of BROS. It is designed to be able to operate in different modes. The surgeon can actually decide to manually do a task if BROS does not succeed to automatically perform it, whether it is facture reduction, blocking the arm or pinning the elbow. Thus, five different operating modes are designed and detailed hereafter: (i) Automatic Mode (AM): The whole surgery is performed by BROS. The surgeon oversees the operation running, (ii) Semi-Automatic Mode (SAM): The surgeon reduces the fracture. BROS performs the remaining tasks, (iii) Degraded Mode for Pining (DMP): BROS only realizes the pinning. It's to the surgeon to insure the rest of the intervention, (iv) Degraded Mode for Blocking (DMB): BROS only blocks the fractured limb. The remaining tasks are manually done by the surgeon, (v) Basic Mode (BM): The whole intervention is manually performed. BROS provides navigation function using the middleware that checks in real time the smooth running of the operation.

### 4.3  Humeral Supracondylar Fracture Treatment

To treat a humeral supracondylar fracture using BROS, the following steps are performed in the automatic mode:

1. the surgeon launches the system and chooses one of the five operating modes;
2. CU asks MW about the fracture coordinates;
3. MW requests an image from BW and the latter sends it;
4. MW determines the different coordinates by image processing and sends them to CU;
5. based on the received coordinates, CU orders B-BROS1 to block the arm at the humerus;
6. B-BROS1 blocks the limb;
7. CU asks B-BROS2 to reduce the fracture based on the latter's line;
8. B-BROS2 reduces the fracture;
9. CU asks MW to ensure that the reduction was successful;
10. MW requests a new image from BW and checks the fracture reduction result. If it is satisfactory, BROS moves to step xi. Steps from vii. to ix. are repeated otherwise;

11. CU orders B-BROS2 to block the arm;
12. under the request of UC, P-BROS performs the first and the second pinning;
13. once the pinning is successful, CU asks B-BROS1 and B-BROS2 to unblock the limb.

---

**Running Example 1**

To test our new robotized platform, we decided to simulate the surgery that would be performed on a real case. Thus, we chose a new patient, a nine-year-old girl, suffering from a fracture similar to the one presented in the case study of Sect. 3 (a a type III fracture). We simulated the whole surgery on June 9th 2014 using the software RobotStudio and the developed middleware and control unit. We will present the obtained results as we introduce these two components in the next sections.

---

## 5   Middleware

We introduce in this section the architecture of the middleware and its image processing module.

### 5.1   Architecture

The Middleware features two important modules: the first performs operations relating to image processing and the second insures the synchronization and communication with the whole robotized platform. Middleware's class diagram is illustrated in Fig. 7. Since the middleware acts as a mediator between the browser and the control unit, several data are exchanged between MW and CU during the surgery. First, the control unit notifies the start of the intervention and the activated operating mode to the middleware. Then, it asks it to compute necessary parameters like fracture's type and spatial coordinates. It also informs MW about the end of reduction and pinning. The middleware and the control unit are connected through an ad hoc network. We illustrate the different exchanges between MW and CU by a sequence diagram as shown in Fig. 8.

The controller is a module that saves the current status reached by the intervention. Indeed, the control unit informs the middleware of each fired transition and the current triggered operating mode. The control unit updates these information as the intervention advances in time. Thus, the middleware is kept aware of the progress of the surgery. This module synchronizes, then, the middleware with the whole operation. The image processing module is deeply detailed in the next section.

### 5.2   Image Processing

Image processing is the most important module of the middleware and provides a number features that we detail below.

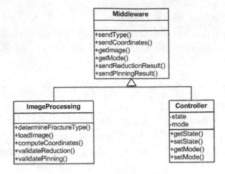

**Fig. 7.** Middleware's class diagram.

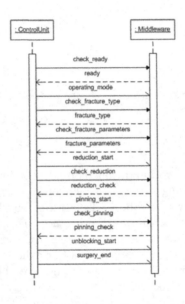

**Fig. 8.** Sequence diagram of communication with CU.

**Locating.** Locating is an important feature that involves setting a spatial reference which is considered during the whole intervention. The middleware and the
control unit must use the same coordinate system since several points coordinates
computed by MW are, firstly, sent to CU so the latter performs a preoperative
simulation and, secondly, to B-BROS and P-BROS to realize the fracture reduction and pinning. We choose to fix the coordinate system origin at the patient's
elbow as illustrated in Fig. 9. The X, Y and Z axes respectively represent the
elbow's rotation axis, the humeral palette length's median and the normal to
(XY) plan.

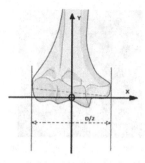

**Fig. 9.** The coordinate system axes.

**Determination of the Fracture Type.** MW starts by receiving from BW a first image of the fracture to determine its type. It compares the acquired image with the ones stored in its database. To do this, the middleware uses two image processing techniques, ensuring, thus, proper detection of the fracture type. The first one is image matching and consists in comparing images in order to obtain a measure of their similarity. It extracts invariant local features for all images, and then uses voting to rank the database images in similarity with the query image [11]. The second used image processing technique is contour comparison. It consists in detecting an image contour by quantifying the presence of a boundary at a given image location through local measurements [1]. The contour comparison is applied on the patient's elbow image acquired from BW and images stored at the database, one at a time.

---

**Running Example 2**
Fig. 10 shows the result of image matching applied on the running example's fractured elbow (on the left) and an image from the MW database (on the right). Figure 11, for its part, shows a contour comparison with another image from the database. The type III is confirmed.

---

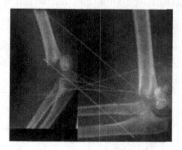

**Fig. 10.** Image matching applied on two fractured elbow images.

**Fig. 11.** Contour comparison performed by MW.

**Coordinates Transformation.** The middleware acquires images from the browser. The latter uses a system camera composed of two lenses to geometrically triangulate the spatial coordinates of each light source on the instrument, reference frame, and C-Arm Target. However, the images it sends to MW are two-dimensional, and MW needs to operate in a three-dimensional environment to properly ensure the different steps of the surgery, such as the fracture reduction and pinning. Thus, we must, first, realize a camera calibration which consists in finding the relationship between the spatial coordinates of a point in space (i.e. the operating theatre) and the associated point in the image taken by the camera [22]. To achieve the desired transformation, two type of parameters must be determined:

- the camera extrinsic parameters which define the position and orientation of camera relative to the space in which we work. Technically, determining these parameters consists in finding the translation vector between the relative positions of the origins of two references: the camera reference and the operating theatre's. A rotation vector aligning the axes of the two references must also be computed.
- the camera intrinsic parameters which are required to bind the image pixels coordinates with the corresponding ones in the camera coordinate system. These parameters present the camera optical, digital and geometric features like the focal length, the geometric distortion and image magnification factors.

Figure 12 illustrates the different used coordinate systems where: (i) (x, y) plan is the image pixels reference, (ii) (x', y', z') is the camera coordinate system, (iii) (x, y, z) is the operating theatre reference.

To translate the coordinates of a point in the image from the latter's reference to the operating theatre's and vice versa, we use the following formula:

$$S \begin{bmatrix} u \\ v \\ 1 \end{bmatrix} = M(R \begin{bmatrix} X \\ Y \\ Z_{const} \end{bmatrix} + T \tag{2}$$

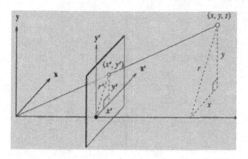

**Fig. 12.** The different coordinate systems.

where: (i) $S\begin{bmatrix} u \\ v \\ 1 \end{bmatrix}$ are the coordinates of a point in the image, (ii) $M$ is the camera matrix, (iii) $R$ represents the rotation vector, (iv) $T$ is the translation vector, (v) $(X, Y, Z_{const})$ are the coordinates corresponding to the point $S$ in the operating theater reference.

**Fracture Reduction Validation.** The validation of fracture reduction consists in checking whether the bone fragments regained their original places or not. Thus, this module detects, based on the acquired image, the bone discontinuity and, then, computes the distance between the displaced bone fragments. We hereafter explain this technique with the most common fracture types of Lagrange classification: II and III.

Validating the reduction of a type II fracture involves calculating the distances $AC$ and $BD$ as illustrated in Fig. 13. A reduction is considered successful when:

$$|AC| = |BD| = 0 \tag{3}$$

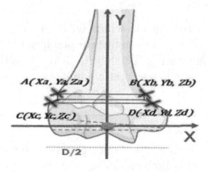

**Fig. 13.** The reduction of a type II fracture.

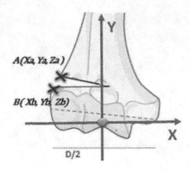

**Fig. 14.** The reduction of a type III fracture.

BROS has only three attempts to achieve a successful reduction before switching to the semi-automatic mode (SAM) to let the surgeon manually perform it. The type III fractures usually present a rotary disorder. Their reduction consists, therefore, in the rotation of the forearm with an $\alpha$ angle which is arcsin $(Z_b$ - $Z_a)$ as illustrated in Fig. 14.

**Pinning Validation.** Pinning validation amounts to checking the respect of the formula 1 introduced in Sect. 4.1 by computing the humeral palette's width and the distance separating the two pins.

## 6   Control Unit

The control unit, the entity responsible of the smooth running and the safety of surgery, is composed of several modules which we detail hereafter. We use RobotStudio to implement it and RobotWare [20] as the robot controller. Both are ABB's products.

### 6.1   Station Definition

This module implements the station which is, in our case, the operating room with all its components. The latter can be grouped into two categories: the mechanisms and the static components. The mechanisms are objects that perform 3D motion during simulations, whereas static components, as their name suggests, remain fixed during all surgery.

---

**Running Example 3**

Fig. 15 shows the implementation of our operating theater with its different robotic arms, the patient's limb modeling and the surgical bed.

---

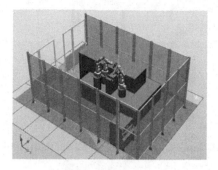

**Fig. 15.** The operating room definition.

**Mechanisms.** Our operating theatre's mechanisms are B-BROS1, B-BROS2 and PBROS. They are all ABB's IRB 120 which we earlier presented in Sect. 2.1. "Blocker 1" is the used tool to block the patient's limb at humerus and lately unblock it according to coordinates computed by the blocking module. To reduce the fracture and block the limb at forearm, "Blocker 2" is used according to coordinates received from the reduction module. Blocker 1 and Blocker 2 have the same 3D modeling illustrated in Fig. 16. "Pinning", as its name suggests, is the used tool to perform pinning at the patient's elbow according coordinates computed by the pinning module. Its 3D modeling is showed in Fig. 17.

**Fig. 16.** Blocker 1 and Blocker 2's 3D modeling.

**Fig. 17.** Pinning's 3D modeling.

**Fig. 18.** Limb's 3D modeling (Color figure online).

To simulate the progress of the surgery on the patient's limb, we model the latter as illustrated in Fig. 18. It is modeled by a mechanism that rotates about the X axis (in red).

**Static Components.** Static components are the different 3D objects which are useful to the simulation like the robotic arms' racks and the surgical bed.

## 6.2  B-BROS1 Module

B-BROS1 module describes the behavior of the robotic arm B-BROS1 and how it blocks the patient's limb at the humerus and unblocks it once the surgery is completed. Thus, this module features two procedures: (i) B_BROS1_humerusBlock (): it blocks the arm at a distance of y+100 mm where y is the coordinate on Y axis of the intersection point of the humeral palette and its median. Figure 19 illustrates how the blocking is performed, (ii) B_BROS1_humerusUnblock (): it releases the patient's limb once the fracture treatment is completed.

**Fig. 19.** Blocking the patient's limb.

### 6.3   B-BROS2 Module

This module features several procedures which allow robotized fracture reduction when the automatic mode is triggered and direct robotized arm blocking when AM, SAM or DMB is triggered. B-BROS2 module releases the patient's limb once the surgery is completed. We, hereafter, detail the procedures: (i) B_BROS2_reduce_II (A, B, C, D): it performs the reduction of a type II fracture and takes into account the parameters that we defined in Sect. 5.2. Figure 20 illustrates a robotized fracture reduction, (ii) B_BROS2_unblock_II (): this procedure unblock the patient's limb suffering from a type II fracture once the surgery is completed, (iii) B_BROS2_reduce_III (A, B): it computes the rotation angle of the rotary disorder in the case of a type III fracture and, then, reduces the latter, (iv) B_BROS2_block (): the procedure blocks the limb at the forearm once a manual reduction is performed during SAM or DMB. Figure 21 shows how this is performed.

### 6.4   P-BROS Module

This section describes the behavior of P-BROS, the robotic arm performing fracture reduction according to its type and the triggered operating mode. We point out that the used pinning technique is Judet's which we mentioned in Sect. 4.1. The orientation of the tool "Pinning" (Sect. 6.1), relatively to the coordinate system defined in Sect. 5.2, depends on the type of the fracture. Thus, the Figs. 22 and 23 respectively shows the orientation of "Pinning" in the case of a type II and a type III fractures.

**Fig. 20.** Robotized fracture reduction.

**Fig. 21.** Blocking the fractured limb at the forearm.

**Fig. 22.** Orientation of "Pinning" in the case of a type II fracture.

**Fig. 23.** Orientation of "Pinning" in the case of a type III fracture.

The P-BROS module features several procedures that we hereafter detail: (i) P_BROS_DoublePin (A, B, C, D, HP): it performs a parallel pinning using two pins inserted from the external condyle to the lateral humeral column in the case of a type II fracture which requires a double pinning. The procedure uses as parameters the four points of the distal dissolution and the width of the humeral palette (HP), (ii) P_BROS_SinglePin_III (A, B, HP): this procedure performs a percutaneous pinning for a type III fracture. The pin is actually inserted from the external condyle throughout the medial column in a rectilinear direction by keeping a fixed (XY) plane, (iii) P_BROS_SinglePin_IV (A, B, HP): it realizes a percutaneous pinning for a type IV fracture. Indeed, for this type of fracture, the pin is inserted in the lateral condyle and makes an angle of 45 relative to the orientation of the pin in the case of a type III fracture. The pin is inserted until reaching the lateral column.

## 6.5    Synchronization Module

We present, in this section, the synchronization module of the control unit. It is the entity that insures the coordination between the tasks of B-BROS1, B-BROS2 and P-BROS modules. To insure this function, we use interruptions through binary logic signals. Indeed, each signal corresponds to a very specific task. The signal is high when the task is running and low when it is idle or finished executing. We note that the used signals represent the steps of a fracture treatment based on the operating mode and regardless to the nature of a given action (robotized or manual).

We define for the control unit the following logic signals which we detail in Table 2:

Table 2. Synchronization logic signals.

| Logic Signal | Description |
| --- | --- |
| HandBlocking | This signal controls the first step of a fracture treatment which is blocking the patient's limb at the humerus. It is the highest priority task. The signal is high when B-BROS1 starts blocking the humerus and it switches to low once blocking is finished |
| HandReduction | The signal controls the fracture reduction and the forearm blocking. It switches to high when HandBlocking is low and either B-BROS2 starts the robotized reduction and/or blocking or the surgeon starts the manual reduction and/or blocking. It is the second priority task |
| HandPinning | HandPinning controls pinning, whether it is manual or robotized. It changes to high when the signal HandReduction changes to low informing, thus, that reduction and blocking are finished. When it switches to high, HandPinning starts pinning and switches to low once it is finished |

## 6.6  CU-MW Communication Module

A good communication between the control unit and the middleware is critical to the smooth functioning of BROS. For example, the control unit cannot start the different processing until it receives key parameters like the fracture type and the coordinates of the points of the distal fragment discontinuity. The module respects the diagram presented in Sect. 5.1.

## 6.7  Surgeon-Robot Interface

It is the graphical interface through which the surgeon communicates with the platform and oversees the progress of the operation. The surgeon can, using this interface, choose the operating mode to start with. Through this GUI, the surgeon consults any medical parameter like the fracture type, the displacement nature or the angle of the rotational trouble in the case of type III fractures. This interface meets the man-machine requirements like: (i) Guidance: All resources used to guide the surgeon during the use of the interface like grouping/distinction, immediate feedback and legibility, (ii) Workload: Minimum and explicit actions ("start reduction", "start pinning"), informational density more or less acceptable for a surgeon, (iii) Error management: This is to protect sensitive actions against errors with error messages, (iv) Ergonomics: The interface must be flexible and adaptable to a surgeon and especially in an operating room.

> **Running Example 4**
> The whole surgery was successfully performed by BROS under the automatic operating mode and simulated using RobotStudio and RobotWare. Only 4 fluoroscopic images were needed, what makes 21 images less than in the study case introduced in Sect. 3. BROS insured all the intervention steps and the surgeon had only to remotely check the smooth running of the surgery and be ready to intervene in the case where the robotized platform would not be able to perform one of the surgery's steps or he would judge that a human intervention is necessary.

## 7   Conclusion and Perspectives

Our work consisted, through this paper, in introducing BROS, this new robotic platform dedicated to the treatment of supracondylar humerus fracture, and its contributions. BROS is a flexible system since it may run under different operating modes to meet the surgeon requirements and the environment constraints: it is reconfigurable. Through the simulation of a real case of BROS-assisted surgery, we proved the usefulness of this robotic platform to avoid the complications that may be generated because of the blind pinning and prevent the danger posed by the recurrent exposition to radiations. We can, now, certify that BROS is an innovating project which will be of a great help to pediatric orthopedic surgeons. The next step is to proceed to the real implementation of BROS using the ABB robotic arms.

**Acknowledgements.** This research work is carried out within a MOBIDOC PhD thesis of the PASRI program, EU-funded and administered by ANPR (Tunisia). The BROS national project is a collaboration between the Children Hospital of Bchir Hamza (Tunis), eHTC and INSAT (LISI Laboratory) in Tunisia. We thank the medical staff, Prof.Dr.med. Mahmoud Smida (Head of Child and Adolescent Orthopedics Service) and Dr.med. Zied Jlalia, for their fruitful collaboration and continuous medical support. A second paper is submitted in the conference for the modeling and verification of BROS.

## References

1. Arbelaez, P., Maire, M., Fowlkes, C., Malik, J.: Contour detection and hierarchical image segmentation. IEEE Trans. Pattern Anal. Mach. Intell. **33**(5), 898–916 (2011)
2. Barton, K.L., Kaminsky, C.K., Green, D.W., Shean, C.J., Kautz, S.M., Skaggs, D.L.: Reliability of a modified Gartland classification of supracondylar humerus fractures. J. Pediatr. Orthop. **21**(1), 27–30 (2001)
3. Cardwell, M.: An IRB 120 robot picks and packs tubes of hair color into boxes for loreal canada. www.abb.com

4. Cleary, K., Nguyen, C.: State of the art in surgical robotics: clinical applications and technology challenges. Comput. Aided Surg. **6**(6), 312–328 (2001)
5. Clein, N.W.: How safe is X-ray and fluoroscopy for the patient andthe doctor? J. Pediatr. **45**(3), 310–315 (1954)
6. Mikaelsson, P., Curtis, M.: Portrait-robot d'un petit prodige: ABB prsente son nouveau robot IRB 120 et son armoire de commande IRC5 Compact. Rev. ABB **4**, 39–41 (2009)
7. Connolly, C.: Technology and applications of ABB RobotStudio. Ind. Robot. Int. J. **36**(6), 540–545 (2009)
8. Emmerson, B.: IRB 120 inserting thermoplastic trays in to boxes for BDMO. www. abb.com
9. Flynn, J.C., Matthews, J.G., Benoit, R.L.: Blind pinning of displaced supracondylar fractures of the humerus in children. J. Bone Joint Surg. **56**(2), 263–272 (1974)
10. Gosens, T., Bongers, K.J.: Neurovascular complications and functional outcome in displaced supracondylar fractures of the humerus in children. Injury **34**(4), 267–273 (2003)
11. Grauman, K., Darrell, T.: Efficient image matching with distributions of local invariant features. In: IEEE Computer Society Conference on Computer Vision and Pattern Recognition, 2005. CVPR 2005, vol. 2, pp. 627–634. IEEE, June 2005
12. Haque, M.U., Shufflebarger, H.L., OBrien, M., Macagno, A.: Radiation exposure during pedicle screw placement in adolescent idiopathic scoliosis: is fluoroscopy safe? Spine **31**(21), 2516–2520 (2006)
13. Judet, J.: Traitement des fractures sus-condyliennes transversales de lhumrus chez lenfant. Rev. Chir. Orthop. **39**, 199–212 (1953)
14. Kwoh, Y.S., Hou, J., Jonckheere, E.A., Hayati, S.: A robot with improved absolute positioning accuracy for CT guided stereotactic brain surgery. IEEE Trans. Biomed. Eng. **35**(2), 153–160 (1988)
15. Lagrange, J., Rigault, P.: Fractures supracondyliennes. Rev. Chir. Orthop. **48**, 337–414 (1962)
16. Landin, L.A.: Fracture patterns in children: analysis of 8,682 fractures with special reference to incidence, etiology and secular changes in a swedish urban population 1950–1979. Acta Orthop. **54**(S202), 3–109 (1983)
17. Landin, L.A., Danielsson, L.G.: Elbow fractures in children: an epidemiological analysis of 589 cases. Acta Orthop. **57**(4), 309–312 (1986)
18. Pirone, A.M., Graham, H.K., Krajbich, J.I.: Management of displaced extension-type supracondylar fractures of the humerus in children. J. Bone Joint Surg. **70**(5), 641–650 (1988)
19. Rampersaud, Y.R., Foley, K.T., Shen, A.C., Williams, S., Solomito, M.: Radiation exposure to the spine surgeon during fluoroscopically assisted pedicle screw insertion. Spine **25**(20), 2637–2645 (2000)
20. Robotics, A.B.B.: Application Manual: Motion Coordination and Supervision, Robot Controller, RobotWare 5.0. Vsters, Sweden (2007)
21. Smida, M., Smaoui, H., Jlila, T.B., Saeid, W., Safi, H., Ammar, C., Ghachem, M.B.: Un index de stabilit pour lembrochage percutan latral parallle des fractures supracondyliennes du coude chez lenfant. Rev. Chir. Orthop. Rparatrice Appar. Mot. **93**(4), 404 (2007)
22. Tsai, R.Y.: A versatile camera calibration technique for high-accuracy 3D machine vision metrology using off-the-shelf TV cameras and lenses. IEEE J. Robot. Autom. **3**(4), 323–344 (1987)

# Real-Time Fuzzy Monitoring of Sitting Posture: Development of a New Prototype and a New Posture Classification Algorithm to Detect Postural Transitions

Leonardo Martins[1,2(✉)], Bruno Ribeiro[1], Hugo Pereira[1],
Rui Almeida[1], Jéssica Costa[1], Cláudia Quaresma[3], Adelaide Jesus[1],
and Pedro Vieira[3]

[1] Department of Physics, Faculdade de Ciências e Tecnologias,
Universidade Nova de Lisboa, Quinta da Torre, 2829-516 Caparica, Portugal
{l.martins,bmf.ribeiro,pl10595,
ja.domingues}@campus.fct.unl.pt,
rui.almeida@ngns-is.com, ajesus@fct.unl.pt
[2] UNINOVA, Institute for the Development of New Technologies,
Quinta da Torre, 2829-516 Caparica, Portugal
[3] LIBPhys-UNL, Department of Physics, Faculdade de Ciências e Tecnologias,
Universidade Nova de Lisboa, 2829-516 Monte da Caparica, Portugal
{q.claudia,pmv}@fct.unl.pt

**Abstract.** In a previous work, a chair prototype was used to detect 11 standardized siting postures of users, using just 8 air bladders (4 in the chair's seat and 4 in the backrest) and one pressure sensor for each bladder. In this paper we describe the development of a new prototype, which is able to classify 12 standard postures with an overall score of 80.9 % (using a Neural Network Algorithm). We tested how this Algorithm worked during postural transitions (frontal and lateral flexion) and in intermediate postures, identifying some limitation of this Algorithm. This prompted the development of a Posture Classification Algorithm based on Fuzzy Logic and is able to determine if the user is adopting a good or a bad posture for specific time periods, using as input the Centre of Pressure, the Posture Adoption Time and the Posture Output from the existing Neural Network Algorithm. This newly developed Classification Algorithms is advancing the development of new Posture Correction Algorithms based on Fuzzy Actuators.

**Keywords:** Intelligent chair · Pressure-distribution sensors · Sitting posture · Posture classification · Fuzzy logic · Neural Networks

## 1 Introduction

Society sedentary behaviours are influenced by many factors and on various domains, including spending extended periods of time in a sitting position in a variety of settings, such as the occupational workspace, transports, leisure activities and household activities [1, 2].

© Springer International Publishing Switzerland 2015
A. Fred et al. (Eds): BIOSTEC 2015, CCIS 574, pp. 424–439, 2015.
DOI: 10.1007/978-3-319-27707-3_26

Low back pain has been identified as one of the leading causes of work-related disability and loss of productivity in industrialized countries [3, 4], but systematic studies haven't been able to establish a causal and independent relationship between occupational sitting and low back pain [5, 6], although an increasing risk of this disorder has been associated with sitting during extended periods of time [7].

Back-related pain have been associated with prolonged sitting behaviours, which can results in a decrease on the lumbar lordosis [8, 9] which then increases physical risk factors related to back, neck and shoulder injuries [10, 11]. This pain is due to anatomical changes and degeneration of the intervertebral disks and joints [12, 13].

Recent studies have shown that a consensus on what comprises a neutral spine posture in a sitting posture has still not been grasped [14], due to the difficulty of doing quantitative clinical studies that target the identification of 'correct' and 'incorrect' postures. Three main approaches have been reported to evaluate the posture: by the use of multiple camera sensors to build a 3D optical model of the body [15], by the use of sensors applied directly to the skin to assess the spinal angles [16] and by the calculus of the relative distance between anatomical landmarks of the spine [17].

In a normal sitting position, the support of the person's weight is transferred to the ischial tuberosities, the thigh muscles and pelvic girdle, while the remaining pressure load is transferred to the ground, by the feet, to the backrest and to the armrests [15]. In most of the 'incorrect' postures the trunk is not totally supported in the backrest due to lateral flexion of the upper part of the trunk, forward or backward inclination or leg crossing. In these positions, weight is incorrectly distributed, which can have an amplifying effect in back and neck associated disorders [16].

To promote a more dynamic sitting posture and increase physical activities, chair manufacturers have developed office chairs with structural elements such as seat pan motors, seat pan suspension and moveable joints that allow movement in the horizontal plane or even freely in all directions. Studies have shown that while the task being performed strongly affects body dynamics, different chairs for the same task do not [17]. This suggests the need to develop an intelligent chair, capable of identifying user posture or the associated tasks and alerting him of a prolonged sitting behaviour. Such intelligent chairs have been developed in recent years by numerous research groups, by implementing sheets of surface-mounted pressure sensors placed in a 2D array or using pure mathematical and statistical approaches to find the best way to place singular force-sensitive resistors in the chair or even conductive textiles [18–27]. These intelligent chairs have been built to be capable of detecting the presence of a person and his sitting posture, alerting the user to improve bad posture habits. They can be used as a physiological monitor and create report tools of everyday activities and for that purpose, some are already being implemented in real homes for year-long tests [28].

In a previous work, a chair prototype was developed to classify 11 standardized seating postures of users, using 8 air bladders (4 in the chair's seat and 4 in the backrest) by using a pressure sensor for each bladder to detect the bladder's interior air pressure [29]. The previous prototype's bladders are shown in Fig. 1-A. The classification algorithm was based on Neural Networks and Decision trees and was able to make a real-time overall classification 93 % (for eight postures) and dropped to 70 % (for the eleven standardized postures) [29].

In this work, we show the classification results of a newly developed prototype for 12 standardized postures, and we show how the classification algorithms work during a lateral and frontal flexion. Two different behaviours were identified, one where the classification algorithms goes through a Transition Zone, where the Classification is not stable, and one where the classification does not have a Transition Zone. The observed results from the classification results coming from the Artificial Neural Network (ANN) Algorithm prompted the integration of Fuzzy Logic to cope with the instability of the ANN during transitions and with the inability of the ANN to accurately differentiate intermediate postures (for example the classification of a lateral flexion of 20° and a lateral flexion of 25°).

## 2   Experimental Section

### 2.1   Equipment

In a previous work, a chair prototype was built with a 2-by-2 matrix of air bladders (see Fig. 1-A). Each individual air bladder was made with thermoplastic polyurethane, was manually sealed and had $20 \times 19$ cm. They were placed in the seat pan and the backrest (as can be seen in Fig. 1). This arrangement covers the most important and distinguishable areas of the seated posture [15], such as the ischial tuberosities, the posterior thigh region, the lumbar region of the spine and the scapula [29].

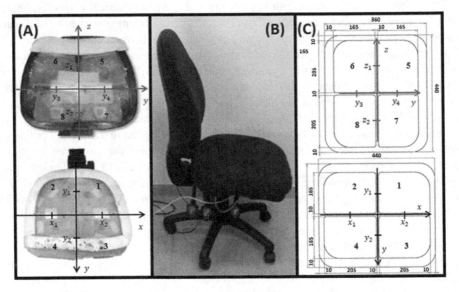

**Fig. 1.** (A) Interior of the previous prototype and Centre of Pressure (COP) measurement (B) New Prototype (C) Schematic of the new air bladder system and COP measurement. Bladder 1 (Back Left – BL), Bladder 2 (Back Right – BR), Bladder 3 (Front Left - FL), Bladder 4 (Front Right - FR), Bladder 5 (Up Left – UL), Bladder 6 (Up Right – UR), Bladder 7 (Down Left – DL) and Bladder 8 (Down Right – DR).

For this project we started by projecting a new set of air bladders, which were industrially manufactured by Aero Tec Laboratories Inc. (that guaranties the volume inside of each bladder is the same when they are fully inflated) and integrated in a different office chair (seen in Fig. 1-B). The new bladder system and its measurements can be seen in Fig. 1-C wherein the bladder placement had the same strategy of the previous work. The original padding foam was also used and placed above the pressure bladders, in order to keep the original anatomical cut of the seat pad and backrest. For the air pressure measurements inside the bladder, it was used the Honeywell 24PC Series piezoelectric gauge pressure sensor, with a max rate of 15 psi and with a sensitivity of 15 mV/psi. The values of $x_1$, $x_2$, $z_1$, $z_2$ correspond to around 10.5 cm and $y_1$, $y_2$, $y_3$, $y_4$ to around 8.5 cm. These values are calculated by halving the bladders lateral dimensions (20.5 and 16.5 cm), adding half of the 10 mm spacing between each bladder, and making rounding adjustments due to the bladders curvature.

Bluetooth communication was also added to this prototype, as we had done in the previous prototype, making it capable to transfer the daily postural information to computers and smartphones, allowing a statistical analysis of the postures taken during a working day in an office environment [29].

## 2.2 Experimental Procedure and Participants

Three experiments were done for this work. The first one (A) followed a similar protocol then in the previous work [29] and served for data acquisition in order to create the Seated Posture Classification Algorithms, based on Neural Networks. The second (B) and third (C) experiments were used to study how the Neural Network behave during standard posture changes (in intermediate postures).

We increased the number of subjects from the previous experiments [29], from 30 to 50 and also tried to have a more indicative sample of office workers (increasing the age of the participants from around 21 to 26). The dataset for both experiments is presented in Table 1. Based on the knowledge from previous experiments [29] and since now the bladders have exactly the same volume, we inflated all the bladders for 5 s (instead of having different inflation times for each bladder) so we could take precise reading of the bladder interior pressure, but not enough to originate discomfort to the users. Before the experiment, participants were asked to empty their pockets and to adjust the stool height to the popliteal height.

**Table 1.** Data of the participants in the experiment, namely, Sex, Age, Weight and Height. Note: values for average ± standard deviation and (M/F) corresponds to (male/female).

| No. of subjects (M/F) | Age (years)[a] | Weight (Kg)[a] | Height (cm)[a] |
|---|---|---|---|
| Experiment A | | | |
| 50 (25/25) | 26,4 ± 9,9 | 66,8 ± 12,1 | 170,9 ± 10,0 |
| Experiment B | | | |
| 12 (6/6) | 25,8 ± 6,6 | 72,8 ± 12,1 | 173,1 ± 10,7 |
| Experiment C | | | |
| 12 (6/6) | 27,3 ± 8,7 | 71,0 ± 16,1 | 170,6 ± 6,0 |

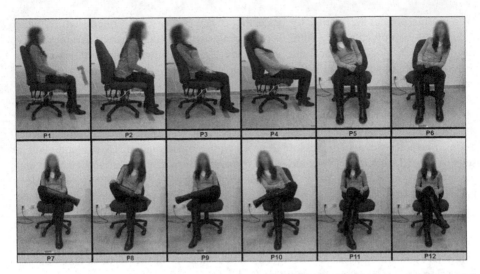

**Fig. 2.** Seated postures their respective class label: (P1) seated upright, (P2) leaning forward, (P3) leaning back, (P4) leaning back with no lumbar support, (P5) leaning left, (P6) leaning right, (P7) right leg crossed, (P8) right leg crossed, leaning left, (P9) left leg crossed, (P10) left leg crossed, leaning right, (P11) left leg over right, (P12) right leg over left.

For the first experiment, we used a similar protocol as in the previous work [29], and showed a presentation of postures P1 to P12 (as can be seen in Fig. 2), each for a duration of 20 s. First we asked the participants to mimic the postures without leaving the chair. Then we asked them to repeat the same postures two more times, but after every postural change we asked the participant to walk out of the chair and move to a certain point in the room and sit down again. Even though most of the chosen postures were the same as the previous work [29], which were based on the most familiar postures observed in office environments [18–21, 23, 30], in this work we have added two new postures (P11 and P12) that are observed as a common posture in women, and we removed a posture which was intermediate between P1 and P2 and had a slight frontal inclination with trunk support (this was because people had trouble mimicking the posture, making it more difficult for the classifier to correctly detect it). We note that the usage of Fuzzy Logic is also one of the reasons for dropping that posture, because we show that we can detect such intermediate postures.

As in previous work, in experiment A we also didn't use the entire 20 s of acquired data, because when a user changes its posture, the sensed pressure values oscillate (Transition zone) and then stabilize (Stable zone) as shown in Fig. 3. The chosen data to be used as input was extracted only from the Stable zone.

Here, using a sampling rate of 8 Hz (which is enough to classify sitting posture behaviour), we took 100 time-points (which correspond to 12.5 s out of 20 for each posture), and divided them in groups of 20 pressure acquisition. Each group was

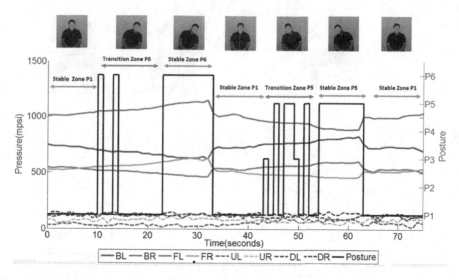

**Fig. 3.** Example of one cycle of Experiment B. The transition from a Stable Zone in P1 to P6 (in green arrows) and then, P1 to P5 goes through a Transition Zone (in red arrow), where the classification is intermittent. Legend of each bladder is according to the Fig. 1. (Color figure online)

averaged, forming 5 pressure maps that serve as input to the Neural Networks, contributing to a total of 4500 pressure maps (50 participants * 12 postures * 5 averages * 3 repetitions).

With experiment B, we wanted to see how the Neural Network algorithm worked during the Transition zone (see Fig. 4) between Posture P1 and Postures related to lateral flexion (which for this first experiment is only done for P5 and P6, as lateral flexion during leg crossing was not tested). For this, we asked the participants to move from P1 to P5, back to P1, then move to P6 and back to P1 (repeating this, 5 times). We observed that during a postural transition, the classification was intermittent (this corresponds to the Transition Zone identified in Fig. 4). So for this experiment we took a picture when the ANN algorithm first changed its value (entering the Transition Zone) and a second picture when the algorithm kept giving the right answer (entering the Stable Zone). Experiment C was done with a similar protocol from experiment B, but here we asked the subjects to perform a frontal flexion (P2 Posture and then doing an extension movement back to P1), repeating 5 times the movement between Posture P1 and Posture P2. During the frontal flexion, no Transition Zone was identified, as the ANN-based algorithm was able to immediately change the classification of Posture P2 after a certain angle between the trunk and the thighs was attained. The observed results for these two experiments (B and C), prompted the integration of Fuzzy Logic to cope with the instability of the ANN during transitions (especially detected during a lateral flexion) and with the inability of the ANN to accurately discern intermediate postures (for example the classification of a frontal flexion of 15° and a frontal flexion of 20°).

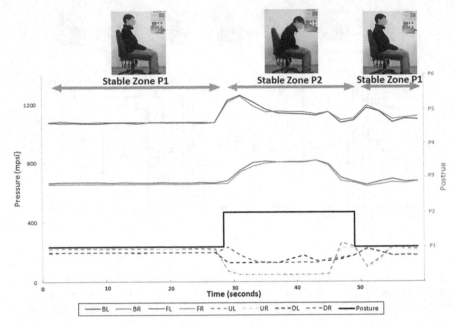

**Fig. 4.** Example of the pressure map for one cycle of Experiment C. The transition from a Stable Zone in P1 to P2 (in green arrows). Legend of each bladder is according to the Fig. 1. (Color figure online)

## 2.3   Classification Methods

Artificial neural network (ANN) based classification algorithms have been shown to be useful in many engineering and biomedical applications [31]. In the preceding works we have used them and since they are able to tackle our multiclass problem in a very satisfactory manner, we decided to continue using them. They also have the advantage of being easily implemented in other systems (by importing weights and bias matrices).

For ANN parameterization, we re-tested the same parameters as in the previous work, by making combinations of layers, neurons, training and transfer functions [29].

Fuzzy logic and specially designed Neuro-fuzzy integrations have also been used in many engineering and biomedical applications [32]. In this case, we opted to use the seat and backrest COPs (Centres of Pressure - calculated through Eqs. 1–4 and using the geometrical information from the bladders represented in Fig. 1), the Neural Network posture output and time interval in that posture output as inputs for the Neuro-Fuzzy algorithm. This algorithm evaluates the posture each 15 min, a time that was chosen based and adapted from ISO standards for max trunk inclination (which has been reported for upwards position, but can be updated to the trunk inclination in a sitting position and without back support) [33].

$$x_{seat} = \frac{(P_2 + P_4) \, x \, x_1 + (P_1 + P_3) \, x \, x_2}{\sum_{i=1}^{4} P_i} \tag{1}$$

$$y_{seat} = \frac{(P_1 + P_2) \, x \, y_1 + (P_3 + P_4) \, x \, y_2}{\sum_{i=1}^{4} P_i} \tag{2}$$

$$y_{back} = \frac{(P_6 + P_8) \, x \, y_3 + (P_5 + P_7) \, x \, y_4}{\sum_{i=5}^{8} P_i} \tag{3}$$

$$z_{back} = \frac{(P_5 + P_6) \, x \, z_1 + (P_7 + P_8) \, x \, z_2}{\sum_{i=5}^{8} P_i} \tag{4}$$

## 3 Experimental Results and Discussion

### 3.1 Results from Experiment A

As in the previous study, the best result was again using the Resilient Back-Propagation algorithm, with the Tansig function, but now with one layer of forty neurons (comparing with one layer of fifteen neurons from the previous work). This result shows that our new bladder system changed slightly the pressure maps, which influenced how the classification algorithms adapted. The confusion matrix for the training, validation and testing data, using the best parameters is shown in Table 2.

**Table 2.** Confusion Matrix for posture classification of the training data, where rows indicate the Output Class and columns indicates the Target Class. The Target Class labels correspond to the respective postures from Fig. 3. The grey boxes in the main diagonal give the output classes that were correctly classified as the target class. The row and column in grey give the percentages of correct classification in the respective class and the overall classification score.

| | P1 | P2 | P3 | P4 | P5 | P6 | P7 | P8 | P9 | P10 | P11 | P12 | (%) |
|---|---|---|---|---|---|---|---|---|---|---|---|---|---|
| P1 | 911 | 7 | 54 | 4 | 1 | 4 | 6 | 2 | 30 | 5 | 0 | 4 | 88.6 |
| P2 | 19 | 989 | 11 | 6 | 21 | 8 | 31 | 9 | 8 | 9 | 8 | 7 | 87.8 |
| P3 | 27 | 9 | 815 | 42 | 0 | 11 | 102 | 4 | 41 | 16 | 0 | 1 | 76.3 |
| P4 | 0 | 3 | 26 | 957 | 13 | 0 | 5 | 9 | 21 | 5 | 19 | 28 | 88.1 |
| P5 | 0 | 6 | 5 | 1 | 860 | 1 | 33 | 5 | 1 | 137 | 16 | 0 | 80.8 |
| P6 | 0 | 0 | 0 | 1 | 5 | 815 | 5 | 115 | 27 | 5 | 0 | 30 | 81.3 |
| P7 | 19 | 0 | 63 | 0 | 14 | 7 | 671 | 27 | 22 | 16 | 17 | 27 | 76.0 |
| P8 | 0 | 0 | 2 | 2 | 0 | 102 | 26 | 777 | 7 | 0 | 3 | 54 | 79.9 |
| P9 | 44 | 1 | 13 | 1 | 2 | 31 | 14 | 20 | 722 | 58 | 30 | 45 | 73.6 |
| P10 | 0 | 1 | 24 | 4 | 86 | 0 | 11 | 0 | 35 | 721 | 23 | 7 | 79.1 |
| P11 | 0 | 1 | 1 | 6 | 23 | 0 | 71 | 14 | 22 | 46 | 900 | 10 | 82.3 |
| P12 | 5 | 8 | 11 | 1 | 0 | 46 | 50 | 43 | 89 | 7 | 9 | 812 | 75.1 |
| (%) | 88.9 | 96.5 | 79.5 | 93.4 | 83.9 | 79.5 | 65.5 | 75.8 | 70.4 | 70.3 | 87.8 | 79.2 | **80.9** |

## 3.2   Results from Experiment B

For experiment B, our first study was related with how the transition between Position P1 and Lateral Flexion (both left and right) influenced the classification algorithm. Figure 3 shows one cycle of this experiment, showing the Transition Zones between postures.

To calculate the angle where these transitions occur, we took a picture of the participants every time they entered a new Zone and then measured the trunk inclination in Position P1 (for reference) and the trunk inclination in the other positions. In order to include the variability among raters, these measurements were performed by 3 different experts, so we also calculated the degree of reliability among raters by using two way Intra Class Correlation (ICC) for each Transition Angle measurement [34].

**Table 3.** Measurement of the lateral angle when the Transition Zone is identified from P1 to the respective Posture, and when the angle at the Stable Zone is identified of the respective Posture. Note: angle values are for average ± standard deviation.

| Subject | 1st – P5 (/°) | 2nd – P5 (/°) | 1st – P6 (/°) | 2nd – P6 (/°) |
|---|---|---|---|---|
| 1 | 10.7 ± 2.2 | 17.5 ± 2.6 | 12.9 ± 3.4 | 20.6 ± 2.4 |
| 2 | 6.5 ± 0.7 | 11.8 ± 1.1 | 6.7 ± 1.2 | 11.4 ± 1.0 |
| 3 | 13.5 ± 2.4 | 19.6 ± 2.7 | 16.7 ± 3.6 | 23.9 ± 2.2 |
| 4 | 8.4 ± 1.4 | 15.1 ± 1.9 | 9.2 ± 1.1 | 14.0 ± 1.6 |
| 5 | 11.6 ± 3.4 | 22.2 ± 2.9 | 13.9 ± 3.0 | 25.7 ± 6.0 |
| 6 | 12.9 ± 2.2 | 21.4 ± 2.5 | 17.3 ± 3.5 | 22.9 ± 5.0 |
| 7 | 16.2 ± 3.6 | 23.2 ± 2.5 | 26.7 ± 4.3 | 35.5 ± 3.3 |
| 8 | 11.3 ± 1.4 | 17.2 ± 1.6 | 10.4 ± 1.5 | 16.8 ± 2.4 |
| 9 | 7.4 ± 1.7 | 12.9 ± 2.7 | 10.4 ± 1.6 | 15.2 ± 1.8 |
| 10 | 11.5 ± 1.4 | 17.3 ± 2.3 | 11.5 ± 1.4 | 17.2 ± 2.1 |
| 11 | 12.0 ± 2.0 | 16.8 ± 2.4 | 12.8 ± 1.8 | 18.1 ± 2.7 |
| 12 | 10.5 ± 2.6 | 16.6 ± 1.7 | 13.7 ± 2.3 | 18.7 ± 2.9 |
| Average | 11.0 ± 3.4 | 17.6 ± 4.0 | 13.5 ± 5.5 | 20.0 ± 6.9 |
| ICC | 0.53 | 0.74 | 0.77 | 0.70 |

The average and standard deviation values for each Transition Angle and the ICC score are presented in Table 3. In the first Transition, we identified an average angle of 11.0° (for P5) and 13.5° (for P6) and in the second Transition we identified an angle of 17.6° and 20.0°, respectively for P5 and P6. These values are in good agreement with the identification of a lateral trunk bending angle, which has normally been defined at 15° [35]. This angle is in our Transition Zone, where we are able to classify as a lateral bending posture (P5 or P6), but there are also mistakes with other postures, due to shear movement (P3 and P1). The ICC values show that the angles measured by each expert rater have good reliability (values between 0.5 and 0.75 have a good reliability, and higher than 0.75 have an excellent reliability), validating our measurements [34].

We use a two-sample Kolmogorov-Smirnov Test (KS Test) to check if the obtained P5 and P6 angles from each subject are drawn from the same distribution [36]. This is because if the user was correctly placed in the chair (during the calibration), the classification algorithm should have similar posture detection for a left inclination and for a right inclination (meaning that it should be an unbiased classification algorithm).

Table 4 shows the results of the KS tests for each Transition and for each subject at a 1 % significance level [36]. The first Test checked the null hypothesis that the data obtained from the 1st detected angle for both posture P5 and P6 are from the same distribution. The second Test checks the same nulls hypothesis but for the angles of the second transition. For a 1 % significance level, the null hypothesis is rejected (value of 1 in Table 4) for p-values inferior to 0.01.

**Table 4.** Two-sample of the KS test at a 1 % significance level. Test 1 checks the null hypothesis that the data from the 1st angles of P5 and P6 are from the same distribution. Test 2 checks the null hypothesis for the 2nd angles.

| Subject | Test 1 | p-value | Test 2 | p-value |
|---------|--------|---------|--------|---------|
| 1 | 0 | 1.6786e-02 | 0 | 1.6786e-02 |
| 2 | 0 | 3.0794e-01 | 0 | 8.8990e-01 |
| 3 | 0 | 5.1467e-02 | 1 | 2.3766e-04 |
| 4 | 0 | 5.8861e-02 | 0 | 1.3586e-01 |
| 5 | 0 | 1.6786e-02 | 0 | 5.1467e-02 |
| 6 | 1 | 4.7152e-03 | 0 | 3.0794e-01 |
| 7 | 1 | 8.7713e-07 | 1 | 1.0054e-07 |
| 8 | 0 | 1.3586e-01 | 0 | 5.8861e-01 |
| 9 | 0 | 5.1467e-02 | 0 | 5.1467e-02 |
| 10 | 0 | 9.9832e-01 | 0 | 9.9832e-01 |
| 11 | 0 | 8.8990e-01 | 0 | 3.0794e-01 |
| 12 | 0 | 5.1467e-02 | 0 | 1.6786e-02 |

The results from the KS Test, indicated that participant 7 had both angle measurements rejected, which is indicative of an unbiased lateral flexion for both Transitions. One of the main reasons for this rejection might be that the participant's has an anthropometric data that was most afar from the mean (a height of 1,58 m compared to 1,73 m and 48,0 kg compared to 72.8 kg). We also detected some postural asymmetry during the calibration phase, which can have a biasing effect during the measurements of trunk lateral inclination. There were 2 other measurements that were rejected (participant 6 – test 1 and participant 3 - test 2), which indicates that the classification system may have a small bias to one of the lateral sides (we detected that the leaning right position had a slightly larger calculated angle in most subjects).

## 3.3   Results from Experiment C

For experiment C, we studied how the classification algorithm acts during a frontal flexion. In this case we didn't observe a Transition Zone as during a lateral flexion, since when the classification changed from P1 to P2, it did not change back to P1. Due to this, only one angle was calculated (the flexion angle in the sagittal plane) which for each subject ascertaining the angle that makes the transition between classified postures. Figure 4 shows an example of one cycle of Experiment C for one subject, and we can check that in this case, we do not observe a Transition Zone.

The angles were also measured by 3 different experts (as in experiment B), in order to calculate ICC of the Transition Angle measurement [34]. Table 5 shows the measured averages angles for the Transition between detected postures and the ICC of the measurements. The ICC obtained value (0.80) show that the angles measured by each expert have good reliability (values higher than 0.75 have an excellent reliability), which also validates the measurements of the obtained angles by 3 different experts [34]. In the first Transition, we identified an average angle of 13,3° for the Transition between P1 and P2. Frontal flexion has normally been defined at 30° [35], so it is necessary that our classification algorithms correctly identifies posture P2 before the 30°. Only subject 4 had a Transition Angle close to 30° (but still inferior to 30°), so our classification algorithm was shown to be useful in identifying a correct flexion of the trunk.

**Table 5.** Measurement of the angles subjects between the trunk and the thighs during a frontal flexion, from which the classification algorithm changes from Posture P1 to Posture P2. Note: angle values are for average ± standard deviation.

| Subject | 1 | 2 | 3 | 4 |
|---|---|---|---|---|
| Frontal flexion angle (/°) | 8,9 ± 0,9 | 16,0 ± 1,6 | 14,9 ± 2,3 | 23,8 ± 3,4 |
| Subject | 5 | 6 | 7 | 8 |
| Frontal flexion angle (/°) | 14,2 ± 2,3 | 7,9 ± 1,0 | 9,6 ± 0,8 | 18,9 ± 1,1 |
| Subject | 9 | 10 | 11 | 12 |
| Frontal flexion angle (/°) | 12,5 ± 4,1 | 11,6 ± 1,8 | 6,8 ± 1,6 | 14,5 ± 1,9 |
| Average angle | 13,3 ± 5,1 | | ICC | 0,80 |

## 4  Classification by Fuzzy Logic

In the Stable Zone of postures P5 and P6, the ANN classification algorithm does not differentiate between smaller lateral flexion and larger flexion (e.g. 20° compared to 30°) or between different frontal flexion movements (e.g. 30° compared to 35°), as the output of the ANN is the same for these postures. Due to this situation, we decided to integrate Fuzzy Logic into our existing Neural Network Posture Classification Algorithm to differentiate between these types of postures and also integrate the adopted time in each sitting posture.

To develop the fuzzy logic algorithm, we created membership functions (shown in Fig. 5) dependent on the COP of the backrest and seat pad and the time spent in each posture. The set of rules is presented in Table 6. Maximum time for the time function is 900 s (15 min), as we will evaluate the sitting posture every 15 min. The interval for each time membership function have been based on ISO standards for trunk inclination [33], although those values are for standing postures instead of sitting postures, so they were increased accordingly. The maximum values for the Centre of Pressure were previously mentioned and are shown in Fig. 1-C.

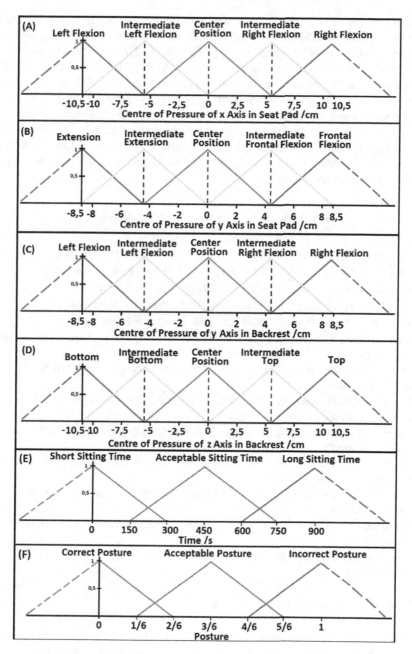

**Fig. 5.** Membership functions of the fuzzy logic algorithm. Dashed lines represent values that are not possible to attain, but we use a full triangle function to avoid asymmetric distributions. (A), (B), (C), (D) and (E) represent the antecedent membership functions and (F) represents the consequent membership functions.

As an example, if in 15 min the user was identified in two postures (e.g. P3 and P6, respectively for 360 and 540 s), with an average value for the COPx_seat, COPy_seat, COPy_back, COPz_back respectively of [2.38; −1.76; 1.22; −2.17] cm for P3 and [−5.61; −1.68; −2.97; −4.48] cm for P6. Using the MATLAB© Fuzzy Logic Toolbox and the Mamdani Centre of Gravity Defuzzification algorithm [37], we can reach a posture value for P3 and P6 respectively of 0.354 and 0.443 (which has 0 for minimum and 1 for maximum). Using these values, we can now try to implement a fuzzy logic actuator system to inflate and deflate specific air bladders depending on the detected postures (by the Neural Network) and the fuzzy output (by the Fuzzy Logic). With this action over the bladders it is expected the user to feel more discomfort in incorrect postures and change its posture.

**Table 6.**  Set of rules for the fuzzy logic algorithm.

| |
|---|
| IF (Time = Long) THEN (Posture = Incorrect) |
| IF (Time = Short) OR (Cxh OR Cxv OR Czv OR Cyh = Center THEN (Posture = Correct) |
| IF (Cxh = Right flexion) OR (Cyh = Frontal Flexion) THEN (Posture = Incorrect) |
| IF (Cxv = Right flexion) OR (Czv = Top) THEN (Posture = Incorrect) |
| IF (Cxh = Left flexion) OR (Cyh = Extension) THEN (Posture = Incorrect) |
| IF (Cxv = Left flexion) OR (Czv = Bottom) THEN (Posture = Incorrect) |
| IF (Cxh = Int. Right Flexion) OR (Cyh = Int. Frontal Flexion)) AND (Time = Long) THEN (Posture = Incorrect) |
| IF ((Cxv = Int. Right Flexion) OR (Czv = Int. Top)) AND (Time = Long) THEN (Posture = Incorrect) |
| IF ((Cxh = Int. Left Flexion) OR (Cyh = Int. Extension)) AND (Time = Long) THEN (Posture = Incorrect) |
| IF ((Cxv = Int. Left Flexion) OR (Czv = Int. Bottom)) AND (Time = Long) THEN (Posture = Incorrect) |
| IF (Cxh = Int. Right Flexion) OR (Cyh = Int. Frontal Flexion)) AND (Time = Accept.) THEN (Posture = Incorrect) |
| IF ((Cxv = Int. Right Flexion) OR (Czv = Int. Top)) AND (Time = Accept.) THEN (Posture = Accept.) |
| IF ((Cxh = Int. Left Flexion) OR (Cyh = Int. Extension)) AND (Time = Accept.) THEN (Posture = Accept.) |
| IF ((Cxv = Int. Left Flexion) OR (Czv = Int. Bottom)) AND (Time = Accept.) THEN (Posture = Accept.) |

## 5  Conclusions and Future Work

In a previous work, a chair prototype was used to classify 11 standardized seating postures of users, using 8 air bladders (4 in the chair's seat and 4 in the backrest). Here we showed that using industrialized air bladders improved the stability of the previously developed Posture Classification Algorithm, which was based on Neural Networks integrated with Decision Trees. One of the identified gaps in that system was the classification behaviour in intermediate postures or during posture changing, as the previous classification was only made in the so called Stable Zone.

In this paper, we first showed the results for the new prototype (of 80.9 % for 12 standardized postures) studied how the classification algorithm handled lateral postural changes, and identified stability and instability zones. During lateral flexion, the stability zone was found to be around an interval of 9°, between 11° and 20°. After 20°, the ANN algorithm is stable in classifying the lateral flexion postures. During frontal flexion, we did not identify an instability zone, and identified an Angle of Posture Transition (between P1 and P2) of 13° (in average). To differentiate between intermediate trunk flexion and extension positions we devised an approach based on integrating Fuzzy logic into the existing Neural Network-based Classification Algorithm that was capable of classifying 12 standard sitting positions.

For future work, we will need to check the how the classification algorithm works during a leaning back movement (which can also be studied by using the backrest reclining) and lateral flexion during leg crossing positions (we need to check if the Transitions have an instable zone.

Based on the output of the Neuro-fuzzy classification algorithm we are now devising a Fuzzy actuation system to create a new Posture Correction algorithm, since the previous algorithm is based on simple Boolean logic [29]. This actuation system is based on giving a discomfort to the user in incorrect postures and lead the user to a correct posture or to change the posture, as staying in the same posture in long periods of time can have a health impact in back and neck related problems.

**Acknowledgements.** This project (QREN 13330 – SYPEC) is supported by FEDER, QREN – Quadro de Referência Estratégico Nacional, Portugal 07/13 and PORLisboa – Programa Operacional Regional de Lisboa. The authors wish to thank Eng. Pedro Duque, Eng. Rui Lucena, Eng. João Belo and Eng. Marcelo Santos for the help provided in the construction of the first prototype.

# References

1. Owen, N., Sugiyama, T., Eakin, E.E., Gardiner, Pa, Tremblay, M.S., Sallis, J.F.: Adults' sedentary behavior determinants and interventions. Am. J. Prev. Med. **41**, 189–196 (2011)
2. Chau, J.Y., Ploeg, H.P.Van, Der, Uffelen, Van, J.G.Z., Wong, J., Riphagen, I., Healy, G.N., Gilson, N.D., Dunstan, D.W., Bauman, A.E., Owen, N., Brown, W.J.: Are workplace interventions to reduce sitting effective ? Sys. Rev. Prev. Med. (Baltim) **51**, 352–356 (2010)
3. Ramdan, N.S.A., Hashim, A.Y.B., Kamat, S.R., Mokhtar, M.N.A., Asmai, S.A.: On lower-back pain and its consequence to productivity. J. Ind. Intell. Inf. **2**, 83–87 (2014)
4. Punnett, L., Wegman, D.H.: Work-related musculoskeletal disorders: the epidemiologic evidence and the debate. J. Electromyogr. Kinesiol. **14**, 13–23 (2004)
5. Hartvigsen, J., Leboeuf-yde, C., Lings, S., Corder, E.H.: Is sitting-while-at-work associated with low back pain? A systematic, critical literature review. Scand. J. Public Health **28**, 230–239 (2000)
6. Roffey, D.M., Wai, E.K., Bishop, P., Kwon, B.K., Dagenais, S.: Causal assessment of occupational sitting and low back pain: results of a systematic review. Spine J. **10**, 252–261 (2010)

7. Todd, A.I., Bennett, A.I., Christie., C.J.: Physical implications of prolonged sitting in a confined posture-a literature review. Ergon. SA J. Ergon. Soc. South Africa Spec. Ed. **19**, 7–21 (2007)

8. Van Dieën, J.H., De Looze, M.P., Hermans, V.: Effects of dynamic office chairs on trunk kinematics, trunk extensor EMG and spinal shrinkage. Ergonomics **44**, 739–750 (2001)

9. Cagnie, B., Danneels, L., Van Tiggelen, D., De Loose, V., Cambier, D.: Individual and work related risk factors for neck pain among office workers: a cross sectional study. Eur. Spine J. **16**, 679–686 (2007)

10. Ariëns, G.A., Bongers, P.M., Douwes, M., Miedema, M.C., Hoogendoorn, W.E., van der Wal, G., Bouter, L.M., van Mechelen, W.: Are neck flexion, neck rotation, and sitting at work risk factors for neck pain? Results of a prospective cohort study. Occup. Environ. Med. **58**, 200–207 (2001)

11. Juul-Kristensen, B., Søgaard, K., Strøyer, J., Jensen, C.: Computer users' risk factors for developing shoulder, elbow and back symptoms. Scand. J. Work Environ. Health **30**, 390–398 (2004)

12. Adams, M., Hutton, W.: The effect of posture on diffusion into lumbar intervertebral discs. J. Anat. **147**, 121–134 (1986)

13. Kingma, I., van Dieën, J.H., Nicolay, K., Maat, J.J., Weinans, H.: Monitoring water content in deforming intervertebral disc tissue by finite element analysis of MRI data. Magn. Reson. Med. **44**, 650–654 (2000)

14. O'Sullivan, K., O'Sullivan, P., O'Sullivan, L., Dankaerts, W.: What do physiotherapists consider to be the best sitting spinal posture? Man. Ther. **17**, 432–437 (2012)

15. Pynt, J., Higgs, J., Mackey, M.: Seeking the optimal posture of the seated lumbar spine. Physiother. Theory Pract. **17**, 5–21 (2001)

16. Lis, A.M., Black, K.M., Korn, H., Nordin, M.: Association between sitting and occupational LBP. Eur. Spine J. **16**, 283–298 (2007)

17. Ellegast, R.P., Kraft, K., Groenesteijn, L., Krause, F., Berger, H., Vink, P.: Comparison of four specific dynamic office chairs with a conventional office chair: impact upon muscle activation, physical activity and posture. Appl. Ergon. **43**, 296–307 (2012)

18. Zhu, M., Mart, A.M., Tan, H.Z.: Template-based recognition of static sitting postures. In: Proceedings of The Workshop on Computer Vision and Pattern Recognition for Human Computer Interaction, held at the IEEE Conference on Computer Vision and Pattern Recognition (CVPR 2003), pp. 1–6. IEEE Computer Society, Madison (2003)

19. Forlizzi, J., Disalvo, C., Zimmerman, J., Mutlu, B., Hurst, A.: The SenseChair : the lounge chair as an intelligent assistive device for elders. In: DUX 2005 Proceedings of the 2005 Conference on Designing for User eXperience, p. Article No. 31 (2005)

20. Tan, H.Z., Slivovsky, L.A., Member, S., Pentland, A.: A sensing chair using pressure distribution sensors. IEEE/ASME Trans. Mechatron. **6**, 261–268 (2001)

21. Zheng, Y., Morrell, J.: A vibrotactile feedback approach to posture guidance. In: IEEE Haptics Symposium, pp. 351–358 (2010)

22. Schrempf, A., Schossleitner, G., Minarik, T., Haller, M., Gross, S.: PostureCare - towards a novel system for posture monitoring and guidance. In: 18th World Congress of the International Federation of Automatic Control (IFAC), pp. 593–598 (2011)

23. Mutlu, B., Krause, A., Forlizzi, J., Guestrin, C., Hodgins, J.: Robust, low-cost, non-intrusive sensing and recognition of seated postures. In: UIST 2007 Proceedings of the 20th Annual ACM Symposium on User Interface Software and Technology, pp. 149–158 (2007)

24. Daian, I., Ruiten, A.M. Van, Visser, A., Zubic, S.: Sensitive chair : a force sensing chair with multimodal real-time feedback via agent. In: ECCE 2007 Proceedings of the 14th European Conference on Cognitive Ergonomics: Invent! Explore!, pp. 163–166 (2007)

25. Griffiths, E., Saponas, T.S.: Health chair : implicitly sensing heart and respiratory rate. In: UbiComp 2014 Proceedings of the 2014 ACM International Joint Conference on Pervasive and Ubiquitous Computing, pp. 661–671 (2014)
26. Faudzi, A., Athif, M., Suzumori, K., Wakimoto, S.: Development of an intelligent chair tool system applying new intelligent pneumatic actuators. Adv. Robot. **24**, 1503–1528 (2010)
27. Goossens, R.H.M., Netten, M.P., van der Doelen, L.H.M.: An office chair to influence the sitting behavior of office workers **41**, 2086–2088 (2012)
28. Palumbo, F., Ullberg, J., Stimec, A., Furfari, F., Karlsson, L., Coradeschi, S.: Sensor network infrastructure for a home care monitoring system. Sens. (Basel) **14**, 3833–3860 (2014)
29. Martins, L., Lucena, R., Almeida, R., Belo, J., Quaresma, C., Jesus, A., Vieira, P.: Intelligent chair sensor: classification and correction of sitting posture. Int. J. Syst. Dyn. Appl. **3**, 65–80 (2014)
30. Vergara, M., Page, A.: System to measure the use of the backrest in sitting-posture office tasks. Appl. Ergon. **31**, 247–254 (2000)
31. Paliwal, M., Kumar, U.A.: Neural networks and statistical techniques: a review of applications. Expert Syst. Appl. **36**, 2–17 (2009)
32. Kar, S., Das, S., Ghosh, P.K.: Applications of neuro fuzzy systems: A brief review and future outline. Appl. Soft Comput. **15**, 243–259 (2014)
33. International Organization for Standardization: Ergonomics—Evaluation of static working postures (2000)
34. Fleiss, J.L.: Reliability of measurement. In: The Design and Analysis of Clinical Experiments, pp. 1–32 (1986)
35. Hobson, D.A.: Comparative effects of posture on pressure and shear at the body-seat interface. J. Rehabil. Res. Dev. **29**, 21 (1992)
36. James, F.: Statistical Methods in Experimental Physics. World Scientific, Singapore (2006)
37. Mamdani, E.H.: Control & science application of fuzzy algorithms for control of simple dynamic plant. Electr. Eng. Proc. Inst. **121**, 1585–1588 (1974)

# TogetherActive - Key Concepts
# and Usability Study

Lamia Elloumi[1]([✉]), Bert-Jan van Beijnum[1], and Hermie Hermens[1,2]

[1] University of Twente, Enschede, The Netherlands
{l.elloumi,b.j.f.vanbeijnum}@utwente.nl,
h.hermens@rrd.nl
[2] Roessingh Research and Development, Enschede, The Netherlands

**Abstract.** Despite the well-known benefits of physical activity on health, the recommended level of physical activity is not reached by everyone. Many interventions are aimed at reducing sedentary behaviour and increasing physical activity. As an intervention, we developed a virtual community system, TogetherActive, aiming at providing the social support to people in their daily life. Typically, virtual communities provide emotional and/or informational support, but our contribution aims mainly the instrumental and appraisal support. The community is coupled with physical activity sensor. In this system we focused on concepts such as individual and group goals, comparison, competition and cooperation in order to increase motivation to meet the daily recommended physical activity level. In this paper presents the design and the key concepts of the TogetherActive system, its implementation and the usability study.

**Keywords:** Physical activity · Virtual support community · Design · Usability evaluation

## 1 Introduction

Physical Inactivity is the fourth leading risk factor for global mortality causing an estimated 3.2 million deaths globally [24]. Moderate regular physical activity has significant benefits on health and can reduce the risk of cardiovascular diseases, diabetes, colon and breast cancer, and depression [24]. Physical activity should not be mistaken with physical exercise. Physical activity is defined as any bodily movement produced by skeletal muscles that requires energy expenditure [24]. Physical activity includes physical exercises, but also can be active transportation, working or house chores, or more generally activities of daily living.

Physical activity is important for all age groups and health conditions. Only the recommendations change depending on the group ages (5–17 years old, 18–64 years old and 65 years and above) and on the health condition (healthy, acute diseases and chronic diseases) [24]. Despite all recommendations and well-known health benefits of regular and sufficient physical activity, physical inactivity remains a global health problem [24].

A. Fred et al. (Eds): BIOSTEC 2015, CCIS 574, pp. 440–457, 2015.
DOI: 10.1007/978-3-319-27707-3_27

The actual focus of many researches and organizations is to reduce the physical inactivity, such as the WHO Member States have agreed to reduce physical inactivity by 10 % in 2025 [24].

In order to reduce physical inactivity and promote this behavioural change, researchers from several fields are involved: social sciences and computer science.

On one hand, in social sciences, researchers base interventions for behavioural change like this on a number of theories and models from social sciences such as [10,11]: classic learning theories, transtheoretical model and social support. These interventions are based on face-to-face meetings and recently implemented in e-coaching systems [14].

On the other hand, in computer sciences, several investigations and researches have been using Information and Communications Technology (ICT) to provide, extend, and enhance interventions to promote the level of physical activity among healthy people and chronic patients [3,7]. They address motivation and monitor physical activity in order to change behaviour regarding physical activity. The assessment of physical activity is important in those interventions and it is either self-reported (for examples with the use of e-diaries and questionnaires) or measured automatically and more objectively (for example with the use of pedometers, actometers, accelerometers and gyroscopes).

Telehealth and Telemonitoring systems couple ICT-based systems with for instance physical activity assessment tools in order to monitor the physical activity and give appropriate feedback taking into account the heath situation of the person [1,21,23]. But some of these ICT-based interventions showed a decrease of physical activity after a period of time compared to the first period of use and assessment (around 2 months in [21]). This decrease may be explained by drop of motivation.

Persuasive technology is also targeting at behavioural changes through persuasion and social influence. PersonA [2], UbiFit [6] and ActiveLifestyle [20] are examples of systems using persuasive technology in order to change physical activity behaviours. These systems are however limited in terms of social support, they are only focusing on the appraisal support.

From another perspective social networks and virtual communities are also used in healthcare in general and in physical activity in particular. These communities mainly provide the needed emotional and informational support. Some examples are: WebMD [22], PatientsLikeMe [18] and MedHelp [17].

Our current work focuses on reducing physical inactivity and we are targeting healthy people in order to improve their health and well-being. In order to overcome drawbacks of previous systems and solutions provided by different research fields, we focus on improving their motivation to be physically active through the use of a virtual community. The virtual community uses physical activity monitoring to assess physical activity. The community aims to provide different forms of social support (informational, emotional, instrumental and appraisal support). The community supports groupings of people using the system. We introduce physical activity goals for groups and individuals. We support competition between groups and cooperation between members within group. Comparison of achievement by members belonging to the same group is also supported. All these functionalities are included in order to increase the awareness and the motivation of users.

The ultimate goal of the virtual community is to motivate people to be physically active and maintain their physical activity level in long term use. It can be used as supporting tool for achieving lifestyle changes in the health prevention and in the management for chronic patients. Lifestyle changes includes physical activity as important facet, but also other facets such as diet and medication.

In this paper we present the design and first steps towards the implementation and evaluation of the TogetherActive community system. We designed TogetherActive and implemented part of the functionalities in order to do a technically evaluation and perform a usability study in order to improve it for the next design iteration.

First we present the architecture and design of TogetherActive system, then we present the implementation and the evaluation of the first prototype. And we finish with discussion and conclusions.

## 2   TogetherActive Overview and Concepts

TogetherActive is a virtual community to support people in their daily physical activity in order to be physically active and/or to maintain an appropriate level of activity. The appropriate level of activity is captured by the activity goal that can be set on an individual basis and depending on the personal context.

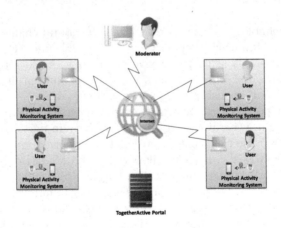

**Fig. 1.** TogetherActive architecture overview.

The TogetherActive architecture consists of a number of subsystems as shown in Fig. 1. First of all there is a physical activity monitoring system, composed by a physical activity sensor and a gateway (which can be a smartphone). The data collected by the physical activity monitoring system is transmitted from the sensor to the user's gateway and then synchronized with the portal.

The second component of the system is the portal. The portal is accessible from an internet-connected device (laptop, tablet or smartphone). Different views on activity and community data are shown to people depending of their roles.

## 2.1  Concepts Used in TogetherActive

In this section we introduce the main concepts used.

**Users and Roles.** A user is a person who is using the TogetherActive portal. We distinguish different users based on their user Role. We defined two different user roles: the user role and the moderator role. In the user role people have a physical activity monitoring system. In the moderator role people are able to configure some features of the system such as setting activity goals for individuals and groups.

**Group and Group Member.** A group is composed of two or more users who interact with each other in order to motivate each other in their daily physical activity. A group is composed of peers (people sharing similar age range, or similar health condition, or similar motivation and goals, or similar physical activity monitoring system) and can include moderator(s).When a user belongs to a group, then he/she is called member of that group.

**Physical Activity and Physical Activity Monitoring Systems.** Assessing the physical activity level is typically done using a physical activity monitoring systems such as a pedometer or accelerometer. Depending on the physical activity sensor used, different outcome measures are generated by these sensors. The four main different types that can be distinguished are: Steps, time, distance, and energy expenditure.

**Goals.** As mentioned in the introduction, the recommended amount of physical activity depends on many factors, including age and health condition. Hence, goals are personal and need to be make known to the system. The formation of the goal depends on the type of physical activity monitoring system used. However, in all cases the target amount of physical activity is always related to a period of time, for instance per day or per week.

Next to the personal goal we introduce the notion of group goal. The group goal is a physical activity level plan that the members intend to achieve together. The period of the goal can vary from one day to multiple days.

## 2.2  Functional View

As the main objective of the TogetherActive community is to provide social support, the functional blocks are described and categorised according to the four types of social support [13]: Informational, Emotional, Appraisal and Instrumental Support.

**Informational Support.** With informational support people can receive and search information on physical activity, sensors to measure these, and general information about the pros and cons of physical activities, new facts published about physical activity importance, and recommendations on physical activity. This information is published on a wiki related to TogetherActive community.

**Emotional Support.** Emotional support involves the provision of empathy, trust and caring. This is done with publishing discussion on a blog in the TogetherActive community. It is also supported by a synchronous communication service (chat) and an asynchronous communication service (private messaging).

**Appraisal Support and Feedback.** Appraisal support and feedback is about encouragement and giving motivational cues to people based on their physical activity achievements. In the TogetherActive system the achievement is measured by the activity sensors in relation to the personal and group goals that have been set. Appraisal and feedback can be given in different modalities and have different origins. For instance, in the TogetherActive system, the system itself may provide appraisal and feedback, and peers in the community can provide appraisal and feedback.

**Instrumental Support.** Instrumental support involves the provision of tangible aids and services that directly assist a person in achieving physically activity goals. It is realised in the community by the functionalities involving physical activity monitoring and self-management:

- Self-measuring of the physical activity: A physical activity monitoring system is provided to users in order to measure the physical activity level.
- Self-monitoring of the physical activity: Users are able to monitor their physical activity them-selves in order to change their physical activity behaviour.
- Self-comparison of physical activity: Users are able to compare their current physical activity level with previous levels; such as the daily level with the previous day level.
- Setting personal goal: It is about setting physical activity level goals. These goals should be realistic and measurable. They are time-targeted, such as daily goals. The users are able to set them-selves the physical activity goals, otherwise the moderator has to set the goals.
- Sharing physical activity with peers: Peers of the same group are able to share their current physical activity level.
- Monitoring other's physical activity: Peers of the same group are able to monitor each other's physical activity level in order to support and motivate them.
- Setting group goal: A group exists in order to motivate each other, have common goal. Each group can set its own goal. It is also possible that the moderator is involved in setting the goal for a group.
- Collaboration: because a group goal is set for each group and this goal is shared among the peers of the group, collaboration is stimulated to reach this group goal.

- Competition: the TogetherActive community is composed by multiple groups. The ability to achieve a common goal in each group creates the possibility to create a competition between groups.
- Comparison: within a group, peers can compare their physical activity achievement with others and give insight in similarities and differences amongst group members.

## 3   TogetherActive Functional Architecture

The architecture of the TogetherActive portal is based on the concepts of a Service Oriented Architecture. Adapting such a flexible architecture allows a flexible integration, developers get accelerated development cycles, reusable services and composite application development.

The TogetherActive functional architecture is shown in (Fig. 2). A portal is generally defined as a software platform for building websites and web applications. The TogetherActive portal is composed by portlets. A portlet implements a reusable independent application component.

In the TogetherActive portal, portlets are divided into three categories: generic, personal level and group level portlets. Portlets may use services. Services are existing services (provided by the platform that will be used) and customised services (implemented depending on the need).

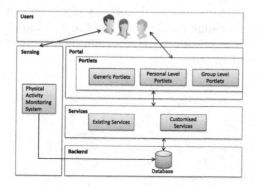

**Fig. 2.** TogetherActive functional architecture.

### 3.1   Portal

The portal is composed of set of pages hosting the portlets. These are categorised into a set of main pages, a set of personal pages and a set of group pages.

- The main pages contain the profile of the logged-in users, their groups' list where they belongs (a user can belong to one or more groups), the portal wiki and portal blog. The pages of the main pages are accessible via the menu bar of the other pages (personal and group pages) for navigation back and forth.

- The personal pages are user-related pages. Via these pages, a logged-in user has access to his/her data such as Daily and History physical activity monitoring data, and personal goals (current and past ones).
- The group pages are group-related pages. Whenever a group is selected from the main pages, the user is redirected to those pages. It contains data and information about the group such as details about the group, the members of the group and the group goal (current and past ones).

Each portal page contains one or more portlets. As part of the design we have defined this relationship as shown in Figs. 3 and 4 for each set of pages respectively.

**Generic Portlets.** The generic portlets are used in the design of set of the main pages. The organization of these portlets on the pages in represented in Fig. 3. The portlets used are:

- User Profile Portlet: It contains the profile of the logged-in user. The content of the portlet can be seen by authorised users/peers.
- Group List Portlet: It shows the list of existing groups. Authorized user can access to some groups pages or all of them, for example a member of the group can access his or her group pages from this portlet.
- Group Leader Board Portlet: it shows the leader board of active groups based on the group goal achievement.
- Group Profile Portlet: It contains a short description about the group.
- Members Activity Timeline Portlet: It shows the list of group members' posts, such as their current physical activity level and their comments on the physical activity achievement of others.

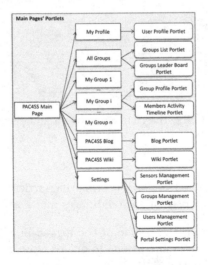

**Fig. 3.** Main pages' portlets.

- Blog Portlet: It is for discussion on the main part of the portal.
- Wiki Portlet: It is for posting information, article and news on the main part of the portal.
- Sensor Management Portlet: The user moderator can manage sensors (add, update or delete sensors) and manage associations between users and sensors (add, update or delete).
- Group Management Portlet: The moderator can manage groups (add, update or delete groups). Management of members and goals of the group is not done in this portlet.
- User Management Portlet: The moderator can manage users (add, update or delete).
- Portal Settings Portlets: for all general settings.


**Personal Level Portlets.** The personal level portlets are used to populate the set of personal pages. The organization of the portlets on the pages is shown in Fig. 4. The following portlets have been designed:

- Daily Physical Activity (PA) Monitoring Portlet: It shows to the logged-in user his or her physical activity level as a graph. The graph shows the recommended level (goal of the day) and the achieved level. The physical activity data is assessed by the physical activity monitoring system. It shows also some system feedback about the achievement of the user. The content of the portlet can be seen by authorised peers such as peers belonging to the same group or the user moderator of the portal.
- Comments portlet: Authorised peers can comment and discuss about the current physical activity achievements of the concerned user.

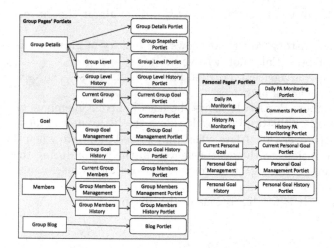

**Fig. 4.** Group and personal pages' portlets.

- History Physical Activity (PA) Monitoring Portlet: It shows to the logged-in user his or her physical activity level history over time. Similar to the Daily Physical Activity (PA) Monitoring Portlet, the content of this portlet can be seen by authorised peers.
- Current Personal Goal Portlet: It shows details about the current personal goal of the logged-in user.
- Personal Goal Management Portlet: Using this portlet th, moderator (and potentially the logged-in user) can manage the goals of the user (create, update or delete).
- Personal Goal History Portlet: It shows the history of goals, both in term of goals set and achievement made.

**Group Level Portlets.** The group level portlets are used to design the set of the group pages. The organization of the portlets on the pages is shown in Fig. 4. The following group level portlets have been designed:

- Group Details Portlet: It is a quick visual overview about the goal achievement. It shows if the member is compliant, over active, under active or if the sensor is not in use for the day.
- Group Snapshot Portlet: It gives an overview of the social activity of the members of the group. This overview is based on the number of messages exchanged such as posted/shared messages, messages replies (on own messages, or on messages from others) and the number of messages received. It measures the interaction between the members.
- Group Level Portlet: It shows the current level of the group. Every new group has a beginner level. Based on the daily group achievement, groups earn points allowing them to level-up (levels are expressed from beginner to expert).
- Group Level History Portlet: It shows the history of the levels earned by the group.
- Current Group Goal Portlet: It shows details about the current goal of the group.
- Group Goal History Portlet: It shows the history of group goals created and achieved (or potentially not achieved).
- Group Goal Management Portlet: The authorised user can create new goals and update existing ones.
- Group Members Portlet: It shows the list of current members of the group.
- Group Members History Portlet: It shows the list of users that were members of the groups over time (date when they joined and date when then left) and their contributions to the goals achieved in the group during that period.
- Group Members Management Portlet: The authorised user can add or remove members of the group.
- Blog Portlet: It is for discussions in the group.

## 3.2   Services

Services are responsible of storing and retrieving data. A portlet asks for data using service, and the service fetches it. The portal can then display this data to the user. The user can, depending on the portlet design, create, read, update or delete the data. If the user chooses to modify (create, update or delete) the data, the portlet passes it back to the service and the service manage and stores it in the database. The portlet doesn't need to know how the services do it. The existing services are services responsible for fetching and modifying the data to the portal in general, and portlets that are provided by the web-based platform. Customised services are implemented services responsible for fetching and modifying the data to the implemented portlets.

The information model in Fig. 5 represents the conceptual classes that are use to implement the services.

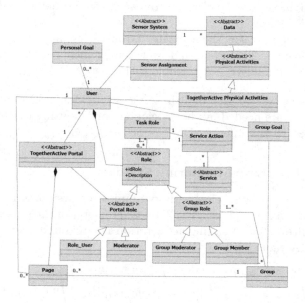

**Fig. 5.** Informational model.

# 4    TogetherActive Implementation

## 4.1    Portal and Portlets

The portal was implemented using Liferay [16]. Liferay is a web based platform supporting features commonly required for the development of websites and portals. It is an open architecture and open source system. Liferay offers you a full choice of application servers, databases, and operating systems to run on. Liferay provides also out of the box portlets such as Liferay CMS and Liferay Collaboration offering web publishing, content management, collaboration and social networking. It offers also a secure single sign on. Liferay users can be intuitively grouped into "user groups" and "roles" providing flexibility and ease of administration.

For implementation we used Liferay Portal 6.1 Community Edition Bundled with Tomcat. We used some of the platform built-in portlets and we implemented a subset of the portlets described in Sect. 3. Priorities were set for developing portlets which would be included in the evaluation (described in Sect. 5). The portlets that were implemented and included for the evaluation are:

- Personal Level Portlets: Daily Physical Activity (PA) Monitoring, History Physical Activity (PA) monitoring and Liferay Comments Portlets (known as Page Comment).
- Group Level Portlets: Group Details, Group Level, Group History, Current Group Goal, Group Goal History and Group Goal Management Portlets. The members' management portlet was replaced by the default administration possible with Liferay.
- Generic Portlets: Sensor Management, Group List and Group Leaderboard were implemented. The group and user management portlets were replaced by the default administration possible with Liferay.

As for the services, we used built-in services provides by Liferay and we built new services that are needed by the developed portlets using the service builder feature provided by the Liferay SDK.

### 4.2   Physical Activity Monitoring System

The current choice for a physical activity monitoring system is ProMove sensor [19] coupled with a smartphone. The 3D-accelerometer of the sensor assesses the energy expenditure. The resulting acceleration is integrated over time, which is referred by IMA value [5]:

$$IMA = \int\limits_{t=t_0}^{t_0+T} |a_x(t)|dt + \int\limits_{t=t_0}^{t_0+T} |a_y(t)|dt + \int\limits_{t=t_0}^{t_0+T} |a_z(t)|dt \tag{1}$$

where x, y and z are the axes of the accelerometer and $a_x$, $a_y$ and $a_z$ the associated accelerations.

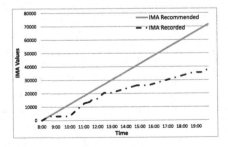

**Fig. 6.** IMA Values - recommended versus recorded.

More details about the sensor and the IMA value in [4]. IMA values allow to capture the physical activity level. Figure 6 presents an example of IMA values recorded and IMA values recommended over a day. The recommended level is the level that a person is supposed to follow.

In order to assess the physical activity achievements, we developed some metrics in [8] based on IMA values.

### 4.3   Personal and Group Goals

**Personal Goals.** The personal goal is a physical activity level that a user plans or needs to achieve during a period of time. This goal is either set by the user him/her-self, or by the moderator. The period of the goal can vary from one day to multiple days depending on the preferences. When a goal is set, the user needs to be compliant to this goal within allowed low and high thresholds. The Fig. 7 shows examples of goals with low and high thresholds. Whenever the physical activity level of the concerned user is in between the low and high thresholds then he/she is considered as compliant to the goal. If the physical activity level is higher than the high threshold then the user is considered as over active, and if it is under the low threshold then he/she is considered as under active. In the current implementation only Fig. 7(a) setup is considered.

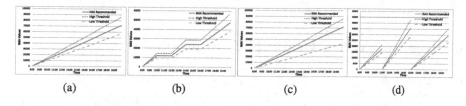

(a)              (b)              (c)              (d)

**Fig. 7.** Physical activity level over time recommended level, high threshold and low threshold. (a) Same percentage of high and low thresholds compared to the goal. (b) No goal for periods of the day (for example in case of office-worker user). (c) Different percentages of high and low thresholds compared to the goal. (d) Different goals for different parts of the day (resetting to 0 for each new goal).

**Group Goals.** The goal of a group is expressed as a physical activity level plan that the members intend to achieve together. The period of the goal can vary from one day to multiple days depending on the settings.

To be able to compute each member's participation to the plan, we take into account the personal achievement regarding a personal goal of each member of the group. We set a rewarding system with points. Points are given every episode of time (example one hour). Points depends on the status of the sensor (off or on) and the current level of activity. Points can vary from a group to another, but also from a goal to another.

To compute the points earned by a group during the period of the group goal, we use the following formula:

$$Points_{group} = \sum_{i=1}^{N}(\sum_{t=1}^{T} Points_{members}(i,t)) \tag{2}$$

With T the total number of the time episode, for example the number of hours for which the goal lasts and N the total number of members in the group.

Then we define the goal ($Goal_{group}$) as percentage of the maximum points that a group can reach:

$$Goal_{group} = Percentage * MaxPoints_{group} \tag{3}$$

With:

$$MaxPoints_{group} = \sum_{i=1}^{N}(Points_{compliance} * T) \tag{4}$$

According to the setting of the goal, a bonus can be attributed. If a member is compliant for the full duration of a goal, $Bonus_{member}$ is attributed.

Then:

$$MaxPoints_{group} = \sum_{i=1}^{N}(Points_{compliance} * T) + \sum_{i=1}^{N}(Bonus_{member}) \tag{5}$$

## 5    TogetherActive Usability Evaluation

### 5.1    Protocol

The goal was to evaluate the usability of the portal in order to acquire feedback from users and improve the portal in the next design cycle.

The evaluation was planned for the duration of 1 week and with 10 participants. A first meeting was planned with the participants. The aim of the meeting was to get an explanation and practice about the physical activity monitoring system ProMove, and to learn how to use and navigate through the TogetherActive Portal. Participants received their credentials to connect the monitoring system and use the community. They were taught how to use the sensor and the TogetherActive Portal. They were asked to sign an informed consent form and a borrowing form of the physical activity monitoring system.

Then, participants were divided into groups so as to test the group-based functionalities. During this week of evaluation, participants were asked to wear the physical activity sensor system from 8:00 to 22:00, and to use the portal.

By the end of the week, a second meeting was organised. The participants returned the physical activity monitoring systems and got a discharge document for borrowing the monitoring system. They received a link for an online questionnaire. The questionnaire was composed by 3 parts:

- General information:
  This part of the questionnaire is to get background information about the participants and their use of social networks and/or apps in general, and for health and well-being purposes.
- Portal usability (based on the Computer System Usability Questionnaire [15]):
  In this part the usability of the portal (computer system in general) was investigated using this standardized questionnaire. This part of the questionnaire has 19 items. The responses to those items would help us understand what aspects of the portal participants are particularly concerned about and the aspects that satisfy them. Items are based on a 7-point scale, where participants can express their opinion from strongly agree to strongly disagree.
- Sensor System:
  It is for measuring the satisfaction with the use of the sensor system and see if there were troubles in its use. This part of the questionnaire is composed by 4 items (with 3 mandatory items with yes or no questions, and one optional item to express extra remarks)

## 5.2   Participants

We recruited 10 participants who had to use the portal and sensor system for a week (5 working day and weekend). Participants were recruited from the University of Twente. Inclusion criterion to participate in the experiment was that participants should have some time for using the physical activity monitoring system and using the portal.

Participants were 9 PhD students and 1 Post-Doc, 2 of them were female participants. The age of participants was between 25 and 35 years old. All participants had an educational background in either technical sciences or social sciences.

## 5.3   Results

After getting all the replies to the questionnaire, we analysed the results. Figures 8(a), (b) and (c) give an overview about the familiarity of the participants in using social networking sites. We can conclude that participants are familiar with social networks (especially the popular ones nowadays), and using them for 0 to 10 hours a week. So we can assume that using our portal/social network will not be difficult for them.

Participants are not or did not use social networking site for health or well-being purposes. But 50 % of the participants used apps (on android, iPhone or other phone operating systems) for health or well-being purposes. All those 5 participants used those apps for exercise/training recording, one of them for exercise/training schedule compliance and 2 of them for informational purposes. So although participants are using social networks in their daily life, adopting these social networks to health or well-being purposes is not yet fully included in their habits, the same holds for the apps.

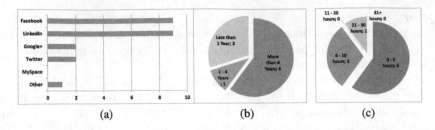

**Fig. 8.** Use of Social Networks (a) Social networking sites. (b) Subscription to the social networking sites. (c) Time spent per week on social networking sites.

After getting the replies from all participants we decided to exclude from the analysis the replies of 2 participants (one replied to all questions with 1 (strongly agree) and one replied to all questions with 7 (strongly disagree)).

Following the guidelines from Lewis [15], the results from user satisfaction (Table 1) are summarized into 4 factors reported as mean values: overall system usability (OVERALL), system usefulness (SYSUSE), information quality (INFOQUAL) and interface quality (INTERQUAL). Based on the result we can say that the interface quality (INTERQUAL) was better judged comparing to the rest.

**Table 1.** Satisfaction results.

| Score | Question items | Average (8 participants) | Standard deviation |
|---|---|---|---|
| OVERALL | 1 to 19 | 3.81 | 1.09 |
| SYSUSE | 1 to 8 | 3.89 | 1.03 |
| INFOQUAL | 9 to 15 | 3.83 | 1.06 |
| INTERQUAL | 16 to 18 | 3.5 | 1.29 |

Regarding the satisfaction with the use of the system sensor, 40 % of the participants were not satisfied and 30 % declared having problems with it. This dissatisfaction is mainly due to (based on their added remarks) large size of the sensor, uncomfortable to wear, sometimes loss of connection to the smartphone and battery charging (battery has to be charged every day). But as shown with the percentages, some participant didn't have concerns with previous list and 30 % of them declared that if the sensor is aimed for real use, they would use it.

## 5.4 Discussion

The results of the scores from the Usability Questionnaire were similar to the results of the other studies. In order to get more insights in the usability, we looked at the background of participants and in particular to their use of social networks. We found no correlation between the use of other social networks and the usability scores of the TogetherActive portal.

The aim of this usability evaluation is to get feedback from users and improve the TogetherActive system in the next design cycle. From the results of the questionnaires we induced two major categories of improvements, these are:

- Suggestions on the improvement on the user interface:
  - Critical remarks were received on the used colour scheme of the portal. So, this is a point of concern in the redesign.
  - Participants expected that the graphs that display daily and history physical activity levels are more interactive and has better quality and more details. So based on this remark, quality and display of the graphs should be improved
  - Some participants experienced difficulties to navigate between the pages. Based on this remark, navigation should be more intuitive.
  - Some participants were looking for help on the portal to understand more some of the functionalities and options of the portal but they could not find it. Basically, the help function was not planned or implemented in this version. So based on this remark help function should be provided in the portal.
- Suggestions on the sensor:
  Many participants complained about the size of the sensor and for some of them they had to change their habitual way of dressing to be able to use the sensor. They also complained about the fact of having to charge the sensor every day. So based on these remarks, we should look for another option for the sensor, satisfying the constraints size, wearability and battery life.

Based on the results from the usability study and suggestions from participants we will update the design theme (colour scheme) of the portal. For the second version of the system we will use the newer version of Liferay Portal: the 6.2 Community Edition Bundled with Tomcat. In the current version of the portal we used the default theme proposed by Liferay but in the new version we will choose a different one from the list of themes developed for Liferay.

For the portlets that show graphs and charts we will use different graphical library (HighCharts [12]) for more interactive and more user-friendly graphics. And we will improve the navigation between different pages of the portal.

A new physical activity monitoring system will be used with the system: Fitbit Zip [9]. This choice is based on the fact that it is relatively inexpensive and it is a reliable measurement device for measuring steps. It offers an open-source API which allows the collection of raw data (steps per minute) from the sensor. Additionally, some updates on the personal and group goals will be introduced.

For the new version of the portal we will implement the remaining portlets that were not included in the first version of the portal. And we will updated portlets that use the concepts related to a new physical activity monitoring system.

# 6    Conclusion

To conclude, in this paper we presented the TogetherActive community system and its design. The community uses physical activity monitoring sensors and aims to provide a full spectrum of social support. We included the concepts of personal goals and group goals. Self-monitoring, competition, cooperation and comparison are included in the system in order to increase the self-awareness and group-awareness, and motivation in order to improve or maintain the physical activity level. We presented the first implementation along with the technical evaluation. Improvements of the current implemented version of the system are planned taking the results of the evaluation into account.

In the current version of the system there is no mechanism to activate people to be physically active. We are thinking about new functionalities such as including "helpers". A helper would be someone that receives notifications from the system to propose a physical activity to the helped subject. Activities would be more as social activities to help in reaching physical activity goals rather than being alone. It will be activities to be active together such as a walk or playing Frisbee during lunch time.

Based on the technical evaluation done for the TogetherActive system, a pilot study is planned. In this evaluation, we will be investigating the added value of the virtual community in improving the physical activity level. The evaluation will include two groups, the control group and the intervention group. The control group will have a simplified version of the system, without the social aspects, and the intervention group will have the full TogetherActive system.

Another evaluation is also planned. The latest is about the usability and usefulness of the TogetherActive system using the mechanism of the social activation of people to be more physically active.

The current version of the system is targeting healthy people and same will be for the two planned evaluation. However, the concepts presented in this paper are easily transferable to other application domains including lifestyle and behaviour change support as needed in many chronic patients and also in the domain of prevention, especially in identified increased risk groups.

# References

1. op den Akker, H., Hermens, H., Jones, V.: A context-aware adaptive feedback system for activity monitoring. In: ICAMPAM 2011 (2012)
2. Ayubi, S.U., Parmanto, B.: PersonA: persuasive social network for physical Activity. 34th Annual International Conference of the IEEE Engineering in Medicine and Biology Society, pp. 2153–2157 (2012)
3. van den Berg, M.H., Schoones, J.W., Vliet Vlieland, T.P.M.: Internet-based physical activity interventions: a systematic review of the literature. J. Med. Internet Res. **9**, e26 (2007)
4. Bosch, S., Marin-Perianu, M., Marin-Perianu, R., Havinga, P., Hermens, H.: Keep on moving! activity monitoring and stimulation using wireless sensor networks. In: Barnaghi, P., Moessner, K., Presser, M., Meissner, S. (eds.) EuroSSC 2009. LNCS, vol. 5741, pp. 11–23. Springer, Heidelberg (2009)

5. Bouten, C.: Assessment of daily physical activity by registration of body movement. In: Eindhoven Univ. Technol. (1995)
6. Consolvo, S., Mcdonald, D.W., Toscos, T., Chen, M.Y., Froehlich, J., Harrison, B., Klasnja, P., Lamarca, A., Legrand, L., Libby, R., Smith, I., Landay, J. A.: Activity Sensing in the Wild : A Field Trial of UbiFit Garden. In: CHI 2008 (2008)
7. Cotter, A.P., Durant, N., Agne, A.A., Cherrington, A.L.: Internet interventions to support lifestyle modification for diabetes management: A systematic review of the evidence. J. Diab. Complications **15**(12), e287 (2013)
8. Elloumi, L., van Beijnum, B., Hermens, H.: Towards physical activity support community. In: 6th International Symposium on Medical Information and Communication Technology, ISMICT 2012 (2012)
9. Fitbit: (2015). Accessed March-2015. https://www.fitbit.com/zip
10. HHS: Theory at-a-glance: a guide for health promotion practice, chap. Theories and Applications (1995)
11. HHS: Physical Activity and Health: A Report of the Surgeon General, chap. Understanding and promoting physical activity (1996)
12. Highcharts: (2015). Accessed March-2015. http://www.highcharts.com/
13. House, J.S.: Social support, and the quality and quantity of life. In: Research on the Quality of Life (1981)
14. Kamphorst, B.A., Klein, M.C.A., Wissen, A.V.: Autonomous E-coaching in the wild: empirical validation of a model-based reasoning system. In: AAMAS 2014 (2014)
15. Lewis, J.R.: Ibm computer usability satisfaction questionnaires: Psychometric evaluation and instructions for use. In: International Journal of Human-Computer Interaction (1995)
16. Liferay: (2000). Accessed March-2015. https://www.liferay.com/
17. MedHelp: (1994). Accessed March-2015. http://www.medhelp.org/
18. PatientsLikeMe: (2004). Accessed March-2015. http://www.patientslikeme.com/
19. ProMove: (2011). Accessed March-2015. http://inertia-technology.com/promove-3d
20. Silveira, P., Daniel, F., Casati, F., van het Reve, E., de Bruin, E. D.: ActiveLifestyle: an application to help elders stay physically and socially active. In: FoSIBLE Workshop at COOP 2012 (2012)
21. Tabak, M.: New treatment approaches to improve daily activity behaviour. Ph.D. thesis, University of Twente (2014)
22. WebMD: (2005). Accessed March-2015. http://www.webmd.com/
23. van Weering, M.: Towards a new treatment for chronic low back pain patients. Ph.D. thesis, University of Twente (2011)
24. WHO: Physical activity (2014). Accessed March-2015. http://www.who.int/topics/physical_activity/en/

# Analysis of Eye Movements with Eyetrace

Thomas C. Kübler[1,3]([✉]), Katrin Sippel[1], Wolfgang Fuhl[1],
Guilherme Schievelbein[1], Johanna Aufreiter[2], Raphael Rosenberg[2],
Wolfgang Rosenstiel[1], and Enkelejda Kasneci[1]

[1] Computer Engineering Department, University of Tübingen, Tübingen, Germany
{thomas.kuebler,katrin.sippel,wolfgang.fuhl,
wolfgang.rosenstiel,enkelejda.kasneci}@uni-tuebingen.de
[2] Department of Art History, University of Vienna, Vienna, Austria
{johanna.aufreiter,raphael.rosenberg}@univie.ac.at
[3] Study Course Ophthalmic Optics/Audiology, University of Applied Sciences Aalen,
Aalen, Germany

**Abstract.** In the time of affordable and comfortable video-based eye
tracking, the need for analysis software becomes more and more important. We introduce Eyetrace, a new software developed for the analysis
of eye-tracking data during static image viewing. The aim of the software is to provide a platform for eye-tracking data analysis which works
with different eye trackers, offering thus the possibility to compare results
beyond the specific characteristics of the hardware devices. Furthermore,
by integrating various state-of-the-art and new developed algorithms for
analysis and visualization of eye-tracking data, the influence of different
analysis steps and parameter choices on typical eye-tracking measures
is totally transparent to the user. Eyetrace integrates several algorithms
to identify fixations and saccades, and to cluster them. Well-established
algorithms can be used side-by-side with bleeding-edge approaches with
a continuous visualization. Eyetrace can be downloaded at http://www.
ti.uni-tuebingen.de/Eyetrace.1751.0.html and we encourage its use for
exploratory data analysis and education.

## 1 Introduction

Eye-tracking technology has found its way into many fields of application and
research during the last years. With ever cheaper and easier to use devices,
the traditional usage in psychology and market investigation was accompanied
by new application fields, especially in medicine and natural sciences. Together
with the number of devices, the number of software for the analysis of eye tracking data increased steadily (e.g., SMI begaze, Tobii Analytics, D-Lab, NYAN,
Eyeworks, ASL Results Plus, or Gazepoint Analysis). Major brands offer their
individual analysis software with ready-to-run algorithms and preset several
parameter settings for their eye-tracker device and typical applications. All of
them share common features (such as visualizing gaze traces, attention maps and
calculating area of interest statistics) and distinguish in minor features. A major

---

Thomas C. Kübler and Katrin Sippel — Contributed equally to this paper.

© Springer International Publishing Switzerland 2015
A. Fred et al. (Eds): BIOSTEC 2015, CCIS 574, pp. 458–471, 2015.
DOI: 10.1007/978-3-319-27707-3_28

restriction in their usability are the licensing regulations. Supplying a class of students with licenses for home use, post-hoc data analysis years after recording the data may be difficult due to the financial effort associated with the licensing regulations. Besides that, these applications usually can not be extended by custom algorithms and specialized evaluation methods. Not few studies reach the point where the manufacturer software is insufficient and its extension is not possible. Thus, the recorded data has to be exported and loaded into other programs, e.g., Matlab, for further processing. Furthermore, individual calculations are often non-opaque or not documented in the necessary detail in order to allow comparison between studies conducted with different eye-tracker devices or even between different recording software versions.

Eyetrace supports a range of common eye-trackers and offers a variety of state-of-the-art algorithms for eye-tracking data analysis. The aim is not only to provide a standardized work-flow, but also to highlight the variability of different eye-tracker devices as well as different algorithms. For example, instead of just finding all fixations and saccades in the data, we enable the data analyst to test whether the choice of parameters for the fixation filter was adequate. Our approach is driven by continuous data visualization such that the result of each analysis step can be visually inspected. Different visualization techniques are available and can be active at the same time, i.e. a scanpath can be drawn over an attention map with areas of interest highlighted. All visualizations are customizable in order to visualize grouping effects, being distinguishable on different backgrounds and for color-blind persons.

We realize that no analysis software can provide all the tools required for every possible study. Therefore, the software is not only extensible but also offers the possibility to export all data and preliminary analysis results for usage with common statistics software such as Gnu R or SPSS.

Eyetrace originates from a collaboration of the department of art history at the University of Vienna [1–3] and the computer science department at the University of Tübingen. It consists of the core analysis component and a preprocessing step that is responsible for compatibility with many different eye-tracker devices as well as data quality analysis. The software bundle, including Eyetrace and EyetraceButler is written in C++, based on the experience of the previous version (EyeTrace 3.10.4, developed by Martin Hirschbühl with Christoph Klein and Raphael Rosenberg) as well as other eye-tracking analysis tools [4]. We are eager to implement state-of-the-art algorithms, such as fixation filters, clustering algorithms, and data-driven area of interest annotation and share the need to understand how these methods work. Therefore, implemented methods as well as their parameters are transparent to the user and well documented including references to the original work introducing them.

Eyetrace is available free of charge for non commercial research and educational purposes. It can be employed for analysis of eye-tracking data in scientific studies, in education and teaching.

## 2    Data Preparation

In order to make the use of different eye-tracker types convenient, recordings are preprocessed and converted into an eye-tracker independent format. The preprocessing software EyetraceButler handles this step and can also be used to split a single recording into subsets (e.g., by task or stimulus) and to perform a data quality check.

EyetraceButler provides a separate plug-in for all supported eye-trackers and converts the individual eye-tracking recordings into a format that holds information common to almost all eye-tracking formats. More specifically, it contains the x and y coordinates for both eyes, the width and height of the pupil as well as a validity bit, together with a joint timestamp. For monocular eye-trackers or eye-trackers that do not include pupil data the corresponding values are set to zero. A quality report is then produced containing information about the overall tracking quality as well as individual tracking losses (Fig. 1). Especially for demanding tracking situations, the quality report enables distinction between an overall low tracking quality and the partial loss of tracking for a time slice (such as at the beginning or end of the recording).

**Fig. 1.** Quality analysis for two recordings with a binocular eye-tracker. The color represents measurement errors (red), successful tracking of both eyes (green) and of only one eye (yellow) over time. It is easy to visually assess the overall quality of a recording as well as the nature of individual tracking failures (Color figure online).

### 2.1    Supplementary Data

In addition to the eye-tracking data, arbitrary supplementary information about the subject or relevant experimental conditions can be added, e.g. gender, age, dominant eye, or (subject's) patient's status. This information is made available to Eyetrace along with information concerning the stimulus viewed. Based on this information, the program is able to sort and group all loaded examinations according to these values.

### 2.2    Supported Eye-Tracker Devices

The EyetraceButler utilizes slim plug-ins in order to implement new eye-tracker profiles. To date, plug-ins for five different eye-trackers are available, among them devices by SMI, Ergoneers, TheEyeTribe as well as a calibration-free tracker recently developed by the Fraunhofer Institute in Ilmenau.

# 3   Data Analysis

## 3.1   Loading, Grouping and Filtering Data

Data files prepared by the EyetraceButler can be batch loaded into Eyetrace together with their accompanying information such as the stimulus image or subject information. Visualization and analysis techniques can handle subjects grouping by any of the arbitrary subject information fields. For example attention maps can be calculated separately for each subject, cumulative for all subjects or by subject groups. This allows to compare subjects with healthy vision to a low vision patient group or to compare the viewing behavior of different age groups. Adaptive filters are provided to select the desired grouping and individual recordings can be included or excluded from the visualization and analysis process.

## 3.2   Fixation and Saccade Identification

One of the earliest and most frequent analysis steps is the identification of fixations and saccades. Their exact identification is essential for the calculation of many eye movement characteristics, such as the average fixation time or saccade length.

Eye-tracking manufacturers often offer the possibility to identify fixations and saccades automatically. However, this filtering step is not as trivial as the automated annotation may suggest. In fact, different algorithms yield quite different results. By offering a variety of calculation methods and making their parameters available for editing, we want to bring to mind the importance of the right choice of parameters. Especially when it comes to identifying the exact first and last point that still belong to a fixation and the merging of subsequent fixations that come to fall to the same location, relevant differences between algorithms and a high sensitivity to parameter changes can be observed.

To date following algorithms are implemented in Eyetrace:

**Standard Algorithm.** The standard algorithm for separating fixations and saccades is based on three adjustable values: The minimum duration of the fixations, the maximum radius of the fixations and the maximum number of points that are allowed to be outside this radius (helpful with noisy data). A time window of the minimum fixation duration is shifted over the measurement points until the conditions of maximum radius and maximum outliers are fulfilled. In the following step the beginning fixation is extended if possible until the number of allowed outliers has been reached. A complete fixation has been identified and the procedure starts anew. Every measurement point that was not assigned to a fixation is assigned to the saccade between its predecessor and successor fixation.

**Velocity-Based Algorithm.** Since saccades show high eye movement speed while fixations and smooth pursuit movements are much slower, putting a threshold on the eye movement speed is a straight forward way of fixation filtering.

Eyetrace currently implements three different variants of velocity based fixation identification. Each of the methods can filter short fixations via a minimum duration in a post-processing step.

*Velocity Threshold by Pixel Speed [px/s].* A simple threshold over the speed between subsequent measurements. If the speed is exceeded, the measurement belongs to a saccade, otherwise to a fixation. While a pixel per second threshold is easy to interpret for the computer, it is often not meaningful to the experimenter and therefore hard to choose.

*Velocity Threshold by Percentile.* Based on the assumption that the velocity is bigger within saccades than within fixations, velocities are sorted by magnitude and a threshold is chosen by a percentile of the data selected by the user (usually 80–90%). An example of sorted distances between measurements:
1 3 7 11 12 13 18 21 21 22

Green distances are supposed to belong to fixations for a 60 % percentile (6 out of 10 distances) and the value 18 would be chosen as velocity threshold.

*Velocity Threshold by Angular Velocity [/s].* This is the representation most common in the literature since it is independent of pixel count and individual viewing behavior. However, it also requires most knowledge about the data recording process in order to be able to convert the pixel distances into angular distances (namely the distance between viewer and screen, screen width and resolution). Suggested values for individual tasks can be found in the literature [5,6].

**Gaussian Mixture Model.** A Gaussian mixture model as introduced in [7] is also implemented in Eyetrace. This method is based on the assumption that distances between subsequent measurement points within a fixation form a Gaussian distribution. Furthermore distances between measurement points that belong

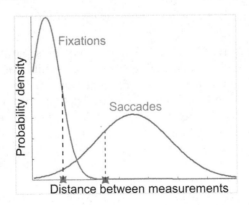

**Fig. 2.** Fit of two Gaussian distributions to the large distances between subsequent measurements within saccades and the short distances between fixations. Two sample points are shown, one with higher probability to belong to a fixation (left) and one with a higher probability for a saccade (right).

to a saccade also form a Gaussian distribution, but with different mean and standard deviation. A maximum likelihood estimation of the parameters of the Mixture of Gaussians is performed. Afterward for each measurement point the probability that it belongs to a fixation or to a saccade can be calculated and fixation/saccade labels are assigned based on these probabilities (Fig. 2). The major advantage of this approach is that all parameters can be derived from the data. One could evaluate data recorded during an unknown experiment without the need to specify any thresholds or experimental conditions. The method has been evaluated in several studies [8–10].

### 3.3   Fixation Clustering

After identification of fixations and saccades the fixations can also be clustered. Clustering fixations either by neighborhood thresholds or mean-shift clustering (as proposed by [11]) results in data-driven, automatically assigned areas of interest.

Fixation clusters can be calculated on the scan patterns of one subject or cumulative on a group of subjects.

**Standard Clustering Algorithm.** This greedy algorithm requires the definition of a minimum number of fixations that will be considered a cluster, the maximum radius of a cluster, and the overlap. Fixations are sorted in descending order of the number of included gaze points. Starting with the longest fixation, the algorithm iterates over all fixations, checking whether they fulfill the conditions of building a cluster with the biggest one. If the number of found fixations is sufficient, all found fixations are assigned to the same cluster and excluded from further clustering. If not, the first fixation cannot be assigned to any cluster and the algorithm starts again from the second longest fixation.

**Mean-Shift Clustering.** The mean-shift clustering method assumes that measurements are sampled from Gaussian distributions around the cluster centers. The algorithm converges towards local point density maxima. The iterative procedure is shown in Fig. 3. One of the main advantages is that it does not require the expected number of clusters in advance but determines an optimal clustering based on the data.

**Cumulative Clustering.** The clustering algorithms mentioned above can also be used on the cumulative data of more than one subject or more than one experiment condition. This way cumulative population clusters can be formed. They are more robust to noise and individual viewing behavior differences. The parameters of the algorithms are adapted for cumulative usage (e.g. the number of minimum fixations for the standard algorithm depends on the number of data sets used for cumulative analysis), but the way the methods work remain the same.

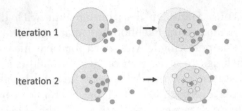

**Fig. 3.** Simplified visualization of the mean-shift algorithm for the first two iterations at one starting point. In each iteration the mean (green square) of all data points (blue circles) within a certain window around a point (big red circle) is calculated. In the next iteration the procedure is repeated with the window shifted towards the previous mean. This is done until the mean convergence (Color figure online).

### 3.4  Areas of Interests (AOIs)

For the evaluation of gaze directed at specific regions, Eyetrace provides both the possibility to annotate top-down AOIs manually (via a graphical editor with only few mouse clicks) or to determine bottom-up clusters of high fixation density. Figure 4 shows an example of manually defined AOIs and bottom-up defined clusters.

**Fig. 4.** Simultaneous overlay of multiple visualization techniques for one scanpath of an image viewing task. The background image is shown together with a scanpath representation of fixed-size fixation markers (small circles) and generated fixation clusters (bigger ellipses) for left (green) and right (blue) eye. AOIs were annotated by hand (marked as white overlay) (Color figure online).

## 4  Data Visualization

The software allows simultaneous visualization of multiple scanpaths. These may represent different subjects, subject groups or distinct experiment conditions.

The scan patterns are rendered in real-time as an overlay to an image or video stimulus. Various customizable visualization techniques are available: Fixations that encode fixation duration in their circular size, elliptical approximations encoding spatial extend as well as attention and shadow maps. Exploratory data analysis can be performed by traversing through the time dimension of the scan patterns as if it was a video. Most of the visualizations are interactive so that placing the cursor over the visualization of e.g. a fixation gives access to detailed information such as its duration and onset time.

## 4.1   Fixations and Fixation Clusters

The visualization of fixations and fixation clusters has to account for their spatial and temporal information. It is common to draw them as circles of either uniform size or to encode the fixation duration in the circle diameter (see Fig. 6). Besides these options, Eyetrace can fit an ellipse to the spatial extend of all measurement points assigned to the fixation. The eigenvectors of these measurements point into the direction of highest variance within the data (see Fig. 5) and can therefore be used as major and minor axis for the ellipse fit. This visualization gives an excellent impression of measurement accuracy, since fixations are supposed to represent a relatively stable eye position and the variance of measurements

**Fig. 5.** The two eigenvectors of Gaussian distributed samples (that correspond to the directions of highest variance). These are used as the major and minor axes for an elliptic fit.

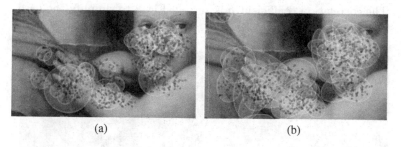

(a)                                                        (b)

**Fig. 6.** (a) Fixation visualization as circles scaled by the respective dwell times, representing location and temporal information. (b) Elliptic fit to the individual measurement points assigned to the fixation. This visualization indicates the measurement accuracy and the quality of the fixation filter. Usually fixations are supposed to be circular, however wrong settings during the fixation identification step may cause adjacent saccade points to be clustered into the fixation and therefore deform the circle towards a more ellipsoid shape.

within a fixation therefore stands representative for the measurement inaccuracy. When evaluating different fixation filters the circularity of resulting ellipses can be used as a quality measure: Realistic fixations are supposed to be approximately circular. Once fixations begin to grow ellipsoid, the choice of parameters for the selected filter is probably inadequate.

## 4.2   Attention and Shadow Maps

Attention maps are one of the most common eye-tracking analysis tools, besides the high number of subjects that have to be measured in order to get reliable results [12]. In order to enable fast attention map rendering even for a large number of recordings and high resolution, the attention map calculation utilizes multiple processor cores. Attention maps can be calculated for gaze points, fixations and fixation clusters. We provide the classical red-green color palette for attention maps as well as blue version for color-blind persons.

For the gaze point attention map each gaze point contributes as a two dimensional Gaussian distribution. The final attention map is calculated as the sum over all Gaussians. The Gaussian distribution is specified by the two parameters size and intensity which are adjustable by the user. This Gaussian distribution is circular because gaze points do not have information about orientation and size. For fixations and fixation clusters the elliptic fit is used to determine the shape and orientation of the Gaussian distribution. Figure 8 shows an example of a circular Gaussian distribution (a) and a stretched, elliptical one (b).

$$P(x,y) = \frac{1}{\sigma_1\sigma_2 2\pi\sqrt{1-\varrho}} e^{-\frac{1}{2(1-\varrho^2)}\left(\frac{x^2}{\sigma_1^2} + \frac{y^2}{\sigma_2^2} - \frac{2\varrho xy}{\sigma_1\sigma_2}\right)} \tag{1}$$

Equation 1 shows the Gaussian distribution in the two dimensional case. $\sigma_1$ and $\sigma_2$ are the variance in horizontal and vertical direction respectively (see Fig. 8). $x$ and $y$ are the offsets to the center of the Gaussian distribution (see Fig. 8). The correlation coefficient $\varrho$ is zero in Fig. 8 to simplify the case.

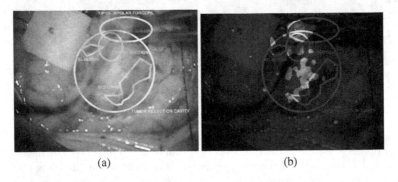

(a)                                    (b)

**Fig. 7.** An attention map calculated for fixation clusters (a) and the corresponding shadow map (b). Data of microsureons during a tumor removal surgery lend from [13].

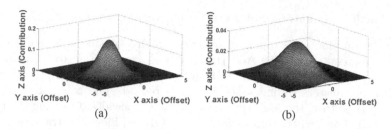

(a)                                    (b)

**Fig. 8.** Two Gaussian distribution calculated with $\sigma_1 = \sigma_2 = 1$ (a) and $\sigma_1 = 1$ and $\sigma_2 = 5$ (b).

A variant of the attention map is the shadow map that reveals only areas frequently hit by gaze (see Fig. 7(b)). Its calculation is identical to that of the attention map with the difference of a smoothing step in order to show the border regions with higher sensitivity. This is done by calculating the n-th root of each map value where n is a user-defined parameter that regulates the desired smoothing.

### 4.3  Saccades

Saccades are typically visualized as arrows or lines connecting two fixations.

Besides this, a statistical evaluation can be visualized as a diagram called anglestar. It consists of a number of slices and a rotation offset. A slice of the anglestar codes in its length the number of saccades with the same angular orientation as the slice (e.g. if the slice represents the angles between $0°$ and $45°$ the number of saccades within that angle range contribute to that slice) to the horizontal axis is considered. The extend of a slice from the center of the star can represent the quantity, summed length or summed duration of the saccades towards that direction. Figure 9 shows a diagram where the extension of the slices is based on the summed length of the saccades.

**Fig. 9.** Representation of an anglestar.

## 4.4  Fixations Clusters and AOI Transitions Diagram

For some evaluation cases it is interesting in which sequence attention is shifted between different areas. The AOI transitions diagram (Fig. 10) visualizes the transition probabilities between AOIs manually annotated and/or converted from fixation clusters during a specific time period. The color of the transition is inherited from the AOI with most outgoing saccades. Hovering the mouse over an AOI shows all transitions from this AOI and hides the transitions from all other AOIs. Hovering the cursor over a specific transition displays an information box containing the number of transitions in both directions. Figure 10(b) visualizes the transitions directly on the image.

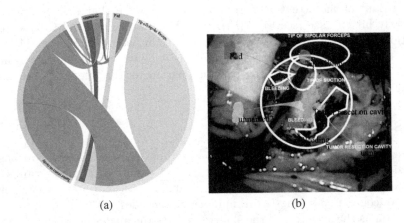

(a)                              (b)

**Fig. 10.** (a) Abstract diagram of the transitions between AOIs. The graphic is interactive and can blend out irrelevant edges if one AOI is selected. (b) AOIs as defined by cumulative fixation clusters and corresponding transition frequencies as an overlay over the original image. Eyetrace offers interaction with the graphic via the cursor in order to show only the currently relevant subset of transitions from and to one AOI.

# 5  Data Export

## 5.1  Statistics

**General Statistics.** Independent of all other calculations it is possible to calculate some general gaze statistics. These include the horizontal and vertical gaze activity, minimum, maximum and average speed of the gaze. These statistics shine a light on the agility and exploratory behavior of the subjects and can be exported in a format ready to use in statistical programs such as JMP or SPSS.

**AOI Statistics.** Numerous gaze characteristics can be calculated for AOIs, such as the total number of glances towards the AOI, the time of the first glance, glance frequency, total glance time, the glance proportion towards the AOI in respect to the whole recording and the minimum, maximum and mean glance duration. Glance in this case means that a sequence of gaze points are located inside the AOI, no matter if they belong to a saccade or a fixation. These statistics are a supplement to the AOI transitions diagram and can also be exported.

## 5.2 Visualization

The transition diagram as well as every visualization can be exported either lossless as vector graphics or as bitmaps (png, jpg). Eyetrace provides the option to export the information about the subject (e.g. age, dominant eye) and the parameters used for calculation and visualization as a footer in the exported image. That way results can be reproduced and understood based solely on the exported image.

## 5.3 Evaluation Results

After calculating fixations, fixation clusters or cumulative clusters Eyetrace provides the possibility to export them as a text file.

Fixations are exported in a table including the running number, the number of included points, x and y coordinate, radius and if calculated the id of the cluster this fixation belongs to. The text file for the clusters and cumulative clusters include an ID number, the number of fixations contained, mean x, mean y and the radius.

# 6 Conclusion and Outlook

Eyetrace is a well-structured software tool, advantageous enough to be employed for academic research in a number of application fields but with its convenient handling nonetheless usable for persons without broad eye-tracking experience, e.g., for teaching students. The major advantages of the software are the flexibility of algorithms and their parameters as well as their actuality with respect to the state-of-the-art.

Eyetrace has already been tested on data of various research projects, ranging from the viewing of fine art [1–3] recorded via a static binocular SMI infrared eyetracker to on-road and simulator driving experiments [14,15] and supermarket search tasks [16,17] recorded via a mobile Dikablis tracker by Ergoneers.

In our future work, we will not only extend the EyetraceButler for applicability to further eye-tracking device, but also extend the general and AOI-based statistics calculations, integrate and implement new calculation and visualization algorithms and make the existing ones more interactive. A special focus will be given to the analysis and processing of saccadic eye movements as well as to the automated annotation of AOIs for dynamic scenarios [18] and non-elliptical AOIs. In addition, we plan on including further automated scanpath comparison metrics, such as MultiMatch [19] or SubsMatch [18].

Another relevant area is for our future work will be the monitoring of vigilance and workload during visual tasks. Especially for medical applications such as reaction or stimulus sensitivity testing, the mental state of the subject is of great importance. Available data, such as the pupil dilation, fatigue waves [20], saccade length differences [21], and blink rate may give important insight into the recorded data and even yield e.g. cognitive workload weighted attention maps.

**Acknowledgements.** We want to thank the department of art history at the university of Vienna for the inspiring collaboration. The project was partly financed by the the WWTF (Project CS11-023 to Helmut Leder and Raphael Rosenberg).

# References

1. Brinkmann, H., Commare, L., Leder, H., Rosenberg, R.: Abstract art as a universal language? Leonardo **47**(3), 256–257 (2014)
2. Klein, C., Betz, J., Hirschbuehl, M., Fuchs, C., Schmiedtová, B., Engelbrecht, M., Mueller-Paul, J., Rosenberg, R.: Describing art-an interdisciplinary approach to the effects of speaking on gaze movements during the beholding of paintings. PloS one **9**, e102439 (2014)
3. Rosenberg, R., Messen, B.: Vorschläge für eine empirische Bildwissenschaft. Jahrbuch der Bayerischen Akademie der Schönen Künste **27**, 71–86 (2014)
4. Tafaj, E., Kübler, T. C., Peter, J., Rosenstiel, W., Bogdan, M., Schiefer, U.: Vishnoo: an open-source software for vision research. In: 24th International Symposium on Computer-Based Medical Systems (CBMS), pp. 1–6. IEEE (2011)
5. Blignaut, P.: Fixation identification: the optimum threshold for a dispersion algorithm. Atten. Percept. Psychophys. **71**, 881–895 (2009)
6. Salvucci, D.D., Goldberg, J.H.: Identifying fixations and saccades in eye-tracking protocols. In: Proceedings of the 2000 Symposium on Eye Tracking Research and Applications, pp. 71–78. ACM (2000)
7. Tafaj, E., Kasneci, G., Rosenstiel, W., Bogdan, M.: Bayesian online clustering of eye movement data. In: Proceedings of the Symposium on Eye Tracking Research and Applications, pp. 285–288. ACM (2012)
8. Kasneci, E.: Towards the automated recognition of assistance need for drivers with impaired visual field. Ph.D. thesis, University of Tübingen, Wilhelmstr. 32, 72074 Tübingen (2013)
9. Kasneci, E., Kasneci, G., Kübler, T. C., Rosenstiel, W.: The applicability of probabilistic methods to the online recognition of fixations and saccades in dynamic scenes. In: Proceedings of the Symposium on Eye Tracking Research and Applications, pp. 323–326. ACM (2014)
10. Kasneci, E., Kasneci, G., Kübler, T.C., Rosenstiel, W.: Online recognition of fixations, saccades, and smooth pursuits for automated analysis of traffic hazard perception. In: Koprinkova-Hristova, P., Mladenov, V., Kasabov, N. (eds.) Artificial Neural Networks. SSBN, vol. 4, pp. 411–434. Springer, Heidelberg (2014)
11. Santella, A., DeCarlo, D.: Robust clustering of eye movement recordings for quantification of visual interest. In: Proceedings of the Eye Tracking Research and Applications Symposium on Eye Tracking Research and Applications - ETRA'2004, pp. 27–34 (2004)

12. Pernice, K., Nielsen, J.: How to conduct eyetracking studies. Nielsen Norman Group (2009)
13. Eivazi, S., Bednarik, R., Tukiainen, M., von und zu Fraunberg, M., Leinonen, V., Jääskeläinen, J.E.: Gaze behaviour of expert and novice microneurosurgeons differs during observations of tumor removal recordings. In: Proceedings of the Symposium on Eye Tracking Research and Applications, pp. 377–380. ACM (2012)
14. Kasneci, E., Sippel, K., Aehling, K., Heister, M., Rosenstiel, W., Schiefer, U., Papageorgiou, E.: Driving with binocular visual field loss? a study on a supervised on-road parcours with simultaneous eye and head tracking. PloS one 9, e87470 (2014)
15. Tafaj, E., Kübler, T.C., Kasneci, G., Rosenstiel, W., Bogdan, M.: Online classification of eye tracking data for automated analysis of traffic hazard perception. In: Mladenov, V., Koprinkova-Hristova, P., Palm, G., Villa, A.E.P., Appollini, B., Kasabov, N. (eds.) ICANN 2013. LNCS, vol. 8131, pp. 442–450. Springer, Heidelberg (2013)
16. Kasneci, E., Sippel, K., Aehling, K., Heister, M., Rosenstiel, W., Schiefer, U., Papageorgiou, E.: Homonymous visual field loss and its impact on visual exploration - a supermarket study. Translational Vision Science and Technology (2014, In Press)
17. Sippel, K., Kasneci, E., Aehling, K., Heister, M., Rosenstiel, W., Schiefer, U., Papageorgiou, E.: Binocular glaucomatous visual field loss and its impact on visual exploration - a supermarket study. PloS one 9, e106089 (2014)
18. Kübler, T.C., Bukenberger, D.R., Ungewiss, J., Wörner, A., Rothe, C., Schiefer, U., Rosenstiel, W., Kasneci, E.: Towards automated comparison of eye-tracking recordings in dynamic scenes. In: EUVIP 2014 (2014)
19. Dewhurst, R., Nyström, M., Jarodzka, H., Foulsham, T., Johansson, R., Holmqvist, K.: It depends on how you look at it: scanpath comparison in multiple dimensions with multimatch, a vector-based approach. Behav. Res. Meth. 44, 1079–1100 (2012)
20. Henson, D.B., Emuh, T.: Monitoring vigilance during perimetry by using pupillography. Invest. Ophthalmol. Vis. Sci. 51, 3540–3543 (2010)
21. Di Stasi, L.L., McCamy, M.B., Macknik, S.L., Mankin, J.A., Hooft, N., Catena, A., Martinez-Conde, S.: Saccadic eye movement metrics reflect surgical residents' fatigue. Ann. Surg. 259, 824–829 (2014)

# Author Index

Printed in the United States
By Bookmasters